Essentials of Bacteriology

Essentials of Bacteriology

Editor: Ricky Parks

www.callistoreference.com

Callisto Reference,
118-35 Queens Blvd., Suite 400,
Forest Hills, NY 11375, USA

Visit us on the World Wide Web at:
www.callistoreference.com

ISBN: 978-1-63239-814-7 (Hardback)

Cataloging-in-publication Data

Essentials of bacteriology / edited by Ricky Parks.
    p. cm.
Includes bibliographical references and index.
ISBN 978-1-63239-814-7
1. Bacteriology. 2. Microbiology. I. Parks, Ricky.
QR74.8 .E87 2017
579.3--dc23

Table of Contents

Permissions

List of Contributors

Index

Preface

As an essential part of microbiology is bacteriology. It is concerned with the study of bacteria. It includes the examination of their structure, their classification, identification and characterization. Also included in bacteriology is the study of the relation between bacteria and its surroundings, like in agriculture and medicine, etc. This book elucidates the concepts and innovative models around prospective developments with respect to bacteriology. It includes some of the vital pieces of work being conducted across the world, on various topics related to this field. This book, with its detailed analyses and data, will prove immensely beneficial to professionals and students involved in this area at various levels.

The main aim of this book is to educate learners and enhance their research focus by presenting diverse topics covering this vast field. This is an advanced book which compiles significant studies by distinguished experts. This book addresses successive solutions to the challenges arising in the area of application, along with it; the book provides scope for future developments.

It was a great honour to edit this book, though there were challenges, as it involved a lot of communication and networking between me and the editorial team. However, the end result was this all-inclusive book covering diverse themes in the field.

Finally, it is important to acknowledge the efforts of the contributors for their excellent chapters, through which a wide variety of issues have been addressed. I would also like to thank my colleagues for their valuable feedback during the making of this book.

Editor

A Novel Model of Chronic Wounds: Importance of Redox Imbalance and Biofilm-Forming Bacteria for Establishment of Chronicity

Sandeep Dhall[1,2], Danh Do[3], Monika Garcia[1], Dayanjan Shanaka Wijesinghe[6,7,8,9], Angela Brandon[10], Jane Kim[4], Antonio Sanchez[11], Julia Lyubovitsky[5], Sean Gallagher[11], Eugene A. Nothnagel[4], Charles E. Chalfant[6,7,8,9], Rakesh P. Patel[10], Neal Schiller[3], Manuela Martins-Green[1,2]*

1 Departments of Cell Biology and Neuroscience, University of California Riverside, Riverside, California, United States of America, 2 Bioengineering Interdepartmental Graduate Program, University of California Riverside, Riverside, California, United States of America, 3 Division of Biomedical Sciences, University of California Riverside, Riverside, California, United States of America, 4 Department of Botany and Plant Sciences, University of California Riverside, Riverside, California, United States of America, 5 Department of Bioengineering, University of California Riverside, Riverside, California, United States of America, 6 Hunter Holmes McGuire Veterans Administration Medical Center, Richmond, Virginia, United States of America, 7 Department of Biochemistry & Molecular Biology, Virginia Commonwealth University, Richmond, Virginia, United States of America, 8 Virginia Commonwealth University Reanimation Engineering Science Center, Richmond, Virginia, United States of America, 9 The Massey Cancer Center, Richmond, Virginia, United States of America, 10 Department of Pathology, University of Alabama at Birmingham, Birmingham, Alabama, United States of America, 11 Department of Product Technology, UVP, LLC, an Analytik Jena Company, Upland, California, United States of America

Abstract

Chronic wounds have a large impact on health, affecting ~6.5 M people and costing ~$25B/year in the US alone [1]. We previously discovered that a genetically modified mouse model displays impaired healing similar to problematic wounds in humans and that sometimes the wounds become chronic. Here we show how and why these impaired wounds become chronic, describe a way whereby we can drive impaired wounds to chronicity at will and propose that the same processes are involved in chronic wound development in humans. We hypothesize that exacerbated levels of oxidative stress are critical for initiation of chronicity. We show that, very early after injury, wounds with impaired healing contain elevated levels of reactive oxygen and nitrogen species and, much like in humans, these levels increase with age. Moreover, the activity of anti-oxidant enzymes is not elevated, leading to buildup of oxidative stress in the wound environment. To induce chronicity, we exacerbated the redox imbalance by further inhibiting the antioxidant enzymes and by infecting the wounds with biofilm-forming bacteria isolated from the chronic wounds that developed naturally in these mice. These wounds do not re-epithelialize, the granulation tissue lacks vascularization and interstitial collagen fibers, they contain an antibiotic-resistant mixed bioflora with biofilm-forming capacity, and they stay open for several weeks. These findings are highly significant because they show for the first time that *chronic wounds* can be generated in an animal model effectively and consistently. The availability of such a model will significantly propel the field forward because it can be used to develop strategies to regain redox balance that may result in inhibition of biofilm formation and result in restoration of healthy wound tissue. Furthermore, the model can lead to the understanding of other fundamental mechanisms of chronic wound development that can potentially lead to novel therapies.

Editor: Vasu D. Appanna, Laurentian University, Canada

Funding: This work was partially supported by research grants from the National Institutes of Health to MMG (5R21AI78208-2), (AI078208-01); to RPP (HL092624); to CEC (HL072925); and to Virginia Commonwealth University for renovation of facilities (NH1C06-RR17393). These studies were also funded by the Department of Veterans Affairs, Veterans Health Administration, Office of Research and Development (Career Development Award CDA1 to DSW, VA), a Merit Award to CEC (BX001792) and a Research Career Scientist Award to CEC. Funding also came from the US-Israel Binational Science Foundation to CEC (BSF#2011380). Services and products in support of the research project were generated by the VCU Massey Cancer Center Lipidomics Shared Resource (Developing Core), supported, in part, with funding from NIH-NCI Cancer Center Support Grant P30 CA016059 as well as a shared resource grant (S10RR031535 to CEC) from the National Institutes of Health. The contents of this manuscript do not represent the views of the Department of Veterans Affairs or the United States Government. Co-authors Antonio Sanchez and Sean Gallagher are employed by UVP, LLC, an Analytik Jena Company. UVP, LLC, an Analytik Jena Company provided support in the form of salaries for authors AS and SG, but did not have any additional role in the study design and decision to publish. They did, however, help in data collection and analysis, and helped write the pertinent Materials and Methods section related to image collection and data analysis. The specific roles of these authors are articulated in the "author contributions" section.

Competing Interests: The authors have the following interests: Co-authors Antonio Sanchez and Sean Gallagher are employed by UVP, LLC, an Analytik Jena Company. There are no patents, products in development or marketed products to declare.

* Email: manuela.martins@ucr.edu

Introduction

Failure of acute wounds to proceed through the normal regulated repair process results in wounds that have impaired healing and/or become chronic [2,3]. Diabetic foot ulcers, venous ulcers, and other similar chronic wounds have a large impact on health, currently affecting ~6.5 M patients and costing ~$25B/year in the US alone [1]. Although great efforts have been made to switch the course of repair from non-healing wounds to healing wounds, success has been limited. This is primarily due to the pathophysiological complexity of changing an acute wound into a chronic wound and the lack of good animal models.

Injury causes the early generation of reactive oxygen species (ROS) in the presence of vascular membrane-bound nicotinamide-adenine-dinucleotide (NADH)-dependent oxidases (NOXs) that are produced by resident endothelial cells and fibroblasts [4]. ROS are required for defense against invading pathogens and low levels of ROS act as essential mediators of intracellular signaling that leads to proper healing [5,6]. However, uncontrolled production of ROS early after injury leads to an altered detoxification process caused by reduction in antioxidant production and activity [7]. Studies have provided evidence that non-healing ulcers in humans have high oxidative and nitrosative stress [8–10]. Furthermore, tissue hypoxia as well as anaerobic glycolysis, contribute to the production of lactate and its accumulation under inflammatory conditions [11,12]. Even in well-oxygenated wounds [11], when the number of neutrophils is high [13], lactate and ROS become significantly elevated as a result of aerobic glycolysis – the so-called "Warburg effect" [14]. This environment leads to a stagnant inflammatory phase. If the inflammatory cells are not removed from the wound tissue, they can promote further tissue damage through excessive production of inflammatory cytokines, proteases, and reactive oxygen intermediates, and increased cell death that, together, result in abnormal granulation tissue development and lead to wounds with impaired healing [15–17].

Nitric oxide (NO) also plays a key role in wound repair [18,19]. The beneficial effects of NO in wound repair relate to its functions in angiogenesis, inflammation, cell proliferation, matrix deposition, and remodeling. However, high levels of NO produced by inducible nitric oxide synthase (iNOS) produce peroxynitrite ($ONOO^-$), a reactive nitrogen species (RNS). $ONOO^-$ causes damage to DNA, lipids and proteins which invariably leads to cell apoptosis and/or necrosis depending on its concentration at the injury site [20].

It is virtually impossible to study the development of chronic wounds in humans. By the time these wounds appear in the clinic, the initial stage of development is well passed. Therefore, animal models to conduct studies on the genesis of non-healing chronic wounds are needed. We recently showed that a mouse in which the Tumor Necrosis Factor Superfamily Member 14 (TNFSF14/LIGHT) gene has been knocked out (LIGHT$^{-/-}$ mice) has impaired healing and that the wounds heal poorly and show many of the characteristics of impaired wounds in humans [21]. When compared to control, the wounds of LIGHT$^{-/-}$ mice show defects in epithelial-dermal interactions, high degree of inflammation, damaged microvessels with virtually no basement membrane or periendothelial cells, the collagen in the granulation tissue is mostly degraded, matrix metalloproteinases (MMPs) are elevated and tissue inhibitors of metalloproteinase (TIMPs) are downregulated. In addition, we also found that sometimes the LIGHT$^{-/-}$ wounds become chronic, and when they do, these defects are highly accentuated. In addition, the wounds become heavily infected with *Staphylococcus epidermidis* [21], a gram-positive bacterium frequently found in human chronic wounds

[22]. All of these characteristics are very similar to those found in chronic wounds in humans [23–25]. Mechanistically, we have shown that LIGHT mediates macrophage cell death induced by vascular endothelial growth factor (VEGF) and that this occurs in a LTβ receptor-dependent manner [26], indicating that LIGHT is involved in the resolution of macrophage-induced inflammation. In addition, LIGHT$^{-/-}$ mice also show increased levels of Forkhead box protein A1 (FOXA1), Cytochrome P450 2E1 (CYP2E1), and Toll-like receptor 6 (TLR6) which are genes involved in oxidative stress [27–30]. Furthermore, Aldehyde oxidase 4 (AOX4) is also elevated in these knockout mice. This enzyme leads to the generation of O_2^- that then aids in release of iron from ferritin [31]. Here we show that by manipulating the microenvironment at wounding we can cause the impaired wounds to become chronic 100% of the time and propose that the same processes are involved in chronic wound development in humans. This model provides an opportunity to understand fundamental mechanisms involved in chronic wound development that can potentially lead to identifying diagnostic molecules and to the discovery of novel treatments.

Results

In order to identify parameters in the wounds with impaired healing that, when changed, may lead these wounds to become chronic, we first characterized the state of ROS/RNS in the early stages of impaired healing by examining a variety of components of the oxidative and nitrosative stress cycle as represented schematically in **Figure S1 in File S1** Superoxide dismutase (SOD) dismutates superoxide anions (O_2^-) to generate H_2O_2, which can then be detoxified by catalase to $H_2O + O_2$ and by glutathione peroxidase (GPx) to H_2O. ROS can also enter the Fenton reaction in the presence of ferrous ions to give rise to.OH+ OH^-. O_2^- can also interact with nitric oxide (NO) produced by nitric oxide synthase (NOS) to give rise to peroxynitrite anion ($ONOO^-$). The effects of oxidative and nitrosative stress are shown in terms of lipid peroxidation, DNA damage, protein modification and cell death. Secondly, we will present the data on the manipulation of the redox balance that leads to development of chronic wounds including the characterization of the polymicrobial environment that favors growth of biofilm-forming aerobic and anaerobic bacteria. For all figures (**Figures 1, 2, 3, and 4**) except (**Figure 2C**), time t = 0 represents unwounded skin.

Characterization of the redox environment in wounds with impaired healing ROS

Oxidative stress. To determine whether the wounds with impaired healing have increased oxidative stress, we measured the levels of SOD. SOD activity was already significantly elevated by 4 hrs post-wounding in the LIGHT$^{-/-}$ wounds compared to the C57BL/6 wounds and remains high through 48 hrs (**Figure 1A**). H_2O_2 levels also were significantly elevated as early at 4 hrs post-wounding in the LIGHT$^{-/-}$ wounds, decreasing to control levels by 48 hrs (**Figure 1B**). Furthermore, we observed that in LIGHT$^{-/-}$ mice, both catalase and GPx activities were similar to control mice, suggesting that accumulation of H_2O_2 was primarily caused by the inability of the antioxidant system to keep up with the oxidative stress (**Figure 1C,D**).

It is well known that, in human wounds, oxidative stress increases with age. We determined that oxidative stress in wounds of old LIGHT$^{-/-}$ mice also increased with age; higher levels of SOD activity were seen in wounds of old LIGHT$^{-/-}$ mice than in their adult counterparts (**Figure 1E**). H_2O_2 levels in the wounds of old LIGHT$^{-/-}$ mice were at least 10 times higher than those in

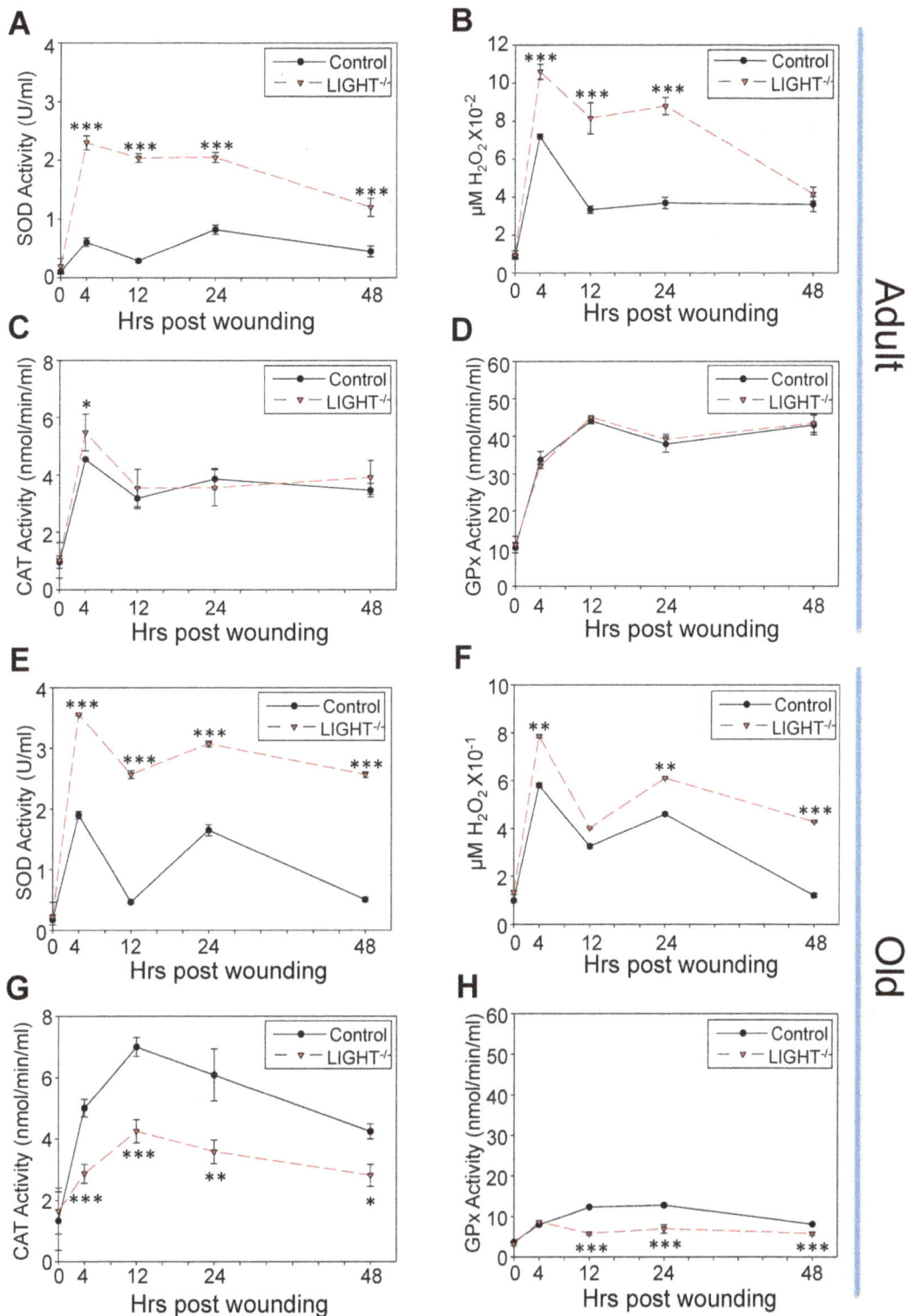

Figure 1. Oxidative stress is elevated in LIGHT$^{-/-}$ wounds. (A) SOD activity was measured using a tetrazolium salt that converts into a formazan dye detectable at 450 nm. SOD activity remains significantly elevated in LIGHT$^{-/-}$ mice in the first 48 hrs post-wounding. $n = 6$. (B) Resofurin formation, detected at 590 nm, was used to determine H_2O_2 levels. Significant increases in H_2O_2 very shortly post-wounding were seen. $n = 8$. (C) Enzymatic reaction of catalase and methanol in the presence of H_2O_2 gives rise to formaldehyde, spectrophotometrically detected with purpald chromogen, at 540 nm. Catalase activity in adult LIGHT$^{-/-}$ and control wounds was similar. $n = 6$. (D) GPx detoxifying activity was measured indirectly at 340 nm by a coupled reaction with glutathione reductase where GPx activity was rate-limiting. The level of GPx activity in the adult LIGHT$^{-/-}$ wounds was essentially identical to that of the controls. $n = 6$. (E-H) The findings in old LIGHT$^{-/-}$ mice were exacerbated in all four parameters when compared to adult LIGHT$^{-/-}$ mice. $n = 6$. Time zero represents unwounded skin. All data are Mean $\pm$ SD. *$p < 0.05$,**$p < 0.01$,***$p < 0.001$.

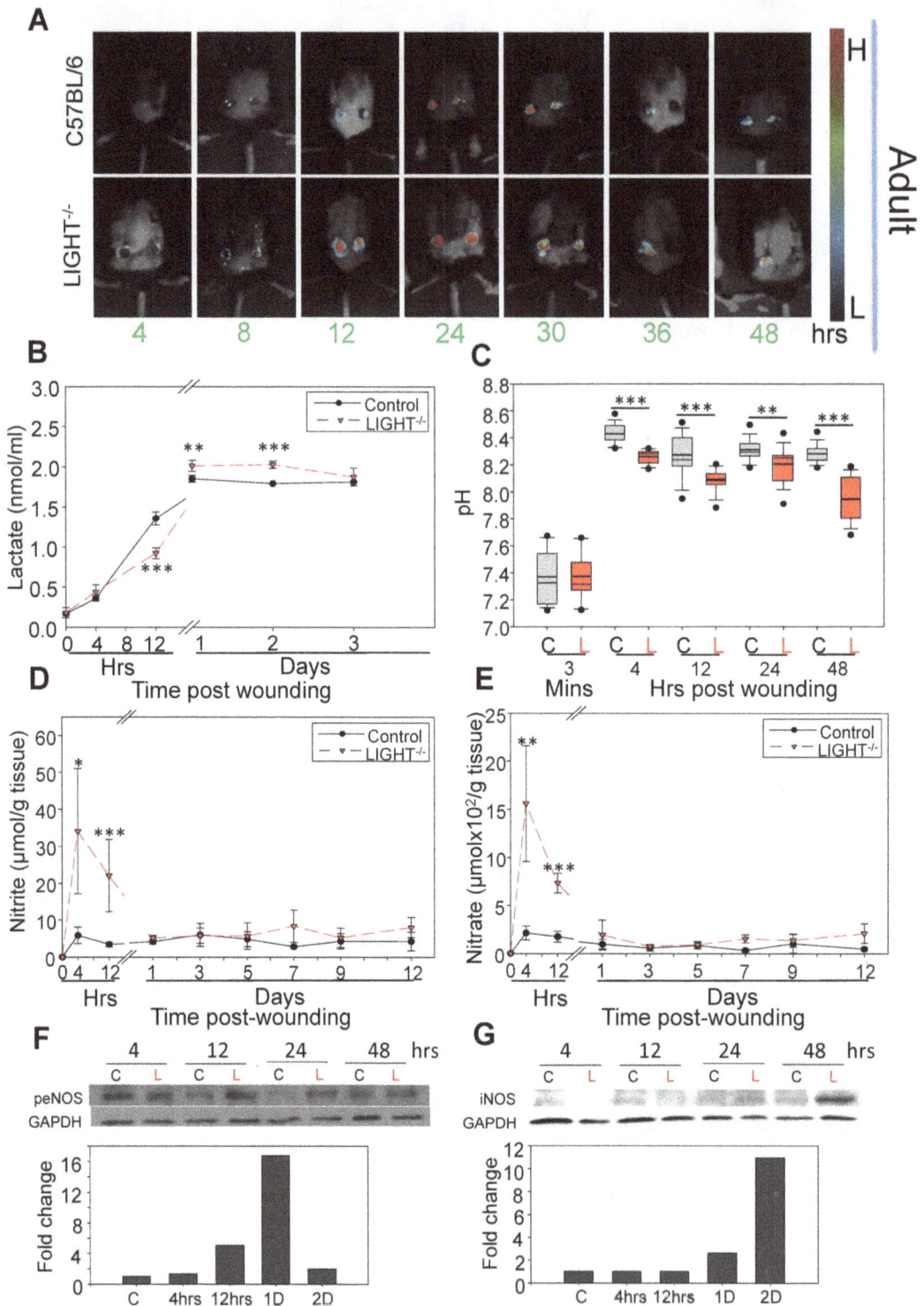

Figure 2. Microscopic, biochemical and chemical markers show imbalanced redox in LIGHT$^{-/-}$ mice. (A) *In vivo* imaging of ROS was carried out using the ImageEM 1K EM-CCD camera with an optical system consisting of a 50 mm f/1.2 lens. Signals were obtained around the periphery of the wound as early as 4 hrs post-wounding in the LIGHT$^{-/-}$ mice and significantly higher signals captured in LIGHT$^{-/-}$ mice peaked at 24 hrs post-wounding. (B) Lactate measurements: An oxidized intermediate was formed when extracted lactate reacted with a probe to give fluorescence detectable at 605 nm. There was significant increase in levels of lactate accumulation in LIGHT$^{-/-}$ mice at 24–48 hrs post wounding. $n = 6$. (C) pH levels were measured using a beetrode microelectrode and micro-reference electrode. The LIGHT$^{-/-}$ wounds were systematically more acidic than controls. $n = 25$. (D,E) Methanolic-extracted nitrite (D) and nitrate (E) were analyzed. Both were greatly increased in LIGHT$^{-/-}$ mice during early response to wounding. $n = 8$. (F-G) Phospho-eNOS levels and iNOS expression in LIGHT$^{-/-}$ wounds were examined by western blotting (representative experiment shown). Analysis by densitometry (normalized to C57BL/6 mouse wound). *Time zero represents unwounded skin except in Figure 2C. All data are Mean $\pm$ SD. *p<0.05,**p<0.01,***p<0.001.*

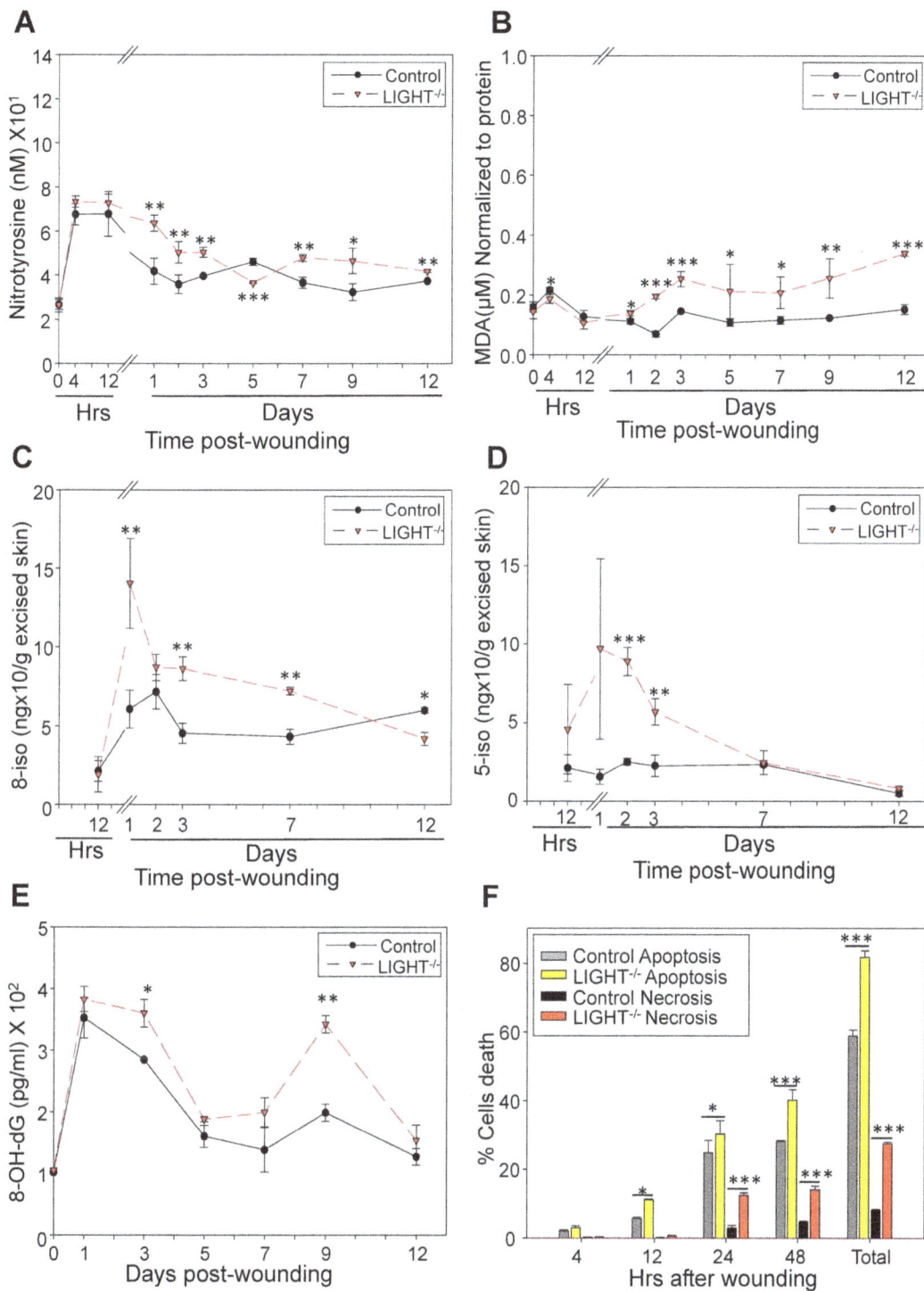

Figure 3. Early oxidative and nitrosative stress in LIGHT$^{-/-}$ wounds have damaging effects on proteins, lipids and DNA and increased cell death. (A) Protein modification measurements were based on a competitive enzyme immunoassay; nitrotyrosine levels in the LIGHT$^{-/-}$ mice were significantly different from control throughout healing. (B) Lipid peroxidation levels were measured fluorometrically at an Ex/Em of 540 nm/590 nm using thiobarbituric acid reactive substances (TBARS); the MDA levels were significantly elevated throughout the course of wound healing in LIGHT$^{-/-}$ mice. $n = 6$. (C, D) F$_2$ isoprostanes, were measured using the approach described in the M&M section; levels of 8- and 5-isoprostanes detected in LIGHT$^{-/-}$ mice were much higher than those in the control mice at early times. This correlates with the MDA levels that are the stable byproducts of lipid peroxidation. $n = 5$. (E) Levels of 8-OH-dG, were based on a competitive enzyme immunoassay; the samples were read spectrophotometrically at 412 nm using Ellman's reagent. 8-OH-dG levels were found to be significantly elevated during the course of healing in LIGHT$^{-/-}$ mice. $n = 4$. (F) Cell death by apoptosis and necrosis was determined by staining with Annexin V-FITC and propidium iodide, respectively, followed by FACS analysis. Cell death was increased significantly in the LIGHT$^{-/-}$ mice. The greatest difference occurred with necrosis, which showed to be much higher in LIGHT$^{-/-}$ mice. *Time zero represents unwounded skin. All data are Mean $\pm$ SD. *$p < 0.05$,**$p < 0.01$,***$p < 0.001$.*

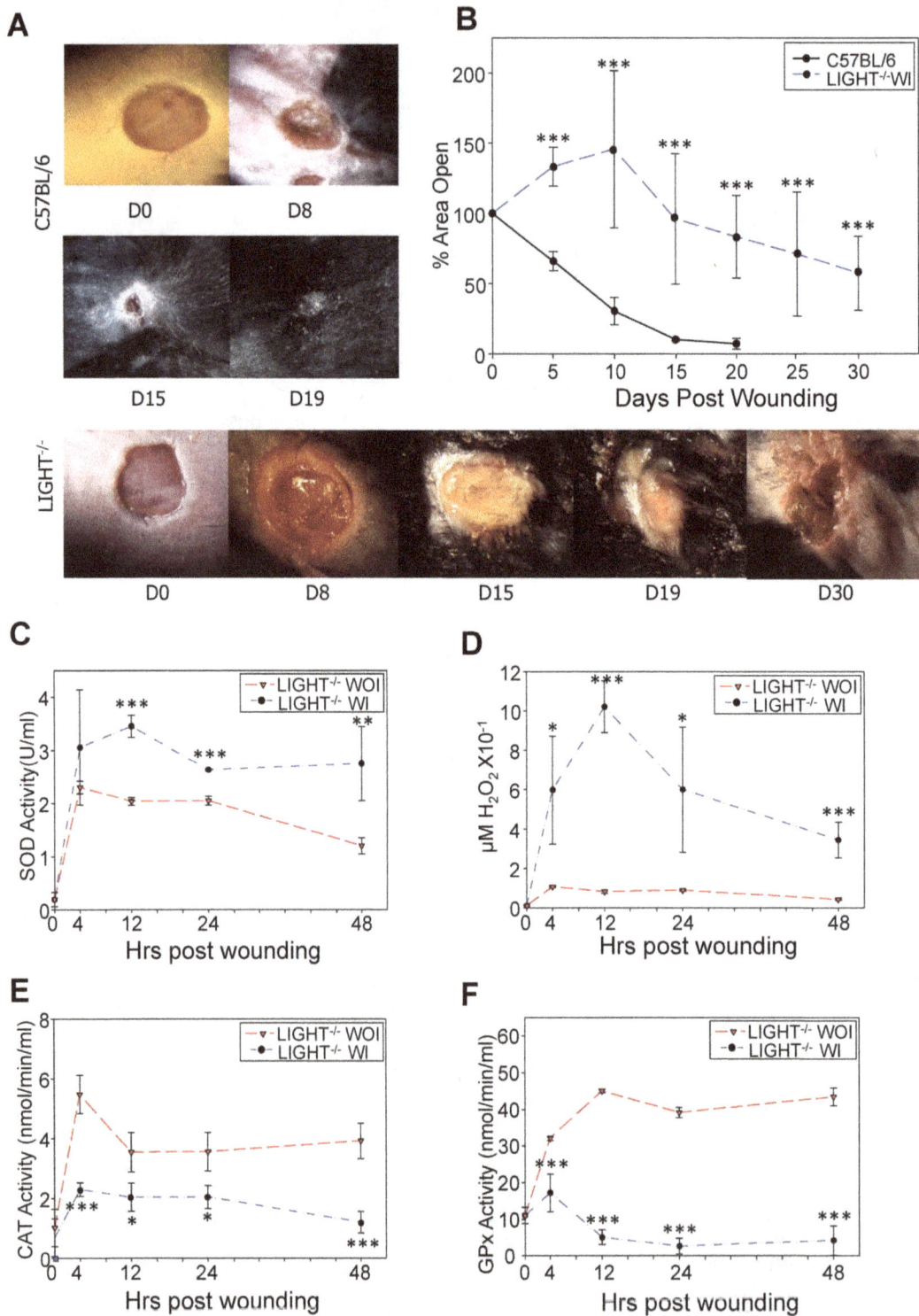

Figure 4. Manipulating redox parameters leads to development of chronic wounds. (A) C57BL/6 and LIGHT$^{-/-}$ mice were wounded and immediately treated with inhibitors for GPx and catalase followed by the application of biofilm-forming bacteria 24 hrs later. The wounds were covered with sterile tegaderm to maintain a moist wound environment and prevent external infection. The LIGHT$^{-/-}$ wounds became chronic and remained open for more than 30 days. $n = 30$. (B) Wound areas were traced using ImageJ and % open wound area was calculated. The LIGHT$^{-/-}$ wounds remained open for significantly longer time than the C57BL/6 wounds with similar treatment. $n = 8$. (C-F) SOD activity (C); H$_2$O$_2$ levels (D); Catalase activity (E); and GPx Activity (F) were measured as described in Figure 1. All were greatly different from controls. For all tests $n = 6$ at minimum. *Time zero in C-F represents unwounded skin. All data are Mean $\pm$ SD. *$p < 0.05$, **$p < 0.01$, ***$p < 0.001$.*

the adult mice (compare **Figure 1F** with **Figure 1B**). In contrast, the level of catalase activity was significantly lower in wounds of old LIGHT$^{-/-}$ mouse than in wounds of old C57BL/6 mice (**Figure 1G**) but was comparable to the wounds of adult LIGHT$^{-/-}$ mice (compare **Figure 1G** with **Figure 1C**). Similarly, GPx activity was significantly lower in the wounds of old LIGHT$^{-/-}$ mice than in old C57BL/6 mice (**Figure 1H**) and was much lower than in either type of adult mice (compare **Figure 1H** with **Figure 1D**). Taken together, these results suggest that adult LIGHT$^{-/-}$ wounds have high levels of oxidative stress and that, much like in humans, these levels are exacerbated with age [24,32].

To further confirm the elevated presence of ROS we performed real time *in vivo* imaging of excision wounds at various time points after wounding. Imaging was initiated immediately after IP injection of luminol that emits light in the presence of an oxidizing agent such as H_2O_2. We detected a signal on the edges of the wound in the LIGHT$^{-/-}$ mouse as early as 4 hrs after wounding. The level of intensity was increased significantly in LIGHT$^{-/-}$ mice compared to C57BL/6 throughout the early hours post-wounding (**Figure 2A**). Similar results were obtained when imaging old LIGHT$^{-/-}$ and C57BL/6 mice (data not shown). These real-time images show for the first time that, *in vivo*, ROS can be detected *in situ* as early as 4 hrs after wounding.

The presence of oxidative stress leads to increase in enzymatic activity of lactate dehydrogenase (LDH), which results in lactate generation. Because ROS-generated oxidative stress is elevated in the LIGHT$^{-/-}$ wounds, we investigated production of lactate in the wound microenvironment [15,33]. Higher levels of lactate production were seen at 12 hrs post-wounding in the control mice whereas LIGHT$^{-/-}$ mice showed a delayed, but significant, accumulation during days 1 and 2 post-wounding (**Figure 2B**). The levels of lactate accumulation in wounds of old LIGHT$^{-/-}$ mice were similar to the wounds of adult LIGHT$^{-/-}$ mice and also were significantly higher than wounds in old C57BL/6 mice (**Figure S2A in File S1**).

The pH in a wound milieu is a dynamic factor that can change rapidly and affect healing. Studies have shown that the presence of acidic pH correlates with compromised, chronic, and infected wounds [34]. pH measurements of the wound bed were collected immediately, within 3 minutes after wounding and then at the indicated hrs. Relative to the control, the pH obtained from LIGHT$^{-/-}$ wounds was more acidic by 4 hrs post-wounding and remained so through at least 48 hrs (**Figure 2C**). Similar results were obtained with old LIGHT$-/-$ mice (**Figure S2B in File S1**). Unwounded skin surface pH was not measured because the glass microelectrodes we used require moisture and the skin is dry. Humidifying the skin with water will alter the pH because of the presence of free fatty acids on the skin that releases H+ ions into the water applied and can give measurements that are not accurate [35]. Correlations between lactate and pH (proton transport) have previously been shown to increase in parallel to each other [36,37]. The same occurs in these wounds.

Nitrosative stress, protein modification, and damage of lipids and DNA.   To determine whether wounds of LIGHT$^{-/-}$ mice have high nitrosative stress, we examined the metabolites of NO, nitrite (NO_2^-) and nitrate (NO_3^-) and found that shortly after wounding the levels of nitrite in the adult LIGHT$^{-/-}$ mice wounds were very much higher than those in the control at 4 and 12 hrs but declined to normal by day 1 (**Figure 2D**). Nitrate levels showed the same pattern of elevation as nitrite (**Figure 2E**). Old mice showed a similar pattern of elevation but the levels were even higher than in adult LIGHT$^{-/-}$ wounds between 4–12 hrs post-wounding (**Figure S2C,D in File S1**). To determine

whether the elevated levels of NO_2^- and NO_3^- early post-wounding were due to changes in nitric oxide synthase (NOS), both endothelial NOS (eNOS) and inducible NOS (iNOS) were examined for phosphorylation/activation of eNOS and elevated expression of iNOS in LIGHT$^{-/-}$ mouse wounds. We found that the levels were significantly elevated but that elevation did not occur until 12 hrs and 24 hrs post-wounding, respectively (**Figure 2F,G**), suggesting that the increase in NO production must be due to activation of other systems/factors occurring very early after wounding.

Modification of tyrosine residues to 3-nitrotyrosine in proteins by ONOO$^-$ or other potential nitrating agents occurs when tissues are subject to nitrosative stress. Because we show the presence of nitrosative stress, we examined the levels of 3-nitrotyrosine (3-NT) to assess the effects of this stress on protein modification during healing of the LIGHT$^{-/-}$ mice. We found that the levels of 3-NT were significantly elevated in LIGHT$^{-/-}$ mouse wounds 1 day post-wounding and, except for day 5, remained significantly elevated throughout the course of healing (**Figure 3A**), confirming the deleterious effects of the presence of nitrosative stress. These effects were almost doubled in the old LIGHT$^{-/-}$ mice (**Figure S3A in File S1**).

It is known that increase in ROS/RNS can cause lipid peroxidation. Lipid peroxides are unstable markers of oxidative stress that decompose to form malondialdehyde (MDA) and 4-hydroxynonenal (4-HNE). We found a significant increase in MDA levels 48 hrs after wounding that remained significantly elevated throughout healing (**Figure 3B**). We also found that the levels of lipid peroxidation were exacerbated in wounds of old LIGHT$^{-/-}$ mice after 48 hr post-wounding (**Figure S3B in File S1**). Furthermore, we used mass spectroscopy to examine whether ROS-induced non-enzymatic peroxidation products of arachidonic acid, such as isoprostanes, were present in the wounds. We found that 8-isoprostane (8-epi-PGF$_{2\alpha}$) and 5-isoprostane were significantly elevated in the LIGHT$^{-/-}$ mouse wounds, suggesting the breakdown of arachidonic acid in the presence of ROS (**Figure 3C,D**). These results confirm that there is lipid damage in the LIGHT$^{-/-}$ wounds.

Another detrimental effect caused by excessive oxidative stress and nitrosative stress is 8-hydroxylation of the DNA guanine base (8OHdG) that results in DNA damage. The overall levels of this stress marker were increased in wounds of adult LIGHT$^{-/-}$ mice (**Figure 3E**), with significant increase at days 3 and 9 post-wounding. We also found that the levels of 8-OHdG in the old LIGHT$^{-/-}$ mouse wounds were significantly more elevated throughout the course of healing (**Figure S3C in File S1**).

Levels of cell death by both apoptosis and necrosis.   Given that excessive redox stress results in damage of DNA, proteins, and lipids that are critical for cell survival and function, we examined cell death both by apoptosis and necrosis (**Figure 3F**). Apoptosis was significantly increased 12 hrs post-wounding and increased even more by 48 hrs post-wounding. Cell death by necrosis was predominantly found at 24 hrs and 48 hrs. Particularly striking is the difference in cell death by necrosis between control and LIGHT$^{-/-}$ mice. This elevated cell death is potentially due to the higher levels of oxidative stress and can lead to chronic inflammation, impaired healing and delayed wound closure.

Manipulation of redox balance and the presence of biofilm-forming bacteria lead to development of chronic wounds

Our previous results [21] and the results presented above strongly suggest that the LIGHT$^{-/-}$ impaired healing is caused by redox imbalance established shortly after injury, resulting in excessive cell death which then creates an environment that increases inflammation and is propitious for the growth of biofilm-forming bacteria, thereby setting the wound on a course that leads to development of chronic ulcers. To test this possibility we significantly increased the oxidative stress in the wound by further inhibiting the antioxidant enzymatic activity and applying the biofilm-forming bacteria, S. epidermidis C2, that we isolated from the spontaneously-developed chronic wounds of the LIGHT$^{-/-}$ mice [21]. Inhibition of catalase by 3-Amino-1,2,4-triazole (ATZ) and GPx by mercaptosuccinic acid (MSA) immediately post-wounding and application of S. epidermidis C2 24 hrs later was sufficient to turn the wounds with impaired healing into chronic wounds 100% of the time (**Figure 4A**). Chronic wounds were successfully created in 30 animals used in 10 different experiments. Wounds of C57BL/6 mice treated under the same conditions closed in 15–19 days whereas the LIGHT$^{-/-}$ wounds remained open for>4 weeks (**Figure 4B**). The wounds were kept covered at all times using sterile tegaderm and changed upon compromised sealant of the bandage.

Following the application of inhibitors post-wounding, and bacteria 24 hours later, we evaluated the levels of ROS to determine whether the levels of oxidative stress increased. With antioxidant inhibitor treatment, SOD (**Figure 4C**) and H_2O_2 levels (**Figure 4D**) were significantly elevated by 12 hours post-wounding. Corresponding to the increase in ROS, antioxidant enzymes catalase and GPx, that were inhibited by ATZ and MSA respectively, were decreased significantly (**Figure 4E,F**). These experiments were conducted simultaneously, under identical conditions.

Histological examination of chronic LIGHT$^{-/-}$ wounds showed that the migrating tongue of the epidermis was blunted and tortuous (**Figure 5A,B**) rather than thin and linear as in the control (**Figure S4A in File S1**). Also, the granulation tissue was poorly developed (**Figure 5A**) when compared to normal granulation tissue (**Figure S4B,C in File S1**). Collagen IV, a component of the basal lamina, was well-formed behind the migrating tongue but was absent under the tortuous migrating edge (**Figure 5C-E**). We also found that these wounds contain macrophages, indicating that inflammation has not been resolved (**Figure 5F,H**; inserts show higher magnification of one macrophage). Furthermore, the interstitial collagen deposition and organization were abnormal in the LIGHT$^{-/-}$ chronic wounds as revealed by Masson trichrome staining (**Figure 5I**) and by second harmonic generation imaging microscopy (SHIM) (**Figure 5J,K**). Although interstitial collagen was present, the collagen fibers were not clearly visible and did not form proper bundles (**Figure 5J**). This is similar to the finding we published on the impaired wounds of LIGHT−/− wounds [21] but much more exaggerated.

To determine whether the application of the bacteria alone or in the presence of a single inhibitor could induce chronicity in the LIGHT$^{-/-}$ wounds, we introduced S. epidermidis C2 24 hrs post-wounding without any inhibitor treatment (**Figure S5A in File S1**) or with just ATZ treatment (**Figure S5B in File S1**) or with just MSA treatment (**Figure S5C in File S1**). In all three cases, both in the C57BL/6 and LIGHT$^{-/-}$ mice, the wounds healed by day 15–19, suggesting that development of chronic wounds requires all three of these elements: inhibition of both catalase and GPx to greatly decrease the antioxidant enzymes in the wound, plus addition of biofilm-forming bacteria.

It has been established that the bioflora that colonize chronic wounds in humans is commonly polymicrobial [38,39]. Therefore, we determined whether the LIGHT$^{-/-}$ chronic wounds also exhibited this polymicrobial phenotype. Wound exudates from both adult and old LIGHT$^{-/-}$ mice were collected and the bacteria genus/species determined as described in Materials and Methods. In addition, staining of adherent cells with Hucker crystal violet, which has been widely used as readout for biofilm-production [40–43], was used as a qualitative measure for biofilm formation in our bacteria isolates. As expected, we found that biofilm-forming (OD570 nm≥0.125) coagulase-negative Staphylococcus epidermidis was present in the wounds throughout healing, given that we infected the wounds with S. epidermidis C2. However, co-colonizing bacteria were also isolated. These co-colonizers were identified as non-biofilm forming hemolytic Streptococcus sp., biofilm-producing oxidase-positive aerobic Gram-negative rods (presumptively Pseudomonas), and Enterobacter cloacae (dotted line in **Figure 6A** defines the minimum optical density for biofilm formation).

Quantification of the relative bacterial prevalence showed that the dynamics of the colonizing bioflora in adult LIGHT$^{-/-}$ mouse wounds changes over time (**Figure 6B**). These changes are marked by the decreased concentration of S. epidermidis populations coupled with the appearance of the oxidase positive Gram-negative rods followed by E. cloacae. As the wound progresses to a non-healing/chronic stage at ~20 days post-wounding, the E. cloacae population dominates the wound with traces of S. epidermidis (**Figure 6B**). Irrespective of the shift in bacterial population of the wounds, the overall degree in biofilm production by these polymicrobial communities (dotted line in **Figure 6C**) did not change significantly over time at least until 22 days. However, the individual contribution to biofilm production varies and is dependent on the time of isolation and is species specific (**Figure 6C**). Eight days post-wounding, biofilm-producing Staphylococcus epidermidis (C2) is significantly different from the non-biofilm-producing negative control, Staphylococcus hominis (SP2, ATCC 35982). SP2 does not adhere to polystyrene plates, does not produce extracellular polysaccharide and is a commensal bacterium found on human skin [44]. Because of these characteristics, this strain has been widely used as a negative control for biofilm production [40,45,46]. Similar observations were made in the old LIGHT$^{-/-}$ mice (**Figure S6A-C in File S1**).

It has been well established that biofilm-associated wound infections are extremely resistant to antimicrobial therapy [47,48]. The community minimal inhibitory concentration (CMIC) of amoxicillin required to inhibit the growth of biofilm-producing microbial flora from LIGHT$^{-/-}$ adult chronic wounds was determined to be 50 µg/mL (day 22/24) compared to the 0.4–0.8 µg/mL required for non-biofilm producing colonizers (day 5) (**Figure 6D**). This suggests that biofilm-producing microbial flora isolated from LIGHT$^{-/-}$ chronic wounds are ~50X more resistant to killing by amoxicillin compared to their non-biofilm producing counterparts.

It has been reported that the majority of chronic wounds in humans have bacterial contamination and high levels of bacterial burden will likely result in impaired healing [49]. At 5 and 8 days post-wounding, colony-forming unit counts (CFU/mL of exudate) from adult and old LIGHT$^{-/-}$ mouse exudates show low levels of bacterial burden (1.6×10^3 CFU/mL and 2.0×10^3 CFU/mL respectively). However, these levels reach 4.0×10^7 CFU/mL and 7.4×10^7 CFU/mL by 22–24 days of healing (**Figure 6E and Figure S6D**).

In order to determine whether the skin of mice contain the bacteria that eventually make biofilm in the chronic wounds, we took skin swabs from unwounded C57 and LIGHT$^{-/-}$ mice and cultured them in vitro (**Figure 6F**). The majority of the cultured

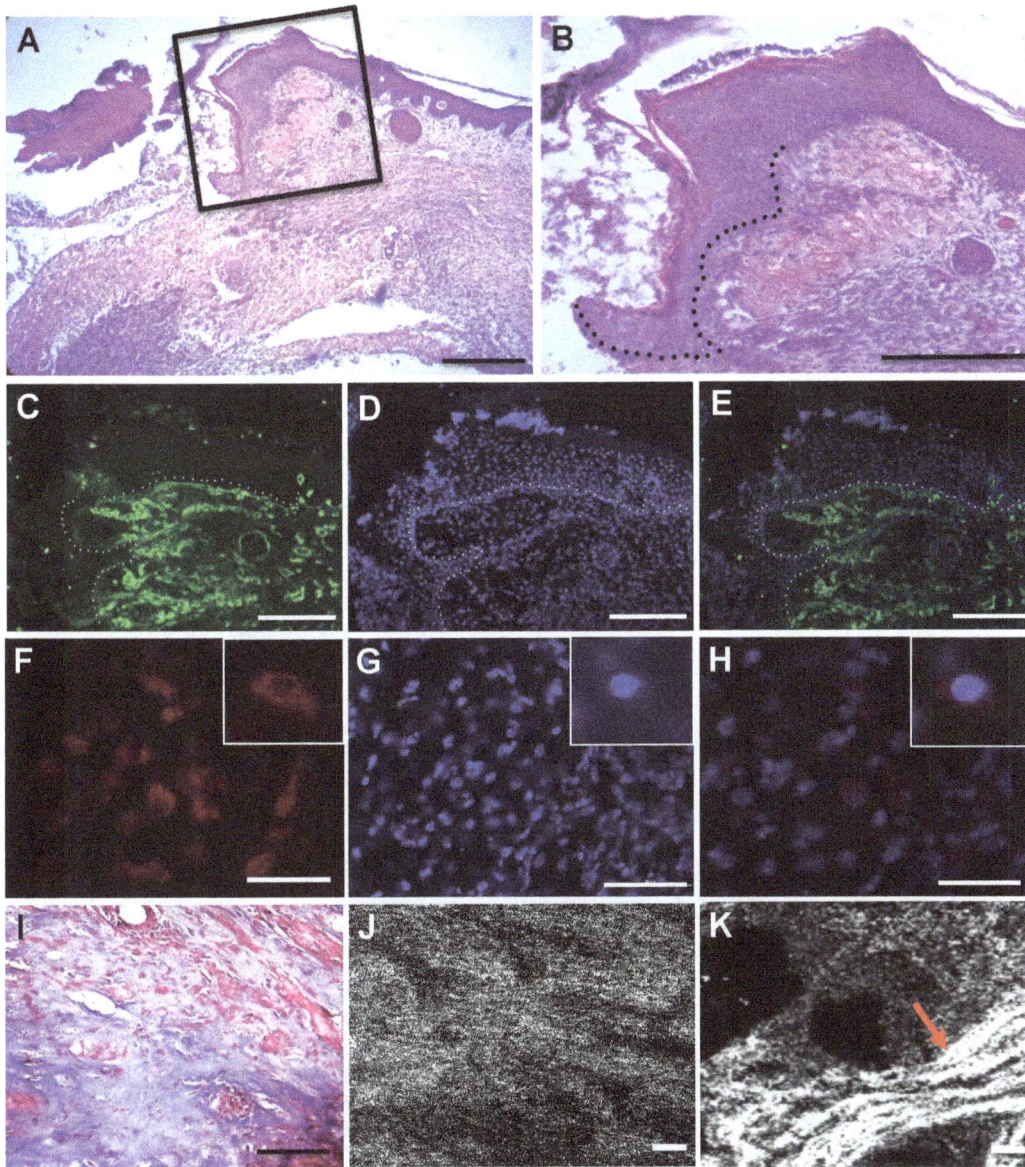

Figure 5. Histological evaluation of chronic wounds. (A) Representative picture of H&E-stained sections of a LIGHT$^{-/-}$ chronic wounds from an animal treated with catalase and GPx inhibitors and the application of bacteria. The epithelium does not cover the wound tissue and the granulation tissue is poorly formed. Scale bar 500 μm. (B) Higher magnification of the boxed area in (A). Epithelial tongue is outlined with a dotted line (compare with Figure S4A). Scale bar 100 μm. (C) Immunolabeling for Collagen IV delineates the presence of basement membrane; dotted line marks where basement membrane is missing in the migrating tongue. (D) propidium iodide staining identifies cell nuclei. (E) Merger of (C) & (D). (F) Immunolabeling for F4/80, a marker for macrophages, to illustrate the presence of inflammation; (G) propidium iodide staining identifies cell nuclei. (H) Merger of (F) & (G). Inserts are high magnifications of a single macrophage. (I) Representative Masson-trichrome (blue color) stained section illustrating loss of collagen bundles; scale bar 100 μm. (J,K) SHIM analysis of a similar section (J) confirms results in (I) and, for comparison, collagen in the granulation tissue of a normal wound similarly analyzed by SHIM (K) showing filamentous collagen (red arrow); scale bar 10 μm.

bacteria belong to the Firmicutes phylum, specifically *Staphylococcus spp. and Streptococcus spp.* We also documented the presence of bacteria that belong to the Proteobacteria phylum (e.g. various Gram-negative rods and *Enterobacter*). These bacteria are all known to be associated with the human skin microbiota [50].

To further confirm the presence of biofilm-forming bacteria in these wounds we performed scanning electron microscopy on LIGHT−/− chronic wounds. An abundance of bacteria was observed in the wound and some of those bacteria were embedded in a biofilm-like matrix (**Figure 7A**), with some of them appearing to reside in a defined niche surrounded by matrix (**Figure 7B**).

Beneath the biofilm we observed the presence of numerous inflammatory cells adherent to extracellular matrix (**Figure 7C**). Furthermore, analysis of the glycosyl composition of the exudate collected from the chronic wounds showed high levels of N-acetylglucosaminyl (GlcNAc), galacturonosyl (GalU), mannosyl, galactosyl and glucosyl residues (data not shown). This glycosyl composition is consistent with the presence of extracellular polysaccharide material, and possibly N-glycoproteins, in the chronic wound. These carbohydrates have also been shown to be present in human chronic wounds during *P. aeruginosa* infections [51] and more recently exopolysaccharides with glycosyl compo-

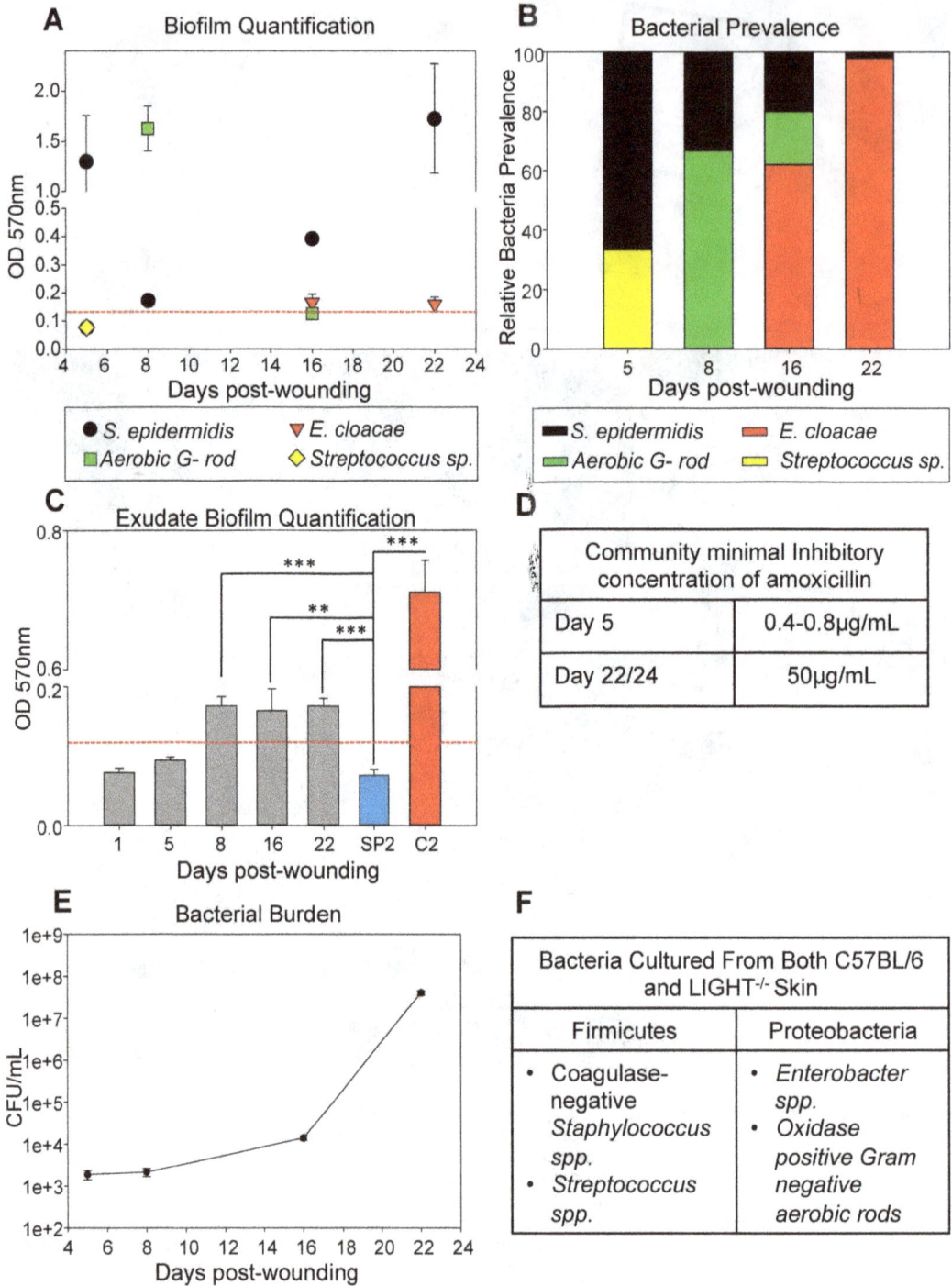

Figure 6. Identification and characterization of the microflora that colonizes the LIGHT$^{-/-}$ chronic wounds. (A) Biofilm production was quantified by measuring the optical densities of stained bacterial films adherent to plastic tissue culture plates. Biofilm forming capacity of *S. epidermidis* was seen throughout the time course of chronic wounds. *n* = 7. (B) Bacterial identification was carried out by growing bacteria on tryptic soy agar. Gram-negative rods were characterized using the API 20E identification kit. *n* = 7. (C) Biofilm quantification of exudate obtained from wounds was performed at OD570 nm. The dynamics of the polymicrobial community in the wounds does not seem to affect the overall degree of biofilm production during the later stages of healing. Controls used were biofilm-negative (OD570 nm<0.125) *S. hominis* SP2 and biofilm-positive *S. epidermidis* C2. *n* = 8. (D) Antibiotic challenge on wound exudates collected from LIGHT$^{-/-}$ mice was done using Amoxicillin. The CMIC of amoxicillin on the bacteria found in the chronic LIGHT$^{-/-}$ wound exudate at day 22/24 was 50 μg/ml, much higher than exudate collected at day 5 when biofilm is not yet abundant. (E) Bacterial burden was evaluated by colony forming unit counts. The CFU/mL was relatively low during the early phases of healing and was highest during the impaired and chronic stages of healing. *n* = 7. (F) Normal skin swabs were collected from LIGHT and C57BL/6 mice to evaluate resident organisms. The microbiota of the skin was similar in both C57BL/6 and LIGHT$^{-/-}$ mice.

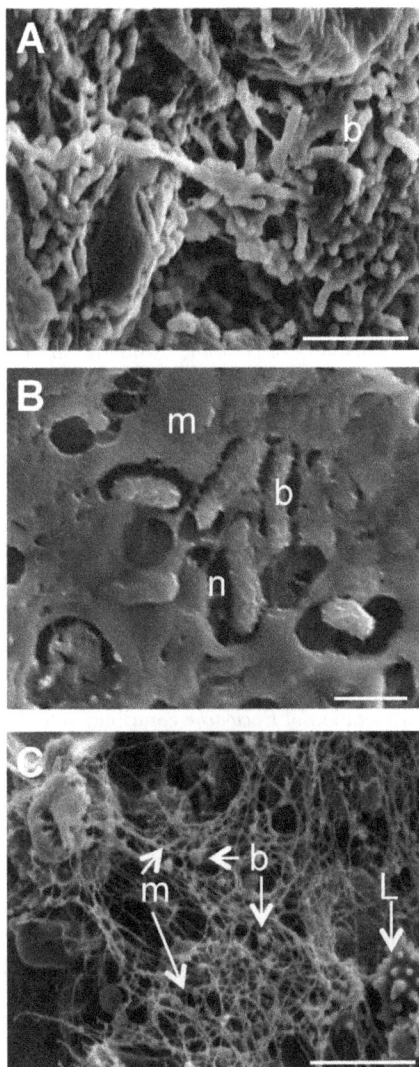

Figure 7. Morphological characterization of biofilm present in LIGHT$^{-/-}$ wounds. Scanning electron microscopy (SEM) images of the Au/Pd sputtered, fixed and dried, chronic wound samples were captured using an FEI XL30 FEG SEM. (A) Image shows the presence of bacterial rods (b) in the wound bed. (B) High magnification image of bacteria embedded in a biofilm-associated matrix (m) in a well-defined niche (n). (C) Matrix beneath the biofilm showing the presence of matrix (m) and of cocci bacteria (b). A Lymphocyte (L/arrow) was highlighted for size references. Scale bars 5 μm (A,C) and 1 μm (B).

sitions including these residues have been characterized in other species such as *Staphylcococcus* and *Enterobacter* which are pathogens commonly found in humans [52].

Discussion

We have shown that we can *create chronic wounds* by manipulation of the impaired wounds of LIGHT$^{-/-}$ mice using antioxidant enzyme inhibitors to further increase ROS/RNS and by adding biofilm-forming bacteria previously isolated from the naturally occurring chronic wounds of these transgenic mice. This approach leads to the generation of chronic wounds 100% of the time. These wounds: (1) Contain high levels of reactive oxygen and nitrogen species and, much like in humans, these levels increase with age; (2) have decreased levels of anti-oxidant enzymes

indicating the buildup of oxidative stress in the wound environment; (3) contain increased peroxynitrite and lipid peroxidation-derived products, increased 3-nitrotyrosine levels, increased DNA damage and high levels of cell death, contributing to redox imbalance in the wound microenvironment; (4) do not heal for weeks.

Our data show that SOD enzymatic activity is highly elevated in the first 48 hrs post-wounding which likely is the cause for continued increase of H_2O_2 at the wound site. This is particularly important because the activities of the antioxidant enzymes, catalase and GPx, are not elevated to compensate for the extra H_2O_2 produced. In old animals, catalase and GPx activity is even lower than in the control, exacerbating the levels of H_2O_2. Furthermore, not only can H_2O_2 cause damage directly, it can also enter the Fenton reaction in the presence of divalent iron ions to produce hydroxyl radicals (.OH) that lead to additional tissue damage [8,53,54].

The LIGHT$^{-/-}$ wounds also have high levels of inflammatory cells early after wounding that persist for a long time [21]. Increase in inflammation in a hypoxic wound tends to drive lactate accumulation that, in turn, leads to an unchecked proton gradient. As a consequence, lactate plays an important role in maintaining the fine acid-base milieu [11–13]. LIGHT$^{-/-}$ wounds showed increases in lactate levels both in adult and old mice, suggesting a pH imbalance. Recent findings on successful acceptance of skin grafts on chronic wounds was higher at elevated pH (alkaline) than at lower (acidic) pH [36,55,56]. In the control wounds, the pH shifted to alkaline at 4 hrs whereas in the LIGHT$^{-/-}$ wounds it shifted to more acidic and the levels remained acidic throughout at least the first 2 days in both adult and old LIGHT$^{-/-}$ mice, potentially contributing to the impaired healing in the wounds of these mice. Although we do not know whether increases in anaerobic metabolism are due to the down regulation of oxidative phosphorylation in an effort to alleviate oxidative stress, we are currently studying the gene expression profiles of the LIGHT$^{-/-}$ wound very early post wounding to obtain in-depth insight into the genes/proteins responsible for such processes.

The levels of nitrite and nitrate, end products of NO metabolism, were significantly elevated very early post-wounding in the adult and old LIGHT$^{-/-}$ mice. This suggests excessive levels of NO production at the wound site that in the presence of O_2^- can generate $ONOO^-$. It has been reported that phosphorylation of eNOS modulates both the production of NO and O_2^- [57] and also that increase in H_2O_2 may exert effects on endothelial cell dysfunction and uncoupling of NOS. Our data show that there is increased phosphorylation/activation of eNOS and increased iNOS levels. However, the elevated levels of phospho-eNOS and of iNOS appear after the increase in nitrite and nitrate levels in LIGHT$^{-/-}$ wounds, hence these enzymes cannot be the reason for the increases in nitrite and nitrate. It is possible that elevation of NO could be the result of either dephosphorylation of Thr495 on eNOS [58] or increases in L-arginine [59] and decrease in endogenous NOS inhibitors [60]. Furthermore, the elevation in eNOS and iNOS at later times after wounding suggests an increase in NO that can combine with O_2^- to give rise to $ONOO^-$, a highly damaging ion species.

Clinical studies on chronic wounds in humans have shown free-radical-induced damage of proteins, lipids and DNA [18,20,61,62]. We found that the levels of malondialdehyde (MDA), a byproduct of lipid peroxidation, were significantly elevated throughout the course of healing in LIGHT$^{-/-}$ mice, indicating lipid damage. We also show the presence of F_2 isoprostanes that are considered to be the gold standard of oxidative stress and lipid peroxidation. Levels of 8- and 5-

isoprostanes detected in $LIGHT^{-/-}$ mice were significantly elevated when compared to the control mice.

DNA damage induced by ROS and RNS can cause modifications that impair DNA repair [63]. Levels of excretion of the free, water soluble, 8-OHdG are reduced, resulting in failing exonuclease activity; this is especially seen with aging, leading to cell damage [64,65]. Our results show two waves of DNA damage in $LIGHT^{-/-}$ mouse wounds. The first wave suggests that the initial accumulation of 8-OHdG in these wounds is potentially controlled by existing exonuclease activity. However, once this activity is exhausted, the cells are no longer able to handle DNA damage. Increased levels of DNA damage in old mice are seen early and remain elevated, underscoring the continuous increase in oxidative and nitrosative stress with aging of this mouse model much as is seen in humans.

NO is a dynamic molecule that reacts with O_2^- at a rate constant three times higher than the rate constant reaction of O_2^- with SOD, giving rise to $ONOO^-$ production [66] that causes nitration of tyrosine residues which damages proteins. Previous studies have shown that changes on tyrosine residues in proteins is an irreversible process that, in turn, severely impairs the regulatory components that undergo phosphorylation or adenylation in signal transduction events [67]. The increased levels of nitrotyrosine in adult and exacerbation in old $LIGHT^{-/-}$ mice suggest an increase in tissue damage.

NO plays a crucial role in bacterial biofilm dispersion [68]. Recent studies have shown that the use of low doses of NO in conjunction with antibiotics can lead to bacterial biofilm dispersion and induce these biofilm-forming bacteria to behave in a planktonic manner, hence reducing biofilm formation [69–71]. Although NO also has been widely considered an important immune cell regulator and a chemoattractant that plays a vital role in signaling events in wound healing [72], the chemistry of excessive NO changes in the presence of oxygen. These two molecules readily react with each other, giving rise to RNS that cause damage in the wound tissue [73,74]. Formation of $ONOO^-$, triggered by combination of NO and O_2^-, has been known to cause nitrosative damage. We show that there are elevated levels of nitrosative damage in $LIGHT^{-/-}$ wounds after we increase the oxidative stress in the wounds. Therefore, we speculate that the complexity of the microbiota and varying levels of NO, due to the presence of ROS, can lead to restricted levels of biofilm dispersion.

Occurrence of oxidative stress and maintenance of redox balance following stress is an essential component for proper wound healing. Stress is triggered by increases in ROS produced by (i) inflammatory cells (termed as oxidative burst), (ii) a family of NADPH oxidase (NOX), consisting of NOX1-5, and DUOX1&2, [73] and (iii) wound fibroblasts when stimulated by pro-inflammatory cytokines [75]. Although moderate increases in ROS regulate various signaling processes and act against invading bacteria, prolonged and excessive presence of ROS and inflammation can lead to hypoxia and tissue damage caused primarily by lipid peroxidation, DNA damage, protein nitrosylation, and cell death. Furthermore, excessive levels of ROS due to tissue damage create an environment that can serve as an inviting substrate for bacteria to thrive upon. The relationship between exacerbated levels of ROS and bacteria has been reported previously [76,77]. Studies have suggested that DNA double-stranded breaks in bacteria caused by oxidative stress lead to mutagenic repair via DNA repair protein RecA, rendering the variants with increased antibiotic resistance and adaptability to the surrounding microenvironment [76]. Furthermore, during excessive oxidative stress, these bacteria upregulate genes that increase their virulence [77]. In an effort to obtain excessive and persistent levels of oxidative stress, we manipulated the early wound microenvironment by inhibiting catalase [78] and GPx activities [79] and introduced our previously isolated S. epidermidis bacterial strain. The intensified levels in redox stress and the presence of biofilm-forming bacteria led to reproducible generation of chronic wounds.

Chronic wounds, and difficult-to-heal wounds are postulated to have an underlying biofilm-associated microbial contribution that is complex and dynamic [23,49,80,81]. We show that chronicity in $LIGHT^{-/-}$ wounds is accompanied by a persistent bacterial infection that is polymicrobial and contains biofilm-producing bacteria. The colonizing bacterial species associated and/or responsible for the formation of chronic wounds in the $LIGHT^{-/-}$ mice, are biofilm-producing and the capacity of these organisms to produce biofilms varies depending on the time of isolation. Furthermore, we showed that the source of infection arises from the $LIGHT^{-/-}$ mouse skin microbiota. Similar observations have been documented for human chronic wounds [82,83].

The presence of biofilm-producing *S. epidermidis* in human chronic wounds [23] and the contribution of *E. cloacae* in nosocomial infections are well known [84]. However, much less is known about *E. cloacae* infection in chronic wounds, although its presence is often reported [85] in diabetic foot infection [86,87], diabetic gangrene [88], and chronic venous leg ulcers [89,90]. Perhaps the lack of consideration of *E. cloacae* contribution to the development of chronic wounds may be in part due to their relatively lower initial abundance compared to the more commonly isolated bacteria *Staphylococcus spp.*, *Enterococcus spp.*, and *Pseudomonas aeruginosa* [49,90,91].

Summary and Conclusions

The chronic wound model we present in this publication is the first to effectively mimic chronic wounds in humans. The model wounds stay open for weeks and capture many of the characteristics of human chronic wounds. $LIGHT^{-/-}$ mice have elevated levels of genes involved in oxidative and nitrosative stress that lead to imbalanced redox levels in the wound tissue that are exacerbated with age. As a consequence, redox burden causes deleterious effects in the wound tissue that lead to impaired healing. Manipulation of the wound microenvironment to increase oxidative stress in the presence of biofilm-forming bacteria leads inevitably to development of chronic wounds, identifying high levels of oxidative stress in the wound tissue as a critical factor for chronic wound development. Furthermore, because of the nature and complexity of the mixed wound microbiota, the model presented here can provide insight into the biology of bacterial dynamics and host interaction and the factors that promote biofilm production. This model system, in which a genetic alteration leads to an imbalance in redox levels in wound tissue, has the potential to lead to the understanding of other fundamental cell and molecular mechanisms of chronic wound development and, by implication, to the development of new therapies. Having established a system where 100% of wounds are chronic, we are currently working on reversing chronicity in $LIGHT^{-/-}$ and other genetically-modified mice.

Materials and Methods

Dermal excisional wound model

Animals were housed at the University of California, Riverside (UCR) vivarium. All experimental protocols were approved by the UCR Institutional Animal Care and Use Committee (IACUC).

Experiments were performed using 12–16 week old mice categorized as adult mice and 85–92 week old mice as old mice. The procedure used was performed as previously described [21].

Superoxide dismutase activity assay

Total tissue superoxide dismutase (SOD) activity was measured by using a commercially available kit (Cayman Chemical, Catalog# 706002, Ann Arbor, USA) that measures all three types of SOD (Cu/Zn-, Mn-, and EC-SOD). One unit of SOD is defined as the amount of enzyme needed to cause 50% dismutation of the superoxide radical. Extracts obtained from tissues collected at 4 hr, 12 hr, 24 hr and 48 hr post-wounding were processed for total SOD activity according to the protocol provided by the assay kit manufacturer. The SOD activities of the samples were calculated from the linear regression of a standard curve that was determined using the SOD activity of bovine erythrocytes at various concentrations run under the same conditions. The SOD activity was expressed as U/ml of tissue extract.

Hydrogen peroxide activity assay

Tissue hydrogen peroxide (H_2O_2) levels were measured by using a commercially available kit (Cell Technology Inc., Catalog# FLOH 100-3, Mountain View, USA) that utilizes a non-fluorescent detection reagent. The assay is based on the peroxidase-catalyzed oxidation by H_2O_2 of the nonfluorescent substrate 10-acetyl-3,7-dihydroxyphenoxazine to a fluorescent resorufin. Fluorescent intensities were measured at 530 nm (excitation)/590 nm (emission) using a Victor 2 microplate reader. The amounts of H_2O_2 in the supernatants were derived from a seven-point standard curve generated with known concentrations of H_2O_2.

Catalase activity assay

Tissue catalase activity was measured by using a commercially available kit (Cayman Chemical, Catalog# 707002, Ann Arbor, USA). The enzyme assay for catalase is based on the peroxidatic function of catalase with methanol to produce formaldehyde in the presence of an optimal concentration of H_2O_2. The formaldehyde produced was measured spectrophotometrically, with 4-amino-3-hydrazino-5-mercapto-1,2,4-triazole (purpald) as the chromogen, at 540 nm in a 96-well place. The catalase activity was expressed as nmol/min/ml of tissue extract.

Glutathione peroxidase activity assay

Tissue glutathione peroxidase (GPx) activity was measured using a commercially available kit (Cayman Chemical, Catalog# 703102, Ann Arbor, USA). The activity was measured indirectly by a coupled reaction with glutathione reductase (GR). GPx reduces H_2O_2 to H_2O and in the process oxidized glutathione (GSSG) is produced that in turn is recycled to its reduced state by GR and NADPH. Furthermore, oxidation of NADPH to NADP$^+$ is accompanied by a decrease in absorbance at 340 nm. Under conditions in which GPx activity is rate limiting, the rate of decrease in the absorbance measured at 340 nm, in a 96-well plate at 1-min interval for a total of 5 min using a Victor 2 microplate reader, is directly proportional to the GPx activity of the sample. GPx activity was expressed as nmol/min/ml of tissue extract.

Lactate measurement assay

Tissue lactate levels were measured using a commercially available kit (Biovision Inc, Catalog# K638-100, Milipitas, USA). Tissue lactate extracts were specifically oxidized to form an intermediate that reacts with a colorless probe to generate fluorescence that was measured at 530 nm (excitation)/590 nm (emission) using Victor 2 microplate reader. The intensity was directly proportional to the amount of lactate measured in nmol/ml.

pH measurements

Wound pH levels were measured using a Beetrode micro pH electrode with a 100 μm tip diameter, 2 mm receptacle (World Precision Instruments, Catalog# NMPH5, Sarasota, USA). A separate reference electrode of 450 μm diameter tip was used (World Precision Instruments, Catalog# DRIREF-450, Sarasota, USA) along with a small, battery-operated compensator (World Precision Instruments, Catalog# SYS-Beecal, Sarasota, USA) to generate mV readings in the range of the standard pH meter used (Beckman Coulter, Catalog# A58754, Brea, USA). The compensator helped adjust the electrode-offset potential. Calibration of the electrodes was done at 37°C (temperature of the mouse body) in pH buffers 4, 7 and 10. A linear Nernstian plot was obtained and was used to convert the mV readings that were obtained from the mouse wound. Measurements on every mouse wound were done at five different locations, four of which were at the periphery of the wound at 90° angle and one in the center.

Nitrate nitrite analysis

Tissues collected were weighed and introduced into eppendorf tubes with equal weights of zirconium oxide beads. Nitrite free methanol at 2 ml/g tissue was added to the tubes. Tissues were homogenized for 10 mins in a bullet blender at 4°C. The extracts were then centrifuged at 10000 rpm for 10 mins at 4°C. The methanolic supernatant was collected and analysis was performed as previously described [92].

Lipid peroxide assay using thiobarbituric acid reactive substances

Tissue thiobarbituric acid reactive substances (TBARS) were measured by using a commercially available kit (Cell Biolabs Inc., Catalog# STA-300, San Diego, USA). Lipid peroxidation forms unstable lipid peroxides that further decompose into natural byproducts such as malondialdehyde (MDA) and 4-hydroxynonenal (4-HNE). MDA forms adducts with TBARS in a 1:2 proportion. These aducts were measured fluorometrically at an excitation of 540 nm and emission at 590 nm. TBARS levels were then calculated in μM by comparison with a predetermined MDA standard curve.

Isolation of DNA and 8-hydroxy-2-deoxy Guanosine (8-OH-dG) analysis

Tissue DNA was extracted by using a commercially available kit (Qiagen, Catalog# 69504, Valencia, USA). Eluted DNA was digested using nuclease P1 and the pH adjusted to 7.5–8.5 using 1 M Tris. The DNA was incubated for 30 min at 37°C with 1U of alkaline phosphatase per 100 μg of DNA and then boiled for 10 min. The 8-OH-dG DNA damage assay was performed by using a commercially available kit (Cayman Chemical, Catalog# 589320, Ann Arbor, USA). The measurements are based on a competitive enzyme immunoassay between 8-OH-dG and an 8-OH-dG-acetylcholinesterase (AChE) conjugate (8-OH-dG tracer) with a limited amount of 8-OH-dG monoclonal antibody. After conjugation, Ellman's reagent (used to quantify the number or concentration of thiol groups) was used as a developing agent and read spectrophotometrically at 412 nm. The intensity measured was proportional to the amount of 8-OH-dG that was expressed in pg/ml.

Nitrotyrosine ELISA

Tissue nitrotyrosine levels were measured by using a commercially available kit (Cell Biolabs Inc., Catalog# STA-305 San Diego, USA). The measurements are based on a competitive enzyme immunoassay. The tissue sample or nitrated BSA were bound to an anti-nitrotyrosine antibody, followed by an HRP conjugated secondary antibody and enzyme substrate. The absorbance was measured spectrophotometrically at 412 nm and the nitrotyrosine content in the unknown sample was then determined by comparing with a standard curve that was prepared from predetermined nitrated BSA standards.

Cell death

Tissue cell death level was measured by using Annexin V apoptosis kits (Southern Biotech, Catalog# 10010-09, Birmingham, USA) according to the manufacturer's instructions and our previously published methodology [26]. Percoll gradients were used to collect the wound cells from the homogenized wound tissue, and the cell stained with the kit reagents. Cells that lose membrane integrity allow propidium iodide to enter and bind to DNA, a phenomenon seen in case of cell death due to necrosis, whereas apoptotic cells only stain for Annexin V. The cells were then separated by FACS analysis to separate the populations staining with propidium iodide from those staining with Annexin V.

Scanning Electron Microscopy

Tissues collected were fixed in 4% paraformaldehyde for 4 hrs at room temperature. Samples were then dehydrated in 25%, 50%, 75%, 95% and 100% ethanol for 20 min each at room temperature. Critical point drying of the tissues was performed using Critical-point-dryer Balzers CPD0202 followed by Au/Pd sputtering for 1 min in the Sputter coater Cressington 108 auto. The coated samples were attached to carbon taped aluminum stubs and were imaged using an XL30 FEG scanning electron microscope.

In Vivo Imaging

Live animal images were captured using the iBox Scientia Small Animal Imaging System (UVP, LLC. Upland, CA, an Analytik Jena Company). Mice were anesthetized and placed on the imaging stage maintained at 37°C for the duration of each imaging experiment. For each time point, age-matched C57BL/6 and LIGHT$^{-/-}$ mice was imaged using the ImageEM 1K EM-CCD (Hamamatsu, Japan), cooled to $-55°C$, and an optical system consisting of a 50 mm f/1.2 lens. Images were captured separately for each time point without an emission filter and at 1×1 binning. Bright field images using a white light channel were captured first at an exposure time of 150 milliseconds followed by a luminescent channel at an exposure time at 10–20 min.

Preparation of tissue extracts

The tissues collected were prepared as previously described [21].

Immunoblotting

Wound tissue extracts were probed for iNOS and phospho eNOS as previously described [21].

Lipidomics

1 ml of LCMS grade ethanol containing 0.05% BHT and 10 ng of each internal standard was added to frozen wound tissues. Internal standards used were, (d_4) 8-iso PGF$_{2\alpha}$, (d_{11}) 5-iso PGF$_{2\alpha}$-VI, (d_4) 6k PGF$_{1\alpha}$, (d_4) PGF$_{2\alpha}$, (d_4) PGE$_2$, (d_4) PGD$_2$, (d_4) LTB$_4$, (d4) TXB$_2$, (d_4) LTC$_4$, (d_5) LTD$_4$, (d_5) LTE$_4$, (d_8) 5-hydroxyeicosatetranoic acid (5HETE), (d_8) 15-hydroxyeicosatetranoic acid (15HETE), (d_8) 14,15 epoxyeicosatrienoic acid, (d_8) arachidonic Acid, and $(d5)$ eicosapentaenoic acid. Samples were mixed using a bath sonicator incubated overnight at $-20°C$ for lipid extraction. The insoluble fraction was precipitated by centrifuging at 12,000xg for 20 min and the supernatant was transferred into a new glass tube. Lipid extracts were then dried under vacuum and reconstituted in of LCMS grade 50:50 EtOH:dH$_2$O (100 µl) for eicosanoid quantitation via UPLC ESI-MS/MS analysis. A 14 min reversed-phase LC method utilizing a Kinetex C18 column (100×2.1 mm, 1.7 µm) and a Shimadzu UPLC was used to separate the eicosanoids at a flow rate of 500 µl/min at 50°C. The column was first equilibrated with 100% Solvent A [acetonitrile:water:formic acid (20:80:0.02, v/v/v)] for two minutes and then 10 µl of sample was injected. 100% Solvent A was used for the first two minutes of elution. Solvent B [acetonitrile:isopropanol (20:80, v/v)] was increased in a linear gradient to 25% Solvent B to 3 min, to 30% by 6 minutes, to 55% by 6.1 min, to 70% by 10 min, and to 100% by 10.1 min. 100% Solvent B was held until 13 min, then decreased to 0% by 13.1 min and held at 0% until 14 min. The eluting eicosanoids were analyzed using a hybrid triple quadrupole linear ion trap mass analyzer (ABSciex 6500 QTRAP,) via multiple-reaction monitoring in negative-ion mode. Eicosanoids were monitored using species specific precursor → product MRM pairs. The mass spectrometer parameters were: curtain gas: 30; CAD: High; ion spray voltage: -3500 V; temperature: 300°C; Gas 1: 40; Gas 2: 60; declustering potential, collision energy, and cell exit potential were optimized per transition.

Antioxidant inhibition and biofilm formation model

Catalase activity was inhibited by intraperitonial injection of 3-Amino-1,2,4-triazole (ATZ) at a concentration of 1 g/kg body weight 20 min prior to creating the excisional wound. GPx activity inhibition was performed by topical application of mercaptosuccinic acid at concentration of 150 mg/kg body weight immediately after wounding and the wound was covered with sterile tegaderm. 24 hrs post-wounding, 20 µl *Staphylococcus epidermidis C2* suspension at a concentration of 1×10^8 CFU/mL was added onto the wound and this covered with sterile tegaderm. The wounds were kept moist at all times and tegaderm was replaced as soon as the sealant of the tegaderm was seen to be compromised to avoid wound contamination. All procedures were carried out in a sterile environment. The inhibitor injection protocol and application of the bacteria were repeated every week.

Bacteria isolation and characterization

Wound exudates from LIGHT$^{-/-}$ mice were collected using sterile cotton swabs and stored at $-80°C$ in 1.0% w/v proteose peptone and 20.0% v/v glycerol solution until analyzed. Samples were thawed on ice, vortexed and cultured for 16–24 hrs at 37°C on tryptic soy agar plates containing 5.0% v/v defibrinated sheep blood and 0.08% w/v Congo red dye. Viable colonies were counted and then differentiated based on size, hemolytic patterns, and Congo red uptake. The cultures were examined for Grams stain reactivity and visualized using a compound light microscope. Grams negative rods were characterized using the API 20E identification kit (Biomerieux, Durham USA), grown on *Pseudo-*

monas Isolation Agar, oxidase activity, growth at 42°C in LB, and motility. Grams positive cocci differentiated based on catalase activity, coagulase activity, growth in 6.5% w/v NaCl tolerance test, and hemolytic activity. Biofilm production was quantified using adherence and staining of extracellular polysaccharide (slime), produced by bacteria, using Congo red staining to deduce whether or not the bacteria was a biofilm former using previously published procedures and criteria [40,46].

Community Minimal antibiotic inhibitory concentration assay

Community minimal inhibitory concentration (CMIC) assay was carried out with amoxicillin as described by DeLoney and Schiller [93] with the following modification. Wound exudates (containing the bacteria) were challenged with antibiotic for 12 hr with concentrations ranges from 100 to 0.78 μg/mL in tryptic soy broth after being seeded at 37oC in a humidified incubator for 4 hr prior to the assay. The CMIC is defined as the lowest antibiotic concentration that resulted in a $\leq 50\%$ increase in the optical density measured at 595 nm compared to the optical density reading prior to the introduction of antibiotic.

Tissue preparation for histology

Tissues collected were prepared as previously described [21].

Statistical analysis

For the statistical analysis of experiments, we used Graphpad Instat Software (Graphpad, La Jolla, CA, USA) and Sigmaplot Software (SigmaPlot, San Jose, USA). Analysis of variance (ANOVA) was used to test the significance of group differences between two or more groups. In experiments with only two groups, statistical analysis was conducted using a Student's t-test.

Supporting Information

File S1 Figure S1, Schematic illustration of oxidative and nitrosative stress cycle. Figure S2, Lactate levels, pH, and nitrosative stress are exacerbated in old LIGHT$-/-$ mice. Figure S3, Detrimental effects of exacerbated stress on protein modification, lipid peroxidation, and DNA damage in old mice. Figure S4, Histology of normal wound healing. Figure S5, Manipulation of LIGHT$-/-$ wounds with bacteria or individual antioxidant inhibitors does not lead to chronic wound development. Figure S6, Identification and characterization of the bioflora that colonized the old LIGHT/- chronic wounds.

Acknowledgments

The authors thank Dr. Carl Ware for generously providing the LIGHT$^{-/-}$ mice. The authors would also like to thank Dr. Devin Binder and Mike Hsu for use of the cryostat, and Dr. Victor G. J. Rodgers for advice on pH measurements *in vivo*.

Author Contributions

Conceived and designed the experiments: SD DD MMG GN. Performed the experiments: SD DD MG DSW AB AS JL GN JK. Analyzed the data: SD DD DSW AS SG RP NS MMG GN JK. Contributed reagents/materials/analysis tools: MMG SG CC RP NS JL GN. Wrote the paper: MMG SD DD AS GN. Edited the manuscript: NS RP SG DSW.

References

1. Sen CK, Gordillo GM, Roy S, Kirsner R, Lambert L, et al. (2009) Human skin wounds: a major and snowballing threat to public health and the economy. Wound Repair Regen 17: 763–771. Available: http://www.pubmedcentral.nih.gov/articlerender.fcgi?artid=2810192&tool=pmcentrez&rendertype=abstract Accessed 2013 January 7.

2. Lazarus GS, Cooper DM, Knighton DR, Margolis DJ, Percoraro RE, et al. (1994) Definitions and guidelines for assessment of wounds and evaluation of healing. Wound Repair Regen: 165–170.

3. Wong VW, Gurtner GC (2012) Tissue engineering for the management of chronic wounds: current concepts and future perspectives. Exp Dermatol 21: 729–734. Available: http://www.ncbi.nlm.nih.gov/pubmed/22742728 Accessed 2013 April 11.

4. Roy S, Khanna S, Nallu K, Hunt TK, Sen CK (2006) Dermal wound healing is subject to redox control. Mol Ther 13: 211–220. Available: http://www.pubmedcentral.nih.gov/articlerender.fcgi?artid=1389791&tool=pmcentrez&rendertype=abstract Accessed 2013 January 5.

5. D'Autréaux B, Toledano MB (2007) ROS as signalling molecules: mechanisms that generate specificity in ROS homeostasis. Nat Rev Mol Cell Biol 8: 813–824. Available: http://www.ncbi.nlm.nih.gov/pubmed/17848967 Accessed 2013 February 27.

6. Sen CK, Roy S (2008) Redox signals in wound healing. Biochim Biophys Acta 1780: 1348–1361. Available: http://www.pubmedcentral.nih.gov/articlerender.fcgi?artid=2574682&tool=pmcentrez&rendertype=abstract Accessed 2012 November 1.

7. Dröge W (2002) Free radicals in the physiological control of cell function. Physiol Rev 82: 47–95. Available: http://www.ncbi.nlm.nih.gov/pubmed/11773609

8. Yeoh-Ellerton S, Stacey MC (2003) Iron and 8-isoprostane levels in acute and chronic wounds. J Invest Dermatol 121: 918–925. Available: http://www.ncbi.nlm.nih.gov/pubmed/14632213

9. Wlaschek M, Scharffetter-Kochanek K (2005) Oxidative stress in chronic venous leg ulcers. Wound Repair Regen 13: 452–461. Available: http://www.ncbi.nlm.nih.gov/pubmed/16176453

10. Yang Q, Phillips PL, Sampson EM, Progulske-Fox A, Jin S, et al. (2013) Development of a novel ex vivo porcine skin explant model for the assessment of mature bacterial biofilms. Wound Repair Regen 21: 704–714. Available: http://www.ncbi.nlm.nih.gov/pubmed/23927831 Accessed 2013 December 2.

11. Hopf HW, Rollins MD (2007) Wounds: an overview of the role of oxygen. Antioxid Redox Signal 9: 1183–1192. Available: http://www.ncbi.nlm.nih.gov/pubmed/17536961 Accessed 2013 August 15.

12. Britland S, Ross-Smith O, Jamil H, Smith AG, Vowden K, et al. (2012) The lactate conundrum in wound healing: clinical and experimental findings indicate the requirement for a rapid point-of-care diagnostic. Biotechnol Prog 28: 917–924. Available: http://www.ncbi.nlm.nih.gov/pubmed/22581665 Accessed 2013 August 14.

13. Fazli M, Bjarnsholt T, Kirketerp-Møller K, Jørgensen A, Andersen CB, et al. (2011) Quantitative analysis of the cellular inflammatory response against biofilm bacteria in chronic wounds. Wound Repair Regen 19: 387–391. Available: http://www.ncbi.nlm.nih.gov/pubmed/21518086 Accessed 2013 August 15.

14. Warburg O (1956) On the Origin of Cancer Cells. Science (80-) 123: 309–314.

15. Hunt TK, Hopf H, Hussain Z (2000) Physiology of wound healing. Adv Skin Wound Care 13: 6–11. Available: http://www.ncbi.nlm.nih.gov/pubmed/11074996 Accessed 2013 August 14.

16. Dovi J V, Szpaderska AM, DiPietro L a (2004) Neutrophil function in the healing wound: adding insult to injury? Thromb Haemost: 275–280. Available: http://www.schattauer.de/index.php?id=1214&doi=10.1160/TH03-11-0720&no_cache=1 Accessed 2013 April 20.

17. McCarty SM, Cochrane C a, Clegg PD, Percival SL (2012) The role of endogenous and exogenous enzymes in chronic wounds: a focus on the implications of aberrant levels of both host and bacterial proteases in wound healing. Wound Repair Regen 20: 125–136. Available: http://www.ncbi.nlm.nih.gov/pubmed/22380687 Accessed 2013 May 19.

18. Schäfer M, Werner S (2008) Oxidative stress in normal and impaired wound repair. Pharmacol Res 58: 165–171. Available: http://www.ncbi.nlm.nih.gov/pubmed/18617006 Accessed 2012 November 1.

19. Luo J, Chen AF (2005) Nitric oxide: a newly discovered function on wound healing. Acta Pharmacol Sin 26: 259–264. Available: http://www.ncbi.nlm.nih.gov/pubmed/15715920 Accessed 2013 March 29.

20. Abd-El-Aleem S a, Ferguson MW, Appleton I, Kairsingh S, Jude EB, et al. (2000) Expression of nitric oxide synthase isoforms and arginase in normal human skin and chronic venous leg ulcers. J Pathol 191: 434–442. Available: http://www.ncbi.nlm.nih.gov/pubmed/10918219

21. Petreaca ML, Do D, Dhall S, McLelland D, Serafino A, et al. (2012) Deletion of a tumor necrosis superfamily gene in mice leads to impaired healing that mimics chronic wounds in humans. Wound Repair Regen 20: 353–366. Available: http://www.ncbi.nlm.nih.gov/pubmed/22564230 Accessed 2012 November 5.

22. Howell-Jones RS, Wilson MJ, Hill KE, Howard a J, Price PE, et al. (2005) A review of the microbiology, antibiotic usage and resistance in chronic skin wounds. J Antimicrob Chemother 55: 143–149. Available: http://www.ncbi.nlm.nih.gov/pubmed/15649989 Accessed 2012 November 9.

23. James G a, Swogger E, Wolcott R, Pulcini E . deLancey, Secor P, et al. (2007) Biofilms in chronic wounds. Wound Repair Regen 16: 37–44. Available: http://www.ncbi.nlm.nih.gov/pubmed/18086294 Accessed 2012 October 29.

24. Guo S, Dipietro L a (2010) Factors affecting wound healing. J Dent Res 89: 219–229. Available: http://www.pubmedcentral.nih.gov/articlerender.fcgi?artid=2903966&tool=pmcentrez&rendertype=abstract Accessed 2014 April 28.

25. Martin JM, Zenilman JM, Lazarus GS (2010) Molecular microbiology: new dimensions for cutaneous biology and wound healing. J Invest Dermatol 130: 38–48. Available: http://www.ncbi.nlm.nih.gov/pubmed/19626034 Accessed 2014 May 8.

26. Petreaca ML, Yao M, Ware C, Martins-Green MM (2008) Vascular endothelial growth factor promotes macrophage apoptosis through stimulation of tumor necrosis factor superfamily member 14 (TNFSF14/LIGHT). Wound Repair Regen 16: 602–614. Available: http://www.ncbi.nlm.nih.gov/pubmed/19128255 Accessed 2014 February 18.

27. Gonzalez FJ (2005) Role of cytochromes P450 in chemical toxicity and oxidative stress: studies with CYP2E1. Mutat Res 569: 101–110. Available: http://www.ncbi.nlm.nih.gov/pubmed/15603755 Accessed 2014 May 8.

28. Song L, Wei X, Zhang B, Luo X, Liu J, et al. (2009) Role of Foxa1 in regulation of bcl2 expression during oxidative-stress-induced apoptosis in A549 type II pneumocytes. Cell Stress Chaperones 14: 417–425. Available: http://www.pubmedcentral.nih.gov/articlerender.fcgi?artid=2728276&tool=pmcentrez&rendertype=abstract Accessed 2014 May 8.

29. Kuhlicke J, Frick JS, Morote-Garcia JC, Rosenberger P, Eltzschig HK (2007) Hypoxia inducible factor (HIF)-1 coordinates induction of Toll-like receptors TLR2 and TLR6 during hypoxia. PLoS One 2: e1364. Available: http://www.pubmedcentral.nih.gov/articlerender.fcgi?artid=2147045&tool=pmcentrez&rendertype=abstract Accessed 2014 May 7.

30. Kundu TK, Velayutham M, Zweier JL (2012) Aldehyde Oxidase Functions as a Superoxide Generating NADH Oxidase: An Important Redox Regulated Pathway of Cellular Oxygen. Biochemistry 51: 2930–2939.

31. Shaw S, Jayatilleke E (1992) The Role of Cellular Oxidases and Catalytic iron in the pathogenesis of Ethanol-Induced Liver Injury. Life Sci 50: 2045–2052.

32. Moor AN, Tummel E, Prather JL, Jung M, Lopez JJ, et al. (2014) Consequences of age on ischemic wound healing in rats: altered antioxidant activity and delayed wound closure. Age (Dordr) 36: 733–748. Available: http://www.ncbi.nlm.nih.gov/pubmed/24443098 Accessed 2014 July 17.

33. Schneider LA, Korber A, Grabbe S, Dissemond J (2007) Influence of pH on wound-healing: a new perspective for wound-therapy? Arch Dermatol Res 298: 413–420. Available: http://www.ncbi.nlm.nih.gov/pubmed/17091276 Accessed 2012 November 5.

34. Schreml S, Szeimies RM, Prantl L, Karrer S, Landthaler M, et al. (2010) Oxygen in acute and chronic wound healing. Br J Dermatol 163: 257–268. Available: http://www.ncbi.nlm.nih.gov/pubmed/20394633 Accessed 2012 December 4.

35. Stefaniak AB, Plessis J Du, John SM, Eloff F, Agner T, et al. (2013) International guidelines for the in vivo assessment of skin properties in non-clinical settings: part 1. pH. Skin Res Technol 19: 59–68. Available: http://www.pubmedcentral.nih.gov/articlerender.fcgi?artid=3747458&tool=pmcentrez&rendertype=abstract Accessed 2014 July 22.

36. Lotito S, Blonder P, Francois AI, Rdmy C (1989) Correlation between intracellular pH and lactate levels in the rat brain during potassium cyanide induced metabolism blockade: a combined 31p-1H in vivo nuclear magnetic spectroscopy study. 97: 91–96.

37. Gethin G (2007) The significance of surface pH in chronic wounds. Wounds 3: 52–56. Available: http://www.woundsinternational.com/pdf/content_124.pdf Accessed 2014 January 29.

38. Bowler PG (2002) Wound pathophysiology, infection and therapeutic options. Ann Med 34: 419–427. Available: http://www.ncbi.nlm.nih.gov/pubmed/12523497

39. Dowd SE, Sun Y, Secor PR, Rhoads DD, Wolcott BM, et al. (2008) Survey of bacterial diversity in chronic wounds using pyrosequencing, DGGE, and full ribosome shotgun sequencing. BMC Microbiol 8: 43. Available: http://www.pubmedcentral.nih.gov/articlerender.fcgi?artid=2289825&tool=pmcentrez&rendertype=abstract Accessed 2014 June 2.

40. Christensen GD, Simpson W a, Younger JJ, Baddour LM, Barrett FF, et al. (1985) Adherence of coagulase-negative staphylococci to plastic tissue culture plates: a quantitative model for the adherence of staphylococci to medical devices. J Clin Microbiol 22: 996–1006. Available: http://www.pubmedcentral.nih.gov/articlerender.fcgi?artid=271866&tool=pmcentrez&rendertype=abstract

41. O'Toole G a, Kolter R (1998) Initiation of biofilm formation in Pseudomonas fluorescens WCS365 proceeds via multiple, convergent signalling pathways: a genetic analysis. Mol Microbiol 28: 449–461. Available: http://www.ncbi.nlm.nih.gov/pubmed/9632250

42. Stepanovic S, Vukovic D, Dakic I, Savic B, Svabic-Vlahovic M (2000) A modified microtiter-plate test for quantification of staphylococcal biofilm formation. J Microbiol Methods 40: 175–179. Available: http://www.ncbi.nlm.nih.gov/pubmed/10699673

43. Kolodkin-Gal I, Cao S, Chai L, Böttcher T, Kolter R, et al. (2012) A self-produced trigger for biofilm disassembly that targets exopolysaccharide. Cell 149: 684–692. Available: http://www.pubmedcentral.nih.gov/articlerender.fcgi?artid=3526955&tool=pmcentrez&rendertype=abstract Accessed 2014 January 21.

44. Kloos W, Schleifer K (1975) Isolation and Characterization of Staphylococci from Human Skin. Int J Syst Bacteriol 25: 62–79.

45. Qu Y, Daley AJ, Istivan TS, Garland SM, Deighton M a (2010) Antibiotic susceptibility of coagulase-negative staphylococci isolated from very low birth weight babies: comprehensive comparisons of bacteria at different stages of biofilm formation. Ann Clin Microbiol Antimicrob 9: 16. Available: http://www.pubmedcentral.nih.gov/articlerender.fcgi?artid=2902406&tool=pmcentrez&rendertype=abstract

46. Cui B, Smooker PM, Rouch D a, Daley AJ, Deighton M a (2013) Differences between two clinical Staphylococcus capitis subspecies as revealed by biofilm, antibiotic resistance, and pulsed-field gel electrophoresis profiling. J Clin Microbiol 51: 9–14. Available: http://www.pubmedcentral.nih.gov/articlerender.fcgi?artid=3536240&tool=pmcentrez&rendertype=abstract Accessed 2014 July 19.

47. Percival SL, Hill KE, Malic S, Thomas DW, Williams DW (2010) Antimicrobial tolerance and the significance of persister cells in recalcitrant chronic wound biofilms. Wound Repair Regen 19: 1–9. Available: http://www.ncbi.nlm.nih.gov/pubmed/21235682 Accessed 2014 January 29.

48. Parsek MR, Singh PK (2003) Bacterial biofilms: an emerging link to disease pathogenesis. Annu Rev Microbiol 57: 677–701. Available: http://www.ncbi.nlm.nih.gov/pubmed/14527295 Accessed 2014 January 22.

49. Siddiqui AR, Bernstein JM (2010) Chronic wound infection: facts and controversies. Clin Dermatol 28: 519–526. Available: http://www.ncbi.nlm.nih.gov/pubmed/20797512 Accessed 2012 December 17.

50. Cho I, Blaser MJ (2012) The human microbiome: at the interface of health and disease. Nat Rev Genet 13: 260–270. Available: http://www.pubmedcentral.nih.gov/articlerender.fcgi?artid=3418802&tool=pmcentrez&rendertype=abstract Accessed 2014 January 22.

51. Stevens D, Lieberman M, McNitt T, Price J (1984) Demonstration of uronic acid capsular material in the cerebrospinal fluid of a patient with meningitis caused by mucoid Pseudomonas aeruginosa. J Clin Microbiol 19: 942–943. Available: http://jcm.asm.org/content/19/6/942.short Accessed 2014 February 8.

52. Bales PM, Renke EM, May SL, Shen Y, Nelson DC (2013) Purification and Characterization of Biofilm-Associated EPS Exopolysaccharides from ESKAPE Organisms and Other Pathogens. PLoS One 8: e67950. Available: http://dx.plos.org/10.1371/journal.pone.0067950 Accessed 2014 January 26.

53. Sindrilaru A, Peters T, Wieschalka S, Baican C, Baican A, et al. (2011) An unrestrained proinflammatory M1 macrophage population induced by iron impairs wound healing in humans and mice. 121: 985–997. doi:10.1172/JCI44490DS1

54. Bryan N, Ahswin H, Smart N, Bayon Y, Wohlert S, et al. (2012) Reactive oxygen species (ROS)–a family of fate deciding molecules pivotal in constructive inflammation and wound healing. Eur Cell Mater 24: 249–265. Available: http://www.ncbi.nlm.nih.gov/pubmed/23007910

55. Messonnier L, Kristensen M, Juel C, Denis C (2007) Importance of pH regulation and lactate/H+ transport capacity for work production during supramaximal exercise in humans. J Appl Physiol 102: 1936–1944. Available: http://www.ncbi.nlm.nih.gov/pubmed/17289910 Accessed 2013 August 15.

56. Shorrock SM, Kun S, Peura RA, Dum RM (2000) Determination of a Relationship between Bacteria Levels and Tissue pH in Wounds: Animal Studies. IEEE: 117–118.

57. Chen C-A, Druhan LJ, Varadharaj S, Chen Y-R, Zweier JL (2008) Phosphorylation of endothelial nitric-oxide synthase regulates superoxide generation from the enzyme. J Biol Chem 283: 27038–27047. Available: http://www.pubmedcentral.nih.gov/articlerender.fcgi?artid=2556006&tool=pmcentrez&rendertype=abstract Accessed 2013 March 24.

58. Sullivan JC, Pollock JS (2006) Coupled and uncoupled NOS: separate but equal? Uncoupled NOS in endothelial cells is a critical pathway for intracellular signaling. Circ Res 98: 717–719. Available: http://www.ncbi.nlm.nih.gov/pubmed/16574911 Accessed 2014 May 8.

59. Erez A, Nagamani SCS, Shchelochkov O a, Premkumar MH, Campeau PM, et al. (2011) Requirement of argininosuccinate lyase for systemic nitric oxide production. Nat Med 17: 1619–1626. Available: http://www.pubmedcentral.nih.gov/articlerender.fcgi?artid=3348956&tool=pmcentrez&rendertype=abstract Accessed 2014 May 8.

60. Miyazaki H, Matsuoka H, Cooke JP, Usui M, Ueda S, et al. (1999) Endogenous Nitric Oxide Synthase Inhibitor: A Novel Marker of Atherosclerosis. Circulation 99: 1141–1146. Available: http://circ.ahajournals.org/cgi/doi/10.1161/01.CIR.99.9.1141 Accessed 2014 May 8.

61. Moseley R, Hilton JR, Waddington RJ, Harding KG (2004) Original Research Articles – Basic Science Comparison of oxidative stress biomarker profiles between acute and chronic wound environments. Wound Repair Regen 12: 419–429.

62. Goel A, Spitz DR, Weiner GJ (2012) Manipulation of cellular redox parameters for improving therapeutic responses in B-cell lymphoma and multiple myeloma. J Cell Biochem 113: 419–425. Available: http://www.pubmedcentral.nih.gov/articlerender.fcgi?artid=3374635&tool=pmcentrez&rendertype=abstract Accessed 2013 May 19.

63. Freeman T a, Parvizi J, Della Valle CJ, Steinbeck MJ (2009) Reactive oxygen and nitrogen species induce protein and DNA modifications driving arthrofibrosis following total knee arthroplasty. Fibrogenesis Tissue Repair 2: 5. Available: http://www.pubmedcentral.nih.gov/articlerender.

fcgi?artid=2785750&tool=pmcentrez&rendertype=abstract Accessed 2013 May 19.

64. Fraga CG, Shigenaga MK, Park JW, Degan P, Ames BN (1990) Oxidative damage to DNA during aging: 8-hydroxy-2′-deoxyguanosine in rat organ DNA and urine. Proc Natl Acad Sci U S A 87: 4533–4537. Available: http://www.pubmedcentral.nih.gov/articlerender.fcgi?artid=54150&tool=pmcentrez&rendertype=abstract

65. Kikuchi S, Kobune M, Iyama S, Sato T, Murase K, et al. (2012) Improvement of iron-mediated oxidative DNA damage in patients with transfusion-dependent myelodysplastic syndrome by treatment with deferasirox. Free Radic Biol Med 53: 643–648. Available: http://www.ncbi.nlm.nih.gov/pubmed/22705364 Accessed 2013 May 19.

66. Beckman JS, Koppenol WH (1996) Nitric oxide, superoxide, and peroxynitrite: the good, the bad, and the ugly. AJP Cell Physiol 271: 1424–1437.

67. Berlett BS, Friguet B, Yim MB, Chock PB, Stadtman ER (1996) Peroxynitrite-mediated nitration of tyrosine residues in Escherichia coli glutamine synthetase mimics adenylylation: relevance to signal transduction. Proc Natl Acad Sci U S A 93: 1776–1780. Available: http://www.pubmedcentral.nih.gov/articlerender.fcgi?artid=39857&tool=pmcentrez&rendertype=abstract

68. Barraud N, Hassett DJ, Hwang S-H, Rice S a, Kjelleberg S, et al. (2006) Involvement of nitric oxide in biofilm dispersal of Pseudomonas aeruginosa. J Bacteriol 188: 7344–7353. Available: http://www.pubmedcentral.nih.gov/articlerender.fcgi?artid=1636254&tool=pmcentrez&rendertype=abstract Accessed 2014 July 17.

69. Schreiber F, Beutler M, Enning D, Lamprecht-Grandio M, Zafra O, et al. (2011) The role of nitric-oxide-synthase-derived nitric oxide in multicellular traits of Bacillus subtilis 3610: biofilm formation, swarming, and dispersal. BMC Microbiol 11: 111. Available: http://www.pubmedcentral.nih.gov/articlerender.fcgi?artid=3224222&tool=pmcentrez&rendertype=abstract Accessed 2014 July 22.

70. Kaplan JB (2010) Biofilm dispersal: mechanisms, clinical implications, and potential therapeutic uses. J Dent Res 89: 205–218. Available: http://www.pubmedcentral.nih.gov/articlerender.fcgi?artid=3318030&tool=pmcentrez&rendertype=abstract Accessed 2014 July 17.

71. Cathie K, Howlin R, Carroll M, Clarke S, Connett G, et al. (2014) Reducing Antibiotic Tolerance using Nitric Oxide in Cystic Fibrosis: report of a proof of concept clinical trial. Arch Dis Child 99: A159–A159. Available: http://adc.bmj.com/cgi/doi/10.1136/archdischild-2014-306237.367 Accessed 2014 July 22.

72. Rizk M, Witte MB, Barbul A (2004) Nitric oxide and wound healing. World J Surg 28: 301–306. Available: http://www.ncbi.nlm.nih.gov/pubmed/14961192 Accessed 2014 July 22.

73. Bedard K, Krause K (2007) The NOX Family of ROS-Generating NADPH Oxidases: Physiology and Pathophysiology. Physiol Rev 87: 245–313. doi:10.1152/physrev.00044.2005

74. Wink D a, Hines HB, Cheng RYS, Switzer CH, Flores-Santana W, et al. (2011) Nitric oxide and redox mechanisms in the immune response. J Leukoc Biol 89: 873–891. Available: http://www.pubmedcentral.nih.gov/articlerender.fcgi?artid=3100761&tool=pmcentrez&rendertype=abstract Accessed 2014 July 22.

75. Clark RAF (2008) Oxidative stress and "senescent" fibroblasts in non-healing wounds as potential therapeutic targets. J Invest Dermatol 128: 2361–2364. Available: http://www.ncbi.nlm.nih.gov/pubmed/18787545 Accessed 2014 July 23.

76. Boles BR, Singh PK (2008) Endogenous oxidative stress produces diversity and adaptability in biofilm communities. Proc Natl Acad Sci U S A 105: 12503–12508. Available: http://www.pubmedcentral.nih.gov/articlerender.fcgi?artid=2527941&tool=pmcentrez&rendertype=abstract

77. Liu X, Sun X, Wu Y, Xie C, Zhang W, et al. (2013) Oxidation-sensing regulator AbfR regulates oxidative stress responses, bacterial aggregation, and biofilm formation in Staphylococcus epidermidis. J Biol Chem 288: 3739–3752. Available: http://www.pubmedcentral.nih.gov/articlerender.

fcgi?artid=3567629&tool=pmcentrez&rendertype=abstract Accessed 2014 July 17.

78. Heck DE, Shakarjian M, Kim HD, Laskin JD, Vetrano AM (2010) Mechanisms of oxidant generation by catalase. Ann N Y Acad Sci 1203: 120–125. Available: http://www.ncbi.nlm.nih.gov/pubmed/20716293 Accessed 2013 May 19.

79. Chaudiere J, Wilhelmsen EC, Tappel a L (1984) Mechanism of selenium-glutathione peroxidase and its inhibition by mercaptocarboxylic acids and other mercaptans. J Biol Chem 259: 1043–1050. Available: http://www.ncbi.nlm.nih.gov/pubmed/6693375

80. Zhao G, Usui ML, Underwood RA, Singh PK, James GA, et al. (2012) Time course study of delayed wound healing in a biofilm-challenged diabetic mouse model. Wound Repair Regen 20: 342–352. Available: http://www.pubmedcentral.nih.gov/articlerender.fcgi?artid=3349451&tool=pmcentrez&rendertype=abstract Accessed 2013 November 6.

81. Costerton JW, Stewart PS, Greenberg EP (1999) Bacterial biofilms: a common cause of persistent infections. Science 284: 1318–1322. Available: http://www.ncbi.nlm.nih.gov/pubmed/10334980 Accessed 2013 September 30.

82. Scales BS, Huffnagle GB (2013) The microbiome in wound repair and tissue fibrosis. J Pathol 229: 323–331. Available: http://www.pubmedcentral.nih.gov/articlerender.fcgi?artid=3631561&tool=pmcentrez&rendertype=abstract Accessed 2014 January 22.

83. Youmans GP, Paterson PY, Sommers HM (1980) The Biologic and clinical basis of infectious diseases. J Am Med Assoc 244: 849. Available: http://books.google.com/books/about/The_Biologic_and_clinical_basis_of_infec.html?id=CaBrAAAAMAAJ&pgis=1 Accessed 2014 January 29.

84. Dalben M, Varkulja G, Basso M, Krebs VLJ, Gibelli M a, et al. (2008) Investigation of an outbreak of Enterobacter cloacae in a neonatal unit and review of the literature. J Hosp Infect 70: 7–14. Available: http://www.ncbi.nlm.nih.gov/pubmed/18632183 Accessed 2013 September 30.

85. Madsen SM, Westh H, Danielsen L, Rosdahl VT (1996) Bacterial colonization and healing of venous leg ulcers. APMIS 104: 895–899. Available: http://www.ncbi.nlm.nih.gov/pubmed/9048868

86. Gerding DN (1995) Foot infections in diabetic patients: the role of anaerobes. Clin Infect Dis 20 Suppl 2: S283–8. Available: http://www.ncbi.nlm.nih.gov/pubmed/7548576 Accessed 2013 September 30.

87. Sapico FL, Witte JL, Canawati HN, Montgomerie JZ, Bessman AN (n.d.) The infected foot of the diabetic patient: quantitative microbiology and analysis of clinical features. Rev Infect Dis 6 Suppl 1: S171–6. Available: http://www.ncbi.nlm.nih.gov/pubmed/6718934 Accessed 2013 September 30.

88. Sharp CS, Bessman AN, Wagner FW, Garland D (n.d.) Microbiology of deep tissue in diabetic gangrene. Diabetes Care 1: 289–292. Available: http://www.ncbi.nlm.nih.gov/pubmed/720182 Accessed 2013 September 30.

89. Hansson C, Hoborn J, Möller A, Swanbeck G (1995) The microbial flora in venous leg ulcers without clinical signs of infection. Repeated culture using a validated standardised microbiological technique. Acta Derm Venereol 75: 24–30. Available: http://www.ncbi.nlm.nih.gov/pubmed/7747531 Accessed 2013 September 30.

90. Gjødsbøl K, Christensen JJ, Karlsmark T, Jørgensen B, Klein BM, et al. (2006) Multiple bacterial species reside in chronic wounds: a longitudinal study. Int Wound J 3: 225–231. Available: http://www.ncbi.nlm.nih.gov/pubmed/16984578 Accessed 2012 November 9.

91. Brook I, Frazier EH (1998) Aerobic and anaerobic microbiology of chronic venous ulcers. Int J Dermatol 37: 426–428. Available: http://www.ncbi.nlm.nih.gov/pubmed/9646126

92. Stapley R, Owusu BY, Brandon A, Cusick M, Rodriguez C, et al. (2012) Erythrocyte storage increases rates of NO and nitrite scavenging: implications for transfusion-related toxicity. Biochem J 446: 499–508. Available: http://www.ncbi.nlm.nih.gov/pubmed/22720637 Accessed 2013 May 22.

93. DeLoney CR, Schiller NL (1999) Competition of Various beta -Lactam Antibiotics for the Major Penicillin-Binding Proteins of Helicobacter pylori: Antibacterial Activity and Effects on Bacterial Morphology. Antimicrob Agents Chemother 43: 2702–2709. Available: http://aac.asm.org/content/43/11/2702.long Accessed 2014 February 17.

2

Synthetic Teichoic Acid Conjugate Vaccine against Nosocomial Gram-Positive Bacteria

Diana Laverde[1,2]**, Dominique Wobser**[1]**, Felipe Romero-Saavedra**[1,2]**, Wouter Hogendorf**[3]**, Gijsbert van der Marel**[3]**, Martin Berthold**[1]**, Andrea Kropec**[1]**, Jeroen Codee**[3]*, **Johannes Huebner**[1,4,5]*

1 Division of Infectious Diseases, Department of Medicine, University Medical Center Freiburg, Freiburg, Germany, **2** EA4655 U2RM Stress/Virulence, University of Caen Lower-Normandy, Caen, France, **3** Bio-organic Synthesis Unit, Faculty of Science, Leiden Institute of Chemistry, Leiden University, Leiden, Netherlands, **4** Division of Pediatric Infectious Diseases, Dr. von Hauner Children's Hospital, Ludwig-Maximilians-University, Munich, Germany, **5** German Center for Infection Research (DZIF), Partnersite Munich, Munich, Germany

Abstract

Lipoteichoic acids (LTA) are amphiphilic polymers that are important constituents of the cell wall of many Gram-positive bacteria. The chemical structures of LTA vary among organisms, albeit in the majority of Gram-positive bacteria the LTAs feature a common poly-1,3-(glycerolphosphate) backbone. Previously, the specificity of opsonic antibodies for this backbone present in some Gram-positive bacteria has been demonstrated, suggesting that this minimal structure may be sufficient for vaccine development. In the present work, we studied a well-defined synthetic LTA-fragment, which is able to inhibit opsonic killing of polyclonal rabbit sera raised against native LTA from *Enterococcus faecalis* 12030. This promising compound was conjugated with BSA and used to raise rabbit polyclonal antibodies. Subsequently, the opsonic activity of this serum was tested in an opsonophagocytic assay and specificity was confirmed by an opsonophagocytic inhibition assay. The conjugated LTA-fragment was able to induce specific opsonic antibodies that mediate killing of the clinical strains *E. faecalis* 12030, *Enterococcus faecium* E1162, and community-acquired *Staphylococcus aureus* strain MW2 (USA400). Prophylactic immunization with the teichoic acid conjugate and with the rabbit serum raised against this compound was evaluated in active and passive immunization studies in mice, and in an enterococcal endocarditis rat model. In all animal models, a statistically significant reduction of colony counts was observed indicating that the novel synthetic LTA-fragment conjugate is a promising vaccine candidate for active or passive immunotherapy against *E. faecalis* and other Gram-positive bacteria.

Editor: Eliane Namie Miyaji, Instituto Butantan, Brazil

Funding: The authors have no support or funding to report.

Competing Interests: The authors have declared that no competing interests exist.

* Email: johannes.huebner@med.uni-muenchen.de (JH); jcodee@chem.leidenuniv.nl (JC)

Introduction

The incidence of infections caused by multidrug resistant enterococci has become a worldwide problem over the last decades, particularly in immunocompromised patients [1]. Acquired resistance to β-lactams and vancomycin has spread almost through all patient populations, not only making nosocomial infections caused by this genus extremely difficult to treat, but also highlighting the necessity to develop alternative treatments [2]. Effective immunotherapies are usually directed against virulence factors like capsular polysaccharides that are present on the outside of the bacterial membrane and which often play a role to evade host responses [3]. Our group has previously identified an enterococcal surface antigen, lipoteichoic acid (LTA), present in nonencapsulated *E. faecalis* strains, that is able to induce opsonic antibodies and protect against *E. faecalis* and *E. faecium* bacteremia [4].

Lipoteichoic acids are amphiphilic glycoconjugate polymers and are important constituents of the cell wall of many Gram-positive bacteria such as staphylococci, streptococci, bacilli, clostridia,

corynebacteria and listeria [3,5]. They play crucial roles in cell division, membrane elasticity, porosity and anchoring of surface proteins [3,6]. The chemical structure of LTAs varies among organisms, but in the majority of Gram-positive bacteria LTA has a relatively conserved poly-1,3-(glycerolphosphate) backbone structure with limited variability, which may be due to its biosynthetic pathway [7,8]. This backbone represents the shared epitope amongst different bacterial strains and variation of the LTA structures between organisms originates from the type and number of carbohydrate appendages and length of the poly-glycerol phosphate chain [3]. The glycolipid anchor of LTA has been reported to be an integral part of the immunostimulatory activity of LTA, although it has also been argued that lipopeptides and lipoproteins that contaminate LTA when isolated from biological sources, are responsible for this activity. The poly-glycerol-phosphate backbone has no innate immunostimulatory activity itself and small teichoic acid fragments are poor immunogens [9–11]. Polysaccharide antigens that are intrinsically poorly immunogenic [12] are often conjugated to a carrier protein to elicit optimal anti-polysaccharide responses, and to induce humoral immune responses with the characteristics of a T-cell dependent antigen [12,13]. Synthetic oligosaccharide-protein conjugate vaccines have emerged recently as an interesting strategy in vaccinology, since they offer two major advantages: a well-defined chemical structure (chain length, nature of the epitope, well-established carbohydrate/protein ratio, single type of linkage between the antigen and the carrier) and lack of impurities present in polysaccharides obtained from bacterial cultures [14,15]. This would apply also for a teichoic acid-based vaccine. The accessibility of the highly conserved LTA polymer on the cell surface, its relatively uniform basic structure and its non-inflammatory nature would be advantages of a synthetic LTA vaccine that targets a wide variety of LTAs in different Gram-positive pathogens. Very recently, a tetanus toxoid conjugate of a 10-mer polyglycerolphosphate (PGP) was evaluated for its potential use as a conjugate vaccine directed against *S. aureus* [11]. Chen and coworkers evaluated the use of this PGP as a prophylactic conjugated vaccine in a mouse model and were able to demonstrate that serum raised against the conjugate elicits specific IgG capable of enhancing in vitro opsonophagocytic killing of *S. aureus* strains and clearance of staphylococcal bacteremia in vivo [11]. Although Chen *et al.* did not evaluate the potential use of this PGP-based conjugate vaccine against other Gram-positive pathogens, they suggest cross-protection against organisms expressing this highly conserved backbone [11].

We have previously shown that opsonic antibodies directed against LTA from *E. faecalis* are cross-reactive against LTA present in *Staphylococcus epidermidis, Staphylococcus aureus* and group B streptococci, and that they mainly bind to the poly-1,3-(glycerolphosphate) backbone, suggesting that this minimal structure may be sufficient for vaccine development against some Gram-positive bacteria [8]. We also have previously demonstrated that short synthetic oligoglycerol phosphates are able to absorb up to 91% of the opsonophagocytic killing of serum raised against LTA from *E. faecalis* 12030 against the homologous strain and also against *S. aureus* strains, indicating that these synthetic antigens could be used as templates for vaccine development [8,16]. In the present study, we evaluated the antigenicity of well-defined synthetic teichoic acid fragments and the development of a synthetic TA-protein conjugate as a vaccine candidate. We show that the semi-synthetic model vaccine modality is capable of eliciting opsonic and protective antibodies that promote *in vitro* opsonophagocytic killing, clearance of bacteria after active and passive immunization and effectively protect against enterococcal bacteremia in a rat endocarditis model.

Materials and Methods

Synthesis of Teichoic Acid Fragments

The short teichoic acid fragments used in this study were synthesized as described earlier using a combination of solution phase, light fluorous supported and automated solid phase techniques [8,17–19]. To scale up the synthesis of the hexaglycerolphosphate with a single glycosyl appendage (WH7, see Figure 1A), a light fluorous supported synthesis was applied [19]. Kojibiosyl [α-Glc-1→2-α-Glc] functionalized hexaglycerol phosphate (WH5, Figure 1B) was assembled using a solution phase synthesis approach [17].

Synthesis of the Model Teichoic Acid Vaccine

Maleimides derivatives of WH7 and WH5 were obtained by treatment of the glycosylated glycerol phosphate hexamers [16] with *N*-succinimidyl-3-maleimido propionate ester (See Supporting Information S1 for full experimental details and characteriza-

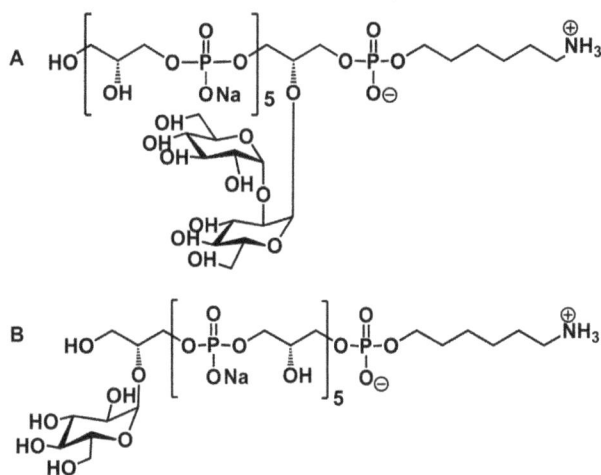

Figure 1. Structure of synthetic teichoic acid hexamers WH5 (A) and WH7 (B).

tion of the new compounds). Bovine serum albumin BSA was thiolated as described by Verez-Bencomo and co-workers [21] by treatment with thiopropionic acid hydroxysuccinimide ester homodisulfide and subsequent reduction of the disulfides. Quantification of the thiols using Ellman's reagent indicated that ±43 thiol groups per BSA molecule were present. The maleimides of WH7 and WH5 were then conjugated with thiolated-BSA through a Michael type addition [20,21] to give conjugates WH7-BSA and WH5-BSA. The teichoic acid conjugates were purified by dialysis and analyzed by SDS-PAGE, which revealed a broad band for the conjugates around 95 KDa and 100 KDa for the WH7 and WH5-conjugates, respectively (Figure 2). The conjugates were also analyzed on protein and sugar content. Combined, these analyses indicated that approximately 20 teichoic acid fragments per BSA molecule were installed.

Purification of LTA

Lipoteichoic acid from *E. faecalis* 12030 was obtained by butanol extraction as described elsewhere [6].

Bacterial Strains and Rabbit Immunizations

The bacterial strains used in this study were the clinical strains *E. faecalis* 12030 [22], *E. faecium* 1162 [23] and community-acquired *S. aureus* MW2 (USA400) [24]. A New Zealand white rabbit was immunized with purified LTA from *E. faecalis* 12030 as described elsewhere (anti-LTA) [4]. Three more rabbits were immunized with synthetic LTA WH5 conjugated with BSA (antiWH5-BSA), synthetic LTA WH7 conjugated with BSA (antiWH7-BSA), and unmodified BSA (anti-BSA). Two subcutaneous injections of 10 µg of conjugate or protein were given 2 weeks apart together with Freund's incomplete antigen; in the fourth and eighth week, three injections of 5 µg were given intravenously; and a final bleed was collected in the ninth week. All sera samples were heat-inactivated at 56°C for 30 min to inactivate complement components. For all experiments, sera from the final bleed of the rabbit were used.

Measurement of antigen (LTA and WH7)-specific IgG titers in rabbit immune sera

Total IgG concentration was determined in each pre-bleed, test bleed (i.e. after 5th antigen injection) and terminal bleed sera with the Easy-Titer Rabbit IgG Assay kit (Thermo Scientific) according

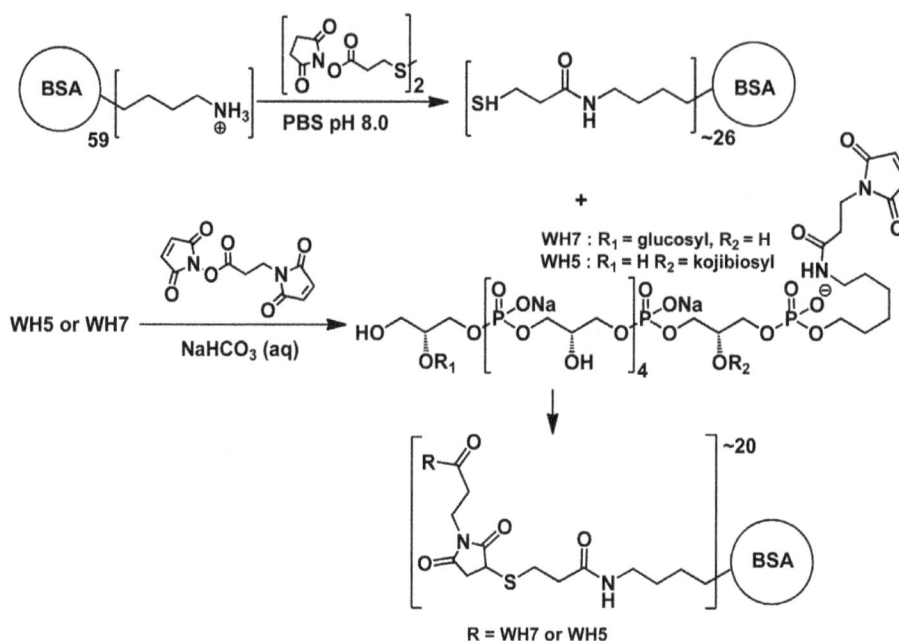

Figure 2. Synthesis of WH5-BSA and WH7-BSA conjugates.

to the manufacturer's instructions. Total IgG concentration in each serum sample was adjusted to 9.3 mg IgG/mL and serum specific IgG titers against WH7 or LTA were measured by ELISA. Nunc-immuno Maxisorp MicroWell 96 well plates were coated either with 0.2 µg of WH7 antigen (previously conjugated with Tetanus Toxoid as described by Hoogerhout and co-workers [25,26]) or 0.125 µg of LTA from *S. aureus* purchased from Sigma (St. Louis, Mo.) in 0.2M carbonate-bicarbonate coating buffer. Plates were incubated overnight at 4°C, washed three times after incubation with PBS containing 0.05% Tween 20 and blocked with 3% cold water fish skin gelatin (Sigma, St. Lois, Mo.) in PBS at 37°C for 2 hours. Rabbit sera were plated in twofold serial dilutions and incubated 1 hour at 37°C, starting with a dilution of 1:12.5 for each serum tested. Alkaline-phosphatase-conjugated anti-rabbit IgG produced in goat (Sigma) diluted 1:1.000 was used as secondary antibody and p-nitrophenyl phosphate (Sigma) was used as substrate (1 mg/mL in 0.1M glycine, 1 mM $MgCl_2$, 1 mM $ZnCl_2$, pH 10.4). After 60 min of incubation at room temperature, the absorbance was measured at 405 nm on a Tecan Infinite 200 PRO (Tecan Group Ltd.). Each experiment was performed twice at different time-points, and wells were measured in quadruplicate, with a coefficient of variation for each measured sample less than 12%. Serum Ig titers were calculated as follows: For each serum sample, a plot of OD value against the reciprocal of the dilution (Log(1/dilution)) was used to calculate the intercept with the specified cutoff value of each test, and this value was taken as the ELISA end point titer. The value extrapolated from the standard curve was then multiplied by the inverse of that dilution to generate the final inverse titer [27].

Opsonophagocytic Assay

An in vitro opsonophagocytic assay (OPA) was performed as described elsewhere [28] using baby rabbit serum as complement source and rabbit sera raised against either purified LTA or the conjugated synthetic LTA antigens. White blood cells (WBCs) were freshly prepared from human blood collected from healthy adult volunteers. Bacterial strains were grown to an

$OD_{650 \text{ nm}} = 0.400$ in tryptic soy broth (TSB). For the assay, the following components were mixed: 100 µl of PMNs; 100 µl of serum dilutions (as indicated), 100 µl of absorbed baby rabbit complement at a dilution of 1:15 and 100 µl of the bacterial suspension adjusted to the desired colony counts (i.e. 1:1 relation PMNs/bacteria). The mixture was incubated on a rotor rack at 37°C for 90 min, and samples were plated in duplicate at time 0 and after 90 min. Percent killing was calculated by comparing the colony forming units (CFUs) surviving in the tubes with bacteria, WBCs, complement and antibody, to the CFUs surviving in the tubes with all these components but lacking WBCs. Each experiment was performed at least twice, under the same conditions and with different blood donors, and measured in quadruplicate samples. Variation in the percentage of killing between the two independent experiments was less than 15%, showing the same trend and indicating the reliability of the procedure and the reproducibility of the experiment.

Opsonophagocytic Inhibition Assay (OPIA)

Synthetic TA mimetics and purified LTA from *E. faecalis* 12030 were used to evaluate its function as inhibitors of the opsonic killing generated by anti-LTA serum at different concentrations (from 0.08 to 100 µg/mL), by incubating them for 60 minutes at 4°C with an equal volume of a 1:200 dilution of anti-LTA serum or the sera raised against WH7. After this incubation step, the opsonophagocytic assay was performed as described above using the mix inhibitor/anti-LTA or inhibitor/anti-WH7-BSA serum instead of the serum dilutions. Inhibition assays were performed at serum dilutions yielding 70 to 80% of opsonic killing of the inoculum without the addition of the inhibitor. The inhibition of killing was calculated as the percentage of CFUs surviving opsonophagocytic killing when the inhibitor was used compared to those surviving when no inhibitor was present. Each experiment was performed at least twice, under the same conditions and with different blood donors, and measured in quadruplicate samples. Variation in the percentage of killing between the two independent experiments was less than 15%, showing the same trend and indicating the reliability of the procedure and the reproducibility of the experiment.

Whole cell ELISA

Bacterial strains *E. faecalis* 12030, *E. faecium* E1162 and *S. aureus* MW2, were grown in tryptic soy agar overnight and resuspended in PBS to an OD_{650} of 0.4 (~5×10^8 CFU/mL). From each bacterial suspension, 100 µL was added to wells of Nunc-immuno Maxisorp MicroWell 96 well plates and incubated overnight at 4°C. The wells were washed three times with PBS containing 0.05% Tween 20. After washing, the wells were blocked with PBS containing 3% cold water fish skin gelatin (Sigma, St. Louis, Mo.) for 2 h at 37°C. Antibodies normal rabbit serum (NRS), anti-LTA, anti WH5-BSA and anti WH7-BSA were tested at 1:100 dilution, with final concentration of 93 µg IgG's/mL. Alkaline-phosphatase conjugated anti-rabbit IgG produced in goat (Sigma) diluted 1:1,000 was used as secondary antibody and p-nitrophenyl phosphate (Sigma) was used as substrate (1 mg/mL in 0.1M glycine, 1 mM $MgCl_2$, 1 mM $ZnCl_2$, pH 10.4). After 60 min of incubation at room temperature, the absorbance was measured at 405 nm on a Tecan Infinite 200 PRO (Tecan Group Ltd.). Each experiment was performed twice, measured in quadruplicate samples, with a coefficient of variation for each measured sample less than 15%. Both antibodies were diluted in PBS with 1% cold water fish skin gelatin.

Mouse Sepsis Model – Active Immunization

Six to eight week-old female BALB/c mice (Charles River Laboratories Germany GmbH) were vaccinated with WH7-BSA conjugate or unmodified BSA antigens. Mice were injected subcutaneously with 10 µg of antigen emulsified in 100 µl complete Freund's adjuvant. Eight days post vaccination mice were boosted subcutaneously with 10 µg of antigen emulsified in 100 µl incomplete Freund's adjuvant. On days 14, 16, 18, 32 and 35 mice were boosted intraperitoneally with 5 µg of antigen suspended in saline. Challenge of the vaccinated mice was performed intravenously seven days after final boost with 3×10^7 CFUs of *E. faecalis* 12030. Mice were sacrificed 24 hours after challenge and colony counts in the liver were determined.

Mouse Sepsis Model – Passive Immunization

The model was performed as previously described [4]. In brief, six to eight week-old female BALB/c mice (Charles River Laboratories Germany GmbH) were immunized i.p. with 100 µL of anti WH7-BSA serum or normal rabbit serum (NRS). After 24 hours, mice were infected i.v. with 9.4×10^6 CFUs of *E. faecalis* 12030 via the tail vein. Mice were sacrificed 24 hours after challenge and colony counts in liver were determined.

Rat Endocarditis Model

A rat endocarditis model was performed as described previously [29]. In brief, female Wistar rats (Charles River Laboratories Germany GmbH), were anesthetized with 5.75% ketamine and 0.2% xylazine. Nonbacterial thrombotic endocarditis was caused by insertion of a plastic catheter (polyethylene tubing; Intramedic PE 10) via the right carotid artery. The catheter was advanced through the aortic valve into the left ventricle, secured properly and left in place. Rats were monitored closely after the catheter implantation. Ten rats were randomly separated in two groups, one received intravenously 500 µL of normal rabbit sera (NRS, Cedarlane, Burlington, Canada) during catheter implantation, and the second group received 500 µL of the anti WH7-BSA serum. An injection of 500 µL of NRS and anti WH7-BSA was made i.p.

after 24 hours of catheter implantation and 4 hours after bacterial challenge. An inoculum of 1.18×10^5 C.F.U.s of *E. faecalis* 12030 was injected via the tail vein in each rat 48 hours after catheter implantation. On postoperative day 6, rats were sacrificed and the correct placement of the catheter was verified. Valve vegetations were removed aseptically, weighed and homogenized in 500 µL TSB. Quantitative assessment was performed by weighing the vegetations as well as culturing serial dilutions on agar plates incubated over night at 37°C. Only animals with correct placement of the catheter were included in the study.

Statistics

The software program GraphPad PRISM version 5.00 was used for statistical analysis. The percentage of opsonophagocytic killing and absorbance in whole-cell ELISA was expressed as the geometrical mean ± the standard errors of the means and statistical significance was determined by one way ANOVA with Dunnett's post-test for multiple comparisons of groups. For animal models statistical significance was determined by nonparametric t-test with Mann-Whitney post-test and P values ≤ 0.05 were considered statistically significant.

Ethics Statement

All animal experiments were performed in compliance with the German animal protection law (TierSchG). The mice and rats were housed and handled in accordance with good animal practice as defined by FELASA and the national animal welfare body GV-SOLAS. The animal welfare committees of the University of Freiburg (Regierungspräsidium Freiburg Az 35/9185.81/G-07/15, Az 35/9185.81/G07-72) approved all animal experiments.

Results

Synthetic antigen selection and synthesis of the teichoic acid carrier protein conjugates

The library of teichoic acid fragments we have synthesized contains fragments varying in length (ranging from 6 to 30 glycerol phosphate monomers), glycosylation pattern (α-glucosyl, α-kojibiosyl, α-glucosamine and α-*N*-acetyl glucosamine) and terminal functionality (either a phosphate or primary alcohol) [8,16,17]. From this library we have selected WH7 as an optimal synthetic antigen because of its ability to inhibit the opsonic killing of antibodies raised against LTA from *E. faecalis* 12030 by more than 80% (See Figure 3) [16]. Furthermore, the synthesis of WH7 can be readily scaled to provide enough material allowing the generation of sufficient amounts of the conjugate [19]. Surprisingly we have found that the analogous WH5, featuring an α-kojibiosyl substituent that is found in native *E. faecalis* LTA is devoid of such activity (Figure 3). Therefore, WH5 was chosen as a negative control and both synthetic TA fragments were conjugated to BSA. To this end, the approach used by Verez-Bencomo *et al.* for the development of the Quimihib-vaccine [20] was employed, because of the structural similarities between the Hib ribitol-based polysaccharide and the synthetic TA fragments at hand. We selected BSA as a carrier protein because of its stability, ease of handling, favorable molecular weight, non-glycosylated nature and modification possibilities. The TA fragments, functionalized with a primary amine spacer, were equipped with a reactive maleimide group and conjugated to thiolated BSA as depicted in Figure 2. The obtained conjugates were characterized by SDS-PAGE and analyzed for carbohydrate and protein content, revealing a carbohydrate-protein ratio of ~1.5: 1, indicating that

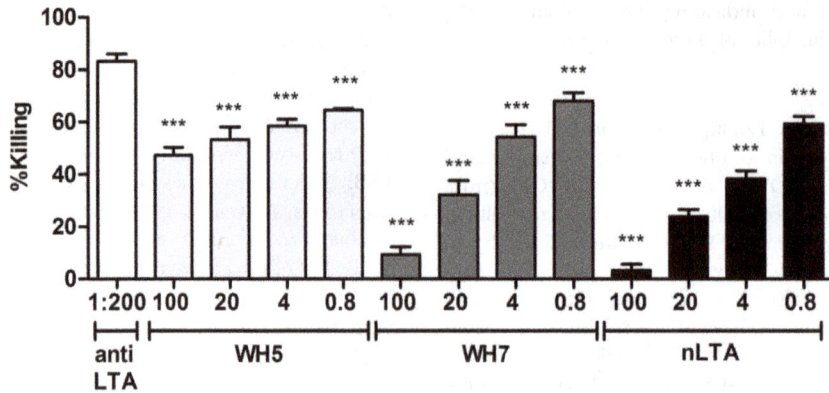

Figure 3. Evaluation of anti-LTA antibodies specificity for synthetic LTA hexamers and native LTA. Antibodies raised against LTA from *E. faecalis* 12030 were evaluated for their ability to bind specifically the different synthetic TA hexamers WH5 and WH7, and native LTA purified from *E. faecalis* 12030 (nLTA). Anti-LTA sera was used at final dilution of 1:200 and the strain tested was *E. faecalis* 12030. Inhibitors WH5, WH7 and nLTA at concentrations of 100, 20, 4 and 0.8 µg/mL were preincubated with anti-LTA antibodies for 1 hour at 4°C prior OPA. Opsonic killing of the target strain with non-absorbed antibodies was used to assess the reduction of opsonic killing produced by each inhibitor. All inhibitors tested showed statistical significance in comparison to control anti-LTA serum (P value <0.001). The corresponding dilutions of inhibitor used in the OPA are indicated in the x-axis and the % killing in the y-axis. Bars represent the mean of data and the error bars represent the standard error of the mean.

approximately 20 copies of antigen were present per carrier molecule.

LTA and teichoic acid fragment-specific IgGs are induced after rabbit immunizations

According to figure 4, LTA-specific IgG antibodies were generated during each immunization procedure using either native LTA from *E. faecalis* 12030 (Figure 4A) or WH7-BSA conjugate (Figure 4B). Antibodies raised against the synthetic hexamer WH7 are able to bind and recognize native LTA, but to a lesser extent than antibodies raised against the whole LTA molecule, while only a very low titer is observed for the terminal bleed of anti WH5-BSA serum. A different behavior is observed for WH7 titers of anti-LTA and anti WH7-BSA sera, where higher

titers were obtained for anti WH7-BSA serum in comparison to anti-LTA serum.

Teichoic acid conjugates WH5-BSA and WH7-BSA induced antibodies with opsonic activity

To confirm that opsonic antibodies were induced by the immunization with the conjugate WH7-BSA but not by immunization with BSA or the WH5-BSA conjugate, the opsonophagocytic activity of the sera was tested by OPA using *S. aureus* MW2 as target strain since this strain was shown to be sensitive to anti-LTA antibodies raised against LTA from *E. faecalis* 12030. At a serum dilution of 1:10 and 1:50, the serum raised against BSA was not able to mediate opsonic killing against *S. aureus* MW2. A similar behavior was observed for the WH5-BSA serum, with very low opsonophagocytic killing (<10%). On the other hand, the

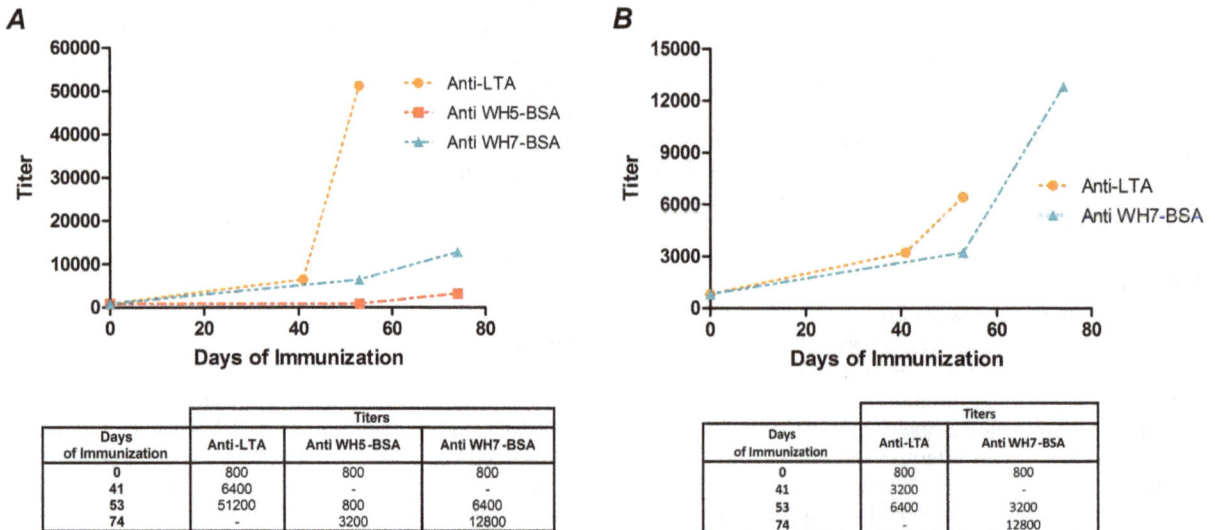

A

Days of Immunization	Titers		
	Anti-LTA	Anti WH5-BSA	Anti WH7-BSA
0	800	800	800
41	6400	-	-
53	51200	800	6400
74	-	3200	12800

B

Days of Immunization	Titers	
	Anti-LTA	Anti WH7-BSA
0	800	800
41	3200	-
53	6400	3200
74	-	12800

Figure 4. LTA and WH7 -specific IgG antibody titer curves. (A) LTA from *S.aureus* specific IgG titers of sera raised against native LTA (Anti-LTA), synthetic WH5-BSA conjugate (Anti WH5-BSA) and synthetic WH7-BSA conjugate (Anti WH7-BSA). **(B)** WH7 specific IgG titers of sera raised against native LTA (Anti-LTA) and synthetic WH7-BSA conjugate (Anti WH7-BSA).

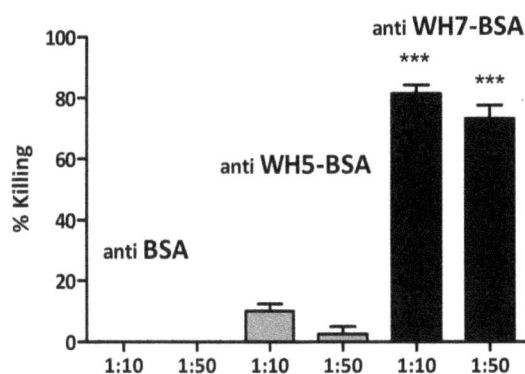

Figure 5. Opsonophagocytic assay. This experiment was used to test the ability of antibodies raised against conjugates with WH5 and WH7, and BSA to mediate killing in the strain *S. aureus* MW2. Anti WH7-BSA sera mediate opsonophagocytic killing of *S. aureus* MW2 in vitro, while the sera anti-BSA and anti-WH5-BSA do not. Comparison of killing percentages of same dilutions (i.e. 1:10 or 1:50) with control anti BSA serum, only showed statistical significance for anti WH7-BSA (*** P value <0.001). The different sera and the corresponding dilutions used in the OPA are indicated on the x-axis and the % killing on the y-axis. Bars represent the mean of data and the error bars represent the standard error of the mean.

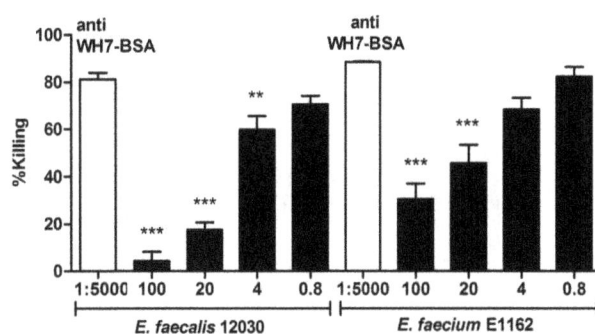

Figure 6. Specificity of anti WH7-BSA serum to WH7-BSA conjugate. Anti WH7-BSA serum was used at final dilution of 1:5,000 and the strains tested were *E. faecalis* 12030 and *E. faecium* E1162. Inhibitor WH7-BSA conjugate at different concentrations 100, 20, 4 and 0.8 µg/mL was preincubated with anti-LTA antibodies for 1 hour at 4°C prior OPA. Opsonic killing of the target strain with non-absorbed antibodies was used to assess the reduction of opsonic killing produced by each inhibitor, asterisks denote statistical significance (** P<0.01, *** P value <0.001). The corresponding dilutions of inhibitor used in the OPA are indicated on the x-axis and the % killing on the y-axis. Bars represent the mean of data and the error bars represent the standard error of the mean.

WH7-BSA serum elicited high opsonophagocytic activity (>60%) for both serum dilutions (Figure 5).

Opsonic antibodies to WH7-BSA conjugate are specific against the teichoic acid conjugate

To confirm the specificity of the opsonic antibodies to WH7-BSA, an opsonophagocytic inhibition assay (OPIA) was performed by preincubation of anti WH7-BSA serum with a range of dilutions of the synthetic WH7-BSA conjugate. The teichoic acid-conjugate was a potent inhibitor of the opsonophagocytic killing activity of the serum, not only against *E. faecalis* 12030, but also against *E. faecium* E1162 (an *E. faecium* strain sensitive to anti-LTA antibodies). In all cases, the inhibition of anti WH7-BSA antibodies is directly proportional to the amount of inhibitor added (Figure 6).

Opsonic antibodies to WH7-BSA conjugate are cross-reactive

To explore the specificity of the antibodies directed to the WH7-BSA conjugate for the LTA expressed by Gram-positive bacteria, the opsonic activity towards the strains *E. faecium* E1162, *S. aureus* MW2 and *E. faecalis* 12030 was evaluated. Different concentrations of the serum were tested to titer out the opsonic killing activity of the antibodies. Although it was observed that anti WH7-BSA serum can mediate opsonic killing of all strains tested (Figure 7), the enterococcal strains are effectively killed at higher dilutions than *S. aureus* MW2.

Antibodies directed against LTA and WH7-BSA conjugate bind bacterial cells

Immunoreactivity of tested sera against whole bacterial cells was analyzed by whole-cell ELISA (Figure 8). Immune sera anti-LTA, anti WH5-BSA and anti WH7-BSA showed statistically significant binding to cells of *E. faecalis* 12030, *E. faecium* E1162 and *S. aureus* MW2, in comparison with control serum (NRS) (* P<0.05, *** P<0.001). Anti-LTA serum showed stronger signal binding to *E. faecalis* 12030 cells, while anti WH5-BSA serum elicited poor

binding to *E. faecalis* and *E. faecium* cells. Serum raised against WH7-BSA-conjugate bound well to all bacterial strains tested.

Vaccination with WH7-BSA conjugate and immunization with anti WH7-BSA serum promotes clearance of bacteria in mice

To demonstrate protection by the teichoic acid conjugate, active immunization with the purified teichoic acid conjugate was performed in mice. After 2 injections of 10 µg/mouse and 5 injections of 5 µg/mouse, animals showed statistically significantly reduced numbers of *E. faecalis* 12030 in livers 24 hours after inoculation (Figure 9A). In another set of experiments, prophylactic treatment with antibodies raised against the WH7-BSA conjugate significantly reduces bacterial counts in mice. Application of 100 µl anti WH7-BSA serum per mouse 24 hours before bacterial challenge resulted in less bacteria (reduction by factor 5) at 24 hours after inoculation (Figure 9B).

Opsonic antibodies to WH7-BSA conjugate are protective in a rat enterococcal endocarditis model

To explore if the opsonophagocytic activity of anti WH7-BSA antibodies correlated with an *in vivo* infection model, a rat endocarditis model was performed. Passive immunization with 3 doses of 500 µL of anti WH7-BSA antibodies resulted in dramatic reductions of CFUs and also a significant reduction in valve vegetations (P<0.05) compared to controls (Figure 10); anti WH7-BSA serum mediated complete clearance of *E. faecalis* 12030 in more than 50% of the rats, while all control rats showed endocarditis.

Discussion

The use of polysaccharide-conjugate vaccines has rapidly emerged as a suitable strategy to combat different pathogenic bacteria. It includes the development of effective vaccines against *Haemophilus influenzae* type b, *Streptococcus pneumoniae*, *Neisseria meningitidis*, *Salmonella typhi* and *S. aureus* [13–15,30]. Many of these approaches have focused on targeting different

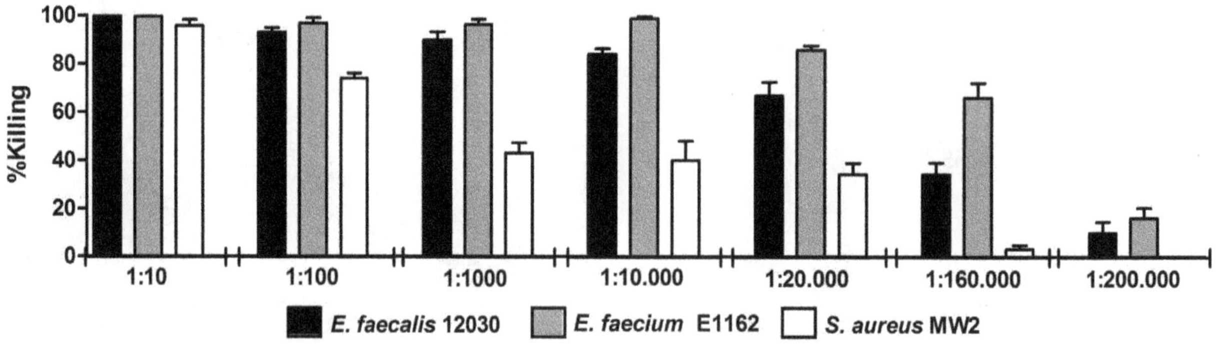

Figure 7. Opsonophagocytic killing activity of anti WH7-BSA antibodies against Gram-positive bacterial strains. The antibodies raised against WH7-BSA conjugate were used at different dilutions 1:10, 1:100, 1:1000, 1:10.000, 1:20.000, 1:160.000 and 1:200.000, as shown in the x-axis, to assess % killing activity of the target strains E. faecalis 12030, E. faecium E1162 and S. aureus MW2. Bars represent the mean of data and the error bars represent the standard error of the mean.

serotypes by the identification of conserved surface carbohydrate structures. LTAs from different common pathogens have been evaluated as possible vaccine candidates against bacterial infections. Intranasal co-administration of LTA from *Streptococcus pyogenes* with cholera toxin subunit B had induced good pharyngeal IgA and systemic IgGs, suggesting its use as an effective approach in prevention of tonsillitis [31]. Rabbit sera raised against conjugated LTA from *Clostridium difficile* have shown immunoreactivity with cells and spores from other *C. difficile* strains; however, no protection studies have been conducted [32]. In a different approach, Goldenberg and coworkers have demonstrated the ability of human antibodies directed against the phosphorylcholine to protect mice from a lethal dose of *S. pneumonia*. This antigenic component of LTA from *S. pneunomiae* and of the lipopolysaccharide from *H. influenzae* has been proposed as a vaccine candidate against pathogenic bacteria residing the upper respiratory tract [33].

Various synthetic approaches have been reported towards the assembly of well-defined teichoic acid structures of different pathogens, including *Clostridium difficile* [34,35], *E. faecalis* [7,8]

and *S. aureus* [11,19]. Martin *et al.* have described the synthesis of oligomers of the LTA antigen in *C. difficile* and used these fragments to define epitopes for vaccine development based on interactions of anti-LTA antibodies with bacteria in the blood of patients. However, no protective efficacy of antibodies directed against the synthetic oligomers has been reported so far [35]. To date, only a chimeric monoclonal antibody developed against LTA from *S. aureus* has been used in clinical trials. Pagibaximab has shown efficacy promoting staphylococcal phagocytosis and survival of animals challenged with coagulase-negative straphylococci [36,37]. This vaccine appeared safe and well tolerated in healthy adult volunteers and very-low-weight neonates; however, no clear protection has been observed against staphylococcal sepsis in this setting [36–39].

Despite the already demonstrated importance of specific antibodies directed against LTA in the treatment of infections caused by many Gram-positive pathogens, the protective efficacy of human pre-existing antibodies, either against LTA structures or phosphorylcholine, is still questionable [33,40]. Hufnagel and coworkers attribute this fact to the low specificity of pre-existing

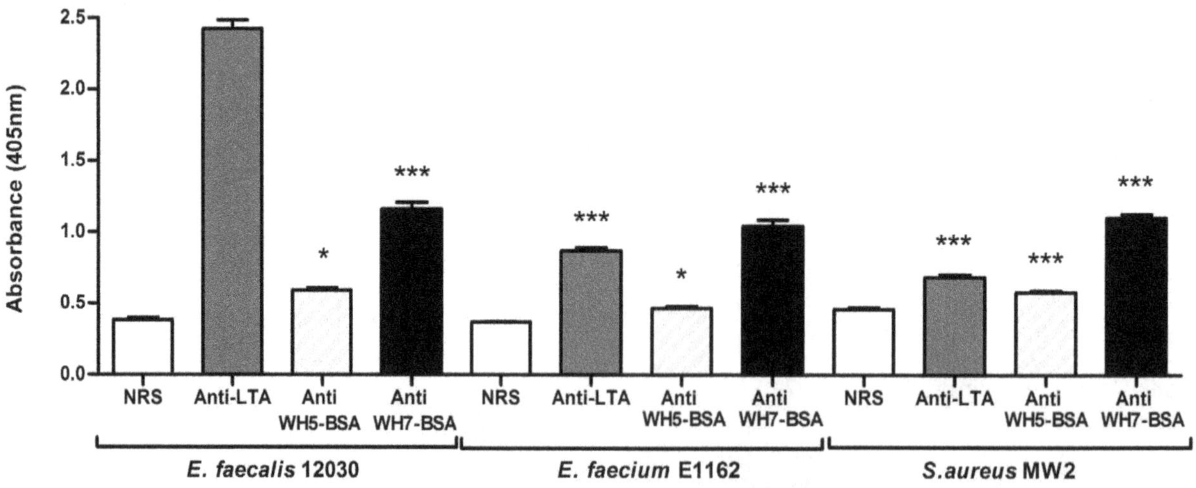

Figure 8. Whole-cell ELISA. Whole cell ELISA of E. faecalis 12030, E. faecium E1162 and S.aureus MW2 using normal rabbit serum (NRS- white bar), anti-LTA serum (grey bar), anti WH5-BSA serum (grey diagonally-striped bar) and anti WH7-BSA serum (black bar). Statistical significance performed by one way ANOVA with Dunnett's post-test for multiple comparison of groups, is denoted in asterisks (* P<0.05, ** P<0.01, *** P<0.001) and was calculated comparing the means of absorbance for each anti sera sample with serum control (NRS).

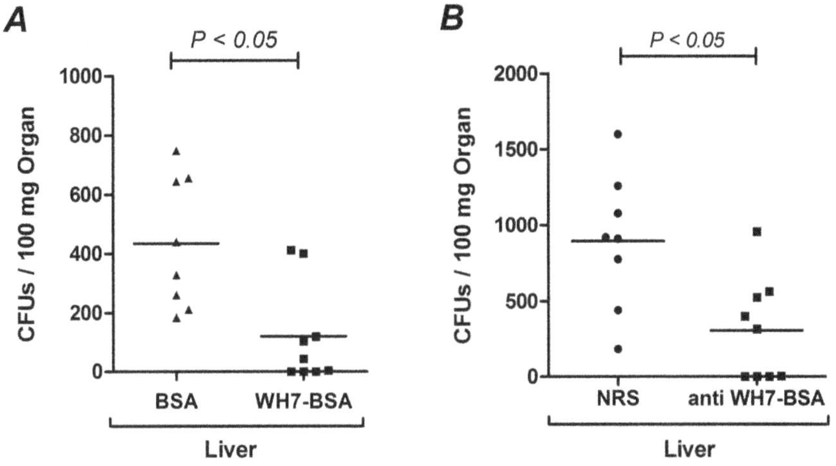

A

B

Figure 9. Mouse Sepsis Models. Active Immunization with WH7-BSA conjugate promotes clearance of *E. faecalis* 12030 in liver in comparison with vaccination with unmodified BSA (**A**). Passive immunization with anti WH7-BSA serum facilitates clearance of *E. faecalis* in liver in comparison with normal rabbit serum (NRS) (**B**). After 24 hours of challenging mice were killed, kidney and liver were removed to assess viable counts. Statistical analysis was done by Students t-test, horizontal bars represent geometric means, and a P value <0.05 was considered statistically significant.

human LTA antibodies and their non-opsonic nature [40]. Others have explain this biological behavior on the basis of individual variation in IgG subclasses, differences in affinity or avidity, or low specificity of naturally acquired antibodies [33,41,42].

We have shown previously that the cross-reactive epitope of anti-LTA antibodies is the poly-1,3-(glycerolphosphate) backbone of LTA in staphylococcal, streptococcal and enterococcal strains; likewise, their ability to opsonize and protect against *E. faecalis*, *E. faecium*, *S. aureus* and *S. epidermidis* bacteremia has been demonstrated [4,6,8,43]. In the current study, two synthetic TA hexamers with different glycosyl substituents were chosen according to their ability to absorb out opsonic anti-LTA antibodies, and their immunogenicity was evaluated after conjugation with a carrier protein. It was found that the kojibiosyl-functionalized TA-conjugate proved ineffective in eliciting opsonic antibodies, which seems to indicate that the kojibiosyl moiety present in the WH5 hexamer probably masks the polyglycerolphosphate backbone. Indeed, we have previously found that the kojibiosyl hexaglycerolphosphate WH5 is unable to inhibit opsonophagocytic killing of

E. faecalis by serum raised against LTA [8]. On the other hand, the ability of hexamer WH7 to inhibit, in a dose-dependent fashion, the opsonic killing activity of anti-LTA antibodies to *E. faecalis* 12030 was confirmed in agreement with previous findings [16]. The conjugate WH7-BSA was able to induce high-titered opsonic and specific antibodies, showing cross-reactivity with two other pathogenic Gram-positive species. The lower opsonophagocytic killing observed in the staphylococcal strain may be explained by the ability of capsular polysaccharides present in the cell wall of some Gram-positive strains to mask components in the cell wall [8,44,45]. Additionally, the opsonic killing activity of the anti WH7-BSA serum observed by OPA, correlates with the ability of the TA conjugate antigen to induce antibodies that promotes clearance of bacteria in the liver of mice, either by active or passive immunization and to protect rats against enterococcal endocarditis by passive immunization. The active immunization schedule used here was chosen because pilot experiments confirmed effective induction of high titers of opsonic antibodies. However, since we did not measure titers against the antigens after

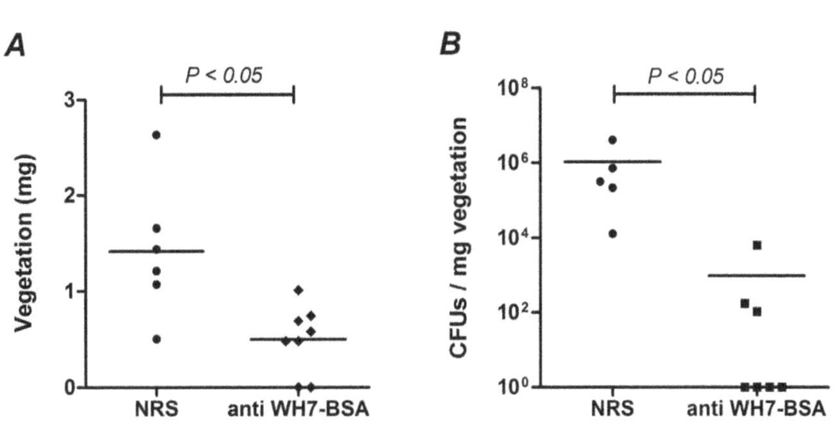

A

B

Figure 10. Passive immunization with anti WH7-BSA serum protects rats against enterococcal endocarditis. Thirteen female Wistar rats were catheterized, passively immunized 2 times with 500 μL of anti WH7-BSA sera and NRS, and challenged after 48 hours with 1.18×10^5 CFUs of *E. faecalis* 12030. After challenging, rats were boosted with either 500 μL of NRS or anti WH7-BSA sera. Three days after challenging, animals were sacrificed and valve vegetations were weighed and quantified. Horizontal lines indicate the geometric means of weighed valve vegetations (**A**) and CFUs per mg of vegetation of *E. faecalis* 12030 (**B**). Significant P values are indicated on the graphs.

all the different injections we cannot assess how many applications of antigen would be necessary to achieve protective titers.

The high affinity against LTA observed in the serum raised against native LTA from *E. faecalis* 12030 is probably caused by the diversity of antibodies against different epitopes of the molecule. Although antibodies directed against the WH7-BSA conjugate show lower binding than anti-LTA serum, these antibodies may have a higher affinity for the specific antigenic determinant present in the WH7 structure that represents the protective epitope, enabling the sera to mediate better in vitro opsonophagocytosis and protect animals against bacteremia or endocarditis. Antibodies directed against different epitopes (i.e. alanine residues present in native LTA but absent in WH7) that are raised through immunization with native LTA do not bind to the synthetic molecule, explaining the relatively low binding of serum raised against native LTA to WH7. The results from the whole cell ELISA confirm these findings since antibodies directed against WH7-BSA are able to bind bacterial cells of *E. faecalis* 12030, *E. faecium* E1162 and *S. aureus* MW2, showing that there is a direct relationship between the antibodies bound to the bacterial cell surface and the ability of antibodies to mediate *in vitro* opsonophagocytic killing or provide protection against bacteremia and endocarditis in animal studies.

In summary, the data presented here describe the potential use of a synthetic polyphosphoglycerol conjugate as vaccine candidate in enterococci. To this end we have used a potent glycerol phosphate hexamer, WH7, selected from a focused library of synthetic TA structures, and conjugated this to BSA, a model carrier protein. We were able to demonstrate that the WH7-BSA conjugate induces specific and opsonic antibodies that promote clearance of *E. faecalis* in the liver of mice and protect against enterococcal endocarditis, suggesting that this teichoic acid conjugate is a promising vaccine candidate against enterococcal infections. Its use as a prophylactic vaccine will require further studies of cross-reactivity against other enterococcal strains and other Gram-positive bacterial species as well as different infection models.

Author Contributions

Conceived and designed the experiments: JH JC GvM. Performed the experiments: DL DW FRS WH MB AK. Analyzed the data: JH JC AK. Wrote the paper: JH JC.

References

1. Theilacker C, Krueger WA, Kropec A, Huebner J (2004) Rationale for the development of immunotherapy regimens against enterococcal infections. Vaccine. Vol. 22. pp. S31–S38. doi:10.1016/j.vaccine.2004.08.014.
2. Leendertse M, Willems RJL, Flierman R, de Vos AF, Bonten MJM, et al. (2010) The complement system facilitates clearance of Enterococcus faecium during murine peritonitis. J Infect Dis 201: 544–552. doi:10.1086/650341.
3. Weidenmaier C, Peschel A (2008) Teichoic acids and related cell-wall glycopolymers in Gram-positive physiology and host interactions. Nat Rev Microbiol 6: 276–287. doi:10.1038/nrmicro1861.
4. Huebner J, Quaas A, Krueger WA, Goldmann DA, Pier GB (2000) Prophylactic and Therapeutic Efficacy of Antibodies to a Capsular Polysaccharide Shared among Vancomycin-Sensitive and -Resistant Enterococci. Infect Immun 68: 4631–4636. doi:10.1128/IAI.68.8.4631-4636.2000.
5. Neuhaus FC, Baddiley J (2003) A continuum of anionic charge: structures and functions of D-alanyl-teichoic acids in gram-positive bacteria. Microbiol Mol Biol Rev 67: 686–723. doi:10.1128/MMBR.67.4.686.
6. Theilacker C, Kaczynski Z, Kropec A, Fabretti F, Sange T, et al. (2006) Opsonic antibodies to Enterococcus faecalis strain 12030 are directed against lipoteichoic acid. Infect Immun 74: 5703–5712. doi:10.1128/IAI.00570-06.
7. Huebner J, Kropec-Huebner A, Van der Marel GA, Hogendorf WFJ, Codee JD (2012) European patent application (51). 1: 1–16. Available: http://www.google.com/patents/EP2500349A1?cl=en.
8. Theilacker C, Kropec A, Hammer F, Sava I, Wobser D, et al. (2012) Protection against Staphylococcus aureus by antibody to the polyglycerolphosphate backbone of heterologous lipoteichoic acid. J Infect Dis 205: 1076–1085. doi:10.1093/infdis/jis022.
9. Morath S (2002) Synthetic Lipoteichoic Acid from Staphylococcus aureus Is a Potent Stimulus of Cytokine Release. J Exp Med 195: 1635–1640. doi:10.1084/jem.20020322.
10. Zähringer U, Lindner B, Inamura S, Heine H, Alexander C (2008) TLR2 - promiscuous or specific? A critical re-evaluation of a receptor expressing apparent broad specificity. Immunobiology 213: 205–224. doi:10.1016/j.imbio.2008.02.005.
11. Chen Q, Dintaman J, Lees A, Sen G, Schwartz D, et al. (2013) Novel synthetic (poly)glycerolphosphate-based antistaphylococcal conjugate vaccine. Infect Immun 81: 2554–2561. doi:10.1128/IAI.00271-13.
12. Mond JJ, Lees A, Snapper CM (1995) T cell-independent antigens type 2. Annu Rev Immunol 13: 655–692. doi:10.1146/annurev.iy.13.040195.003255.
13. Finn A (2004) Bacterial polysaccharide-protein conjugate vaccines. Br Med Bull 70: 1–14. doi:10.1093/bmb/ldh021.
14. Phalipon A, Tanguy M, Grandjean C, Guerreiro C, Bélot F, et al. (2009) A synthetic carbohydrate-protein conjugate vaccine candidate against Shigella flexneri 2a infection. J Immunol 182: 2241–2247. doi:10.4049/jimmunol.0803141.
15. Chu K-C, Wu C-Y (2012) Carbohydrate-based synthetic vaccines: does the synthesis of longer chains of carbohydrates make this a step ever closer? Future Med Chem 4: 1767–1770. doi:10.4155/fmc.12.102.
16. Hogendorf WFJ, Meeuwenoord N, Overkleeft HS, Filippov DV, Laverde D, et al. (2011) Automated solid phase synthesis of teichoic acids. Chem Commun (Camb) 47: 8961–8963. doi:10.1039/c1cc13132j.
17. Hogendorf WFJ, Bos LJ vd, Overkleeft HS, Codée JDC, Marel GA vd (2010) Synthesis of an α-kojibiosyl substituted glycerol teichoic acid hexamer. Bioorganic Med Chem 18: 3668–3678. doi:10.1016/j.bmc.2010.03.071.
18. Hogendorf WFJ, Lameijer LN, Beenakker TJM, Herman S, Filippov DV, et al. (2012) Fluorous Linker Facilitated Synthesis of Teichoic Acid Fragments: 2010–2013. doi:10.1021/ol2033652.
19. Hogendorf WFJ, Kropec A, Filippov D V, Overkleeft HS, Huebner J, et al. (2012) Light fluorous synthesis of glucosylated glycerol teichoic acids. Carbohydr Res 356: 142–151. doi:10.1016/j.carres.2012.02.023.
20. Verez-Bencomo V, Fernández-Santana V, Hardy E, Toledo ME, Rodríguez MC, et al. (2004) A synthetic conjugate polysaccharide vaccine against Haemophilus influenzae type b. Science 305: 522–525. doi:10.1126/science.1095209.
21. Fernández-Santana V, González-Lio R, Sarracent-Pérez J, Verez-Bencomo V (1998) Conjugation of 5-azido-3-oxapentyl glycosides with thiolated proteins through the use of thiophilic derivatives. Glycoconj J 15: 549–553. doi:10.1023/A:1006903524007.
22. Huebner J, Wang Y, Krueger WA, Madoff LC, Martirosian G, et al. (1999) Isolation and Chemical Characterization of a Capsular Polysaccharide Antigen Shared by Clinical Isolates of Enterococcus faecalis and Vancomycin-Resistant Enterococcus faecium. Infect Immun 67: 1213–1219.
23. Heikens E, Bonten MJM, Willems RJL (2007) Enterococcal surface protein Esp is important for biofilm formation of Enterococcus faecium E1162. J Bacteriol 189: 8233–8240. doi:10.1128/JB.01205-07.
24. Voyich JM, Otto M, Mathema B, Braughton KR, Whitney AR, et al. (2006) Is Panton-Valentine leukocidin the major virulence determinant in community-associated methicillin-resistant Staphylococcus aureus disease? J Infect Dis 194: 1761–1770. doi:10.1086/509506.
25. Van der Ley P, Heckels JE, Virji M, Hoogerhout P, Poolman JT (1991) Topology of outer membrane porins in pathogenic Neisseria spp. Infect Immun 59: 2963–2971.
26. Drijfhout JW, Bloemhoff W, Poolman JT, Hoogerhout P (1990) Solid-phase synthesis and applications of N-(S-acetylmercaptoacetyl) peptides. Anal Biochem 187: 349–354. doi:10.1056/NEJMra011223.
27. Chen Q, Cannons JL, Paton JC, Akiba H, Schwartzberg PL, et al. (2008) A novel ICOS-independent, but CD28- and SAP-dependent, pathway of T cell-dependent, polysaccharide-specific humoral immunity in response to intact Streptococcus pneumoniae versus pneumococcal conjugate vaccine. J Immunol 181: 8258–8266. doi:10.4049/jimmunol.181.12.8258.
28. Kropec A, Sava IG, Vonend C, Sakinc T, Grohmann E, et al. (2011) Identification of SagA as a novel vaccine target for the prevention of Enterococcus faecium infections. Microbiology 157: 3429–3434. doi:10.1099/mic.0.053207-0.
29. Haller C, Berthold M, Wobser D, Kropec A, Lauriola M, et al. (2014) Cell-Wall Glycolipid Mutations and Their Effects on Virulence of E. faecalis in a Rat

Model of Infective Endocarditis. PLoS One 9: e91863. doi:10.1371/journal.-pone.0091863.

30. Mond JJ, Lees A, Snapper CM (1995) T CELL-INDEPENDENT. Annu Rev Immunol 13: 655–692.

31. Yokoyama Y, Harabuchi Y (2002) Intranasal immunization with lipoteichoic acid and cholera toxin evokes specific pharyngeal IgA and systemic IgG responses and inhibits streptococcal adherence to pharyngeal epithelial cells in mice. Int J Pediatr Otorhinolaryngol 63: 235–241. doi:10.1016/S0165-5876(02)00021-6.

32. Cox AD, St. Michael F, Aubry A, Cairns CM, Strong PCR, et al. (2013) Investigating the candidacy of a lipoteichoic acid-based glycoconjugate as a vaccine to combat Clostridium difficile infection. Glycoconj J 30: 843–855. doi:10.1007/s10719-013-9489-3.

33. Goldenberg HB, McCool TL, Weiser JN (2004) Cross-reactivity of human immunoglobulin G2 recognizing phosphorylcholine and evidence for protection against major bacterial pathogens of the human respiratory tract. J Infect Dis 190: 1254–1263. doi:10.1086/424517.

34. Oberli MA, Hecht M-L, Bindschädler P, Adibekian A, Adam T, et al. (2011) A possible oligosaccharide-conjugate vaccine candidate for Clostridium difficile is antigenic and immunogenic. Chem Biol 18: 580–588. doi:10.1016/j.chembiol.2011.03.009.

35. Martin CE, Broecker F, Eller S, Oberli MA, Anish C, et al. (2013) Glycan arrays containing synthetic Clostridium difficile lipoteichoic acid oligomers as tools toward a carbohydrate vaccine. Chem Commun (Camb) 49: 7159–7161. doi:10.1039/c3cc43545h.

36. Weisman LE, Fischer GW, Thackray HM, Johnson KE, Schuman RF, et al. (2009) Safety and pharmacokinetics of a chimerized anti-lipoteichoic acid monoclonal antibody in healthy adults. Int Immunopharmacol 9: 639–644. doi:10.1016/j.intimp.2009.02.008.

37. Weisman LE, Thackray HM, Garcia-Prats JA, Nesin M, Schneider JH, et al. (2009) Phase 1/2 double-blind, placebo-controlled, dose escalation, safety, and

pharmacokinetic study of pagibaximab (BSYX-A110), an antistaphylococcal monoclonal antibody for the prevention of staphylococcal bloodstream infections, in very-low-birth-weight neon. doi:10.1128/AAC.01565-08.

38. Weisman LE, Thackray HM, Steinhorn RH, Walsh WF, Lassiter HA, et al. (2011) A randomized study of a monoclonal antibody (pagibaximab) to prevent staphylococcal sepsis. Pediatrics 128: 271–279. doi:10.1542/peds.2010-3081.

39. Percy MG, Gründling A (2014) Lipoteichoic Acid Synthesis and Function in Gram-Positive Bacteria. Annu Rev Microbiol. doi:10.1146/annurev-micro-091213-112949.

40. Hufnagel M, Kropec A, Theilacker C, Huebner J (2005) Naturally acquired antibodies against four Enterococcus faecalis capsular polysaccharides in healthy human sera. Clin Diagn Lab Immunol 12: 930–934. doi:10.1128/CDLI.12.8.930-934.2005.

41. Kelly-Quintos C, Kropec A, Briggs S, Ordonez CL, Goldmann DA, et al. (2005) The role of epitope specificity in the human opsonic antibody response to the staphylococcal surface polysaccharide poly N-acetyl glucosamine. J Infect Dis 192: 2012–2019. doi:10.1086/497604.

42. Romero-Steiner S, Musher DM, Cetron MS, Pais LB, Groover JE, et al. (1999) Reduction in functional antibody activity against Streptococcus pneumoniae in vaccinated elderly individuals highly correlates with decreased IgG antibody avidity. Clin Infect Dis 29: 281–288. doi:10.1086/520200.

43. Theilacker C, Kaczyński Z, Kropec A, Sava I, Ye L, et al. (2011) Serodiversity of opsonic antibodies against Enterococcus faecalis–glycans of the cell wall revisited. PLoS One 6: e17839. doi:10.1371/journal.pone.0017839.

44. Graveline R, Segura M, Radzioch D, Gottschalk M (2007) TLR2-dependent recognition of Streptococcus suis is modulated by the presence of capsular polysaccharide which modifies macrophage responsiveness. Int Immunol 19: 375–389. doi:10.1093/intimm/dxm003.

45. Riordan KO, Lee JC (2004) Staphylococcus aureus capsular polysaccharides. Clin Microbiol Rev 17: 218–234. doi:10.1128/CMR.17.1.218.

3

The *Porphyromonas gingivalis* Ferric Uptake Regulator Orthologue Binds Hemin and Regulates Hemin-Responsive Biofilm Development

Catherine A. Butler◗, Stuart G. Dashper◗, Lianyi Zhang◗, Christine A. Seers, Helen L. Mitchell, Deanne V. Catmull, Michelle D. Glew, Jacqueline E. Heath, Yan Tan, Hasnah S. G. Khan, Eric C. Reynolds*

Oral Health Cooperative Research Centre, Melbourne Dental School, Bio21 Institute, The University of Melbourne, Victoria, Australia

Abstract

Porphyromonas gingivalis is a Gram-negative pathogen associated with the biofilm-mediated disease chronic periodontitis. *P. gingivalis* biofilm formation is dependent on environmental heme for which *P. gingivalis* has an obligate requirement as it is unable to synthesize protoporphyrin IX *de novo*, hence *P. gingivalis* transports iron and heme liberated from the human host. Homeostasis of a variety of transition metal ions is often mediated in Gram-negative bacteria at the transcriptional level by members of the Ferric Uptake Regulator (Fur) superfamily. *P. gingivalis* has a single predicted Fur superfamily orthologue which we have designated Har (heme associated regulator). Recombinant Har formed dimers in the presence of Zn^{2+} and bound one hemin molecule per monomer with high affinity (K_d of 0.23 µM). The binding of hemin resulted in conformational changes of Zn(II)Har and residue ^{97}Cys was involved in hemin binding as part of a predicted -^{97}C-^{98}P-^{99}L-hemin binding motif. The expression of 35 genes was down-regulated and 9 up-regulated in a Har mutant (ECR455) relative to wild-type. Twenty six of the down-regulated genes were previously found to be up-regulated in *P. gingivalis* grown as a biofilm and 11 were up-regulated under hemin limitation. A truncated Zn(II)Har bound the promoter region of *dnaA* (PGN_0001), one of the up-regulated genes in the ECR455 mutant. This binding decreased as hemin concentration increased which was consistent with gene expression being regulated by hemin availability. ECR455 formed significantly less biofilm than the wild-type and unlike wild-type biofilm formation was independent of hemin availability. *P. gingivalis* possesses a hemin-binding Fur orthologue that regulates hemin-dependent biofilm formation.

Editor: Benfang Lei, Montana State University, United States of America

Funding: Financial assistance was received from The Australian National Health and Medical Research Council (https://www.nhmrc.gov.au/) project grant 1008055 and Australian Dental Research Foundation (http://www.ada.org.au/about/adrfgrants.aspx) grant 22/2002. The funders had no role in study design, data collection and analysis, decision to publish, or preparation of the manuscript.

Competing Interests: The authors have declared that no competing interests exist.

* Email: e.reynolds@unimelb.edu.au

◗ These authors contributed equally to this work.

Introduction

Chronic periodontitis is an inflammatory disease of the supporting tissues of the teeth associated with specific bacteria in a biofilm and is a major cause of tooth loss [1]. *Porphyromonas gingivalis* is considered to be a principal pathogen in chronic periodontitis due to its close association with the disease in humans and its virulence in animal models [1–5]. *P. gingivalis* and other oral bacterial species exist *in vivo* as a polymicrobial biofilm called subgingival plaque accreted onto the surface of the tooth root. *P. gingivalis* has recently been described as a 'keystone pathogen' that manipulates the host response to allow proliferation of the subgingival plaque community to produce dysbiosis and disease progression [6]. Sessile *P. gingivalis* cells release antigens, toxins and hydrolytic enzymes such as proteinases into the surrounding tissue that stimulate and dysregulate the host immune response causing tissue destruction [7].

Like most bacteria, *P. gingivalis* has an essential growth requirement for iron but unlike most bacteria *P. gingivalis* cannot synthesize protoporphyrin IX, a porphyrin derivative that complexes ferrous iron (Fe^{2+}) to form heme, a cofactor used with various enzymes and in electron transport systems [8]. Thus *P. gingivalis* must acquire protoporphyrin IX from the environment, which may explain the reported preferential utilisation of heme as an iron source by this bacterium [9]. *P. gingivalis* also utilises manganese especially for protection from oxidative stress and intracellular survival in host cells [10,11]. Vascular disruption and bleeding are characteristics of periodontitis, providing an iron/heme rich environment for bacterial growth. However, *P. gingivalis* would also be exposed to low iron/heme environments and oxidative stress during colonization and periods of disease quiescence. In response to this dynamic environment, *P. gingivalis* must tightly regulate iron homeostasis gene expression to survive. We have characterised the *P. gingivalis* W50 response to hemin-limitation in continuous culture, using proteomic and transcriptomic approaches that identified 160 genes and 70 proteins that are differentially regulated by hemin availability [12]. We have also demonstrated the importance of ferrous iron uptake in *P. gingivalis* W50 using the ferrous iron transporter mutant W50FB1, which has half the iron content of the wild-type, and was avirulent in an animal model of disease [13].

Iron homeostasis is mediated in most Gram-negative bacteria and in Gram-positive bacteria with low GC content by the transcriptional repressor protein Fur, using ferrous iron as co-factor [14,15]. During iron-rich conditions Fur binds intracellular Fe^{2+}, acquiring a conformation able to bind target DNA sequences, known as Fur boxes that are found overlapping the promoters of Fur-regulated genes and thereby inhibits transcription of these genes. When iron is scarce, the equilibrium shifts to release Fe^{2+}, Fur dissociates from the Fur box and allows access to RNA polymerase and the genes are expressed [16]. Fur is dimeric, and in addition to the labile Fe^{2+} binding site (S2) it also binds zinc in a structurally important site (S1) and can have a further metal binding site per monomer (S3) [17,18]. The molecular mechanisms of transcriptional control by Fur appear to be shared by many bacterial species, as Fur orthologues from numerous species are able to complement an *E. coli fur* mutant [16]. The genes regulated by *E. coli* Fur encode proteins that are not only involved in iron uptake [19], but also in cellular processes such as defence against oxygen radicals [20], metabolic pathways [21], chemotaxis [22], and the production of toxins and other virulence factors [23,24]. Iron responsive Fur is the best characterized member of a larger Fur superfamily, with Fur-like proteins responding to manganese (Mur), zinc (Zur), nickel (Nur), hydrogen peroxide (PerR) and iron via heme (Irr) [18].

Characterisation of the *P. gingivalis* Fur orthologue we have designated Har (**h**eme **a**ssociated **r**egulator) showed a hemin and iron-binding transcriptional regulator that plays a role in hemin-responsive biofilm development. We have further demonstrated a relationship between Har, hemin availability and biofilm development with the Har regulon overlapping previously identified *P. gingivalis* hemin-responsive and biofilm adaptation regulons [12]. *P. gingivalis* is unique as an iron-dependent Gram-negative bacterium with a single Fur superfamily orthologue, Har that regulates hemin-dependent biofilm formation.

Materials and Methods

Bacterial strains and culture conditions

The bacterial strains and plasmids used in this study are listed in Table 1. *P. gingivalis* ATCC 33277 was obtained from the culture collection of the Oral Health CRC, Melbourne Dental School, The University of Melbourne. Strains ECR455 and ECR475 were derived from ATCC 33277 during this study. *P. gingivalis* strains were routinely maintained on Horse Blood Agar (HBA) plates (HBA; 40 g/L Blood Agar Base No. 2 (Oxoid), 100 mL/L Defibrinated Horse Blood (Equicell, Bayles, Victoria, Australia) containing 5 μg/mL vitamin K and antibiotic selection of 10 μg/mL erythromycin or 5 μg/mL ampicillin where appropriate. Batch cultures of all *P. gingivalis* strains were grown without antibiotics in Brain Heart Infusion (BHI; Oxoid) or Mycoplasma Basal Broth (MBB; Becton, Dickinson and Company BBL) supplemented with 0.5 g/L cysteine, 5 μg/mL hemin and 5 μg/mL vitamin K. All *P. gingivalis* cultures were incubated anaerobically at 37°C in a MACS MG500 anaerobic workstation (Don Whitley Scientific). *P. gingivalis* ATCC 33277 and ECR455 were grown in continuous culture in Bioflo 110 biofermentors (New Brunswick Scientific) as previously described [12], with a 400 mL working volume in BHI supplemented with 0.5 g/L cysteine, 5 μg/mL hemin and 5 μg/mL vitamin K.

DNA analysis and manipulations

Oligonucleotide primers used in this study are listed in Table 2. Genomic DNA from *P. gingivalis* strains was prepared using the DNeasy Blood and Tissue kit (Qiagen) and plasmid DNA was extracted from *E. coli* strains using the QIAprep spin miniprep kit (Qiagen). The Herculase II DNA Polymerase (Stratagene) and Platinum Taq DNA Polymerase High Fidelity (Invitrogen) were used according to manufacturer's instructions in PCR reactions. SOE PCRs (gene splicing by overlap extension PCRs) were performed essentially as previously described [25]. PCR products were purified using the NucleoSpin Extract II purification kit (Macherey Nagel) according to manufacturer's instructions. Ligations were transformed into *E. coli* alpha-select gold competent cells (Bioline) by heat shock according to manufacturer's instructions. DNA was sequenced by Applied Genetics Diagnostics, The University of Melbourne.

Expression and purification of recombinant Har proteins

The full-length *har* gene PGN_1503 (501 bp) was PCR amplified from *P. gingivalis* 33277 using primers HarNterm and HarCterm which had *Eco*RI and *Xho*I restriction sites respectively. The HarNterm primer also encoded a FactorXa cleavage site (IEGR) immediately before the start of the *har* gene. The *har* gene was also amplified with a C-terminal truncation (450 bp) using primers Har150_Nterm and Har150_Cterm which had *Bam*HI and *Sma*I restriction sites respectively. The full length *har* gene and the expression vector pGEX-4T-1 were each digested with *Eco*RI/*Xho*I, whilst the truncated *har* gene and pGEX-4T-1 were each digested with *Bam*HI/*Sma*I, then ligated and transformed into *E. coli* alpha gold (Bioline). The resulting plasmids pGEX-4T-Har and pGEX-4T-Har150, encoding full-length and truncated Har respectively, were transformed into the expression strain *E. coli* BL21(DE3) (Bioline). Site-directed mutagenesis was performed on the pGEX-4T-Har plasmid using the QuikChange Lightning Site Directed Mutagenesis kit (Agilent) and the HarC97A SDM primer as per manufacturer's instructions.

Recombinant *P. gingivalis* Har, truncated Har150 and the site-directed mutant C97A were over-expressed as glutathione S-transferase (GST) fusion proteins with an engineered N-terminal Factor Xa cleavage site. The recombinant genes were expressed in *E. coli* BL21(DE3) by 5 h induction with 200 μM IPTG at 32°C from cell culture OD_{600} around 0.8. After lysis, the GST-Har fusion proteins were applied to GSTrap HP columns (GE Healthcare) at pH 6.2 in 20 mM phosphate binding buffer containing 500 mM NaCl, 1% Triton X-100 and protease inhibitors (Roche). The GST tag was cleaved on column with Factor Xa (GE Healthcare) at pH 8.0 in 50 mM Tris buffer containing 2 mM $CaCl_2$ and 150 mM NaCl. Har and its variants bound to a cation exchange HiTrap HP column (GE Healthcare) in 20 mM phosphate buffer at pH 6.8 and were eluted at an ionic strength of 450 mM NaCl. After further purification with size exclusion chromatography, the Har, Har150 and C97A proteins had a purity of over 95%, with a yield of ~6 mg/L culture.

Zinc binding by Har

The zinc contents of recombinant Har and variants were determined by electrospray ionization mass spectrometry (ESI-MS) in the presence and absence of formic acid or by inductively coupled plasma mass spectrometry (ICP-MS) of protein solutions treated with Chelex-100 resin and EDTA with or without 8 M urea. Chelex-100 treatment was carried out with a 100 μM protein solution being extensively dialysed at 4°C against 1 g Chelex-100 resin in 50 mM acetate buffer containing 100 mM NaCl and 5 mM DTT at pH 5.0 [26]. EDTA treatment was

Table 1. Bacterial strains and plasmids used in this study.

Bacterial strain or plasmid	Description[a]	Reference or source
Strains		
Escherichia coli		
Alpha-Select Gold	F⁻ *deoR endA1 recA1 relA1 gyrA96 hsdR17*(r$_K$⁻, m$_K$⁺) *supE44 thi-1 phoA* Δ(*lacZYA-argF*)U169 Φ80*lacZ*ΔM15 λ⁻	Bioline
BL21(DE3)	F– *ompT hsd*SB(rB–, mB–) *gal dcm* (DE3)	Bioline
Porphyromonas gingivalis		
ATCC 33277	Wild-type	Oral Health CRC
ECR455	*P. gingivalis* 33277 *har::ermF*, Em^r	This study
ECR475	ECR455 *ermF::har cepA*, Ap^r	This study
Plasmids		
pBluescript II	Cloning vector; Ap^r	Stratagene
pGEM-TEasy	Cloning vector linearized with T overhangs for ligation of PCR products generated with A overhangs by Taq polymerase; Ap^r	Promega
pGEX-4T-Har	pGEX-4T-1 containing *har* and an engineered Factor Xa cleavage site, for cleavage of the expressed full-length Har from an N-terminal GST tag.	This study
pGEX-4T-Har150	pGEX-4T-1 containing a 3′ shortened *har* gene and an engineered Factor Xa cleavage site, for cleavage of the expressed C-terminally truncated Har150 from an N-terminal GST tag.	This study
pGEX-4T-HarC97A	pGEX-4T-Har which had the TGT codon for Cys97 mutated to GCA for Ala.	This study
pVA2198	*E. coli-Bacteroides* shuttle vector carrying *ermF-ermAM* cassette; Em^r	[76]
pHarSOE1-4	Recombination cassette for the deletion of *har* from *P. gingivalis* 33277, cloned blunt into the SmaI site of pBluescript II; Ap^r in *E. coli*	This study
pEC474	*cepA* in pBR322; Ap^r	[77]
pHarComp	Recombination cassette for the insertion of *har* into *P. gingivalis* ECR455, cloned via A-overhangs into pGEM-T Easy; Ap^r	This study

[a]Ap^r, ampicillin resistant; Em^r, erythromycin resistant.

performed by overnight incubation of a protein solution (100 μM) at 4°C with 50 mM EDTA at pH 8.0 in 10 mM HEPES containing 150 mM NaCl and 5 mM DTT, with or without 8 M urea, followed by EDTA removal with buffer exchange. Lysozyme was used as a control protein known to bind zinc ions very weakly [27]. To characterise Cys involvement in zinc binding, Zn^{2+} released from Har and Har150 by incubation with the Cys oxidising Ellman's reagent, 5,5′-dithiol-2,2′-nitrobenzoic acid (DTNB) [28,29] was separated by centrifugation through a 3 kDa MWCO filter and determined by ICP-MS.

Thiol assay

Free sulfhydryl groups in Har were determined in air with DTNB (120 μM final concentration) in sodium phosphate buffer (0.1 M, pH 8.0) [28,29]. Protein samples (1.3–5.2 μM) were mixed with DTNB and incubated for 15 min before the absorbance at 412 nm was recorded.

Determination of oligomeric states of Har and variants

The dimerization of Har, Har150 and C97A was determined by analytical size exclusion chromatography performed on a Superdex 75 HR 10/300 column (GE Healthcare).

Hemin binding by Har

Hemin solutions were prepared freshly by dissolving porcine hemin chloride (≥98% HPLC, Sigma-Aldrich) in 0.1 M NaOH and then diluted into the TBS buffer (50 mM TrisHCl, 150 mM NaCl, pH 8.0). The stock hemin solution was filtered through a 0.22 μm filter unit (Millipore) and kept cold on ice in the dark before use. Concentrations of hemin were determined with the extinction coefficient of 58,400 $M^{-1}.cm^{-1}$ on a Cary50 UV-vis spectrometer [30]. Hemin (1 – 2 μM) was incubated with the purified Har and C97A proteins at 0 to 4 protein to hemin molar ratios. The spectra (700 – 250 nm) of these solutions were collected after 1 h incubation. Dissociation constants of hemin binding by Har and C97A were estimated by fitting the absorbance changes at 419 nm from the spectrophotometric titrations using the biochemical analysis program Dynafit [31]. The hemin binding affinity of each protein was estimated from separate titrations using three different hemin concentrations. Lysozyme was used as a negative control in the hemin binding determinations.

Spectroscopic and affinity estimation of divalent metal cation binding by Har and C97A

Fluorescence spectra were collected on a Cary eclipse fluorescence spectrophotometer (Varian) at room temperature. The Har or C97A mutant concentration was 8.0 μM in 10 mM HEPES buffer (pH 7.0) containing 250 mM NaCl and the reducing agent tris(2-carboxyethyl) phosphine (TCEP) (2 mM). The cation (Fe^{2+} or Mn^{2+}) was added at 0–8 molar equivalents to protein. Metal binding affinities were estimated by fitting the titration curves with the biochemical analysis program Dynafit [31].

Secondary structural analysis

Circular dichroism (CD) data of Har and Har150 in the presence and absence of one equivalent of Fe^{2+} (as $(NH_4)_2Fe(SO_4)_2$) or 0.8 equivalents of hemin were acquired from

Table 2. Oligonucleotides used in this study.

Oligonucleotide Primers	Sequence (5'-3')[a]
Recombinant expression of GST-Har and variants	
HarNterm	CGGAATTCATCGAAGGTCGTATGATAGTCACATCA
HarCterm	CGCTCGAGCCATATCGGATCAATGTTATATGTCT
Har150_Nterm	CCGCGTGGATCCATGATAGTCACATCACTG
Har150_Cterm	CGTGGATCCCGGGTTACTGCTTCTTCCTGCATTT
HarC97A SDM	GCTTCATTTGCAGAGCAGGCACCGCTGCTTTTCTGTACC
DNA target for EMSA	
PGN_0001_240bp_For	GGTGTTGATAACTCGGTCGCGCCTT
PGN_0001Rev	CTAAAAAAATATCGTTTTGAGAGCAGT
Construction of *har* mutant	
PGN_1502-Fwd	ATGTCGCCTTCCGAGGCTAT
ErmF-PGN_1502-Rev	GCAATAGCGGAAGCTATCGGTTATCTTTTCGATCCATTCTTGC
PGN_1502-ErmF-Fwd	AGAATGGATCGAAAAGATAACCGATAGCTTCCGCTATTGC
Term-ErmF-Rev	GTCTTTCGACTGAGCCTTTCGTTTTAGCATCTAATTTAACTTCAATTCC
Term-Prom-region-Fwd	GCTCAGTCGAAAGACTGGGCCTTTCGTTTTACGGAGTGAAAAAGGAGCCG
PGN_1504-Prom-region-Rev	CAATGTTATATGTCTGTGTTATCTCTCTTTTACATCATATTTTCC
Prom-region-PGN_1504-Fwd	AATATGATGTAAAAGAGAGATAACACAGACATATAACATTGATCC
PGN_1504-Rev	GCAGATATTTTGTAGCCTCCATC
Construction of *har* complement	
PGN_1502-Fwd	ATGTCGCCTTCCGAGGCTAT
CepA-Har-Rev	ACTTTCCTTAACTCTTTTGACGTCTTATTTTTTCTTCTTGGGAGCGGCT
Har-CepA-Fwd	AGCCGCTCCCAAGAAGAAAAAATAAGACGTCAAAAGAGTTAAGGAAAGT
ErmF-CepA-Rev	TGTGTAGGTTCTAATTGAAGGACAGACGTCTCAAGTCACCGATAG
CepA-ErmF-Fwd	CTATCGGTGACTTGAGACGTCTGTCCTTCAATTAGAACCTACACA
ErmF-Rev	GATACTGCACTATCAACACACTC
qRT-PCR	
PGN_0287 RT Fwd	CCAGCAGCACTTTCCATACAAA
PGN_0287 RT Rev	CCACTGATTACGGCCTCATTT
PGN_0448 RT Fwd	AGTAAAGGGGTAGGGCAACG
PGN_0448 RT Rev	ATCGGATTCGTGTTCCAAAGC
PGN_1296 RT Fwd	CCTGCAAGAGCGTGAAGTTG
PGN_1296 RT Rev	GGATCGGAAAGCCGTATAAGC
PGN_1578 RT Fwd	TGTTGTGGAAAGGAGTGTGG
PGN_1578 RT Rev	AGAAGGAATGAAGTCGGTTGTT
PGN_2083 RT Fwd	GCATTCTTTTCTGGCGTAGCA
PGN_2083 RT Rev	TTTGCGAAACGGCACTCCCT

[a]Underlined sequence of SOE primers indicates the part of the primer that is complementary to the target sequence, with the remainder of the primer providing complementarity with a second PCR product for splicing.

260 to 190 nm on a Jasco J815 spectropolarimeter [32]. Secondary structures of the proteins were estimated by analysing the CD data using the DichroWeb online server [33,34].

Har DNA binding

All EMSA reactions were performed using 50 mM TrisHCl pH 7.0, 40 mM MgCl$_2$, 100 mM NaCl, 5 mM DTT and various concentrations of Zn(II)Har150 protein (2.85 – 14 µM), *dnaA* promoter DNA (0.2 µM) or hemin (0 – 140 µM) in a total reaction volume of 20 µL. A negative control EMSA was performed using Zn(II)Har150 and PGN_1308 promoter DNA that is bound by a different *P. gingivalis* transcriptional regulator. FAM-labeled

DNA was generated via PCR using 5′FAM-labeled PCR primers (Geneworks). EMSA reactions were incubated 25°C for 2 h then 4°C for 1 h before gently adding DNA loading buffer and loading each EMSA reaction onto a 1% agarose gel, with the wells cast centrally in the gel, in 2 x TA buffer. After electrophoresis DNA was visualized by staining gels in a SYBR Safe DNA gel stain bath (Life Technologies) or FAM fluorescence was visualized with a LAS-3000 Imager. Proteins were visualized by staining gels with SimplyBlue SafeStain (Life Technologies).

Construction of *P. gingivalis har* mutant and complemented strains

The recombination cassette for deletion of the *har* gene consisted of the final 600 bp of the PGN_1502 gene, the *ermF* gene encoding erythromycin resistance in *P. gingivalis*, followed by a transcriptional terminator then a copy of the promoter region that drives the operon containing PGN_1503, followed by the first 614 bp of the PGN_1504 gene. The truncated PGN_1502 and PGN_1504 genes were used as the sites of homologous recombination with the *P. gingivalis* ATCC 33277 chromosome and would result in replacement of PGN_1503 with *ermF* followed by a transcriptional terminator then the promoter to drive transcription of the remaining genes downstream of PGN_1503 in the operon. This cassette was constructed from four separate PCR products that were spliced together to form the final cassette. The following PCR products were amplified from ATCC 33277 genomic DNA: PGN_1502' (primers PGN_1502-Fwd and *ErmF*-PGN_1502-Rev), the promoter region (primers Term-Prom-region-Fwd and PGN_1504-Prom-region-Rev) and PGN_1504' (primers Prom-region-PGN_1504-Fwd and PGN_1504-Rev). The *ermF* gene was amplified from pVA2198 with primers PGN_1502-*ErmF*-Fwd and Term-*ErmF*-Rev. The transcriptional terminator was included in the Term-*ErmF*-Rev (used to amplify *ermF*) and Term-Prom-region-Fwd (used to amplify the promoter region) primers so that when the *ermF* and promoter PCR products were joined by SOE PCR, the terminator sequence would be between them. All PCRs were performed with Herculase II and the final product cloned into the SmaI site of pBluescript to form pHarSOE1-4. The recombination cassette was released from the plasmid with EcoRV/XbaI and 200 ng electroporated into *P. gingivalis* ATCC 33277 in a 0.1 cm gap cuvette at 1.8 kV, 200 Ohms resistance. The resulting mutant was called ECR455.

The recombination cassette for complementation of the *har* mutant consisted of the final 600 bp of the PGN_1502 gene plus the *har* gene PGN_1503, followed by the *cepA* gene encoding ampicillin resistance, then the final 578 bp of the *ermF* gene. The truncated PGN_1502 and *ermF* genes were used as the sites of homologous recombination with the *P. gingivalis* ECR455 chromosome and would result in the insertion of the PGN_1503 gene and the ampicillin resistance gene. This cassette was constructed from three separate PCR products that were spliced together to form the final cassette. PGN_1502' through to the end of PGN_1503 was amplified from ATCC 33277 chromosome (primers PGN_1502-Fwd and CepA-Har-Rev), the *cepA* gene was amplified from pEC474 (primers Har-CepA-Fwd and ErmF-CepA-Rev) whilst *ermF'* was amplified from pVA2198 (primers CepA-ErmF-Fwd and ErmF-Rev). All PCRs were performed with Herculase II except for the final SOE PCR which was amplified with Platinum Taq DNA Polymerase High Fidelity, then cloned into pGEM-TEasy to produce pHarComp. This plasmid was electroporated into ECR455 resulting in the *har*-complemented strain, ECR475.

Western blot analyses of Har expression

Bacterial whole cell lysates or cytoplasmic protein extracts (25 μg) were separated on 4–12% Bis-Tris polyacrylamide gels (Invitrogen) in MES buffer before Western transfer and immunoblotting with rabbit anti-rHar serum diluted 1:2500.

Determination of cellular metal content

Three biological replicates of each strain of *P. gingivalis* ATCC 33277, ECR455 and ECR475 were grown in MBB supplemented with 0.5 g/L cysteine, 5 μg/mL hemin and 5 μg/mL vitamin K and cell lysates were prepared as previously described [13]. Measurements were made using an Agilent 7700 series ICP-MS instrument under operating conditions suitable for routine multi-element analysis in Helium Reaction Gas Cell mode.

Extraction of RNA for transcriptomic analyses

Extraction of total RNA was performed as previously described [12].

Microarray hybridization and analyses

Porphyromonas gingivalis W83 microarray slides version 1 were obtained from the Pathogen Functional Genomics Resource Centre of the J. Craig Venter Institute. cDNA synthesis, labeling and microarray hybridization were all performed as previously described except that 5 μg total RNA was reverse transcribed instead of 10 μg [12]. Paired samples were compared on the same microarray using a two-colour system. A total of 6 paired microarray hybridizations were performed representing 6 biological replicates, where a balanced dye design was used, with the overall analyses including three microarrays where *P. gingivalis* ATCC 33277 samples were labeled with Cy3 and the paired ECR455 samples were labeled with Cy5 and three other microarrays where samples were labeled with the opposite combination of fluorophores. Image analysis was also performed as previously described except that print tip loess normalization was used [12].

qRT-PCR

Three biological replicates of *P. gingivalis* ATCC 33277, ECR455 and ECR475 were grown in batch culture to an OD_{650} of 1.0 in BHI supplemented with 0.5 g/L cysteine, 5 μg/mL hemin and 5 μg/mL vitamin K. cDNA was generated from RNA isolated from these strains using a NucleoSpin RNA II Total RNA Isolation kit (Macherey-Nagel) and SuperScript III Reverse Transcriptase First-Strand Synthesis SuperMix for qRT-PCR kit with random hexamers (Invitrogen), then qPCR analysis was performed using 0.3 ng cDNA per reaction and Power SYBR Green PCR master mix (Applied Biosystems), all according to manufacturer's instructions. cDNA was quantified relative to standard curves generated by amplification of ATCC 33277 gDNA with the same primer pair used for each cDNA. Primer pairs used are listed in Table 2, with primers for PGN_1296 representing genes that did not change in the microarrays, PGN_0287 and PGN_1578 representing genes that increased in transcription in ECR455, and PGN_0448 and PGN_2083 representing genes that decreased in transcription in ECR455 relative to ATCC 33277.

Biofilm assays

Parental and mutant strains of *P. gingivalis* grown in BHI supplemented with 0.5 g/L cysteine and 5 μg/mL vitamin K containing either 7 μM hemin (hemin excess) or no added hemin (hemin limitation) for 24 h were diluted in the same medium to a density of 5×10^7 cfu/mL, and incubated in CultureWell imaging chambers (Invitrogen) for 24 h under anaerobic conditions at 6 rpm on a rocking platform. Four wells were inoculated per strain per experiment. The supernatant was removed and the biofilms washed carefully with 0.85% NaCl to remove any non-adherent cells and stained with the LIVE/DEAD BacLight Bacterial Viability kit (Invitrogen) for 20 min according to the manufacturer's instructions. Wells were then washed a final time with 0.85% NaCl post-staining prior to imaging.

```
PgHar    1   M-IVTSLEDLRSRLRAYVSENGLRHTPERYSILEVAYNLK-KIFTPDDLFDLTR-ENGLPVSLSTVYNTLTLLERCGIVL   77
BfFur    1   METQNVKDTVRQIFTEYLNANGHRKTPERYAILDTIYSID-GHFDIDMLYSQMMNQENFRVSRATLYNTIILLINARLVI   79
HpFur    1   MKRLETLESILERLRMSIKKNGLKNSKQREEVVSVLYRSG-THLSPEEITHSIR-QKDKNTSISSVYRILNFLEKENFIC   78
VcFur    1   ---------MSDNNQALKDAGLKVTLPRLKILEVLQQPECQHISAEELYKKLI-DLGEEIGLATVYRVLNQFDDAGIVT   69
PaFur    1   ---------MVENSE-LRKAGLKVTLPRLKILQMLDSAEQRHMSAEDVYKALM-EAGEDVGLATVYRVLTQFEAAGLVV   68
EcFur    1   ---------MTDNNTALKKAGLKVTLPRLKILEVLQEPDNHHVSAEDLYKRLI-DMGEEIGLATVYRVLNQFDDAGIVT   69
MtZur    1   ----------------MSAAGVRSTRQRAAISTLLETLD-DFRSAQELHDELR-RRGENIGLTTVYRTLQSMASSGLVD   61
ScNur    1   M--------VSTDWKSDLRQRGYRLTPQRQLVLEAVDTLE--HATPDDILGEVR-KTASGINISTVYRTLELLEELGLVS   69
BsPerR   1   -----MAAHELKEALETLKETGVRITPQRHAILEYLVNSM-AHPTADDIYKALE-GKFPNMSVATVYNNLRVFRESGLVK   73
RlMur    1   MT--DVAKT----LEELCTERGMRMTEQRRVIARILEDSED-HPDVEELYRRSV-KVDAKISISTVYRTVKLFEDAGIIA   72
RlIrr    1   M----MTGAFPIAIEVRLRGAGLRPTRQRVALGDLLFAKGDRHLTVEELHEEAV-AAGVPVSLATVYNTLHQFTEAGLIR   75
                      *               *                 *                                 *
```

```
PgHar    78   RLPSPETKYQYLMASFAEQCPLLFCTECAQFSTYYRRNVKSILADKDLRPPRFSYRQALICLYGICDKCRKKQSALKKAA   157
BfFur    80   KHQF-GTSSQYEKSYNRETHHQICTQCGKVTEFQNEALQN--AIENTKLSKFQLSHYSLYIYGICSKC---DRANKRKR   153
HpFur    79   VLETSKSGRRYEIAA-KEHHDHIICLHCGKIIEFADPEIEN-RQNEVVKKYQAKLISHDMKMFVWCKEC--------QE   147
VcFur    70   RHHFEGGKSVFELST-QHHHDHLVCLDCGEVIEFSDDVIEQ-RQKEIAAKYNVQLTNHSLYLYGKCGSDGSC------KD   141
PaFur    69   RHNFDGGHAVFELAD-SGHHDHMVCVDTGEVIEFMDAEIEK-RQKEIVRERGFELVDHNLVLYVRK--------------   132
EcFur    70   RHNFEGGKSVFELTQ-QHHHDHLICLDCGKVIEFSDDSIEA-RQREIAAKHGIRLTNHSLYLYGHCAEGDC-------RE   140
MtZur    62   TLHTDTGESVYRRCS-EHHHHLVCRSCGSTIEVGDHEVEA-WAAEVATKHGFSDVSHTIEIFGTCSDC--------RS   130
ScNur    70   HAHLGHGAPTYHLAD-RHHHIHLVCRDCTNVIEADLSVAAD-FTAKLREQFGFDTDMKHFAIFGRCESC-----SLKGST   142
BsPerR   74   ELTYGDASSRFDFVT-SDHYHAICENCGKIVDFHYPGLDE-VEQLAAHVTGFKVSHHRLEIYGVCQEC--------SK   141
RlMur    73   RHDFRDGRSRYETVP-EEHHDHLIDLKTGTVIEFRSPEIEA-LQERIAREHGFRLVDHRLELYGVP-------------   136
RlIrr    76   VLAVESAKTYFDTNV--SDHHHFFVEGDNEVLDIPVS---N-LTIANLPEPPEGMEIAHVDVVIRL---RA-------KQ   139
```

```
PgHar   158   DKAAPKKKK    166
BfFur   154   VNNNNKKEK    162
HpFur   148   SEC------    150
VcFur   142   NPNAHKPKK    150
PaFur   133   -------KK    134
EcFur   141   DEHAHEGK-    148
MtZur         ---------
ScNur   143   TDS------    145
BsPerR  142   KENH-----    145
RlMur   137   ----LKKEDL   142
RlIrr   140   G--------    140
```

Figure 1. Alignment of Fur family proteins with PGN_1503 (Har) from *P. gingivalis* ATCC 33277. Fur family proteins were aligned using COBALT [75] with *P. gingivalis* Har (PgHar; UniProt B2RKX7). Seven of these proteins have structures in the Protein DataBank: HpFur is the iron responsive Fur from *Helicobacter pylori* (UniProt B9XY52), VcFur is the iron-responsive Fur from *Vibrio cholerae* (UniProt P0C6C8), PaFur is the iron-responsive Fur from *Pseudomonas aeruginosa* (UniProt Q03456), EcFur is the iron-responsive Fur from *Escherichia coli* (UniProt P0A9A9), MtZur is the zinc-responsive FurB from *Mycobacterium tuberculosis* (UniProt O05839), ScNur is the nickel-responsive Nur from *Streptomyces coelicolor* (UniProt Q9K4F8) and BsPerR is the peroxide-responsive PerR from *Bacillus subtilis* (UniProt P71086). The other three Fur family proteins are BfFur, an iron-responsive Fur from *Bacteroides fragilis* NCTC9343 (UniProt Q64QR6), RlMur and RlIrr, the manganese-responsive Mur (UniProt Q1MMB4) and the iron response regulator Irr (UniProt Q1MN49) respectively, from *Rhizobium leguminosarum* bv. *viciae* (strain 3841). * indicates identical amino acids in all 11 proteins. Shading indicates residues experimentally confirmed to be involved in the three distinct metal binding sites: S1 in blue; S2 in red, S3 in yellow as reported in Dian *et al.* [17], except for those in PgHar, BfFur and RlMur which were inferred by similarity to the other sequences. The five residues underlined in RlIrr show the amino acids that would make up S2 and although Irr has been shown to bind metals *in vitro*, the metal binding site is still unknown. The principal heme binding site of RlIrr is the HxH motif shaded green [47], which is part of the S2 motif. The putative heme regulatory motif of PgHar is boxed.

Biofilms were imaged as previously described [35] using a Zeiss LSM 510 META Confocal Laser Scanning Microscope with a C-Apochromat 63x/1.2 numerical aperture, water immersion objective lens fitted with a correction collar. SYTO 9 fluorescence was detected by excitation with a 488 nm Argon Ion laser and emission collected with a 500–550 nm bandpass filter. Propidium Iodide (PI) fluorescence was detected by excitation with a 543 nm Helium Neon laser and emission collected by 560 nm longpass filter. Biofilm images were analysed using the COMSTAT 3D biofilm structure quantifying software [36].

Statistical analyses

All statistical analyses were performed using Minitab16 statistical software. Following a Levene's test for equal variances, data were analysed using one-way ANOVA with Tukey multiple comparison tests or the Kruskal-Wallis test with pairwise Mann Whitney tests with a Bonferroni correction. The p-value was set at 0.05 and 95% confidence intervals were calculated.

Microarray data accession number

The microarray data presented in this paper have been entered into the NCBI GEO databank (www.ncbi.nlm.nih.gov/projects/geo) with the accession number GSE37099.

Results

P. gingivalis Fur orthologue

The *P. gingivalis* ATCC 33277 genome sequence contains a single predicted Fur orthologue encoded by PGN_1503 [37], which we have designated Har, for **H**eme **a**ssociated **r**egulator. Strikingly, the predicted pI is high at 9.47 due to the presence of numerous lysine and arginine residues, particularly in the lysine-

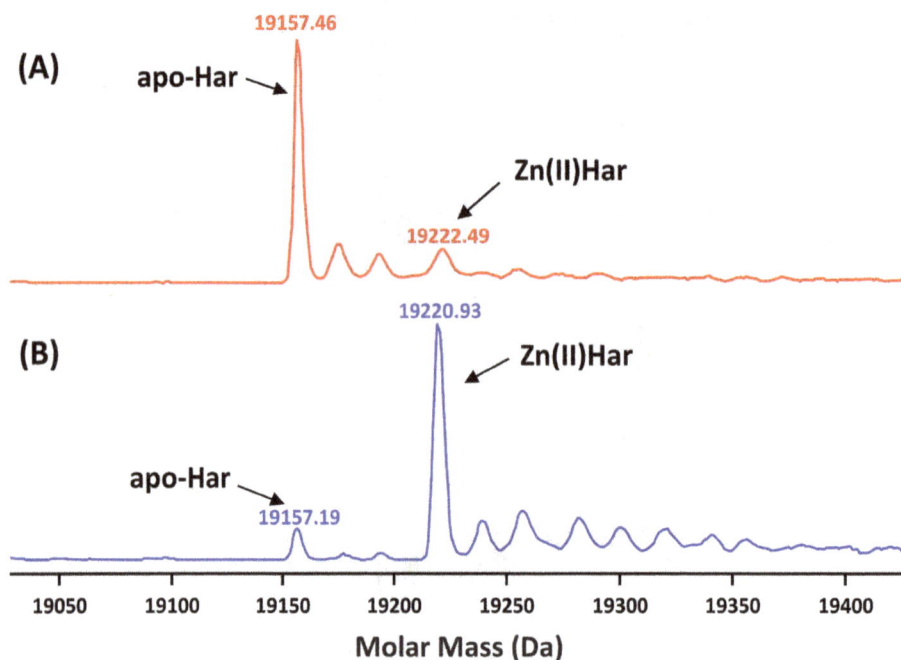

Figure 2. Zinc binding by recombinant Har. Purified Har (5 µM) was buffer exchanged into ammonium acetate (10 mM) via extensive dialysis at 4°C and subjected to ESI-MS analysis on a Quadropole-Time of Flight mass spectrometer (Agilent) in the positive mode with a fragmentor voltage of 200-300 V and a skimmer voltage of 65 V at a flow rate of 500 µL/h for direct syringe infusion delivery to the electrospray probe in the presence (**A**) and absence (**B**) of 0.1% v/v formic acid in the mobile phase. The average molar masses were obtained by application of a deconvolution algorithm to the recorded spectra and were calibrated with horse heart myoglobin (16951.5 Da).

rich C-terminal tail of 20 residues. The predicted amino acid sequence of *P. gingivalis* PGN_1503 was aligned with FurA from *Bacteroides fragilis* NCTC9343 (BfFur) which is the closest related species to *P. gingivalis* that has an experimentally determined Fur orthologue (Fig. 1) [38]. Har and BfFur share 27% identity and 85% similarity and both have C-terminal tails rich in lysine and arginine residues. Seven other members of the Fur family for which there is structural data were also aligned with PGN_1503 as well as two other representatives of the Fur superfamily, Mur and Irr, from *Rhizobium leguminosarum* (Fig. 1). None of these other Fur family members have the highly positively charged C-terminal tail that is found in Har and BfFur. Har contains the dual -C-X-X-C- motifs involved in binding zinc in the S1 structural site which is found in some but not all Fur family members (Fig. 1). The S2 site is the Fur metal sensory site and metallation of this site is essential for specific DNA binding. The S2 site is conserved in Fur superfamily proteins [17] and can be inferred in all sequences in

the alignment except Har (Fig. 1). A predicted heme regulatory motif (HRM) [39] -[97]C-[98]P-[99]L- which RlIrr uses for hemin binding (Fig. 1) was identified in the *P. gingivalis* Har sequence in place of the -H-X-H- S2 motif. Furthermore the S3 site, a supplementary metal binding site that strengthens DNA-binding affinity of the Fur protein [17], was not detectable in the Har sequence. *P. gingivalis* Har and two variants, Har150 which was truncated by 16 amino acids at the C-terminus to remove the lysine rich tail and C97A where Cys97 was substituted with Ala to mutate the proposed hemin binding motif, were then expressed and purified as recombinant proteins from *E. coli* for characterisation.

Characterisation of zinc binding by recombinant Har and C97A

Purified *P. gingivalis* Har and C97A were subjected to ESI-MS analysis in ammonium acetate in the presence of 0.1% v/v formic

Table 3. Zinc content of purified recombinant Har and C97A with and without metal chelator treatment.

Protein sample	Har				C97A	Lysozyme
	Untreated	Treated with 50 mM EDTA	Treated with Chelex-100 resin	Treated with 50 mM EDTA in 8 M urea	Treated with 50 mM EDTA	
Protein concentration (µM)	23.2	23.3	22.9	21.4	14.2	25.5
Zn bound (µM)	22.0	22.0	21.2	0.1	13.4	nd[a]
Zn/Har molar ratio	0.95	0.94	0.93	0.005	0.94	-

Protein solutions were treated with the strong metal chelator Chelex-100 at pH 5.0 or EDTA with or without 8 M urea at pH 8.0. After removal of the metal chelators and urea, the zinc ions in the protein solutions were determined by ICP-MS, with lysozyme as a negative control protein.
[a]not detected.

Table 4. Zinc ions released from Zn(II)Har and Zn(II)Har150 by the cysteine oxidising reagent DTNB.

	Without DTNB		With DTNB	
Protein sample	Zn(II)Har	Zn(II)Har150	Zn(II)Har	Zn(II)Har150
Protein conc. (µM) [a]	13.21	7.31	13.21	7.31
Zn in filtrate (µM)	nd [c]	nd	11.39	5.88
Zn/P [b]	-	-	0.86	0.81

Zinc content in the filtrates of Zn(II)Har and Zn(II)Har150 after 15 min incubation of the proteins with and without DTNB (160 µM) in 100 mM sodium phosphate pH 8.0 as determined by ICP-MS.
[a] before centrifugation through the 3 kDa MWCO filter.
[b] the ratio of zinc content in the filtrate to the protein concentration before centrifugation.
[c] not detected.

acid. Har showed a major peak with a molar mass of 19157.46 Da that is consistent with the theoretical mass of apo-Har, 19157.20 Da (Fig. 2). In contrast, in the absence of formic acid the major peak was at 19220.93. This difference in mass corresponds to the atomic mass of zinc minus two protons suggesting that one zinc ion was bound to the protein monomer (Fig. 2). ESI-MS analysis also confirmed the identity of C97A with the measured molar mass of 19125.8 Da consistent with the theoretical mass of 19125.3 Da (data not shown). ICP-MS analysis of Har and C97A confirmed that a single zinc ion was bound to each Har or C97A monomer (Table 3). EDTA could not remove the zinc ions from the proteins however when Har was treated with EDTA under denaturing conditions (8 M urea), negligible levels of zinc were detected after buffer exchange to remove the EDTA and urea (Table 3). As removal of zinc required denaturation of Har the zinc loaded forms (Zn(II)Har and Zn(II)C97A) were used for characterisation studies.

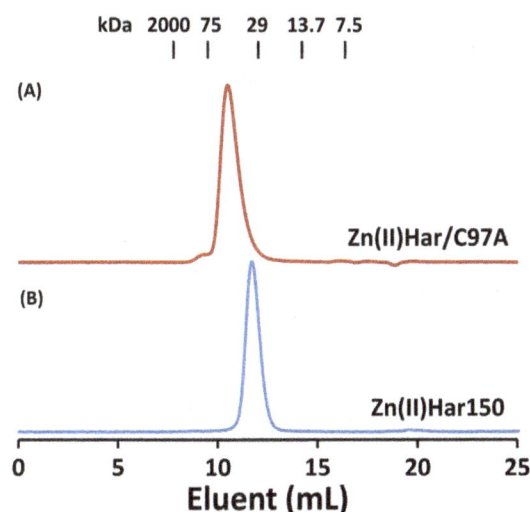

Figure 3. Dimerization of Zn(II)Har, Zn(II)C97A and Zn(II)-Har150. Representative elution profiles of Zn(II)Har/Zn(II)C97A (**A**) and Zn(II)Har150 (**B**) from a Superdex 75 analytical gel filtration column at 4°C in an AKTA FPLC Chromatographic System (GE Healthcare). Proteins (100 µg) were applied to the column pre-equilibrated with 500 mM NaCl containing buffers (20 mM) at pH 6.0 (MES), pH 7.0 (HEPES), 7.4 (KPi) and 8.5 (borate). The molar masses were calculated against a calibration curve of retention volumes (Ve) of the protein standards, which are indicated at the top of the chart. The elution profiles at pH 7.0 are presented.

Free thiol assay and release of zinc from Har/Har150 by DTNB

Ellman's reagent detected seven (experimentally 6.6–6.8) free sulfhydryl (-SH) groups in each monomer of Zn(II)Har and Zn(II)Har150 under oxidative conditions, suggesting that all cysteine residues in both proteins exist in a reduced form. This indicates that the protein structure allows all the free thiol groups in the seven cysteine residues to be readily accessible to DTNB but, interestingly, not to be oxidised to disulfide by air. Therefore, neither intramolecular nor intermolecular disulfide bonds were formed in the proteins even in the presence of oxygen.

ICP-MS detected over 0.8 equivalents of zinc in the filtrates of both Zn(II)Har and Zn(II)Har150 in a concentrator of 3 kDa MWCO after incubation with DTNB for 15 min but no released zinc could be detected in the filtrate when in the absence of DTNB (Table 4). DTNB therefore was able to release the bound zinc from Har/Har150, indicating a zinc binding site with cysteines being involved as ligands, which was disabled by oxidation with DTNB. ICP-MS analyses of Har150 treated with DTNB also showed one zinc ion binding stoichiometry of the protein (Table 4), thus truncation of the C-terminal lysine rich tail did not affect the zinc binding ability of Har.

Oligomerisation states of Zn(II)Har, Zn(II)C97A and Zn(II)Har150

Size exclusion chromatography of Zn(II)Har, Zn(II)C97A and Zn(II)Har50 in reducing or non-reducing environments showed a single major peak in each elution profile with an apparent molar mass of 41.0 kDa which corresponded to dimeric Har or C97A with two zinc ions, or 35.0 kDa which corresponded to dimeric Zn(II)Har150 (Fig. 3). The Zn(II)Har homodimer was stable over the pH range 6.0–8.5 and no variation in the elution profile of the dimer was found at 4°C and ambient temperature (22°C) suggesting a stable dimeric form of Zn(II)Har (data not shown). Since all the seven thiol groups are unpaired, such dimerizaton was not caused by intermolecular disulfide formation.

Hemin binding by Zn(II)Har and Zn(II)C97A

Addition of Zn(II)Har or Zn(II)C97A to hemin induced a solution spectrum change that indicated the formation of a complex between hemin and each protein. UV-visible spectra of the Zn(II)Har-hemin and Zn(II)C97A-hemin complexes showed a blue shift of the typical hemin absorption peak at 388 nm in the near UV range by ~16 nm (372 nm; Fig. 4A) consistent with hemin binding to HRMs [39] and indicating both proteins had affinity for hemin. Difference absorption spectra of the Zn(II)Har-hemin and Zn(II)C97A-hemin complexes showed a second

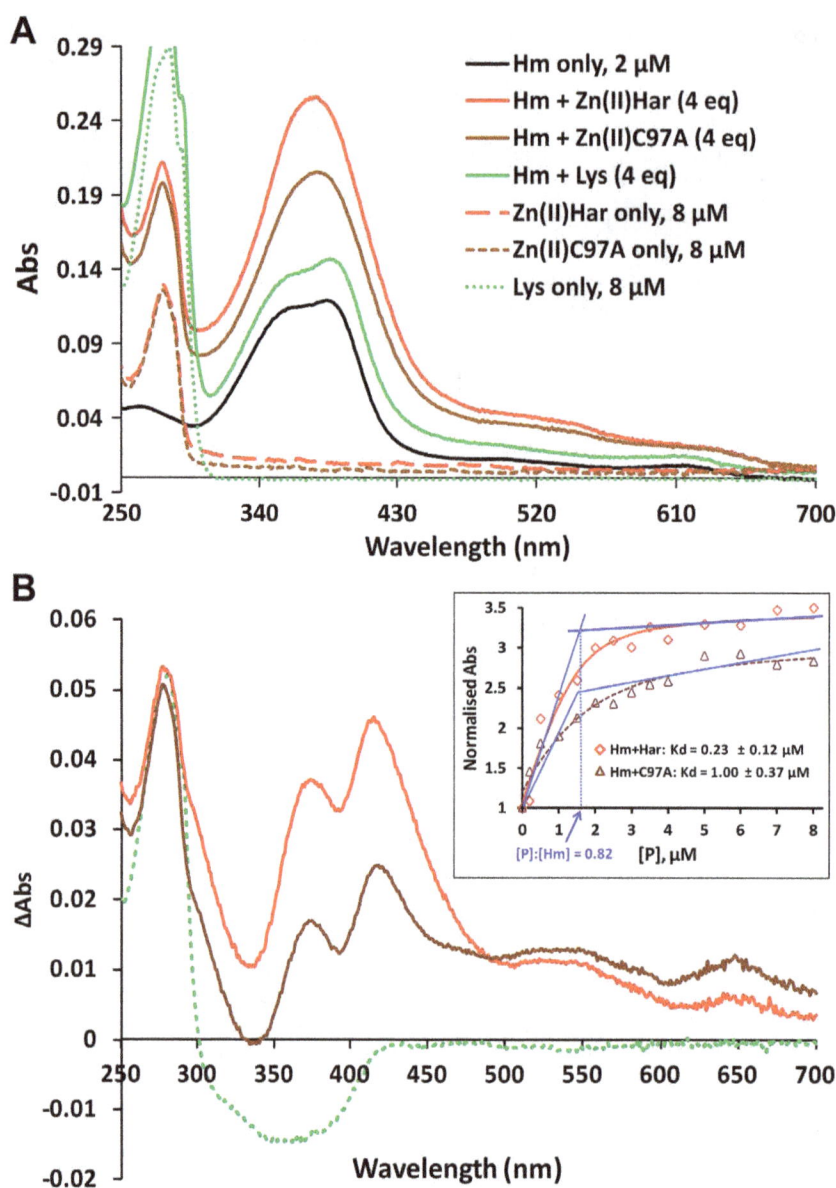

Figure 4. Spectrometric determination of hemin binding by Zn(II)Har and Zn(II)C97A. Hemin (Hm, 1–2 µM) in TBS was incubated with Zn(II)Har and Zn(II)C97A at protein to hemin molar ratios of 0:1 to 4:1 for 1 h and solution spectra collected on a Cary 50 UV-visible spectrometer (Varian). Lysozyme (Lys) was used as a negative control (green lines). (**A**) Absorption spectra of 2 µM free hemin, 8 µM Zn(II)Har or Zn(II)C97A, and hemin plus four equivalents of Zn(II)Har or Zn(II)C97A. Based on the hemin binding affinities of Zn(II)Har and Zn(II)C97A estimated in (B), at the starting protein:hemin molar ratio of 4:1, free hemin in the equilibrium solution was 3.8% and 15.3% of the total hemin after reaction with Zn(II)Har and Zn(II)C97A, respectively. (**B**) Spectra of 1:1 protein to hemin (2 µM) molar ratio are presented as a subtraction from the spectrum of hemin only (red line for Zn(II)Har, brown line for Zn(II)C97A, green line for lysozyme). The hemin binding affinity of the protein was estimated by fitting the absorbance changes at 419 nm for Zn(II)Har and Zn(II)C97A against protein concentrations (inset) using the biochemical analysis program Dynafit [31]. Inset: Fitted titration curves, apparent dissociation constants (K_d) and the titration data point sets of the normalised absorbance at 419 nm for Zn(II)Har and Zn(II)C97A. Estimation of binding stoichiometry is shown in blue. P: protein.

absorption maxima at ~420 nm (Fig. 4B). The intensities of the Soret maxima for Zn(II)C97A were reduced with respect to that of the wild-type spectrum. Addition of lysozyme to hemin resulted in no obvious change to the hemin spectrum. Titration of a hemin solution with Zn(II)Har or Zn(II)C97A showed that Zn(II)Har bound one hemin molecule per monomer with high affinity (K_d of 0.23±0.12 µM), and Zn(II)C97A had a four-fold lower affinity for hemin with a K_d of 1.00±0.37 µM (Fig. 4B, inset).

Divalent metal cation binding by Zn(II)Har

There are ten tyrosine residues in Zn(II)Har which act as fluorophores allowing fluorescence to be used as a probe to determine divalent cation binding activity. The protein exhibited intrinsic fluorescence to a maximum intensity at 305 nm upon excitation at 275 nm. Fluorescence spectroscopic titration of the protein at micromolar concentration with ferrous ions showed a linear decrease in fluorescence intensity at 305 nm up to one molar equivalent of ferrous ions. Further addition of ferrous ions

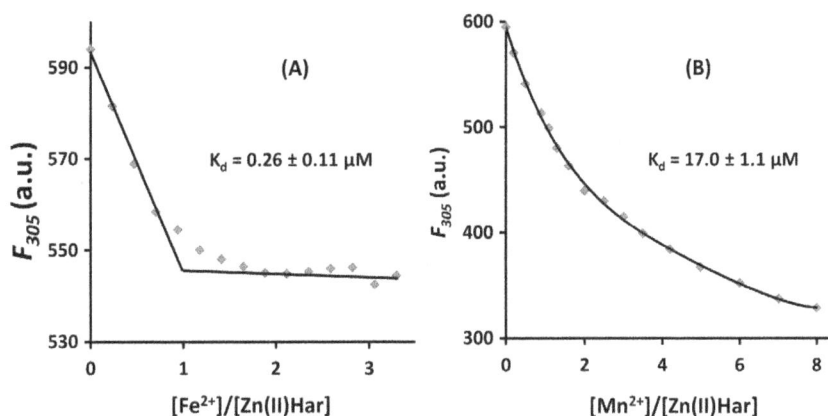

Figure 5. Divalent metal cation binding by Zn(II)Har. Fluorescence spectroscopic titration of Zn(II)Har with ferrous ions and manganese ions in HEPES (5 mM, pH 7.0) containing 250 mM NaCl in the presence of TCEP (2 mM). Change in fluorescence emission intensity of Zn(II)Har (8 μM) at 305 nm upon addition of 0 – 3.3 molar equivalent Fe^{2+} (**A**) and 0 – 8 molar equivalent Mn^{2+} (**B**), with each set of presented data being averaged from three individual titrations. Apparent dissociation constants (K_d) were estimated by fitting the titration data using the biochemical analysis program Dynafit [31]. $\lambda_{ex} = 275$ nm.

essentially did not alter the fluorescence intensity, suggesting a 1:1 molar ratio of iron to protein monomer stoichiometry with a K_d of 0.26 μM (Fig. 5). In contrast Zn(II)C97A interacted with Fe^{2+} non-specifically, resulting in a small fluorescence change with no end point on addition of the metal ion under the same conditions (data not shown). Fluorescence titration of Zn(II)Har with Mn^{2+} showed a nonlinear decrease in fluorescence intensity indicating a relatively lower binding affinity with a K_d of 17 μM (Fig. 5).

Secondary structure analyses

Addition of Fe^{2+} or hemin to Zn(II)Har caused significant changes to secondary structure (Table 5). Addition of hemin (0.8 eq) to Zn(II)Har in PBS caused a decrease in the α-helical content from 46% to 36% and β-strand content from 20% to 16% with an increase in unordered content. Addition of Fe^{2+} (1.0 eq) to Zn(II)Har in 10 mM HEPES buffered saline (75 mM NaCl) at pH 7.0 also resulted in a significant change in secondary structure by decreasing the α-helical content from 44% to 35%. While the β-strand content remained essentially the same, the content of turns decreased and unordered structure increased. Similarly, the presence of Fe^{2+} or hemin had a significant effect on the secondary structure of Zn(II)Har150. The addition of either Fe^{2+} or hemin to

Zn(II)Har150 caused a significant decrease in its α-helical content, accompanied by an increase in percentage of β-strands and turns (Table 5).

Har DNA binding

Zn(II)Har150 without its lysine-rich tail was used for EMSA analyses to minimize non-specific DNA interactions as it has a lower predicted pI of 8.98 compared with Zn(II)Har (pI of 9.47). The promoter region of *dnaA* (PGN_0001) was used as the Zn(II)Har binding target because *dnaA* has been identified from the microarray results of the current study (*vide infra*) as being negatively regulated by Har. The EMSAs were run on agarose gels with the wells cast centrally to enable electrophoresis of both protein and DNA which are oppositely charged [40]. Zn(II)-Har150 bound with the promoter region of *dnaA* forming a Zn(II)Har150-DNA complex preventing the DNA from entering the gel under the applied electric field. The negative control BSA which does not bind DNA did not affect the migration of the *dnaA* promoter region (Fig. 6A). The positively charged Zn(II)Har150 migrated towards the cathode (-) but this movement was retarded in the presence of the specific DNA due to the formation of the Zn(II)Har150-DNA complex (Fig. 6B). Unlabeled specific DNA

Table 5. Secondary structure of Zn(II)Har and Zn(II)Har150 in the absence or presence of Fe^{2+} or hemin.

	α-helix	β-strand	Turns	Unordered	Total
Zn(II)Har only	0.44	0.13	0.19	0.23	0.99
+ Fe^{2+} (1 eq)	0.35	0.15	0.12	0.37	0.99
Zn(II)Har only	0.46	0.20	0.14	0.20	1
+ Hemin (0.8 eq)	0.36	0.16	0.14	0.35	1.01
Zn(II)Har150 only	0.34	0.23	0.21	0.22	1
+Fe^{2+} (1 eq)	0.27	0.28	0.26	0.20	1.01
Zn(II)Har150 only	0.33	0.18	0.23	0.26	1
+ Hemin (0.8 eq)	0.17	0.26	0.26	0.31	1

Secondary structures were estimated by DichroWeb [33,34] analyses of the CD data of the proteins (5 μM) at 20°C in the absence or presence of Fe^{2+} (1 eq) and hemin (0.8 eq) in 10 mM HEPES reducing buffer (75 mM NaCl, 0.5 mM TCEP, pH 7.0) and 10 mM PBS (75 mM NaCl, pH 7.4), respectively.

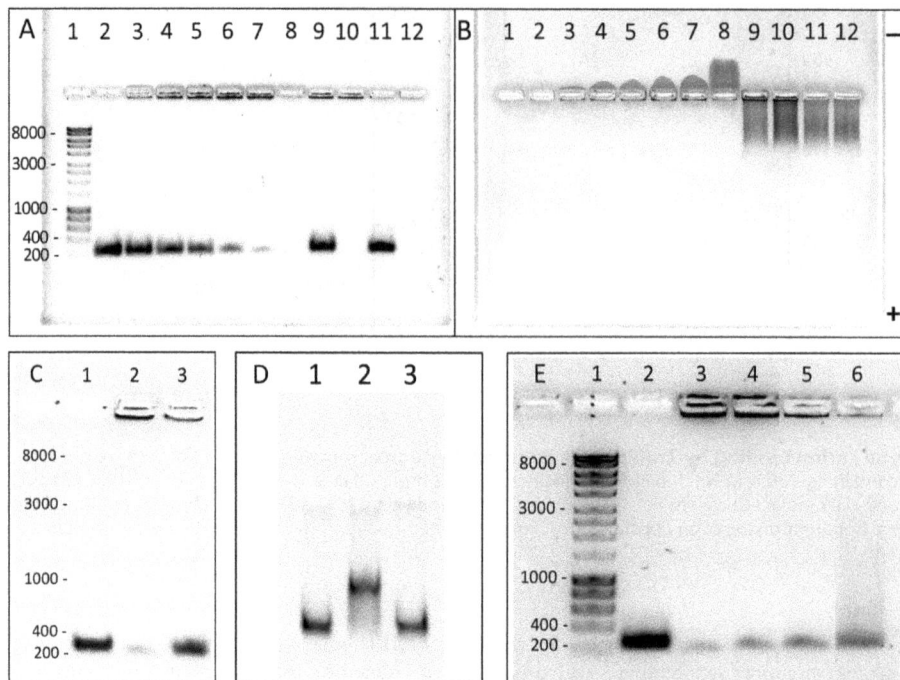

Figure 6. EMSA of Zn(II)Har150 binding to DNA. (A). Agarose gel electrophoresis stained with SYBR Safe DNA gel stain (Life Technologies) for visualizing DNA. Lane 1, Hyperladderl (Bioline) DNA size markers in bp. Lanes 2–7, 9 & 11 all contain 500 ng (0.2 µM) of a 240 bp PCR product encompassing the 33277 *dnaA* promoter sequence (PGN_0001). Additionally, lane 3 has 1 µg (2.8 µM) Zn(II)Har150; lane 4, 2 µg (5.6 µM) Zn(II)Har150; lane 5, 3 µg (8.4 µM) Zn(II)Har150; lane 6, 4 µg (11.2 µM) Zn(II)Har150 and lane 7, 5 µg (14 µM) Zn(II)Har150. Lane 2 contains DNA only whereas Lane 8 contains 5 µg (14 µM) Zn(II)Har150 only. Lanes 9 – 12 contain the negative control protein for DNA binding, BSA, where there is 5 µg BSA in lanes 9 & 10, and 3 µg BSA in lanes 11 & 12. (B). Agarose gel electrophoresis stained with SimplyBlue SafeStain (Life Technologies) for visualizing protein following DNA visualisation. Lanes are as described in (A). The position of the anode (+) and cathode (-) are noted. (C). EMSA competition experiment where an excess of unlabeled *dnaA* promoter DNA (1250 ng) competed with 250 ng (0.1 µM) FAM-labeled *dnaA* promoter DNA for binding to 3 µg (8.4 µM) Zn(II)Har150 (lane 3). Lane 1 contains 250 ng FAM-labeled DNA only, whereas lane 2 contains 250 ng FAM-labeled DNA bound to 3 µg Zn(II)Har150. Visualised is the fluorescence of the FAM-labeled DNA after agarose gel electrophoresis (D) EMSA experiment where the promoter-containing DNA of PGN_1308 (lanes 1–3) was shifted by its cognate transcriptional repressor (lane 2) but not by Zn(II)Har150 (lane 3). (E). Inhibition of Zn(II)Har150 DNA binding by hemin. The addition of increasing concentrations of hemin (lane 3, 0 µM; lane 4, 14 µM; lane 5, 70 µM; lane 6, 140 µM) to a constant amount of DNA (500 ng, lanes 2–6) and Zn(II)Har150 (14 µM, lanes 3–6) resulted in increasing inhibition of DNA binding by Zn(II)Har150. Lane 1, Hyperladderl (Bioline) DNA size markers in bp. Agarose gel electrophoresis stained with SYBR Safe DNA gel stain (Life Technologies) for visualizing DNA.

competed for binding of Zn(II)Har150 to the FAM-labeled specific DNA (Fig. 6C), however non-specific DNA at a similar concentration to the specific DNA did not compete for binding indicating that the binding of Zn(II)Har150 was specific for the *dnaA* promoter. Zn(II)Har150 did not bind to the promoter region of PGN_1308 a DNA target for a different transcriptional regulator of *P. gingivalis* (PgMntR), indicating that Zn(II)Har150 does not bind DNA non-specifically (Fig. 6D). Increasing concentrations of hemin resulted in increasing dissociation of the Zn(II)Har150-DNA complex showing that the binding of Zn(II)Har150 to the *dnaA* promoter region was specifically inhibited by hemin at molar excess concentrations (Fig. 6E).

Construction of *P. gingivalis* *har* mutant and complemented strains

RT-PCR analysis showed that the *P. gingivalis* *har* gene is in the midst of an operon (data not shown), thus *har* was deleted such that there was minimal effect on the transcription of genes downstream of *har* (Fig. 7). A recombination cassette was designed where *har* was replaced by *ermF* followed by a strong Rho-independent transcriptional terminator, then a copy of the intergenic region containing the *har* operon promoter. The genes

(PGN_1502 and PGN_1504) on either side of *har* were used as the flanking DNA for homologous recombination of the cassette with the *P. gingivalis* chromosome (Fig. 7). Deletion of PGN_1503 from ATCC 33277 to produce strain ECR455 was confirmed by Southern blot and PCR analyses (data not shown). Reverse transcription-PCR showed no *har* transcript, but amplification of PGN_1504 cDNA confirmed transcription of the genes downstream of the deleted *har* gene (Fig. 7).

Complementation of the *har* deletion involved the insertion of the *har* ORF back into position downstream of PGN_1502 followed by *cepA* that was inserted into the *ermF* gene. PGN_1502 and part of the *ermF* sequences were used as flanking DNA for homologous recombination of the cassette into the *har* deletion mutant ECR455 (Fig. 7). The resulting strain ECR475 was shown to have the correct chromosomal arrangement by PCR and the *har* gene amplified from the chromosome was sequenced and found to be correct (data not shown). Furthermore RT-PCR showed transcription of the complemented *har* gene (Fig. 7). Western blot analysis using rHar antisera showed that *P. gingivalis* ATCC 33277 and ECR475 whole cell lysates contained Har whilst the ECR455 *har* deletion mutant did not (Fig. 7).

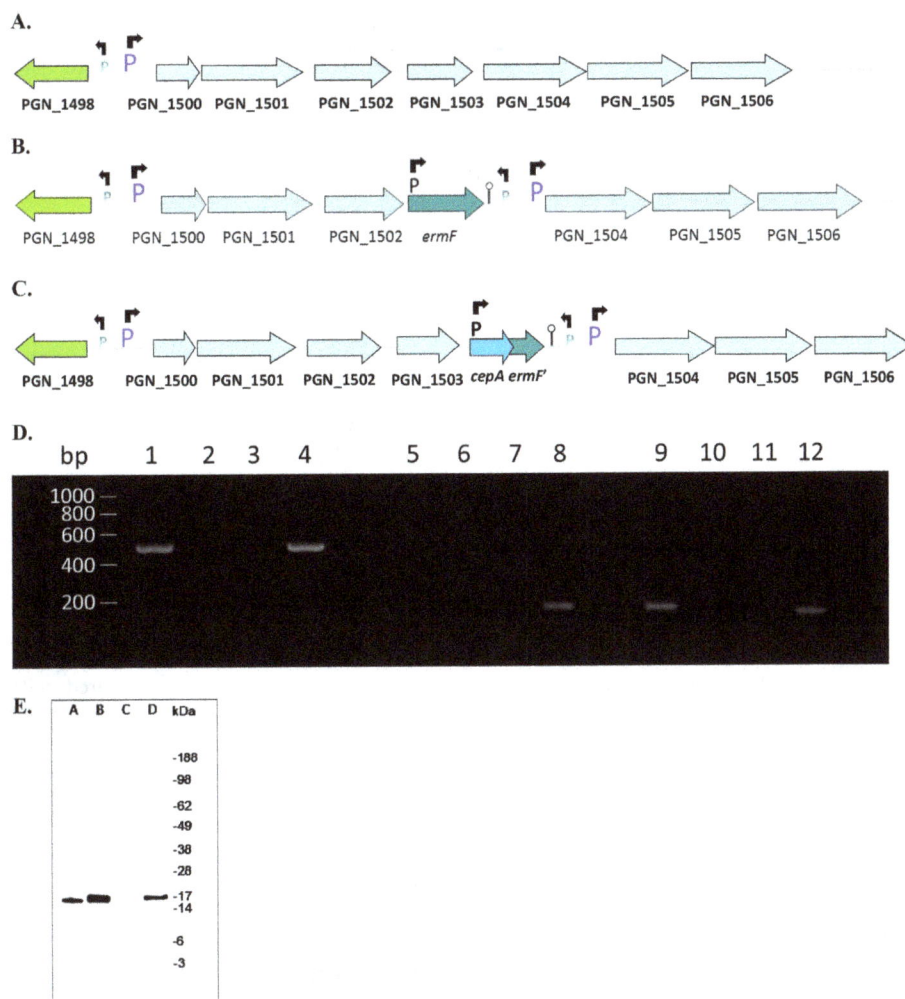

Figure 7. Genomic arrangement of *P. gingivalis* **ATCC 33277 in (A) the wild-type strain, (B)** *har* **mutant strain ECR455 and (C)** *har* **complemented strain ECR475.** 'P' denotes promoter positions, the arrows above 'P' denote the direction of transcription whilst the stem loop following *ermF* indicates a Rho-independent transcriptional terminator. Not drawn to scale. **(D) RT-PCR analysis of** *PGN_1504* **and** *PGN_1503* **(***har***).** Reverse transcription of ECR455 and ECR475 RNA was performed using random hexamers. PCR was then performed using oligonucleotide primers specific for *PGN_1504* (lanes 1–4) or *PGN_1503* (*har*) (lanes 5–8 and 9–12). The templates used for PCR were: reverse transcribed ECR455 RNA (lanes 1 and 5), reverse transcribed ECR475 RNA (lane 9), RNA that was not reverse transcribed (lanes 2, 6 and 10), no template (lanes 3, 7 and 11) and *P. gingivalis* ATCC 33277 genomic DNA (lanes 4, 8 and 12). PGN_1504 transcript was detected in the *har* mutant ECR455 (lane 1), whilst PGN_1503 (*har*) transcript was not detected in the *har* mutant strain ECR455 (lane 5), but was detected in the *har* complemented strain ECR475 (lane 9). **(E) Western blot detection of Har expression in** *P. gingivalis* **33277, ECR455 and ECR475.** Cytoplasmic protein extracts (25 µg) from *P. gingivalis* strains 33277 (B), ECR455 (C), ECR475 (D) and 5 ng purified Har (A) were separated on a 4–12% Bis-Tris polyacrylamide gel (Invitrogen) before Western transfer and blotting with anti-rHar sera. Har protein was detected in the 33277 wild-type and ECR475 complement, but not the ECR455 mutant strain.

Table 6. Elemental content of *P. gingivalis* ATCC 33277, ECR455 and ECR475 as determined by ICP-MS.

P. gingivalis	Fe	Mn	Zn	Ni	Mg
33277	8,514±251[a]	87±6	1,250±152	2.9±1.1	58,476±2,284
ECR455	9,523±199	78±7	1,088±121	4.1±1.5	62,592±1,448
ECR475	9,751±286	93±11	1,680±354	3.3±1.0	60,793±3,388

[a]All values are presented as pmol/mg cellular dry weight and represent the mean of three biological replicates for each strain. Metal content was statistically analysed using a one-way ANOVA with Tukey multiple comparison tests. In total 34 different elements: Li, B, Na, Mg, Al, P, K, Ca, Ti, V, Cr, Mn, Fe, Co, Ni, Cu, Zn, Ga, Ge, As, Se, Rb, Sr, Zr, Mo, Rh, Ru, Cd, Sn, Sb, Cs, Ba, W and Pb, were measured.

Table 7. Gene transcripts significantly up-regulated in the *P. gingivalis* ECR455 mutant compared with ATCC 33277 wild-type.

PGN_ID[a,b]	JCVI Probe Name	Gene	Annotation	Cellular Role	Fold Change	p-Value
PGN_0001	PG0001	*dnaA*	Chromosomal replication initiator protein DnaA	DNA replication, recombination and repair	1.69	9.3E-10
PGN_0287	PG0176	*mfaI*	MfaI fimbrillin	Cell envelope surface structure	1.81	1.5E-4
PGN_1578	PG0387	*tuf*	Translation elongation factor Tu	Translation factor	1.78	3.0E-3
PGN_1851	PG1921	*rpsE*	30S ribosomal protein S5	Translation – ribosomal protein	1.52	2.5E-3
PGN_1853	PG1923	*rplF*	50S ribosomal protein L6	Translation – ribosomal protein	1.68	1.7E-2
PGN_1858	PG1928	*rplN*	50S ribosomal protein L14	Translation – ribosomal protein	1.72	1.8E-2
PGN_1860	PG1930	*rpmC*	50S ribosomal protein L29	Translation – ribosomal protein	1.51	1.5E-3
PGN_2088	PG2224	*husD*	Hypothetical protein	Unknown, part of operon encoding hemophore HusA	1.77	2.6E-2
PGN_2089	PG2225	*husC*	Transcriptional regulator MarR family	Proposed regulator of HusA hemophore expression	1.68	3.6E-2

[a]Results are sorted by ascending PGN_ID (locus ID in 33277).
[b]Predicted operons: PGN_1851-1860; PGN_2088-2089.

Cellular metal content

The cellular metal content of three biological replicate cell lysates of *P. gingivalis* ATCC 33277, ECR455 and ECR475 was determined for 34 elements using ICP-MS. No significant differences were identified in the metal contents of ECR455 (Har mutant) relative to both ATCC 33277 and ECR475 (Har complemented) for any of the 34 elements analysed (Table 6).

Transcriptomic analyses of ECR455 versus wild-type

Total RNA was extracted from six biological replicates of *P. gingivalis* ATCC 33277 and ECR455 grown in continuous culture under defined conditions with a fixed generation time of 8.6 h and used in microarray analyses to identify the Har regulon. A total of 44 genes had significantly altered expression in ECR455 ($\geq$1.5 fold change, p<0.05), compared with the wild-type. Nine genes were up-regulated including two operons (Table 7) whereas 35 genes were down-regulated including 10 operons and 16 genes encoding hypothetical proteins (Table 8). No genes encoding known iron homeostasis or storage proteins had altered expression in the *har* mutant. However the gene transcript encoding the hemophore HmuY decreased ~2-fold and interestingly 11 of the 35 down-regulated genes have previously been shown to have increased expression in *P. gingivalis* grown under hemin-limitation (Table 8) [12], thus suggesting a relationship between Har and hemin availability. A clear relationship between Har and biofilm growth was also seen, as 26 of the 35 down-regulated genes had previously been found to be up-regulated when *P. gingivalis* was grown as a biofilm compared with planktonic growth (Table 8) [41]. Three operons that were down-regulated have been proposed to play roles in aerotolerance (PGN_0527-31), potassium uptake (PGN_2082-3) and efflux of proteins and small molecules (PGN_0446-9). Quantitative real time PCR (qRT-PCR) analysis of five selected genes confirmed the changes in expression showed by the microarray analysis (Table 9).

Biofilm assay

Biometric analysis of the biofilms grown in either hemin-limited or non-limiting growth conditions showed that *P. gingivalis* ATCC 33277 wild-type and the *har* complemented ECR475 produced biofilms that were not statistically different (Fig. 8). The *har* mutant ECR455 produced biofilms that had significantly reduced biovolume and average thickness and an increased surface area (SA):biovolume ratio compared with the wild-type or ECR475 (Fig. 8). Comparison of the biofilm formed by each strain under hemin-limitation or non-limitation showed that hemin availability had no effect on the biovolume or average thickness of the biofilm formed by ECR455 whereas there was a significant reduction in the biovolume and average thickness of the wild-type and ECR475 biofilms grown under hemin limitation (Fig. 8). Planktonic growth of the three strains under the same growth conditions was similar (data not shown).

Discussion

This study demonstrates a novel function for a Fur superfamily protein in *P. gingivalis* regulating hemin-dependent biofilm formation, a prerequisite for colonization and virulence of this bacterium within the host.

Sequence comparison indicated that Har, like other Fur superfamily members has the two -C-X-X-C- motifs associated with Zn^{2+} binding to the S1 structural site that is required for dimerization [18]. A recombinant *P. gingivalis* Har protein tightly bound one Zn^{2+} ion per monomer and oxidation of Cys residues demonstrated they were involved in metal coordination. Zn(II)Har formed a stable dimer in the absence of other divalent metal cations and therefore the results are consistent with Zn^{2+} binding to the two -C-X-X-C- motifs in the predicted S1 structural site as reported for other bacterial species [23,24]. Mutation of Cys97 did not result in any change in Zn^{2+} binding or dimerization state of the protein indicating that this Cys residue was not a component of the S1 structural site. The need to denature Har to release Zn^{2+} is consistent with the high affinity of Zn^{2+} binding, unlike *H. pylori* Fur where EDTA-treatment alone was sufficient to remove Zn^{2+} [42].

The conserved metal binding residues (-H-X-H-) that constitute the S2 divalent metal cation sensory binding site in other Fur sequences are absent in the *P. gingivalis* Har, making it unusual amongst characterized Fur superfamily proteins (Fig. 1). *P. gingivalis* recombinant Zn(II)Har bound hemin/Fe^{2+} in a 1:1 ratio with high affinity, indicating a novel binding site in this Fur orthologue. A Soret shift to 372 nm and 420 nm was observed upon the addition of Zn(II)Har to hemin. The Soret band shift to

Table 8. Gene transcripts significantly down-regulated in the *P. gingivalis* ECR455 mutant compared with ATCC 33277 wild-type.

PGN_ID[a,b]	JCVI Probe Name	Gene	Annotation	Cellular Role	Fold Change	p-Value	HL[c] Fold Change	Biofilm[d] Fold Change
PGN_0300	PG0192	*ompH-1*	cationic outer membrane protein OmpH	Cell wall/membrane biogenesis	0.57	7.4E-05	1.62	1.64
PGN_0301	PG0193	*ompH-2*	cationic outer membrane protein OmpH	Cell wall/membrane biogenesis	0.63	1.4E-03	1.50	1.74
PGN_0320	PG0215	-	hypothetical protein	Unknown	0.66	1.8E-06	1.45	1.87
PGN_0321	PG0216	-	hypothetical protein	Unknown	0.66	2.3E-04		1.88
PGN_0400	PG1715	-	hypothetical protein	Unknown	0.51	7.5E-06	1.56	2.06
PGN_0444	PG1667	-	outer membrane efflux protein PG52	Intracellular trafficking, secretion, and vesicular transport	0.66	1.1E-04		1.06
PGN_0446	PG1665	-	putative ABC transporter permease protein	Defence mechanisms	0.61	9.1E-07		1.15
PGN_0447	PG1664	-	putative ABC transporter permease protein	Defence mechanisms	0.54	1.8E-06		1.15
PGN_0448	PG1663	-	ABC transporter ATP-binding protein	Defence mechanisms	0.47	6.7E-06		1.21
PGN_0449	PG1662	-	hypothetical protein	Unknown	0.58	4.1E-07		1.13
PGN_0449_b	PG1661	-	hypothetical protein	Unknown	0.50	7.0E-05		
PGN_0451	PG1659	-	hypothetical protein	Unknown	0.57	8.1E-05		1.28
PGN_0485	PG1634	-	hypothetical protein	Unknown	0.64	4.8E-04	1.60	2.08
PGN_0486	PG1635	-	hypothetical protein	Unknown	0.66	1.4E-04	1.51	1.7
PGN_0527	PG1584	*batC*	probable aerotolerance-related exported protein BatC	Unknown	0.66	4.8E-03		1.13
PGN_0529	PG1582	*batA*	aerotolerance-related membrane protein BatA	Coenzyme transport and metabolism	0.60	1.4E-05		1.10
PGN_0531	PG1580	-	conserved hypothetical protein	Unknown	0.63	1.4E-05		1.73
PGN_0558	PG1551	*hmuY*	HmuY	Heme binding and transport	0.51	4.8E-03	10.1	1.17
PGN_0968	PG0987	-	hypothetical protein	Unknown	0.65	3.0E-02		1.08
PGN_0970	PG0985	-	RNA polymerase sigma-70 factor ECF subfamily	Transcription	0.59	3.0E-03		
PGN_1019	PG0928	-	response regulator	Signal transduction mechanisms	0.59	2.8E-04		1.64
PGN_1021	PG0926	-	hypothetical protein	Unknown	0.59	5.2E-06	1.8	1.64
PGN_1104	PG1314	*aroC*	chorismate synthase	Amino acid transport and metabolism	0.40	2.3E-06		2.19
PGN_1105	PG1315	*slyD*	peptidyl-prolyl cis-trans isomerase SlyD, FKBP-type	Posttranslational modification, protein turnover, chaperones	0.57	1.9E-02	1.47	1.49
PGN_1106	PG1316	-	hypothetical protein	Unknown	0.53	9.1E-07		1.57
PGN_1107	PG1317	-	hypothetical protein	Unknown	0.66	9.5E-03		1.82

Table 8. Cont.

PGN_ID[a,b]	JCVI Probe Name	Gene	Annotation	Cellular Role	Fold Change	p-Value	HL[c] Fold Change	Biofilm[d] Fold Change
PGN_1272	PG2188	lysA	diaminopimelate decarboxylase	Amino acid transport and metabolism	0.54	4.1E-07		
PGN_1273	PG2187	menA	1,4-dihydroxy-2-naphthoate octaprenyl-transferase	Coenzyme transport and metabolism	0.61	5.6E-07		
PGN_1351	PG1002	-	hypothetical protein	Unknown	0.66	3.3E-04		
PGN_1503	PG0465	har	Fur family transcriptional regulator	Inorganic ion transport and metabolism	0.15	1.8E-13		
PGN_1548	PG0419	-	hypothetical protein	Unknown	0.53	1.1E-03		2.59
PGN_1622	PG0339	-	hypothetical protein	Unknown	0.64	1.7E-03	1.83	1.27
PGN_1791	PG1858	-	flavodoxin FldA	Energy production and conversion	0.60	7.1E-04	15.25	1.34
PGN_2082	PG2218	trkA	potassium transporter peripheral membrane component	Inorganic ion transport and metabolism	0.55	4.3E-07		
PGN_2083	PG2219	trkH	potassium uptake protein TrkH	Inorganic ion transport and metabolism	0.66	2.8E-07		

[a]Results are sorted by ascending PGN_ID (locus ID in 33277).
[b]Predicted operons: PGN_0300-0301, PGN_0320-0321, PGN_0444-0449_b; PGN_0485-0486; PGN_0527-0531; PGN_0968-0970; PGN_1019-1021; PGN_1104-1107; PGN_1272-1273; PGN_2082-2083.
[c]HL - hemin-limited compared with hemin-excess as reported in Dashper et al. [12]. These data had p-values <0.05.
[d]Biofilm - biofilm compared with planktonic growth as reported in Lo et al. [41] and ArrayExpress E-TABM-467. These data had p-values <0.01.

Table 9. Validation of microarray data using qRT-PCR.

Gene	Microarray fold ratio (ECR455/33277)	qRT-PCR fold ratio (ECR455/33277)	qRT-PCR fold ratio (ECR455/ECR475)
PGN_0287	1.81	1.6	1.6
PGN_0448	0.47	0.5	0.6
PGN_1296	1	0.9	0.8
PGN_1578	1.78	1.6	1.5
PGN_2083	0.66	0.2	0.15

RNA from three biological replicates of *P. gingivalis* ATCC 33277, ECR455 and ECR475 grown in batch culture was extracted, then 800 ng RNA was reverse transcribed using random hexamers, before 0.3 ng cDNA was used as template in real time PCR with Power SYBR Green PCR master mix (Applied Biosystems) for 35 cycles. The number of copies of transcript was quantified relative to a standard curve amplified from 33277 genomic DNA for each primer pair. The mean number of transcripts from each bacterial strain was calculated for each primer pair and then used to calculate the qRT-PCR fold ratios.

420 nm has previously been used to demonstrate protein interaction via an axial Cys ligand with the central ferric ion of hemin [43]. The Soret band shift to 372 nm is typical of hemin binding to the Cys-Pro motif where pentacoordination of the central Fe^{3+} in hemin appears to be the preferred binding mechanism [44]. This motif, also known as a heme regulatory motif (HRM) features invariant Cys and Pro residues followed by a hydrophobic residue [39]. *P. gingivalis* Har has a putative HRM, -[97]C-P-[99]L- that replaces the S2 -H-X-H- motif conserved in other species, and mutation of Cys97 to Ala reduced the hemin binding affinity four-fold. This is consistent with mutation of HRMs in other hemin binding proteins such as the hemin iron sensing eukaryotic initiation factor 2α kinase (HRI) which had a five-fold decrease in hemin affinity when Cys409 of its -C-P- motif was mutated to Ser [45]. The -C-P- motif in HRI binds hemin via the coordination of Cys to the hemin iron center [45]. Given the Zn(II)Har Soret shift data and the similar effect of the Cys mutation on Zn(II)Har hemin binding compared to HRI, it is likely that Cys97 of Har binds hemin through the iron centre. This is supported by the lack of specific iron binding by the Zn(II)C97A Har protein.

Binding hemin or Fe^{2+} caused changes to the secondary structure of Zn(II)Har and Zn(II)Har150. Based on previous studies of Fur superfamily proteins conformational changes upon metal binding can have various consequences, such as inducing DNA binding [46], reversible dissociation from DNA [47] or rapid protein degradation [48].

The high pI of *P. gingivalis* Har made the study of specific DNA binding challenging due to nonspecific charge interactions. Thus we used Zn(II)Har150 with the lysine-rich tail removed for the DNA binding studies. This lysine-rich tail is only found amongst the Bacteroidetes Fur homologues. The *dnaA* promoter DNA was used as the Zn(II)Har150 binding target because in the microarray analysis, transcription of the *dnaA* gene was significantly increased in the *har* mutant ECR455 compared with wild-type, suggesting that *dnaA* is repressed by Har. The EMSA results showed that Zn(II)Har150 bound specifically to the *dnaA* promoter and that in the presence of high concentrations of hemin the binding of Har was decreased being consistent with the microarray data. These results therefore suggest that apo-Har (without its co-factor heme) is a repressor of the *dnaA* gene with upregulation of DNA replication being linked with heme availability which would better support metabolism and virulence. The repressor function of apo-Har in *P. gingivalis* is similar to the recently reported repressor function of apo-Fur in *Helicobacter pylori* [49].

The absence of the lysine-rich tail of Zn(II)Har150 may have removed some of the complexity of DNA binding regulation as the lysine-rich tail could serve as a site for lysine acetylation, a regulatory post-translational modification commonly found in eukaryotes and more recently in bacteria [50,51]. There is precedence for this type of regulation, with *in vitro* evidence that reversible lysine acetylation modulates the DNA-binding activity of the bacterial transcriptional regulator RcsB [52].

P. gingivalis Har does not appear to play a role in metal ion homeostasis unlike in other bacterial species [53–55] as suggested by the lack of difference in the cellular content of 34 metals including iron, manganese, zinc and nickel between the *P. gingivalis har* mutant ECR455, the wild-type parental strain ATCC 33277 and the *har* complemented strain ECR475 (Table 6). DNA microarray analysis of the *P. gingivalis har* mutant ECR455 also suggested that Har may not be involved in metal ion homeostasis as only one gene known to play a role in iron/heme uptake or iron homeostasis, PGN_0558 encoding the hemophore HmuY, was differentially regulated in ECR455 (Table 7 and 8). PGN_2089 (*husC*) that was up-regulated in ECR455 may also play a role in heme uptake as it is the proposed transcriptional repressor of the HusA hemophore found only under hemin-limited conditions, however there has been no experimental characterization of HusC [30]. The results of our study are consistent with a recent report suggesting that the *P. gingivalis* Fur orthologue does not play a role in iron homeostasis [56]. Interestingly, 11 of the 35 genes that were down-regulated in ECR455 were previously seen to increase in expression under hemin-limitation (Table 8) [12] and 26 of the 35 down-regulated genes increased in expression when *P. gingivalis* was grown as a homotypic biofilm compared with planktonic growth (Table 8) [41]. This suggests that Har is a transcriptional regulator associated with heme homeostasis and biofilm formation in *P. gingivalis*. The regulation of iron homeostasis in *P. gingivalis* is likely to be complex given the importance of heme and iron to *P. gingivalis* and the complex interplay of metals in this organism [13]. James *et al.* [57] have shown that LuxS was required for a 1.5-fold increase in transcript levels of the ferrous ion transport system but negative regulators of this system have not yet been identified.

Iron availability is known to variably affect bacterial biofilm formation and development [58–63]. The effect of hemin availability on bacterial biofilm development is less well known but we have shown that hemin-limitation decreases homotypic *P. gingivalis* ATCC 33277 biofilm formation and development (Fig. 8) [64]. In this study we have shown that Har was required for maximal *P. gingivalis* biofilm development as demonstrated by

A.

B.

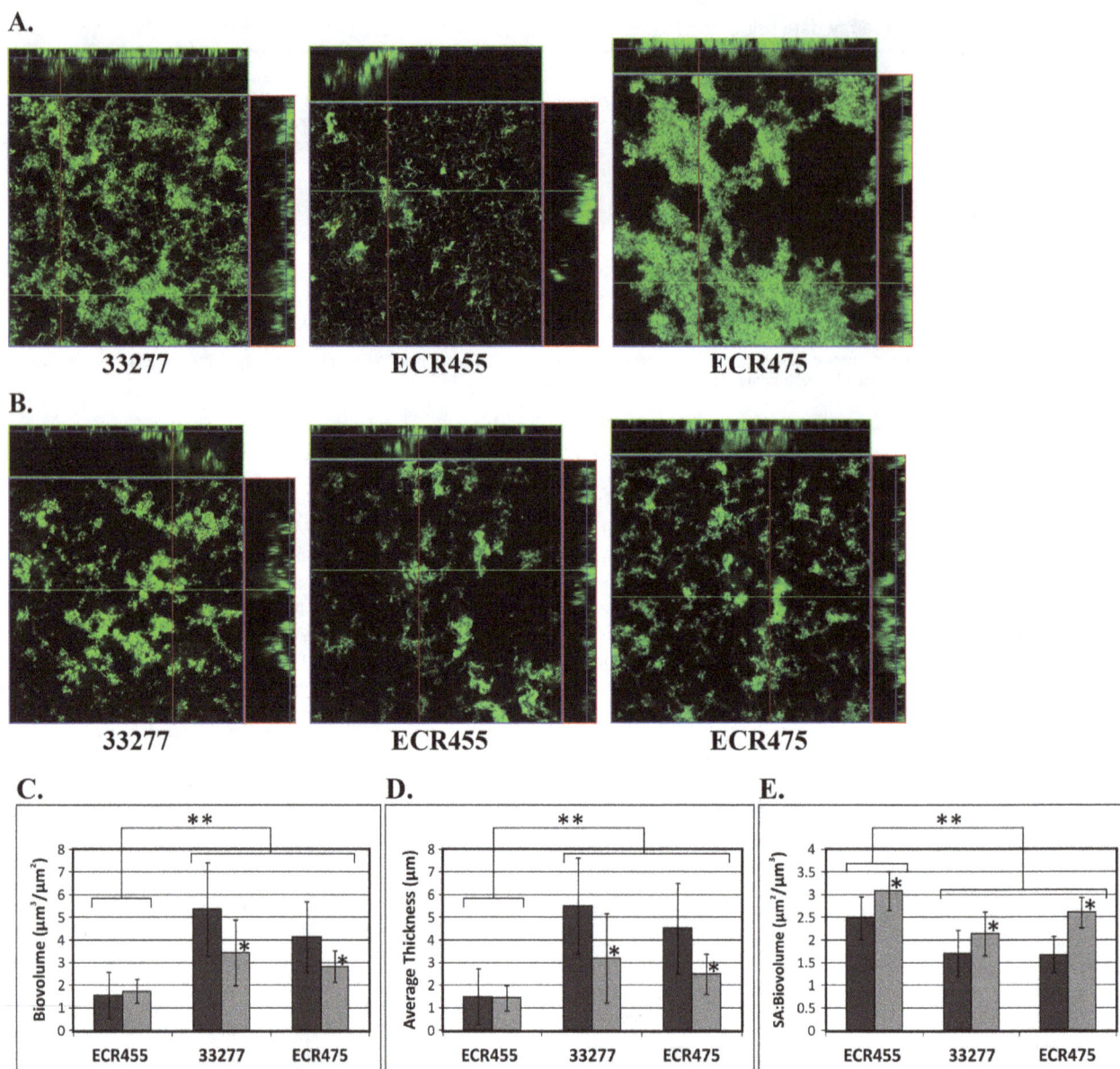

Figure 8. *P. gingivalis* **biofilm development.** Orthogonal projections of CLSM images showing a representative region of the x-y plane over the depth of the biofilm in both xz and yz dimensions of the ATCC 33277 wild-type, *har* mutant ECR455 and *har* complement ECR 475 strains grown in excess hemin (**A**) or hemin-limitation (**B**). Comparison of the Biovolume (**C**), Average Thickness (**D**) and SA:Biovolume (**E**) calculated for each strain's biofilm growth in either excess hemin (dark bars) or limited hemin (light bars) over three independent experiments. All biometric parameters analysed for the biofilms formed by ATCC 33277 and ECR475 were significantly ($p < 0.005$) altered when hemin was limited. * indicates a significant difference in excess hemin versus limited hemin ($p < 0.005$); ** indicates a significant difference in ECR455 biovolume, average thickness and SA:biovolume when compared to the same biometric parameters of both the ATCC 33277 and ECR475 biofilms ($p < 0.001$).

the significant decrease in biovolume and average thickness of ECR455 biofilms under both hemin-limited and non-limited conditions compared with the ATCC 33277 wild-type and Har-complemented strain ECR475 (Fig. 8). This indicates that Har is acting as a positive regulator of biofilm formation, which is consistent with the microarray data.

The biometric analysis of the biofilms formed by the *har* mutant in hemin-limited versus non-limited growth conditions were not significantly different as they were in the wild-type indicating that Har regulated biofilm development in a hemin-responsive manner. Both ATCC 33277 and ECR475 (Har⁺) produced homotypic biofilms with significantly less biovolume under hemin-

limitation compared with non-hemin limited growth conditions whilst the biovolume of the ECR455 biofilms was the same under both growth conditions. Har controls the expression of two genes important for homotypic biofilm formation, *hmuY* and *mfaI* (Table 7 and 8). Down-regulation of *hmuY* expression in ECR455 would contribute to the reduced ability of the Har mutant to form a hemin-responsive biofilm as HmuY plays a role in both hemin uptake and homotypic biofilm development in *P. gingivalis* [65,66]. This is consistent with our previous work showing HmuY was over 2.5 times more abundant in *P. gingivalis* homotypic biofilms than in planktonic cultures [67] and *hmuY* transcription was up-regulated ten-fold under conditions of hemin-limitation in continuous culture [12]. The *mfaI* transcript encoding the *P. gingivalis* minor fimbrillin was more abundant in ECR455 and may also contribute to the reduced biovolume of the *P. gingivalis har* mutant biofilm. Although *mfaI* transcription can fluctuate over time in mature biofilms the minor fimbriae have a suppressive regulatory role on initial attachment and organization of homotypic *P. gingivalis* ATCC 33277 biofilms [68,69].

As the Har regulon contained genes that were either positively- or negatively-regulated, it suggests that Har can act as both repressor and activator. Usually Fur orthologues act as repressor proteins, achieving positive regulation by repressing transcription of the sRNA RyhB [70]. The RNA chaperone Hfq mediates pairing between the RhyB and target mRNA, resulting in promoted degradation of mRNAs by RNase E [71]. Based on genomic sequences, *P. gingivalis* does not appear to have an Hfq orthologue, so positive and negative regulation may be achieved directly by the Har protein itself. There are examples of Fur proteins positively regulating gene expression directly [72,73] and binding DNA in the absence of a co-factor [74]. In fact *H. pylori* Fur can function as a repressor and activator both with and without its iron cofactor [49].

Conclusion

P. gingivalis is an iron and protoporphyrin IX-dependent Gram-negative bacterium that utilizes its only Fur superfamily orthologue, Har, to regulate hemin-responsive biofilm development. The ability to respond to environmental hemin and develop biofilms are key components of the *in vivo* survival and pathogenicity of *P. gingivalis*.

Acknowledgments

We are grateful to J. Patricia Lissel for technical assistance, Irene Volitakis and Robert Cherny for ICP-MS, Jian-Guo Zhang for valuable discussions, Wei Hong Toh for design of the RT-PCR primers and Glenn Walker for statistical advice.

Author Contributions

Conceived and designed the experiments: CAB SGD LZ HSGK ECR. Performed the experiments: CAB LZ CAS HLM DVC MDG JEH YT HSGK. Analyzed the data: CAB SGD LZ CAS HLM DVC MDG JEH YT ECR. Wrote the paper: CAB SGD LZ ECR.

References

1. Loesche WJ, Syed SA, Morrison EC, Laughon B, Grossman NS (1981) Treatment of periodontal infections due to anaerobic bacteria with short-term treatment with metronidazole. J Clin Periodontol 8: 29–44.
2. Slots J (1977) Microflora in the healthy gingival sulcus in man. Scand J Dent Res 85: 247–254.
3. Spiegel CA, Hayduk SE, Minah GE, Krywolap GN (1979) Black-pigmented *Bacteroides* from clinically characterized periodontal sites. J Periodontal Res 14: 376–382.
4. Van Dyke TE, Offenbacher S, Place D, Dowell VR, Jones J (1988) Refractory periodontitis: mixed infection with *Bacteroides gingivalis* and other unusual *Bacteroides* species. A case report. J Periodontol 59: 184–189.
5. White D, Mayrand D (1981) Association of oral *Bacteroides* with gingivitis and adult periodontitis. J Periodontal Res 16: 259–265.
6. Hajishengallis G, Liang S, Payne MA, Hashim A, Jotwani R, et al. (2011) Low-abundance biofilm species orchestrates inflammatory periodontal disease through the commensal microbiota and complement. Cell Host Microbe 10: 497–506.
7. Lamont RJ, Jenkinson HF (1998) Life Below the Gum Line: Pathogenic Mechanisms of *Porphyromonas gingivalis*. Microbiol Mol Biol Rev 62: 1244–1263.
8. Roper JM, Raux E, Brindley AA, Schubert HL, Gharbia SE, et al. (2000) The enigma of cobalamin (Vitamin B12) biosynthesis in *Porphyromonas gingivalis*. Identification and characterization of a functional corrin pathway. J Biol Chem 275: 40316–40323.
9. Shizukuishi S, Tazaki K, Inoshita E, Kataoka K, Hanioka T, et al. (1995) Effect of concentration of compounds containing iron on the growth of *Porphyromonas gingivalis*. FEMS Microbiol Lett 131: 313–317.
10. He J, Miyazaki H, Anaya C, Yu F, Yeudall WA, et al. (2006) Role of *Porphyromonas gingivalis* FeoB2 in Metal Uptake and Oxidative Stress Protection. Infect Immun 74: 4214–4223.
11. Lewis JP (2010) Metal uptake in host-pathogen interactions: role of iron in *Porphyromonas gingivalis* interactions with host organisms. Periodontol 2000 52: 94–116.
12. Dashper SG, Ang C-S, Veith PD, Mitchell HL, Lo AWH, et al. (2009) Response of *Porphyromonas gingivalis* to Heme Limitation in Continuous Culture. J Bacteriol 191: 1044–1055.
13. Dashper SG, Butler CA, Lissel JP, Paolini RA, Hoffmann B, et al. (2005) A Novel *Porphyromonas gingivalis* FeoB Plays a Role in Manganese Accumulation. J Biol Chem 280: 28095–28102.
14. Hantke K (2001) Iron and metal regulation in bacteria. Curr Opin Microbiol 4: 172–177.
15. Lee J-W, Helmann J (2007) Functional specialization within the Fur family of metalloregulators. BioMetals 20: 485–499.
16. Escolar L, Perez-Martin J, de Lorenzo V (1999) Opening the Iron Box: Transcriptional Metalloregulation by the Fur Protein. J Bacteriol 181: 6223–6229.
17. Dian C, Vitale S, Leonard GA, Bahlawane C, Fauquant C, et al. (2011) The structure of the *Helicobacter pylori* ferric uptake regulator Fur reveals three functional metal binding sites. Mol Microbiol 79: 1260–1275.
18. Fleischhacker AS, Kiley PJ (2011) Iron-containing transcription factors and their roles as sensors. Curr Opin Chem Biol 15: 335–341.
19. Hantke K (1981) Regulation of ferric iron transport in *Escherichia coli* K12: Isolation of a constitutive mutant. Mol Gen Genet 182: 288–292.
20. Niederhoffer EC, Naranjo CM, Bradley KL, Fee JA (1990) Control of *Escherichia coli* superoxide dismutase (*sodA* and *sodB*) genes by the ferric uptake regulation (*fur*) locus. J Bacteriol 172: 1930–1938.
21. Hantke K (1987) Ferrous iron transport mutants in *Escherichia coli* K12. FEMS Microbiol Lett 44: 53–57.
22. Karjalainen TK, Evans DG, Evans Jr DJ, Graham DY, Lee C-H (1991) Iron represses the expression of CFA/I fimbriae of enterotoxigenic *E. coli*. Microb Pathog 11: 317–323.
23. Calderwood SB, Mekalanos JJ (1987) Iron regulation of Shiga-like toxin expression in *Escherichia coli* is mediated by the *fur* locus. J Bacteriol 169: 4759–4764.
24. Lebek G, Gruenig HM (1985) Relation between the hemolytic property and iron metabolism in *Escherichia coli*. Infect Immun 50: 682–686.
25. Horton R (1995) PCR-mediated recombination and mutagenesis. Mol Biotechnol 3: 93–99.
26. Althaus EW, Outten CE, Olson KE, Cao H, O'Halloran TV (1999) The Ferric Uptake Regulation (Fur) Repressor Is a Zinc Metalloprotein. Biochemistry (Mosc) 38: 6559–6569.
27. Ostroy F, Gams RA, Glickson JD, Lenkinski RE (1978) Inhibition of lysozyme by polyvalent metal ions. Biochim Biophys Acta 527: 56–62.
28. Ellman GL (1959) Arch Biochem Biophys 82: 70–77.
29. Hansen RE, Winther JR (2009) An introduction to methods for analyzing thiols and disulfides: Reactions, reagents, and practical considerations. Anal Biochem 394: 147–158.
30. Gao JL, Nguyen KA, Hunter N (2010) Characterization of a hemophore-like protein from *Porphyromonas gingivalis*. J Biol Chem 285: 40028–40038.
31. Kuzmic P (1996) Program DYNAFIT for the analysis of enzyme kinetic data: application to HIV proteinase. Anal Biochem 237: 260–273.
32. Lobley A, Whitmore L, Wallace BA (2002) DICHROWEB: an interactive website for the analysis of protein secondary structure from circular dichroism spectra. Bioinformatics 18: 211–212.
33. Whitmore L, Wallace BA (2004) DICHROWEB, an online server for protein secondary structure analyses from circular dichroism spectroscopic data. Nucleic Acids Res 32: W668–W673.
34. Whitmore L, Wallace BA (2008) Protein secondary structure analyses from circular dichroism spectroscopy: Methods and reference databases. Biopolymers 89: 392–400.
35. Pidot SJ, Porter JL, Tobias NJ, Anderson J, Catmull D, et al. (2010) Regulation of the 18 kDa heat shock protein in *Mycobacterium ulcerans*: an alpha-crystallin orthologue that promotes biofilm formation. Mol Microbiol 78: 1216–1231.
36. Heydorn A, Nielsen AT, Hentzer M, Sternberg C, Givskov M, et al. (2000) Quantification of biofilm structures by the novel computer program COMSTAT. Microbiology 146: 2395–2407.
37. Naito M, Hirakawa H, Yamashita A, Ohara N, Shoji M, et al. (2008) Determination of the Genome Sequence of *Porphyromonas gingivalis* Strain ATCC 33277 and Genomic Comparison with Strain W83 Revealed Extensive Genome Rearrangements in *P. gingivalis*. DNA Res 15: 215–225.
38. Robertson KP, Smith CJ, Gough AM, Rocha ER (2006) Characterization of *Bacteroides fragilis* Hemolysins and Regulation and Synergistic Interactions of HlyA and HlyB. Infect Immun 74: 2304–2316.
39. Zhang L, Guarente L (1995) Heme binds to a short sequence that serves a regulatory function in diverse proteins. EMBO J 14: 313–320.

40. Leiros I, Timmins J, Hall DR, McSweeney S (2005) Crystal structure and DNA-binding analysis of RecO from *Deinococcus radiodurans*. EMBO J 24: 906–918.

41. Lo A, Seers C, Boyce J, Dashper S, Slakeski N, et al. (2009) Comparative transcriptomic analysis of *Porphyromonas gingivalis* biofilm and planktonic cells. BMC Microbiol 9: 18.

42. Vitale S, Fauquant C, Lascoux D, Schauer K, Saint-Pierre C, et al. (2009) A ZnS$_4$ Structural Zinc Site in the *Helicobacter pylori* Ferric Uptake Regulator. Biochemistry (Mosc) 48: 5582–5591.

43. Ishikawa H, Kato M, Hori H, Ishimori K, Kirisako T, et al. (2005) Involvement of Heme Regulatory Motif in Heme-Mediated Ubiquitination and Degradation of IRP2. Mol Cell 19: 171–181.

44. Kuhl T, Wissbrock A, Goradia N, Sahoo N, Galler K, et al. (2013) Analysis of Fe(III) Heme Binding to Cysteine-Containing Heme-Regulatory Motifs in Proteins. ACS Chem Biol 8: 1785–1793.

45. Igarashi J, Murase M, Iizuka A, Pichierri F, Martinkova M, et al. (2008) Elucidation of the Heme Binding Site of Heme-regulated Eukaryotic Initiation Factor 2α Kinase and the Role of the Regulatory Motif in Heme Sensing by Spectroscopic and Catalytic Studies of Mutant Proteins. J Biol Chem 283: 18782–18791.

46. Ahmad R, Brandsdal BO, Michaud-Soret I, Willassen N-P (2009) Ferric uptake regulator protein: Binding free energy calculations and per-residue free energy decomposition. Proteins 75: 373–386.

47. Singleton C, White GF, Todd JD, Marritt SJ, Cheesman MR, et al. (2010) Heme-responsive DNA Binding by the Global Iron Regulator Irr from *Rhizobium leguminosarum*. J Biol Chem 285: 16023–16031.

48. Yang J, Panek HR, O'Brian MR (2006) Oxidative stress promotes degradation of the Irr protein to regulate haem biosynthesis in *Bradyrhizobium japonicum*. Mol Microbiol 60: 209–218.

49. Carpenter BM, Gilbreath JJ, Pich OQ, McKelvey AM, Maynard EL, et al. (2013) Identification and Characterization of Novel *Helicobacter pylori* apo-Fur-Regulated Target Genes. J Bacteriol 195: 5526–5539.

50. Kim S, Sprung R, Chen Y, Xu Y, Ball H, et al. (2006) Substrate and functional diversity of lysine acetylation revealed by a proteomics survey. Mol Cell 23: 607–618.

51. Zhang J, Sprung R, Pei J, Tan X, Kim S, et al. (2009) Lysine Acetylation Is a Highly Abundant and Evolutionarily Conserved Modification in *Escherichia coli*. Mol Cell Proteomics 8: 215–225.

52. Thao S, Chen C, Zhu H, Escalante Semerena J, Chen C-S, et al. (2010) Nε-lysine acetylation of a bacterial transcription factor inhibits Its DNA-binding activity. PLoS ONE 5: e15123.

53. Abdul-Tehrani H, Hudson AJ, Chang Y-S, Timms AR, Hawkins C, et al. (1999) Ferritin Mutants of *Escherichia coli* Are Iron Deficient and Growth Impaired, and *fur* Mutants are Iron Deficient. J Bacteriol 181: 1415–1428.

54. Ahn B-E, Cha J, Lee E-J, Han A-R, Thompson CJ, et al. (2006) Nur, a nickel-responsive regulator of the Fur family, regulates superoxide dismutases and nickel transport in *Streptomyces coelicolor*. Mol Microbiol 59: 1848–1858.

55. Vajrala N, Sayavedra-Soto L, Bottomley P, Arp D (2011) Role of a Fur homolog in iron metabolism in *Nitrosomonas europaea*. BMC Microbiol 11: 37.

56. Anaya-Bergman C, Rosato A, Lewis JP (2014) Iron- and hemin-dependent gene expression of *Porphyromonas gingivalis*. Mol Microbiol July 8 [doi:10.1111/omi.12066].

57. James CE, Hasegawa Y, Park Y, Yeung V, Tribble GD, et al. (2006) LuxS Involvement in the Regulation of Genes Coding for Hemin and Iron Acquisition Systems in *Porphyromonas gingivalis*. Infect Immun 74: 3834–3844.

58. Francesca B, Ajello M, Bosso P, Morea C, Andrea P, et al. (2004) Both lactoferrin and iron influence aggregation and biofilm formation in *Streptococcus mutans*. BioMetals 17: 271–278.

59. Hancock V, Ferrières L, Klemm P (2008) The ferric yersiniabactin uptake receptor FyuA is required for efficient biofilm formation by urinary tract infectious *Escherichia coli* in human urine. Microbiology 154: 167–175.

60. Hindré T, Brüggemann H, Buchrieser C, Héchard Y (2008) Transcriptional profiling of *Legionella pneumophila* biofilm cells and the influence of iron on biofilm formation. Microbiology 154: 30–41.

61. Johnson M, Cockayne A, Williams PH, Morrissey JA (2005) Iron-Responsive Regulation of Biofilm Formation in *Staphylococcus aureus* Involves Fur-Dependent and Fur-Independent Mechanisms. J Bacteriol 187: 8211–8215.

62. Mey AR, Craig SA, Payne SM (2005) Characterization of *Vibrio cholerae* RyhB: the RyhB Regulon and Role of ryhB in Biofilm Formation. Infect Immun 73: 5706–5719.

63. Toney JH, Koh ML (2006) Inhibition of *Xylella fastidiosa* Biofilm Formation via Metal Chelators. JALA 11: 30–32.

64. Dashper SG, Pan Y, Veith PD, Chen Y-Y, Toh ECY, et al. (2012) Lactoferrin Inhibits *Porphyromonas gingivalis* Proteinases and Has Sustained Biofilm Inhibitory Activity. Antimicrob Agents Chemother 56: 1548–1556.

65. Lewis JP, Plata K, Yu F, Rosato A, Anaya C (2006) Transcriptional organization, regulation and role of the *Porphyromonas gingivalis* W83 hmu haemin-uptake locus. Microbiology 152: 3367–3382.

66. Olczak T, Wojtowicz H, Ciuraszkiewicz J, Olczak M (2010) Species specificity, surface exposure, protein expression, immunogenicity, and participation in biofilm formation of *Porphyromonas gingivalis* HmuY. BMC Microbiol 10: 134.

67. Ang C-S, Veith PD, Dashper SG, Reynolds EC (2008) Application of ^{16}O/^{18}O reverse proteolytic labeling to determine the effect of biofilm culture on the cell envelope proteome of *Porphyromonas gingivalis* W50. Proteomics 8: 1645–1660.

68. Kuboniwa M, Amano A, Hashino E, Yamamoto Y, Inaba H, et al. (2009) Distinct roles of long/short fimbriae and gingipains in homotypic biofilm development by *Porphyromonas gingivalis*. BMC Microbiol 9: 105.

69. Yamamoto R, Noiri Y, Yamaguchi M, Asahi Y, Maezono H, et al. (2011) Time Course of Gene Expression during *Porphyromonas gingivalis* Strain ATCC 33277 Biofilm Formation. Appl Environ Microbiol 77: 6733–6736.

70. Massé E, Gottesman S (2002) A small RNA regulates the expression of genes involved in iron metabolism in *Escherichia coli*. Proc Natl Acad Sci U S A 99: 4620–4625.

71. Massé E, Escorcia FE, Gottesman S (2003) Coupled degradation of a small regulatory RNA and its mRNA targets in *Escherichia coli*. Genes Dev 17: 2374–2383.

72. Delany I, Spohn G, Rappuoli R, Scarlato V (2001) The Fur repressor controls transcription of iron-activated and -repressed genes in *Helicobacter pylori*. Mol Microbiol 42: 1297–1309.

73. Lee H-J, Bang SH, Lee K-H, Park S-J (2007) Positive Regulation of *fur* Gene Expression via Direct Interaction of Fur in a Pathogenic Bacterium, *Vibrio vulnificus*. J Bacteriol 189: 2629–2636.

74. Bsat N, Helmann JD (1999) Interaction of *Bacillus subtilis* Fur (Ferric Uptake Repressor) with the *dhb* Operator *In Vitro* and *In Vivo*. J Bacteriol 181: 4299–4307.

75. Papadopoulos JS, Agarwala R (2007) COBALT: constraint-based alignment tool for multiple protein sequences. Bioinformatics 23: 1073–1079.

76. Fletcher HM, Schenkein HA, Morgan RM, Bailey KA, Berry CR, et al. (1995) Virulence of a *Porphyromonas gingivalis* W83 mutant defective in the *prtH* gene. Infect Immun 63: 1521–1528.

77. Seers CA, Slakeski N, Veith PD, Nikolof T, Chen Y-Y, et al. (2006) The RgpB C-Terminal Domain Has a Role in Attachment of RgpB to the Outer Membrane and Belongs to a Novel C-Terminal-Domain Family Found in *Porphyromonas gingivalis*. J Bacteriol 188: 6376–6386.

4

Exopolysaccharide Biosynthesis Enables Mature Biofilm Formation on Abiotic Surfaces by *Herbaspirillum seropedicae*

Eduardo Balsanelli, Válter Antonio de Baura, Fábio de Oliveira Pedrosa, Emanuel Maltempi de Souza, Rose Adele Monteiro*

Department of Biochemistry and Molecular Biology, Universidade Federal do Paraná, Curitiba, Paraná, Brazil

Abstract

H. seropedicae associates endophytically and epiphytically with important poaceous crops and is capable of promoting their growth. The molecular mechanisms involved in plant colonization by this microrganism are not fully understood. Exopolysaccharides (EPS) are usually necessary for bacterial attachment to solid surfaces, to other bacteria, and to form biofilms. The role of *H. seropedicae* SmR1 exopolysaccharide in biofilm formation on both inert and plant substrates was assessed by characterization of a mutant in the *espB* gene which codes for a glucosyltransferase. The mutant strain was severely affected in EPS production and biofilm formation on glass wool. In contrast, the plant colonization capacity of the mutant strain was not altered when compared to the parental strain. The requirement of EPS for biofilm formation on inert surface was reinforced by the induction of *eps* genes in biofilms grown on glass and polypropylene. On the other hand, a strong repression of *eps* genes was observed in *H. seropedicae* cells adhered to maize roots. Our data suggest that *H. seropedicae* EPS is a structural component of mature biofilms, but this development stage of biofilm is not achieved during plant colonization.

Editor: Michael Otto, National Institutes of Health, United States of America

Funding: CNPq - Conselho Nacional de Desenvolvimento Científico e Tecnológico (484309/2012-9). The funders had no role in study design, data collection and analysis, decision to publish, or preparation of the manuscript.

Competing Interests: The authors have declared that no competing interests exist.

* Email: rose.adele@ufpr.br

Introduction

H. seropedicae is a nitrogen-fixing, plant-growth-promoting Betaproteobacterium found attached to and within tissues of important crops such as maize (*Zea mays*), rice (*Oryza sativa*), sorghum (*Sorghum bicolor*) sugar-cane (*Saccharum officinarum*) and wheat (*Triticum aestivum*) [1]. The molecular mechanisms of plant recognition, attachment, penetration and endophytic colonization of this microrganism are not well known [1]. EPS are carbohydrate polymers of highly variable composition and structure found outside cells [2]. Bacterial EPS are usually responsible for attachment to solid surfaces and to other bacteria, thus forming microscopic and macroscopic cell aggregates [3]. When the aggregates are neatly organized, they are called biofilms [4]. In these communities the surface-associated microorganisms grow in matrix-enclosed microcolonies separated by a network of open-water channels [5,6]. The presence of a matrix between cells confers a series of selective advantages, such as protection against environmental variations, nutrient and ions retention, resistance to desiccation and mechanical protection [4,7,8].

Most of microorganisms do not occur naturally in planktonic communities, being generally found attached to biological and non-biological surfaces forming biofilms [9]. Initial stages of biofilm formation involves the redistribution of attached cells by surface motility [10–12], binary division of attached cells [13] or recruitment of cells from the surrounding fluid to the developing biofilm [14]. The individual adherent cells that initiate biofilm formation on a surface are capable of independent movement [12] before they begin to exude exopolysaccharide and adhere irreversibly [5]. Biofilm maturation results in the generation of a complex architecture with channels, pores, and redistribution of bacteria away from the substrate [15]. As the biofilm matures many cells alter their physiological processes in response to the conditions in their particular niches. The biofilm cells express genes in a pattern that deeply differs from that of their planktonic counterparts [16]. Finally, individual cells or whole microcolonies may detach from the biofilm and colonize other surfaces [17].

EPS and biofilm formation have been associated with the capacity of bacteria to colonize plants in symbiotic, neutral or pathogenic associations. One of the EPS functions in plant-bacterial interaction is to permit epiphytic colonization of the plant host [18]. Also, in plant-pathogen interaction EPS helps to create a favorable environment for pathogen survival and growth inside the infected plant, acting as a protective barrier against plant metabolic defenses [19]. The knockout of EPS biosynthesis genes (*exo* or *eps*) resulted in loss of virulence by *Erwinia stewartii* and *Xanthomonas axopodis* [20]. The mutation of *Xanthomonas campestris gumD*, which codes for a glucosyltransferase, drastically decreased the pathogenicity of this organism [21]. Also, EPS was the main factor required for bacterial wilt caused by *Ralstonia solanacearum* [22], where it seems to interfere with plant water transport by clogging the xylem [23]. In the case of diazotrophic symbionts, EPS seems to be indispensable for functional nodule establishment [24,25]. *Ensifer meliloti* mutant strains deficient in the production of one kind of EPS induce nodule formation, but they do not contains bacteroids [26,27]. The knockout of acidic EPS biosynthesis genes of *Ensifer* sp. NGR234 also results in pseudonodule formation [28]. The infection and subsequent nodulation of legumes by *R. leguminosarum* requires bacterial attachment onto root hair, a process that involves EPS production [29]. In the case of associative diazotrophs such as *Azospirillum brasilense* and *Gluconacetobacter diazotrophicus*, EPS seems to influence cellular aggregation and biofilm formation on plant root surface [30–32]. The knockout of rhamnose biosynthesis in *A. brasilense* led to a decrease in EPS production, and a decrease in maize colonization [33]. In *G. diazotrophicus*, exopolysaccharides seem to have a more dramatic effect, where knockout of *gumD* abolished attachment to rice root surface and endophytic colonization [32].

There is no evidence of the role of *H. seropedicae* EPS in plant colonization, although scanning electron microscopy revealed production of mucilaginous and fibrillar materials by *H. seropedicae* during colonization of maize, rice and sorghum root surfaces [34,35]. This material might be EPS. In this work we knocked out the *epsB* gene which codes for a putative glucosyl-transferase of the EPS biosynthesis gene cluster of *H. seropedicae*. The mutant strain has diminished EPS production and biofilm formation on abiotic surfaces, but showed no alterations on maize colonization profile compared to the wild type.

Materials and Methods

Growth of bacterial strains, DNA manipulations and mutagenesis

Bacterial strains and their relevant characteristics are listed in Table 1. *Herbaspirillum seropedicae* strains were grown at 30°C and 120 rpm in NFbHPN medium [41]. *Escherichia coli* strains were grown at 37°C in LB medium [42]. Antibiotics were added at the following concentrations when required: ampicillin (Ap) 10 $\mu g.mL^{-1}$; kanamycin (Km) 50 $\mu g.mL^{-1}$; chloramphenicol (Cm) 30 $\mu g.mL^{-1}$; tetracycline (Tc) 10 $\mu g.mL^{-1}$; streptomycin (Sm) 80 $\mu g.mL^{-1}$. The plasmids used in this study are listed in Table 1. Plasmid and total DNA preparations, agarose gel electrophoresis, restriction endonuclease digestion and cloning were performed according to standard protocols [42].

For *epsB* mutagenesis the primers HSepsB-F (5′- gctggaaccgca-tatgatcgt-3′) and HSepsB-R (5′- ccaggtggatccggtcaataa-3′) were used to amplify the *epsB* gene from *H. seropedicae* genomic DNA, and the amplicon was cloned in pTZ57R/T. The generated plasmid pTZHSwaaL was disrupted in the EcoRV site by the *nptI* cassette isolated from pKIXX that confers resistance to kanamycin (Km). The disrupted gene was transferred to pSUP202. This construction was electro-transformed in *E. coli* S17.1, and the transformants were conjugated into *H. seropedicae* SmR1. The mutant strains were selected and named *H. seropedicae* EPSEB (*epsB⁻*). Insertion of the cassette in the genome of the mutant strain by double crossover event was confirmed by PCR analyses. Wild-type and EPSEB mutant strains were GFP-marked through conjugation with *E. coli* S17.1 harboring the pHC60 plasmid.

EPS and LPS analyses

For EPS extraction, the *H. seropedicae* wild type and EPSEB mutant strains were grown in 10 mL of NFbHPN medium [41] at 30C and 120 rpm in the presence of 50 mg of sterile glass fiber. After 12hours, the bacterial cultures together with the glass fiber were transferred to a 50 mL centrifuge tube, and vortexed vigorously for 1 minute to remove glass fiber attached bacteria. The cells and the glass fiber were then removed by centrifugation (15 min, 3000 g) and the supernatant was filtered through a 0.22 µm membrane to remove residual cells. Exopolysaccharides in the filtered supernatant were precipitated with 3 volumes of cold ethanol for 24hours at 20C and centrifuged for 10 minutes at 4°C and 3000 g. The precipitate was vacuum dried, resuspended in MilliQ water and dialyzed against MilliQ water. Ten microliters of dialyzed samples were mixed with sample buffer (120 mM Tris pH 6.8; 3% SDS; 9% β-mercaptoethanol; 30% glycerol; 0.03% bromophenol blue), separated by SDS-PAGE (12% acrylamide) and visualized by silver periodate oxidation staining [43]. Total sugar concentration of the samples was determined with phenol/sulfuric acid [44], using glucose as standard.

LPS extraction for electrophoretic analysis was performed according to Balsanelli *et al.* [45] by the proteinase K – SDS method. Four microliters of final mixture were separated by SDS-PAGE (16% acrylamide) and visualized by silver periodate oxidation staining [43].

Biofilm formation on glass fiber

H. seropedicae strains were grown as described for EPS isolation, and biofilm formation was evaluated according to Balsanelli et al. [45]. Briefly, twelve hours after inoculation glass fiber was removed from the medium, stained with 20 µL of crystal violet 1% for 2 minutes, and washed three times with 0.9% saline solution. Then, 1 mL of absolute ethanol was added to remove the dye, and the alcoholic solution was used to determine the OD_{550}. The values are expressed as OD_{550} of the samples subtracted from the OD_{550} of the fiber glass treated culture medium. The results reported represent the average of three independent experiments. Purified wild type EPS (100 µg of glucose equivalents.mL^{-1}) was added to the system during incubation with glass fiber to test complementation of the mutant strain phenotype. Samples of stained glass fibers were analyzed by light microscopy for visualization of biofilm structure.

Plant interaction assays

Assays of maize colonization by *H. seropedicae* strains were performed according to Balsanelli *et al.* [46]. Briefly, seeds of *Zea mays* cv. 30F53, *Oryza sativa* cv Nipponbare or *Sorghum bicolor* cv A07 were surface-sterilized, germinated and each seedling was inoculated with 10^5 CFU of *H. seropedicae* strains. The inoculated seedlings were transferred to a hydroponic system containing 30 mL of plant medium [47] and 10 g of sterile culture beads in 100 mL glass tubes. Bacterial counts were made immediately after inoculation to access attached bacteria and 1, 4, 7 or 10 days after inoculation to access endophytic and epiphytic bacteria. The results reported represent the average of at least three independent experiments.

The GFP-marked strains were used as inoculants as described above, and longitudinal root cuts were freshly prepared for visualization. Root attached and 7 d.a.i. epiphytic bacteria were visualized by confocal laser scanning microscopy (CLSM) on a Nikon Ti Microscope. Plant tissues showed DAPI autofluorescence. Snapshots of the tridimensional images were obtained with the NIS-Elements software (Nikon).

Table 1. Bacterial strains and plasmids used in this study.

Strains	Relevant characteristics [a]	Reference
E. coli Top 10	F⁻ mcrA Δ(mcrr-hsdRMS-mcrBC) φ80lacZΔM15 ΔlacX74 ara Δ139 Δ(ara,leu) 7697 nupG λ⁻	Invitrogen
E. coli S17.1	RP4-2-Tc::Mu-Km::Tn7	[36]
H. seropedicae SmR1	Spontaneous Smʳ derived from strain Z78 (ATCC 35893)	[37]
H. seropedicae EPSEB	epsB mutant, Smʳ, Kmʳ	This work
H. seropedicae MHS01	epsG::lacZ chromosomal reporter fusion, Smʳ, Kmʳ	[38]
H. seropedicae SmR1+pHC60	H. seropedicae SmR1 constitutively expressing GFP from pHC60, Smʳ, Tcʳ	This work
H. seropedicae EPSEB+pHC60	H. seropedicae EPSEB constitutively expressing GFP from pHC60, Smʳ, Kmʳ, Tcʳ	This work
Plasmids and vectors		
pTZHSepsB	pTZ57 containing H. seropedicae SmR1 epsB gene, Apʳ	This work
pTZHSepsBKM	pTZHSepsB with epsB gene disrupted by Tn5 Kan cassette, Apʳ, Kmʳ	This work
pSUPHSepsBKM	epsB gene disrupted by Tn5 Kan cassette inside Tc gene of pSUP202, Apʳ; Kmʳ; Cmʳ;	This work
pUC4-KIXX	Apʳ; Kmʳ; cassette Tn5 Kan	[39]
pSUP202	Apʳ; Tcʳ; Cmʳ; mob site	[36]
pTZ57R/T	Apʳ, TA cloning vector	Fermentas
pHC60	Tcʳ; constitutive GFP (GFP-S65T) expression	[40]

[a]Ap = ampicillin; Km = kanamycin; Sm = streptomycin; Tc = tetracycline; Cm = chloramphenicol; and the superscript r = resistant.

Competition assays were performed using as inoculant a mixture of *H. seropedicae* wild type and *epsB* strains in 1:1 proportion, with a total of approximately 10, 10^2, 10^3, 10^4, or 10^5 bacteria per seedling. Total bacterial counts were made as described before, and the strains were identified by antibiotic resistance.

Chemical resistance assays

Resistance to chemical compounds by *H. seropedicae* strains was determined by serial dilution of liquid cultures and microdrop plating on solid NFbHPN medium containing naringenin (0–250 μM), quercetin (0–250 μM), jasmonic acid (0–10 μM), salicylic acid (0–50 μg.mL^{-1}), sodium dodecyl sulphate (0–0.01% w/v) or phenol (0–1% w/v). Data were expressed as percentage of colony forming units in the test plates compared to the control after 24 hours of growth at 30°C.

EPS biosynthesis gene expression during rhizoplane colonization and biofilm formation

To evaluate *eps* gene expression during rhizosphere colonization, the *H. seropedicae* MHS01 [38] (*epsG::lacZ*) reporter strain was grown in NFbHPN medium for 16 h. After adjusting the culture to $OD_{600} = 1.0$ in saline buffer, 10^8 cells (1 mL) were inoculated onto maize in the hydroponic system described above and incubated at 28°C. After 24 h, bacterial cells were recovered from the liquid medium by centrifugation and attached cells were recovered from root surface and polypropylene spheres by vortexing and centrifugation.

To evaluate *eps* gene expression during biofilm formation, the *H. seropedicae* MHS01 reporter strain was grown in the presence of glass fiber as described. After 12 h of growth the free living cells were recovered by centrifugation and attached cells were recovered from glass fiber by vortexing and centrifugation. The β-galactosidase activity of the recovered cells was then measured [48]. Protein determination was carried out according to Bradford [49]. The β-galactosidase activity is reported as nmol of o-nitrophenol produced per minute and mg of protein. The results reported represent the average of at least three independent

experiments. The control containing uninoculated maize seedlings did not show any detectable β-galactosidase activity.

Results

Genomic organization of *H. seropedicae* EPS biosynthesis genes and mutagenesis

Analyses of *H. seropedicae* SmR1 genome sequence (CP002039) showed a cluster of 28 genes that code for proteins probably involved in the biosynthesis and secretion of EPS (Fig. S1). The organization of these genes is highly similar to the *eps* cluster of *Herminiimonas arsenicoxydans* [50] and *Methylobacillus* sp. 12S [51], and the encoded proteins share high identity to the homologous proteins of all three microorganisms (Table S1). The EPS produced by *Methylobacillus* sp. 12S, named metanolan, is a heteropolymer composed of glucose, galactose and mannose in a 3:1:1 molar proportion [52]. The analyses of *H. seropedicae eps* genes that code for glycosyltransferases and sugar modifying proteins (such as epimerases and phosphatases in Table S1) suggest that the EPS is composed of these same monosaccharides. Indeed, monosaccharide composition analysis of *H. seropedicae* Z67^T EPS showed galactose, glucose and mannose as constituents at a proportion of 4:3:1, with possible substitutions with tetracarboxylic acids [53].

Knockout of *epsB* strongly reduces EPS production by *H. seropedicae*

The production of EPS was initially evaluated by precipitation of *H. seropedicae* strain culture supernatant with 3 volumes of cold ethanol. When the wild type and EPSEB (*epsB*) strains were grown in liquid NFbHPN for 24h no EPS was produced in the culture supernatant. Since in many bacteria EPS biosynthesis is induced during biofilm formation [54], the supernatant of *H. seropedicae* wild type culture grown for 12hours in the presence of glass fiber was processed as above and 0.8 mg.mL^{-1} of EPS was obtained. In contrast with the wild type strain, no EPS could be detected from the EPSEB strain. The samples were then analyzed by a 12% SDS-PAGE (Fig. 1). Exopolysaccharide from the wild type strain

WT   EPSEB

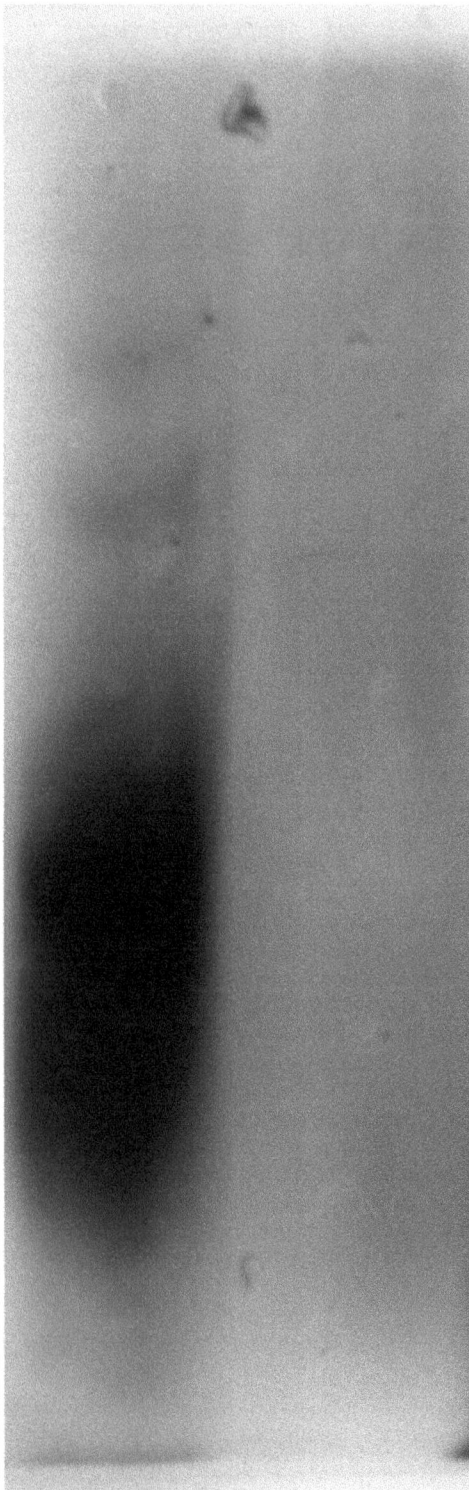

Figure 1. Electrophoretic pattern of EPS isolated from *H. seropedicae* strains SmR1 (wild type) and EPSEB (*epsB* mutant). SDS-PAGE was performed with EPS extracted by cold ethanol precipitation of the supernatant of biofilm growing bacteria in glass fiber submersed in NFbHPN medium.

showed three poorly defined bands of different molecular weight/charge, while supernatant of EPSEB strain had no polysaccharide band.

The EPSEB strain LPS electrophoretic profile did not differ from that of the wild type (Fig. S2), suggesting that this glucosyltransferase is specific for EPS biosynthesis.

H. seropedicae EPS is necessary for biofilm formation on glass fiber

To evaluate the role of EPS in biofilm formation, the strains were grown in the presence of glass fiber and biofilm formation was evaluated quantitatively by staining attached bacteria (Table 2), and qualitatively by light microscopy (Fig. 2A–D). After twelve hours of growth the EPSEB strain showed a 45% reduction in biofilm formation compared to the wild type. Furthermore, microscopic observation showed that the wild type strain formed large tridimensional structures, considered as mature biofilms (Fig. 2A). On the other hand, the mutant strain did not form mature biofilms, with only few attached cells (Fig. 2B). This phenotype was partially restored by the addition of purified *H. seropedicae* EPS (Fig. 2D), suggesting that this polysaccharide is required for biofilm development.

The reporter strain MHS01 (*epsG::lacZ*) was used to determine the regulation of *eps* genes in glass fiber biofilm formation (Fig. 2E). After 12hours of growth in the above-described system, *epsG* expression in glass fiber attached cells was about 3 times higher than in planktonic cells. The *eps* genes up-regulation on bacteria adhered to glass fiber suggests the involvement of EPS in biofilm formation on inert matrix.

Maize colonization by *H. seropedicae* is not dependent on EPS production

Colonization of *H. seropedicae* strains on maize roots was followed to evaluate the role of EPS in this interaction. The colonization profile of the EPSEB strain was very similar to that of the wild type (Fig. 3), suggesting that attachment, epiphytic and endophytic colonization are not dependent on *epsB* gene. Colonization of rice and sorghum by the EPSEB strain was also very similar to that of the wild type strain (Fig. S3), suggesting that EPS production is not required for interaction with poaceous plants. The maize colonization profile of MHS01 was also similar to the wild type one [48], indicating that the *eps* gene cluster and its product are not involved in plant interaction. The use of smaller numbers of wild type and EPSEB cells in attachment assays on maize roots did not show differences of colonization between the strains (Fig. 4).

CLSM analyses showed that both wild type and *epsB* mutant strains attach onto the maize root epidermis and root hair as individual cells and in similar numbers (Fig. 5A). Seven days after inoculation (Fig. 5B), the epiphytic population of both strains was still formed of individualized cells, not comprising tridimensional biofilm structures. These results indicate that *H. seropedicae* do not develop mature biofilms on roots as observed on glass fiber, stressing that EPS production is not required in plant colonization.

H. seropedicae EPS is required for resistance to abiotic stress

EPS production has been associated with protection against chemical stress [3,8,33,55]. We tested the resistance of the mutant and parental strain to the flavonoids naringenin and quercetin, to the plant immune metabolites jasmonic and salicylic acids, to phenol and SDS (Fig. 6). The mutant strain's resistance to plant bactericidal compounds was not different from that of the wild

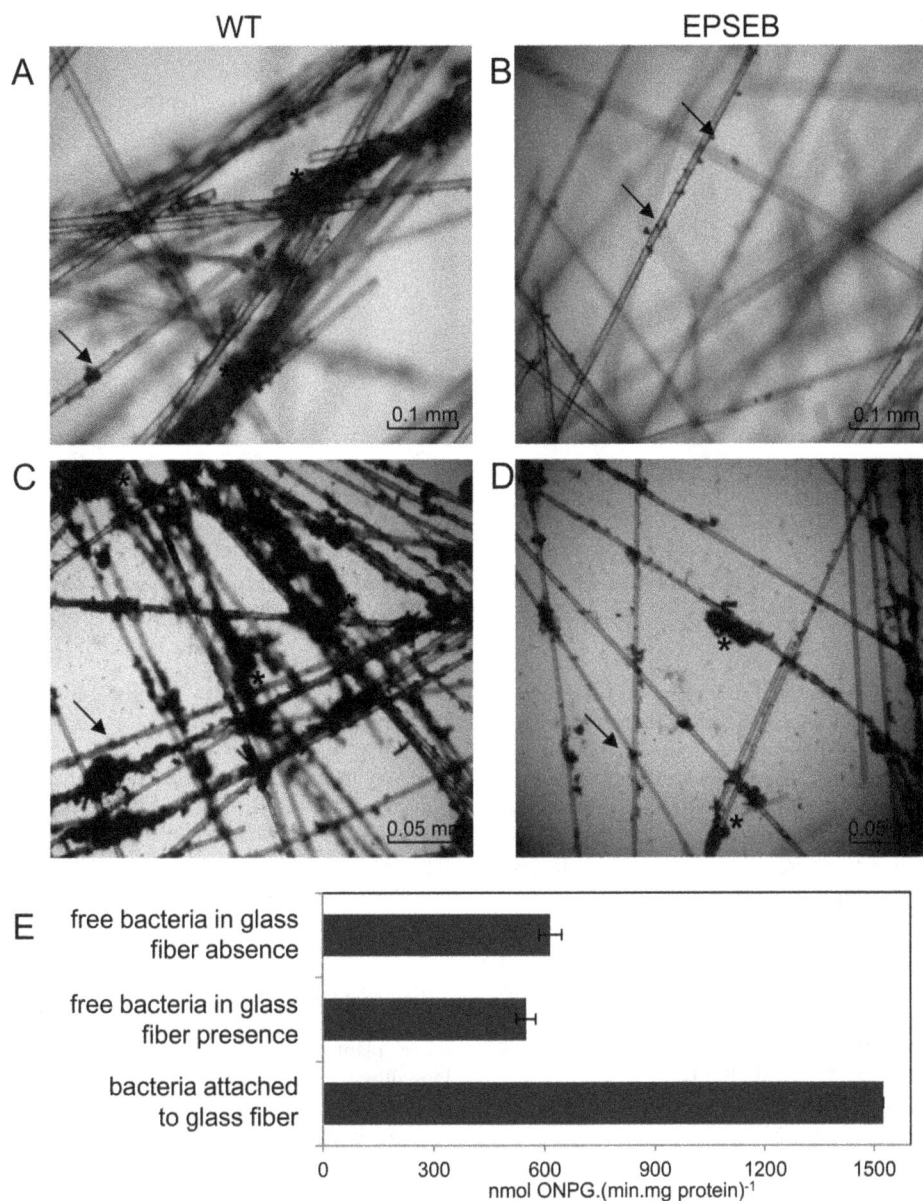

Figure 2. *H. seropedicae* biofilm formation on glass fiber. Light microscopy was performed with *H. seropedicae* SmR1 and EPSEB (*epsB* mutant) grown in the presence of glass fiber for 12 hours, without (A,B) and with (C,D) addition of purified wild-type EPS (100 µg.mL^{-1}). Arrows indicate attached bacteria. Asterisks indicate mature biofilm colonies. For biofilm expression analyses (E), *H. seropedicae* MHS-01 cells were grown for 12 h in the presence or absence of glass fiber, the free living bacteria were directly used and biofilm bacteria were recovered from glass fiber by vortex. β-galactosidase activity was determined, standardized by total protein concentration, and expressed as nmol ONP.(min.mg protein)$^{-1}$± standard deviation. Different letters indicate significant differences (p<0.01, Duncan multiple range test) in *epsG* expression between the tested conditions.

Table 2. *H. seropedicae* EPS is required for biofilm formation on glass fiber.

Strains	Biofilm in glass fiber (O.D.550$_{nm}$)	Biofilm in glass fiber+wild-type EPS (O.D.550$_{nm}$)
H. seropedicae SmR1	0.66±0.02 a	0.67±0.02 a
H. seropedicae EPSEB	0.30±0.01 b	0.54±0.03 c

H. seropedicae strains were grown in the presence of glass fiber and purified wild type EPS (100 µg.mL^{-1}) when indicated. After 12 hours, bacteria attached to the fiber were stained with crystal violet, washed and de-stained with absolute ethanol. The absorbance of the ethanol (550 nm) was determined and subtracted from the absorbance of the control without bacteria. Different letters indicate significant difference (p<0.001, Duncan multiple range test) between biofilm formation by the strains.

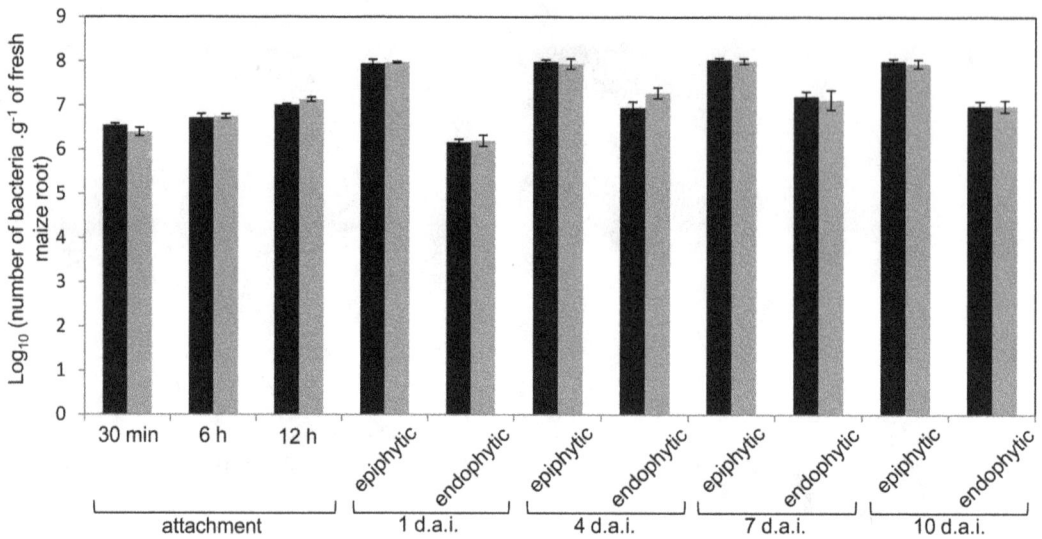

Figure 3. Maize root colonization by *H. seropedicae* wild type (black bars) and *epsB* (gray bars) mutant strain. Results are shown as average of Log_{10} (number of bacteria.g^{-1} of fresh root) $\pm$ standard deviation. d.a.i. = days after inoculation.

type. On the other hand, the parental strain showed resistance to low concentrations of phenol and SDS, while mutation in *epsB* gene reduced the survival of the mutant strain by 95%. These results suggest that *H. seropedicae* EPS is involved in resistance to non-biochemical stress, but not in resistance to plant basal defense.

H. seropedicae eps genes expression is down-regulated during maize colonization

Tadra-Sfeir and coworkers [38] showed by RT-PCR that the expression of *epsB* and *epsG* (code for glucosyltransferases) was repressed in the presence of the flavonoid naringenin. The reporter strain MHS01 (*epsG::lacZ*) was used to determine if the *eps* genes were regulated during maize colonization (Fig. 7). The results show that *epsG* is repressed during the first steps of interaction with maize, suggesting that EPS biosynthesis is diminished under this condition. Such repression was observed

both in planktonic bacteria free in the hydroponic medium in the presence of the plant roots and in root-attached bacteria, suggesting that *H. seropedicae* EPS is not required for the attachment on root surface. On the other hand, *eps* genes were induced (2.5-fold) in the bacteria adhered to the polypropylene spheres of the hydroponic system compared to planktonic bacteria, regardless the plant presence. This result stress the involvement of *H. seropedicae* EPS in biofilm formation on inert matrices.

Discussion

Exopolysaccharides are important factors that enable cellular aggregation and biofilm formation on solid surfaces. As shown for other plant associative bacteria [30–32], mutation of EPS biosynthesis genes in *H. seropedicae* SmR1 decrease EPS production and consequently biofilm formation, but surprisingly, did not alter maize colonization profile.

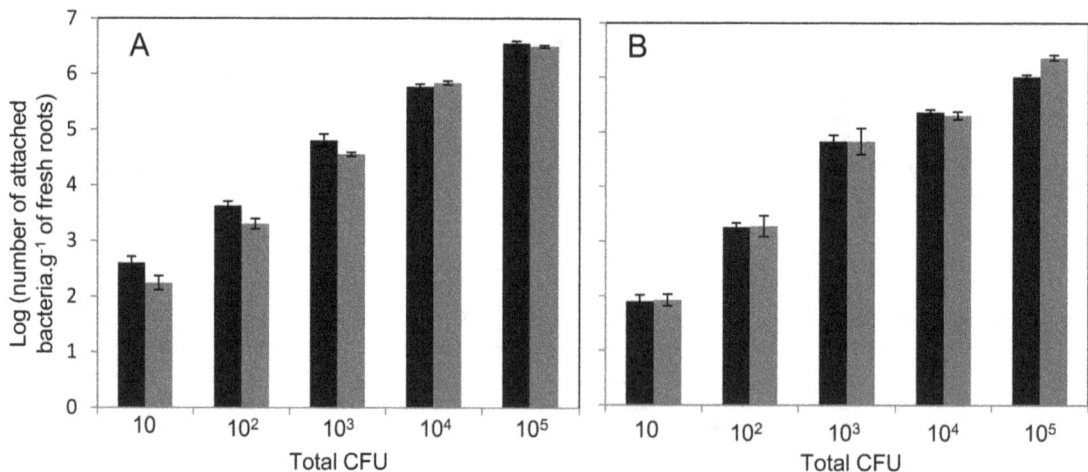

Figure 4. *H. seropedicae* strains competition for attachment on maize roots. *H. seropedicae* wild type (black bars) and *epsB⁻* (gray bars) strains were inoculated on maize separately (A) or co-inoculated in a 1:1 proportion (B), with the total of bacteria inoculated per plantlet indicated in the x axis. Results are shown as average of Log_{10} (number of recovered attached bacteria.g^{-1} of fresh root) $\pm$ standard deviation, CFU = colony forming units.

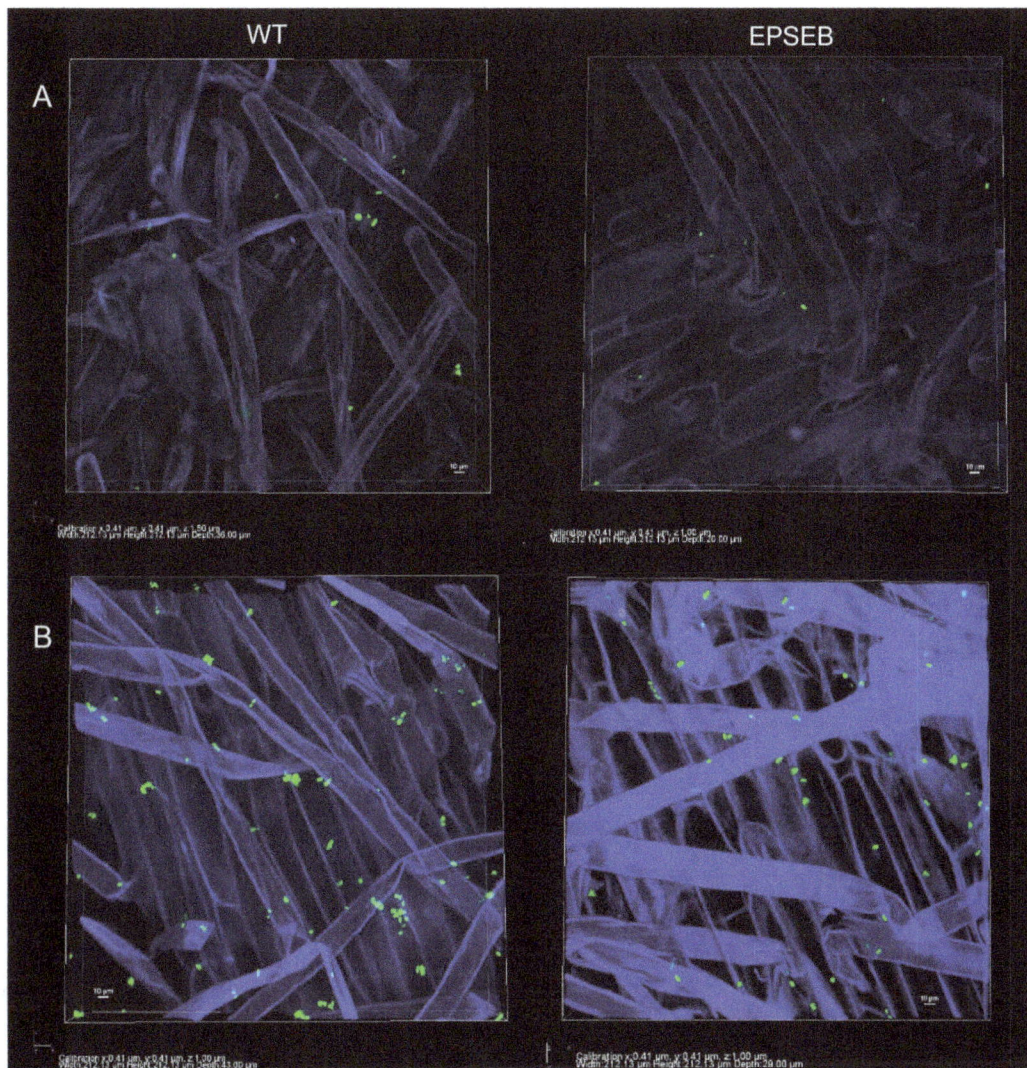

Figure 5. *H. seropedicae* attachment and epiphytic colonization of maize roots. *H. seropedicae* SmR1+pHC60 (GFP- wild type) and EPSEB+ pHC60 (GFP- *epsB* mutant) strains were inoculated on maize, and immediately after inoculation (A) or 7 days after inoculation (B), longitudinal samples of the roots were analyzed by laser scan confocal microscopy. Legends under the figures show positioning coordinates of the tridimensional images.

The importance of EPS in biofilm formation is supported by the induction of *eps* genes in the presence of inert substrates such as glass fiber and the polypropylene spheres. On the other hand, no difference was observed between the wild type and mutant strains in maize, rice or sorghum epiphytic colonization capacity. Even when lower numbers of bacteria were used to inoculate maize plants, both strains had similar root attachment patterns. Moreover, the increase and maintenance of the root epiphytic population seemed not to be dependent on EPS production. In agreement with those results, *eps* gene expression was repressed in *H. seropedicae* cells colonizing maize root surfaces. A huge impact in attachment and epiphytic colonization was observed by the lack of EPS production in *G. diazotrophicus* [32], but that seems not to be the case in *H. seropedicae* SmR1.

EPS can contribute to survival of bacteria within the plant by acting as a barrier against plant defense mechanisms, and creating a favorable microenvironment [55,56]. EPS production seems to be important for *H. seropedicae* resistance to chemical stress caused by phenol and SDS, but not required for resistance to plant

defense metabolites such as flavonoids, jasmonic and salicylic acids. Indeed, the mutant strain was able to cope with the plant chemical defense and endophytically colonize maize roots to the same extend than the wildtype. These results indicate that the product of the *eps* gene cluster is not necessary for maize root endophytic colonization by *H. seropedicae*.

The results lead us to propose a model for the early steps of *H. seropedicae* maize colonization. Upon contact with the rhizosphere environment *eps* genes are down-regulated, decreasing EPS biosynthesis. On the other hand, LPS biosynthesis is up-regulated, which allows the bacteria to bind to plant lectins on the root surface [46]. In accordance with this suggestion, scanning electron microscopy [34,35] and the CLSM results showed that *H. seropedicae* cells form a monolayer on maize root surface, not developing to mature biofilm. It seems that *H. seropedicae* biofilm development is arrested on roots by the reduced biosynthesis of EPS. The loosely attached bacterial cell can then penetrate inner root tissues and colonize them. By avoiding permanent attachment

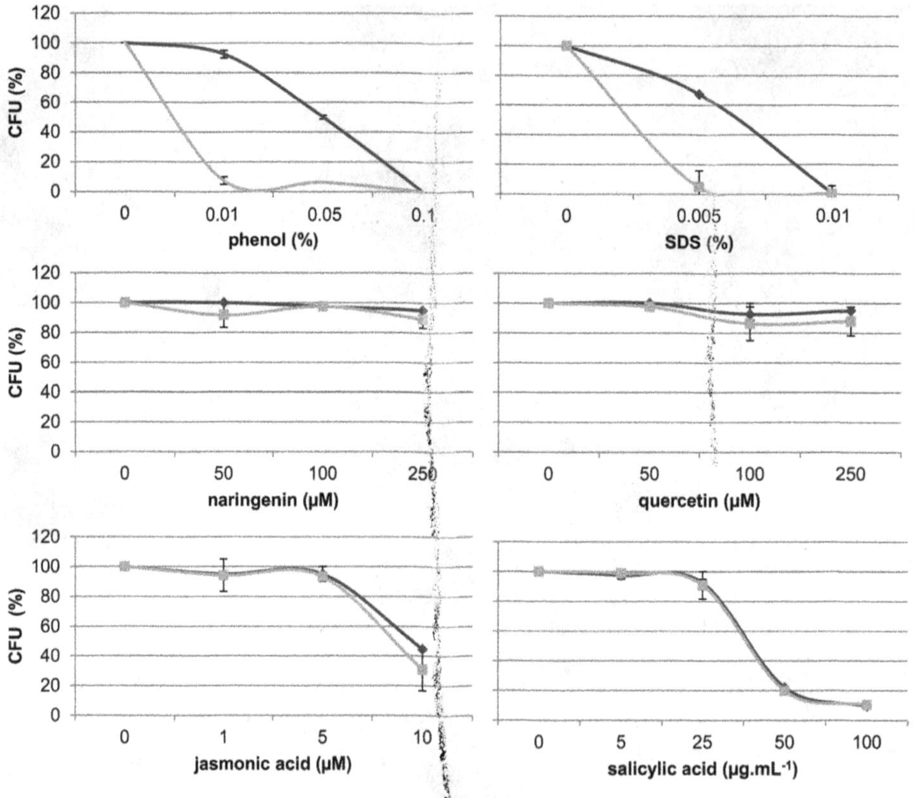

Figure 6. Resistance of *H. seropedicae* strains to chemical stress. *H. seropedicae* wild type (black lines) and EPSEB (gray lines) strains were plated on solid NFbHPN medium containing the compounds. Data expressed as percentage of colony forming units (CFU) in the test plates compared to the control after 24 hours of growth at 30°C.

and biofilm maturation *H. seropedicae* would remain available to seek penetration sites and nutrient sources.

In most plant-interacting bacteria studied so far, including associative, symbiotic or pathogenic, whenever the EPS is involved in biofilm formation it is also required for plant colonization or acts as a virulence factor [32,55–73]. In a stark contrast, *H. seropedicae* SmR1 EPS is necessary for biofilm formation but EPS synthesis is repressed during maize root colonization.

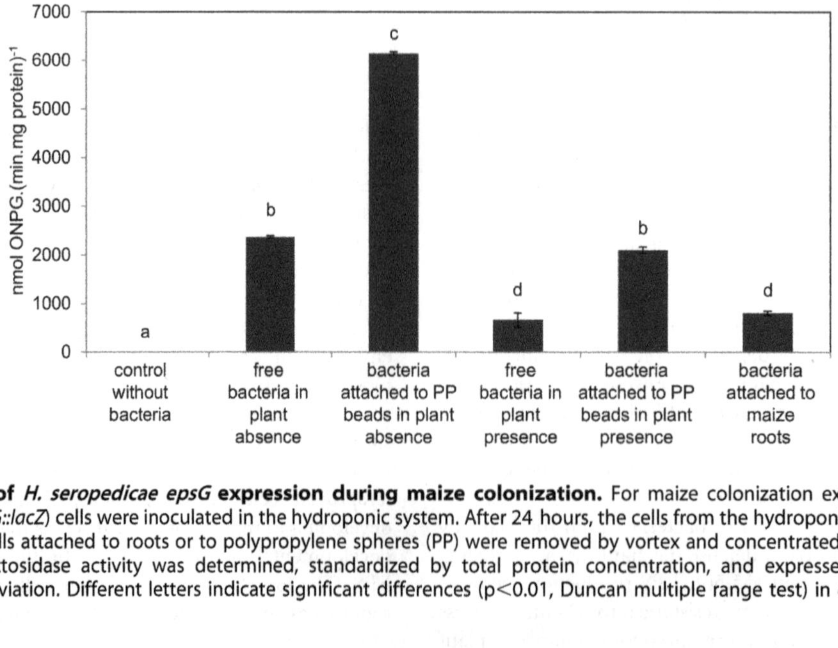

Figure 7. Regulation of *H. seropedicae* epsG expression during maize colonization. For maize colonization expression analyses, 10^8 *H. seropedicae* MHS-01 (*epsG::lacZ*) cells were inoculated in the hydroponic system. After 24 hours, the cells from the hydroponic medium were collected by centrifugation. The cells attached to roots or to polypropylene spheres (PP) were removed by vortex and concentrated by centrifugation. For all the samples the β-galactosidase activity was determined, standardized by total protein concentration, and expressed as nmol ONP.(min.mg protein)$^{-1}$± standard deviation. Different letters indicate significant differences (p<0.01, Duncan multiple range test) in *epsG* expression between the tested conditions.

Supporting Information

Figure S1   *H. seropedicae* SmR1 *eps* gene cluster. The proteins coded by the showed genes were analyzed in Table S1. The indicated probable promoter regions were identified with the BPROM software (SoftBerry).

Figure S2   Electrophoretic pattern of LPS isolated from *H. seropedicae* SmR1 (A) and EPSEB (B). SDS-PAGE was performed with total LPS extracted from 10^7 cells grown in NFbHPN medium by the SDS/proteinase K method, and visualized with silver periodate oxidation staining.

Figure S3   Rice (A) and sorghum (B) root colonization by *H. seropedicae* wild type (black bars) and *epsB* (gray bars) mutant strain. Results are shown as average of Log_{10}

(number of bacteria.g^{-1} of fresh root) $\pm$ standard deviation. d.a.i. = days after inoculation.

Table S1   *H. seropedicae* Eps proteins.

Acknowledgments

We thank Roseli Prado, Marilza Lamour, and Alexsandro Albani for technical support.

Author Contributions

Conceived and designed the experiments: EB EMS RAM. Performed the experiments: EB VB. Analyzed the data: EB EMS RAM. Contributed reagents/materials/analysis tools: FOP EMS RAM. Contributed to the writing of the manuscript: EB FOP EMS RAM.

References

1. Monteiro RA, Balsanelli E, Wassen R, Marin AM, Brusamarello-Santos LCC, et al. (2012) *Herbaspirillum*-plant interactions: microscopical, histological and molecular aspects. Plant Soil 356: 175–196.
2. Sutherland IW (1980) Biosynthesis of microbial exopolysaccharides. Ann Rev Microbiol 34: 79–150.
3. Flemming HC, Wingender J (2001) Relevance of microbial extracellular polymeric substances (EPSs)–Part I: structural and ecological aspects. Water Science Technol 43: 1–8.
4. Costerton JW, Lawandowski Z, Caldwell DE, Korber DR, Lappin-Scott HM (1995) Microbial biofilms. Ann Rev Microbiol 49: 711–745.
5. Stoodley P, Sauer K, Davies DG, Costerton JW (2002) Biofilms as complex differentiated communities. Annu Rev Microbiol 56: 187–209.
6. Stewart PS, Franklin MJ (2008) Physiological heterogeneity in biofilms. Nature Rev Microbiol 6: 199–210.
7. Sutherland IW (1982) Biosynthesis of microbial polysaccharides. In: AH Rose and JG Morris, editors. Advances in Microbial Physiology: Academic Press, London. 79–150.
8. Roberson E, Firestone M (1992) Relationship between desiccation and exopolysaccharide production by soil *Pseudomonas* sp. Appl Environ Microbiol 58: 1284–91.
9. Sutherland IW (2001) The biofilm matrix: an immobilized but dynamic microbial environment. Trends Microbiol 9: 222–227.
10. Dalton HM, Goodman AE, Marshall KC (1996) Diversity in surface colonization behavior in marine bacteria. J Ind Microbiol 17: 228–34.
11. Korber DR, Lawrence JR, Lappin-Scott HM, Costerton JW (1995) Growth of microorganisms on surfaces. In: Lappin-Scott HM, Costerton JW, editors. Microbial Biofilms. Cambridge, UK: Cambridge Univ. Press. 15–45.
12. O'Toole GA, Kolter R (1998) Flagellar and twitching motility are necessary for *Pseudomonas aeruginosa* biofilm development. Mol Microbiol 30: 295–304.
13. Heydorn A, Nielsen AT, Hentzer M, Sternberg C, Givskov M, et al. (2000) Quantification of biofilm structures by the novel computer program COMSTAT. Microbiol 146: 2395–407.
14. Tolker-Nielson T, Brinch UC, Ragas PC, Andersen JB, Jacobsen CS, et al. (2000) Development and dynamics of *Pseudomonas* sp. biofilms. J Bacteriol 182: 6482–89.
15. Davies DG, Parsek MR, Pearson JP, Iglewski BH, Costerton JW, et al. (1998) The involvement of cell-to-cell signals in the development of a bacterial biofilm. Science 280: 295–98.
16. Sauer K, Camper AK, Ehrlich GD, Costerton JW, Davies DG (2002) *Pseudomonas aeruginosa* displays multiple phenotypes during development as a biofilm. J Bacteriol, 184: 1140–54.
17. Hall-Stoodley L, Costerton JW, Stoodley P (2004) Bacterial biofilms: from the environment to infectious disease. Nat Rev Microbiol 2: 95–108.
18. Bogino PC, Oliva MM, Sorroche FG, Giordano W (2013) The Role of Bacterial Biofilms and Surface Components in Plant-Bacterial Associations. Int J Mol Sci 14: 15838–15859.
19. Leigh JA, Coplin DL (1992) Exopolysaccharides in plant-bacteria interaction. Annu Rev Microbiol 46: 307–346.
20. Denny T (1995) Involvement of bacterial exopolysaccharides in plant pathogenesis. Ann Rev Phytopathol 33: 173–197.
21. Chou FL, Chou HC, Lin YS, Yang BY, Lin NT, et al. (1997) The *Xanthomonas campestris gumD* gene required for synthesis of xanthan gum is involved in normal pigmentation and virulence in causing black rot. Biochem Biophys Res Commun 233: 265–269.
22. Hayward AC (1991) Biology and epidemiology of bacterial wilt caused by *Pseudomonas solanacearum*. Annu Rev Phytopathol 29: 65–108.
23. Hussain A, Kelman A (1958) Relation of slime production to mechanism of wilting and pathogenicity of *Pseudomonas solanacearum*. Phytopathol 48: 155–165.
24. Leigh JA, Walker GC (1994) Exopolysaccharides of *Rhizobium*: synthesis, regulation and symbiotic function. Trends in Genet 10: 63–67.
25. Skorupska A, Janczarek M, Marczak M, Mazur A, Król J (2006) Rhizobial exopolysaccharides: genetic control and symbiotic functions. Microbial Cell Fact 5: 7.
26. Leigh JA, Reed JW, Hanks JF, Hirsch AM, Walker GC (1987) *Rhizobium meliloti* mutants that fail to succinilate their calcofluor-binding exopolysaccharide are defective in nodule invasion. Cell 51: 579–587.
27. Long S, Reed JW, Himawan J, Walker GC (1988) Genetic analysis of a cluster of genes required for synthesis of the Calcofluor-binding exopolysaccharide of *Rhizobium meliloti*. J Bacteriol 170: 4239–4248.
28. Staehelin C, Forsberg LS, D'Haeze W, Gao M, Carlson RW, et al. (2006) Exo-Oligosaccharides of *Rhizobium* sp strain NGR234 Are Required for Symbiosis with Various Legumes. J Bacteriol 188: 6168–6178.
29. Laus MC, Van Brussel AAN, Kijne JW (2005) Exopolysaccharide structure is not a determinant of host-plant specificity in nodulation of *Vicia sativa* roots. Mol Plant Microbe Interac 18: 1123–1129.
30. Burdman S, Okon Y, Jurkevitch E (2000) Surface characteristics of *Azospirillum brasilense* in relation to cell aggregation and attachment to plant roots. Crit Rev Microbiol 26: 91–110.
31. Steenhoudt O, Vanderleyden J (2000) *Azospirillum*, a free-living nitrogen-fixing bacterium closely associated with grasses: genetic, biochemical and ecological aspects. FEMS Microbiol Rev 24: 487–506.
32. Meneses CHSG, Rouws LFM, Simões-Araújo JL, Vidal MS, Baldani JI (2011) Exopolysaccharide production is required for biofilm formation and plant colonization by the nitrogen-fixing endophyte *Gluconacetobacter diazotrophicus*. Mol Plant Microbe Interac 24: 1448–1458.
33. Jofré E, Lagares A, Mori G (2004) Disruption of dTDP-rhamnose biosynthesis modifies lipopolysaccharide core, exopolysaccharide production and root colonization in *Azospirillum brasilense*. FEMS Microbiol 231: 267–275.
34. Roncato-Macari LDB, Ramos HJO, Pedrosa FO, Alquini Y, Yates MG, et al. (2003) Endophytic *Herbaspirillum seropedicae* expresses *nif* gene in gramineous plants. FEMS Microbiol Ecol 45: 39–47.
35. Gyaneshwar P, James EK, Reddy PM, Ladha JK (2002) *Herbaspirillum* colonization increases growth and nitrogen accumulation in aluminium-tolerant rice varieties. New Phytol 154: 131–145.
36. Simon R, Priefer U, Pühler A (1983) A broad host range mobilization system for *in vivo* genetic engineering: transposon mutagenesis in gram negative bacteria. Nat Biotechnol 1: 784–791.
37. Pedrosa FO, Teixeira KRS, Machado IMP, Steffens MBR, Klassen G, et al. (1997) Structural organization and regulation of the *nif* genes of *Herbaspirillum seropedicae*. Soil Biol Biochem 29: 843–846.
38. Tadra-Sfeir MZ, Souza EM, Faoro H, Müller-Santos M, Baura VA, et al. (2011) Naringenin regulates expression of genes involved in cell wall synthesis in *Herbaspirillum seropedicae*. Appl Environ Microbiol 77: 2180–2183.
39. Barany F (1985) Two-codon insertion mutagenesis of plasmid genes by using single-stranded hexameric oligonucleotides. Proc Natl Acad Sci USA 82: 4202–4206.
40. Cheng HP, Walker GC (1998) Succinoglycan is required for initiation and elongation of infection threads during nodulation of alfalfa by *Rhizobium meliloti*. J Bacteriol 180: 5183–5191.
41. Klassen G, Pedrosa FO, Souza EM, Funayama S, Rigo LU (1997) Effect of nitrogen compounds on nitrogenase activity in *Herbaspirillum seropedicae* SmR1. Can J Microbiol 43: 887–891.
42. Sambrook J, Fritsch EF, Maniatis T (1989) Molecular cloning: a laboratory manual. 2 ed. Cold Spring Harbor, New York: Cold Spring Harbor Laboratory Press.
43. Tsai C, Frisch CE (1982) A sensitive silver stain for detecting lipopolysaccharides in polyacrylamide gels. Anal Biochem 119: 115–119.

44. Dubois M, Gilles KA, Hamilton JK, Rebers PA, Smith F (1956) Colorimetric Method for Determination of Sugars and Related Substances. Analytic Chem 8: 350–356.

45. Balsanelli E, Serrato RV, de Baura VA, Sassaki G, Yates MG, et al. (2010) *Herbaspirillum seropedicae rfbB* and *rfbC* genes are required for maize colonization. Environ Microbiol 12: 2233–2244.

46. Balsanelli E, Tuleski TR, de Baura VA, Yates MG, Chubatsu LS, et al. (2013) Maize Root Lectins Mediate the Interaction with *Herbaspirillum seropedicae* via N-Acetyl Glucosamine Residues of Lipopolysaccharides. PLoS ONE 8: e77001.

47. Egener T, Hurek T, Reinhold-Hurek B (1999) Endophytic expression of *nif* genes of *Azoarcus sp.* strain BH72 in rice roots. Mol Plant-Microbe Interact 12: 813–819.

48. Miller JH (1972) Experiments in molecular genetics. Cold Springer Horbor: NY.

49. Bradford MM (1976) A rapid and sensitive method for the quantification of microgram quantities of protein utilizing the principle of protein dye binding. Anal Biochem 72: 248–254.

50. Muller D, Medigue C, Koechler S, Barbe V, Barakat M, et al. (2007) A tale of two oxidation states: bacterial colonization of arsenic-rich environments. PLoS Genet 3: e53.

51. Yoshida T, Ayabe Y, Yasunaga M, Usami Y, Habe H, et al. (2003) Genes involved in the synthesis of the exopolysaccharide methanolan by the obligate methylotroph *Methylobacillus* sp strain 12S. Microbiol 149: 431–444.

52. Yoshida T, Ayabe T, Horinouchi M, Habe H, Nojiri H, et al. (2000) Saccharide production from methanol by Tn5-mutants derived from the extracellular polysaccharide-producing bacterium *Methylobacillus* sp. strain 12S. Appl Microbiol Biotechnol 54: 341–347.

53. Smol'Kina ON, Shishonkovaa NS, Yurasovb NA, Ignatov VV (2012) Capsular and Extracellular Polysaccharides of the Diazotrophic Rhizobacterium *Herbaspirillum seropedicae* Z78. Microbiol 81: 317–323.

54. Karatan E, Watnick P (2009) Signals, regulatory networks, and materials that build and break bacterial biofilms. Microbiol Mol Biol Rev 73: 310–347.

55. Janczarek M, Kutkowska J, Piersiak T, Skorupska A (2010) *Rhizobium leguminosarum* bv. *trifolii rosR* is required for interaction with clover, biofilm formation and adaptation to the environment. BMC Microbiol doi:10.1186/1471-2180-10-284.

56. Rinaudi LV, Giordano W (2010) An integrated view of biofilm formation in rhizobia. FEMS Microbiol Lett 304: 1–11.

57. Fujishige NA, Kapadia NN, De Hoff PL, Hirsch AM (2006) Investigations of *Rhizobium* biofilm formation. FEMS Microbiol Ecol 56: 195–206.

58. González JE, Reuhs BL, Walker GC (1996) Low molecular weight EPS II of *Rhizobium meliloti* allows nodule invasion in *Medicago sativa*. Proc Natl Acad Sci USA 93: 8636–8641.

59. Saile E, McGarvey JA, Schell MA, Denny TP (1997) Role of extracellular polysaccharide and endoglucanase in root invasion and colonization of tomato plants by *Ralstonia solanacearum*. Phytopathol 87: 1264–1271.

60. Rinaudi LV, González JE (2009) The low-molecular-weight fraction of exopolysaccharide II from *Sinorhizobium meliloti* is a crucial determinant of biofilm formation. J Bacteriol 191: 7216–7224.

61. Rolfe BG, Carlson RW, Ridge RW, Dazzo RW, Mateos FB, et al. (1996) Defective infection and nodulation of clovers by exopolysaccharide mutants of *Rhizobium leguminosarum* bv. *trifolii*. Aust J Plant Physiol 23: 285–303.

62. Van Workum WAT, Van Slageren S, Van Brussel AAN, Kijne JW (1998) Role of exopolysaccharides of *Rhizobium leguminosarum* bv. *viciae* as host plant specific molecules required for infection thread formation during nodulation of *Vicia sativa*. Mol Plant Microbe Interact 11: 1233–1241.

63. Pérez-Giménez J, Mongiardini EJ, Althabegoiti MJ, Covelli J, Quelas JI, et al. (2009) Soybean lectin enhances biofilm formation by *Bradyrhizobium japonicum* in the absence of plants. Int J Microbiol doi:10.1155/2009/719367.

64. Wang P, Zhong Z, Zhou J, Cai T, Zhu J (2008) Exopolysaccharide biosynthesis is important for *Mesorhizobium tianshanense*: plant host interaction. Arch Microbiol 189: 525–530.

65. Tomlinson AD, Ramey-Hartung B, Day TW, Merritt PM, Fuqua C (2010) *Agrobacterium tumefaciens* ExoR represses succinoglycan biosynthesis and is required for biofilm formation and motility. Microbiol 156: 2670–2681.

66. Roper MC, Greve LC, Labavitch JM, Kirkpatrick BC (2007) Detection and visualization of an exopolysaccharide produced by *Xylella fastidiosa in vitro* and *in planta*. Appl Environ Microbiol 73: 7252–7258.

67. Torres PS, Malamud F, Rigano LA, Russo DM, Marano MR, et al. (2007) Controlled synthesis of the DSF cell-cell signal is required for biofilm formation and virulence in *Xanthomonas campestris*. Environ Microbiol 9: 2101–2109.

68. Koutsoudis MD, Tsaltas D, Minogue TD, Von Bodman SB (2006) Quorum-sensing regulation governs bacterial adhesion, biofilm development, and host colonization in *Pantoea stewartii* subspecies *stewartii*. Proc Natl Acad Sci USA 103: 5983–5988.

69. Koczan JM, Mcgrath MJ, Zhao Y, Sundin GW (2009) Contribution of *Erwinia amylovora* exopolysaccharides amylovoran and levan to biofilm formation: Implications in pathogenicity. Phytopathol 99: 1237–1244.

70. Chapman MR, Kao CC (1998) EpsR modulates production of extracellular polysaccharides in the bacterial wilt pathogen *Ralstonia (Pseudomonas) solanacearum*. J Bacteriol 180: 27–34.

71. Janczarek M, Rachwal K (2013) Mutation in the *pssA* Gene Involved in Exopolysaccharide Synthesis Leads to Several Physiological and Symbiotic Defects in *Rhizobium leguminosarum* bv. *trifolii*. Int J Mol Sci 14: 23711–23735.

72. Killiny N, Hernandez-Martinez R, Dumenyo CK, Cooksey DA, Almeida RPP (2013) The exopolysaccharide of *Xylella fastidiosa* is essential for biofilm formation, plant virulence and vector transmission. Mol Plant Microbe Interact 26: 1044–1053.

73. Zhang Y, Wei C, Jiang W, Wang L, Li C, et al. (2013) The HD-GYP Domain Protein RpfG of *Xanthomonas oryzae* pv.*oryzicola* Regulates Synthesis of Extracellular Polysaccharides that Contribute to Biofilm Formation and Virulence on Rice. Plos One 8: e59428.

5

Coincidental Loss of Bacterial Virulence in Multi-Enemy Microbial Communities

Ji Zhang[1,2]*, Tarmo Ketola[1], Anni-Maria Örmälä-Odegrip[2], Johanna Mappes[1], Jouni Laakso[1,2]

1 Centre of Excellence in Biological Interactions, Department of Biological and Environmental Science, University of Jyväskylä, Jyväskylä, Finland, 2 Department of Biological and Environmental Science, University of Helsinki, Helsinki, Finland

Abstract

The coincidental virulence evolution hypothesis suggests that outside-host selection, such as predation, parasitism and resource competition can indirectly affect the virulence of environmentally-growing bacterial pathogens. While there are some examples of coincidental environmental selection for virulence, it is also possible that the resource acquisition and enemy defence is selecting against it. To test these ideas we conducted an evolutionary experiment by exposing the opportunistic pathogen bacterium *Serratia marcescens* to the particle-feeding ciliate *Tetrahymena thermophila*, the surface-feeding amoeba *Acanthamoeba castellanii*, and the lytic bacteriophage Semad11, in all possible combinations in a simulated pond water environment. After 8 weeks the virulence of the 384 evolved clones were quantified with fruit fly *Drosophila melanogaster* oral infection model, and several other life-history traits were measured. We found that in comparison to ancestor bacteria, evolutionary treatments reduced the virulence in most of the treatments, but this reduction was not clearly related to any changes in other life-history traits. This suggests that virulence traits do not evolve in close relation with these life-history traits, or that different traits might link to virulence in different selective environments, for example via resource allocation trade-offs.

Editor: Boris Alexander Vinatzer, Virginia Tech, United States of America

Funding: This work was supported by the Finnish Academy to JL, 1130724 and 1255572 (URL: www.aka.fi), which had a role in study design, data collection and analysis, decision to publish and preparation of the manuscript. This work was also supported by the CoE in Biological Interactions to JM, 252411 (URL: https://www.jyu.fi/bioenv/en/divisions/coe-interactions), which had a role in study design, data collection and analysis, decision to publish and preparation of the manuscript. This work was also supported by the Finnish Cultural Foundation to JZ (URL: www.skr.fi), which had a role in data collection and analysis, decision to publish and preparation of the manuscript. This work was also supported by the Ellen and Artturi Nyyssö nen Foundation to JZ (URL: www.eans.fi), which had a role in study design, data collection and analysis.

Competing Interests: The authors have declared that no competing interests exist.

* Email: Ji.Zhang@Helsinki.Fi

Introduction

Compared to the vast knowledge on the prevention and treatment of bacterial infectious disease, relatively little is known about how the virulence of bacteria has evolved. Virulence evolution is often exemplified as a tug of war between the multicellular host and the pathogen, where the virulence (the degree of host damage or mortality caused by the pathogen) [1] evolves solely through host-pathogen interaction [2–4]. Contrary to this idea, the "coincidental evolution of virulence hypothesis" suggests that virulence evolves indirectly due to selection forces that are not related to the host-pathogen interaction *per se*, but because of selection that occurs outside host environments [2,3,5,6]. This is a plausible expectation when considering opportunistic, environmentally growing bacterial pathogens because they typically live in a complex web of interactions with biotic and abiotic selection pressures that might not be directly connected to their potential hosts [7].

In the natural environment, top-down regulation by bacteriophages and protozoans are two major biotic causes of bacterial mortality [8,9]. In order to survive, bacteria have evolved wide arrays of defence mechanisms against their natural enemies [10,11]. These adaptations have also been suggested to alter the virulence of the bacteria [11–13]. For example, a biofilm-forming ability can effectively lower predation pressure by ciliate predators that prey in the open water. However, the biofilm-forming ability of many bacteria can also be directly linked to the virulence of bacteria as it can prevent macrophage phagocytosis inside the multicellular host [11,14–16].

In addition to the means that prevent predator ingestion in the first place, bacteria have evolved ways to survive the ingestion process and even benefit from it [11]. Survival and reproduction inside protozoan predators, especially in amoebae, may have even contributed to the evolution of several bacterial pathogens [11]. Therefore, virulence could have evolved via adaptations to survive inside protozoan food vacuoles, which could then promote survival within phagocytes in the immune system [17,18]. Perhaps the most typical example of this type of evolution is *Legionella pneumophila* causing Legionnaires' disease. This species is sometimes found as a parasite of free-living amoeba [19–21]. However, infection of the human body is an evolutionary dead end for *L. pneumophila* because human-to-human transmission is unlikely [22,23]. This suggests that the virulence traits of *L. pneumophila* are not evolved from human-bacteria interaction, but rather "coincidentally" via amoeba-bacteria interaction

[2,24]. In fact this linkage is assumed strong enough that the virulence of bacterial clones are frequently assayed indirectly via amoebae resistance tests [25–31].

Bacteriophages can also have a profound impact on the evolution of bacterial virulence. Bacteriophages are known to carry important virulence genes [32–34]. For example, they have been found to contain genes encoding exotoxins and other virulence factors that can be horizontally transferred into the bacterial genome [35,36]. Moreover, bacteria can alter their cell surface antigens to evade phage adsorption [10], whilst host immune systems rely on bacterial surface antigens to identify bacterial invaders [37,38]. Thus bacteriophage selected bacterial surface antigens could indirectly affect host entry, either positively or negatively.

Although protozoan predators and bacteriophages could potentially contribute to elevated bacterial virulence, outside-host defensive adaptations can also be costly and traded off with virulence related traits [39,40]. For example, when bacteria experience protozoan predation, the motility of bacteria that is sometimes positively linked to virulence [41,42] can trade off with anti-predator traits resulting in lowered virulence [39]. It has also been shown that elevated outside-host temperature can select for higher virulence in *Serratia marcescens*, while coevolution with phage can counteract this effect [43]. Moreover high virulence in *Salmonella typhimurium* can be costly in terms of reduced growth in the outside host environment because of the expression of virulence factor (type III secretion system) in a non-host environment [44]. The nutritional conditions of the bacterial growth environment can also significantly affect bacterial metabolism and the expression of virulence factors [45–47]. For example, it has been found that the virulence of the pathogenic fungi was negatively correlated to the carbon-to-nitrogen (C:N) ratio of the culturing medium [48–50]. Therefore, if a similar correlation occurs for bacteria, then the costly virulence traits might be selected against during a prolonged period in a non-host environment. In conclusion, the environmental lifestyle can attenuate or strengthen the virulence depending on the selection forces in the system [7].

Although predators are supposed to play an important role in the evolution of virulence, experiments testing this theory are rare and the studies that do exist only consider a single predator system [40,51,52]. However, in a natural environment it is more conceivable that several predators are present simultaneously, potentially complicating the picture considerably. To test how virulence and other life-history traits evolve in complex enemy communities, we cultured the facultative pathogen *S. marcescens* either alone or with three types of common bacterial predators (amoeba, ciliate and bacteriophage in all seven possible combinations) in a simulated pond water environment for 8 weeks. *S. marcescens* is a gram-negative opportunistic pathogen infecting a broad spectrum of hosts, including plants, corals, nematodes, insects, fish and mammals [57,58]. They can also be found free-living in soil, freshwater, and marine ecosystems [59,60] making it likely that *S. marcescens* frequently encounters parasitic and predatory organisms. Notably, *S. marcescens* is also capable of re-entering the environment after decomposing the host. This creates the possibility that the pathogen virulence is selected in nature by both environmental and host-pathogen interactions. During the experiment we followed the population dynamics of the prey bacterium. Due to the presumed importance of predators on the evolution of the bacterial virulence [28,53,54], the amoeba densities were also followed throughout the experiment. After the evolution experiment, a library containing 384 differentially evolved clones was built to detect changes in virulence, growth

ability, biofilm-forming ability and amoeba resistance. The virulence of the ancestor and the evolved bacteria, *S. marcescens* Db11 was quantified in the fruit fly (*Drosophila melanogaster*) hosts via an oral infection model [55]. Since phagocytes play a vital role in the clearance of the Db11 from the hemolymph in this animal model [55], we believed that choosing the bacterial strain and infection model was relevant to our study. We hypothesized that if *S. marcescens* Db11 gained amoeba-resistance in the presence of amoeba predation, this resistance could be used to fight against phagocytes in the hemolymph, and thus gain higher virulence. With data from a multi-predator experiment we can test if the bacterial virulence is selected for, a result that is expected in the presence of bacterial enemies (phage, ciliate and amoebae), especially amoebae. However, it is also plausible that selection pressures by bacterial enemies could select against virulence [39,40].

Methods

Study species

Serratia marcescens Db11 [56,57] was initially isolated from a dead fruit fly and was kindly provided by Prof. Hinrich Schulenburg. The predatory particle feeding ciliate, *Tetrahymena thermophila* (strain ATCC 30008) has a short generation time of ca. 2 h [61] and was obtained from American Type Culture Collection. It is routinely maintained in PPY (Proteose Peptone Yeast Medium) at 25°C [51,62]. The free-living amoeba, *Acanthamoeba castellanii* (strain CCAP 1501/10) has a generation time ca. 7 h [62] and was obtained from Culture Collection of Algae and Protozoa (Freshwater Biological Association, The Ferry House, Ambleside, United Kingdom) and routinely maintained in PPG (Proteose Peptone Glucose Medium) [63] at 25°C. Obligatory lytic bacteriophage Semad11, capable of infecting *S. marcescens* Db11, was isolated from a sewage treatment plant in Jyväskylä, Finland in 2009. No specific permission was required for collection or location of the bacteriophage. Semad11 is a T7-like bacteriophage belonging to Podoviridae (A.-M. Örmälä-Odegrip, unpublished data).

The evolution experiment was performed in New Cereal Leaf - Page's Amoeba Saline Solution (NAS) medium which was prepared as follows: 1 g of cereal grass powder (Aldon Corp., Avon, NY) was boiled in 1 liter of dH_2O for 5 minutes, and then filtered through a glass fiber filter (GF/C, Whatman). After cooling, 5 ml of both PAS stock solutions I and II were added before being made up to a final volume of 1 litre with deionized water [64,65].

Before the experiment started, the organisms were cultured separately and prepared as follows: bacterial culture, a single colony of *S. marcescens* was seeded to 80 ml of NAS medium in a polycarbonate Erlenmeyer flask capped with a membrane filter (Corning). The flask was incubated at 25°C on a rotating shaker (120 rpm) for 48 hours. The amoeba and ciliate cells were harvested and washed twice in 40 ml of PAS (Page's Amoeba Saline) with centrifugation at $1200 \times$ g for 15 min to pellet the cells. After the centrifugation, cells were suspended in PAS and adjusted to a final concentration of ca. 10 cells μl^{-1}. To prepare the bacteriophage stock, LB-Soft agar (0.7%) from semi-confluent plates was collected and mixed with LB (4 ml per plate), and incubated for 3.5 h at 37°C. Debris was removed by centrifugation for 20 min at $9682 \times$ g at 5°C. Stock was filtered with 0.2 µm Acrodisc Syringe Filters (Pall). The bacteriophage stock was diluted 1:100,000 in NAS medium, giving approximately 10^6 plaque-forming unit (PFU) ml^{-1}.

Evolution experiment

The bacterium *S. marcescens* was either cultured alone or in a co-culture with the ciliates, amoebae and bacteriophages enemies 8 combinations (B, BA, BC, BP, BAC, BAP, BCP and BACP; B: bacteria; A: amoebae; C: ciliate; P: phage; Figure 1) for 8 weeks. Each treatment was replicated in 8 flasks. The experiment was initiated in 25 cm^2 polystyrene flasks with 0.2 μm hydrophobic filter membrane caps (Sarstedt). Each flask was inoculated with 1 ml of the appropriate microorganism suspension and then the total volume was adjusted to 15 ml with NAS medium. The static liquid cultures were incubated at 25°C and 50% of the medium were replaced weekly with fresh NAS medium, making the system a pulsed resource type [66,67]. Static liquid culture would create a spatial structuration that was similar to the pond water environment. All samples were taken just before the weekly medium renewal (Figure 1).

Measurements during the evolution experiment

Bacterial biomass dynamics. Bacterial biomass in the free water phase was measured from 5 separate 200 μl samples from each flask on 100-well Honeycomb plates (Oy Growth Curves Ab Ltd). The amount of biomass was measured as optical density (OD) at 460–580 nm wavelength using Bioscreen C spectrophotometer (Oy Growth Curves Ab Ltd). The measurements were repeated 10 times at 5 min intervals. The mean of the measurements was used in the data analysis. To measure the amount of *S. marcescens* biofilm attached to the flask walls after 8 weeks had expired, 15 ml of 1% crystal violet solution (Sigma-Aldrich) was injected to the flasks. After 10 minutes, the flasks were rinsed 3 times with distilled water, and then 15 ml of 96% ethanol was added to flasks to dissolve crystal violet from the walls for 24 hours [68]. The amount of biofilm was quantified with the OD of the crystal violet-ethanol solution at 460–580 nm with Bioscreen C spectrophotometer [66].

Amoeba population dynamics. To follow the population dynamics of the amoeba, we measured the density of amoeba cells attached on the flask well. This measurement largely reflects the amoeba population dynamics in the flasks since the proportion of floating cells and cysts would be minimal after 7 days culture in the static cultures. In brief the flasks were carefully flipped and images (total area 5.23 mm^2) of the flask wall were digitized with an Olympus SZX microscope (32× magnification). The amoeba cells attached to the flask wall were counted with a script developed in our lab for the Image Pro Plus software (v. 7.0) (Material S1). To determine the ciliate density by the end of the experiment, 250 μl of open water sample was mixed with 10 μl Lugol solution and injected into a glass cuvette rack (depth 2.34 mm). For each sample, 8 randomly placed images (total area 41.84 mm^2) were digitized with an Olympus SZX microscope (32× magnification). The cell numbers in each image were counted with an Image Pro Plus script [69].

Detecting phage presence. To detect if the bacteriophages were present in the microcosms throughout the experiment and to detect possible contamination, we took 3 independent 500 μl samples from all flasks at the end of evolution experiment. The samples were treated with chloroform and centrifuged to remove bacteria, amoebas and ciliates. 10 μl supernatant drops were then added to 1.5% agar plates. The upper layer of the each plate was covered with 0.7% LB-agar that was mixed with 200 μl of overnight grown *S. marcescens* Db11 ancestor cells. The plates were incubated overnight in 25°C and the presence of phage plaques were checked.

Measurements after the evolution experiment

Amoeba plaque test. After the evolution experiment was finished, half of the flasks from each treatment were randomly sampled to test for any resistance of the bacteria to amoeba predation. The test was adapted from a Wildschutte *et al.* [70] briefly, the flasks from the evolution experiments were shaken vigorously before 1 ml of the culture was transferred to a new tube containing 7 ml of dH$_2$O. The tubes were mixed thoroughly and then centrifuged at 250 g for 10 min to bring down the suspended protozoan cells. 1 ml of the supernatant was spread evenly on to LN agar plates (PAS with 0.2% peptone, 0.2% glucose and 1% agar). A total of 10^5 predatory amoeba cells (washed twice in PAS) suspended in 15 μl PAS solution were added to a sterile paper disk, and then placed in the middle of the plate. All the plates were incubated at 25°C for 8 days and then photographed. The images of the plates were used to measure plaque sizes with Image Pro Plus software (v. 7.0). A large plaque size indicates a small amoeba predation resistance.

Growth and biofilm forming ability of the individual clones. After the evolution experiment was complete, liquid samples from each replicate population of all treatments were

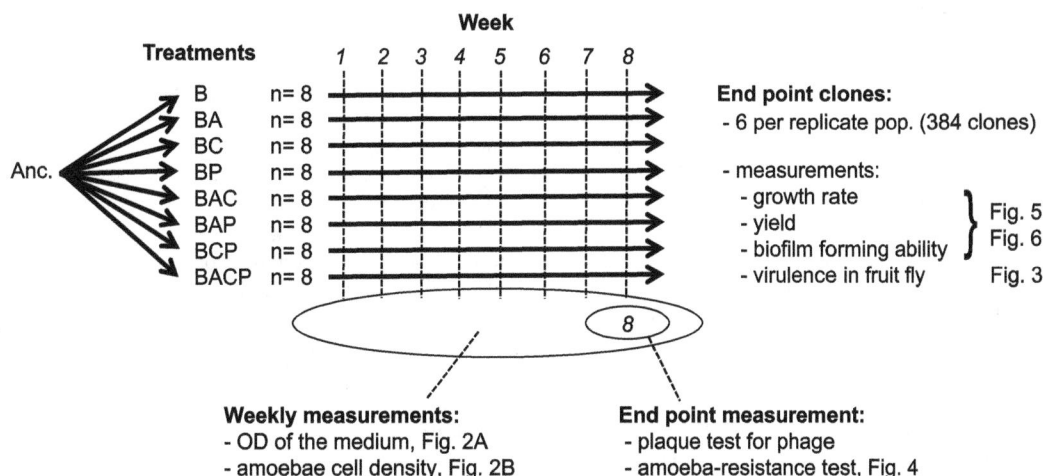

Figure 1. Schematic overview of our experimental evolution study, number of replicate populations, legends for treatments (Anc.: ancestral bacterial strain DB 11; B: bacteria; A: amoebae; C: ciliate; P: phage), and descriptions of different measurements.

streaked on three Luria–Bertani (LB) agar plates. They were incubated for 48 hours at 25°C, two bacterial colonies were randomly picked from each plate and inoculated to 5 ml LB liquid medium. The clones were grown at 25°C overnight on a shaker (120 rpm). To make stock cultures of the clone library, 200 μl of the liquid culture of each strain was mixed with 200 μl of 80% glycerol on 100-well Honeycomb plates in a randomized order and stored at −80°C. Prior to clonal growth measurements stock cultures from the clone library were inoculated to 100-well Honeycomb plates, directly from the freezer, with a plate replicator (EnzyScreen). Each well of the 100-well Honeycomb plates contained 400 μl fresh LB liquid medium. The OD of each well was measured continuously without shaking for 30 hours in 5 min intervals to estimate the maximum growth rate and maximum population size of the clones. After 30 hours, 100 μl of 1% crystal violet solution (Sigma-Aldrich) was added to each well to quantify amount of biofilm that was produced. After 10 minutes incubation in 25°C, the plates were rinsed with distilled water 3 times and then 400 μl of 96% ethanol was added to each well and left for 24 hours to dissolve the crystal violet from the walls [68]. The amount of biofilm was quantified by measuring the OD of the crystal violet-ethanol solution at 460–580 nm with Bioscreen C spectrophotometer [66].

Identical measurements were also recorded with NAS medium either with or without the protozoan predators (amoeba or ciliate). The abovementioned clones were first grown in 400 μl LB liquid medium in 100-well Honeycomb plates. After incubation at 25°C for 24 h in static cultures, 10 μl of the bacterial culture was transferred to 100-well Honeycomb plates. Each well contained 390 μl NAS mixed with or without NAS washed amoeba (5–10 cells/μl) or ciliate cells (0.5–1 cells/μl). Subsequent measurements for growth and biofilm assay were performed as described above.

Estimation of growth rate and yield was based on the Matlab script written by TK that fits linear regression to 25 time-points along a sliding data window with background correction, using ln-transformed OD data. The maximal growth rate is determined by finding the largest slope of linear regression within all fitted regressions for the particular clone. The yield was determined as the highest average OD over the 25 data point window.

Virulence of the evolved clones. Stock cultures of the clone library were inoculated to 100-well Honeycomb plates, filled with 400 μl fresh LB liquid medium with plate replicator (EnzyScreen). For a positive control the ancestor *S. marcescens* Db11 was added to two separate wells in each plate and used in the subsequent infection experiment. After 24 h incubation at 25°C without shaking, 800 μl of the bacterial culture was mixed with the 800 μl of 100 mM sucrose solution. The mixture was absorbed to cotton dental roll (Top Dent, Lifco Dental, Enköping, Sweden) folded on the bottom of a standard 75×23 mm fly vial (Sarstedt, Nümbrecht, Germany). 1600 μl of 100 mM sucrose solution was used as a negative control. Ten *D. melanogaster* adults (2–3 days old) from a large laboratory colony (Oregon R, kindly provided by Christina Nokkala from the University of Turku) were transferred to each vial and plugged with cotton. This was done for all the bacterial clones. Deaths of flies were monitored over next 4 days at 3–6 h intervals.

Statistical analysis

Changes in bacterial density and amoeba density were compared using repeated measurements ANOVA. The effects of the evolutionary treatments on bacterial virulence were quantified with Cox regression by fitting evolutionary treatment and identity of population as categorical covariates. The amoeba plaque test and the amount of biofilm at the end of the evolution experiment

were compared using ANOVA. All the analyses were done with SPSS v. 19 (IBM).

Life history and defensive traits of evolved clones were tested with ANOVA including treatment (all possible combinations of predators) as a fixed factor and population identity as a random factor. From the data we tested effects of treatments on growth rate, yield and biofilm, as well as growth rate and yield under the influence of ciliate or amoebae presence. Coevolution of traits was studied with MANOVA and subsequent eigenanalysis to reveal if changes in certain traits would lead to corresponding changes in other traits across the treatments. Thus, this analysis allows pinpointing strongly interconnected traits [71]. MANOVA and eigenanalysis was performed with MATLAB function manova1 (R2012a, Mathworks; Statistics toolbox) for population averaged trait values. Values used for MANOVA for virulence were hazard function coefficients averaged over the populations within each treatment.

Two replicated populations (one in the treatment BC and one in BAC) were found contaminated by Semad11 phages. Moreover, in two replicates of the treatment BACP phages were not detected by plaque assay. All the other samples from phage containing treatments formed phage plaques on the ancestor Db11 bacterial lawn. This confirms that phages did not go extinct during the experiment. The aforementioned 4 flasks were excluded from the data analysis.

Results

The population dynamics of bacterial prey and amoeba predators during the experiment

The presence of predators generally reduced the bacterial biomass in free water phase (OD of the medium: $F_{7, 52} = 674.620$, p<0.001; Figure 2A). The ciliates reduced the biomass most dramatically: on average by 27% during the weeks 1–8 when compared to the control (B). Biomass reduction by ciliate and phage (BCP), and amoeba and ciliate (BAC) communities was 25%. Amoeba (BA) and amoeba and phage (BAP) communities reduced bacteria biomasses by 20%. The bacteriophage (BP) reduced the biomass only by 2%. The pairwise-comparisons of the rest of the treatments were significant after Bonferroni correction, except BCP vs. BC, and BACP vs. BAC.

The amount of biofilm produced in each treatment was different at week 8 when measured directly from the microcosm walls (ANOVA, $F_{7, 52} = 39.101$, p<0.001). The highest amount of biofilm was found in the presence of ciliates (BC) and the lowest amount of biofilm was found in the treatments BAC and BACP. Detailed pairwise comparison can be found in Table S1.

The amoeba population sizes declined in all treatments after the initial increase during the first week ($F_{3, 28} = 280.257$, p<0.001; Figure 2B). The amoeba population sizes were higher in amoeba (BA) and amoeba and phage (BAP) treatments (on average 30 cells μl^{-1}) throughout the 8-week evolution experiment. Adding phage to the amoeba treatment did not change the population dynamics of the amoeba (Fisher's LSD: BA vs. BAP, p = 0.485). However, ciliates reduced the amoeba population sizes: on average only 5 cells μl^{-1} were found throughout the experiment in treatment BAC, and on average 8 cells/μl in treatment BACP.

Virulence

In order to explore if past selection with predators had influenced virulence we utilized the *Drosophila* oral infection assay. The treatment group that had evolved with ciliates and phages (BCP) had clearly lower virulence than the rest of the evolved treatment groups (p<0.01 in all pairwise comparisons;

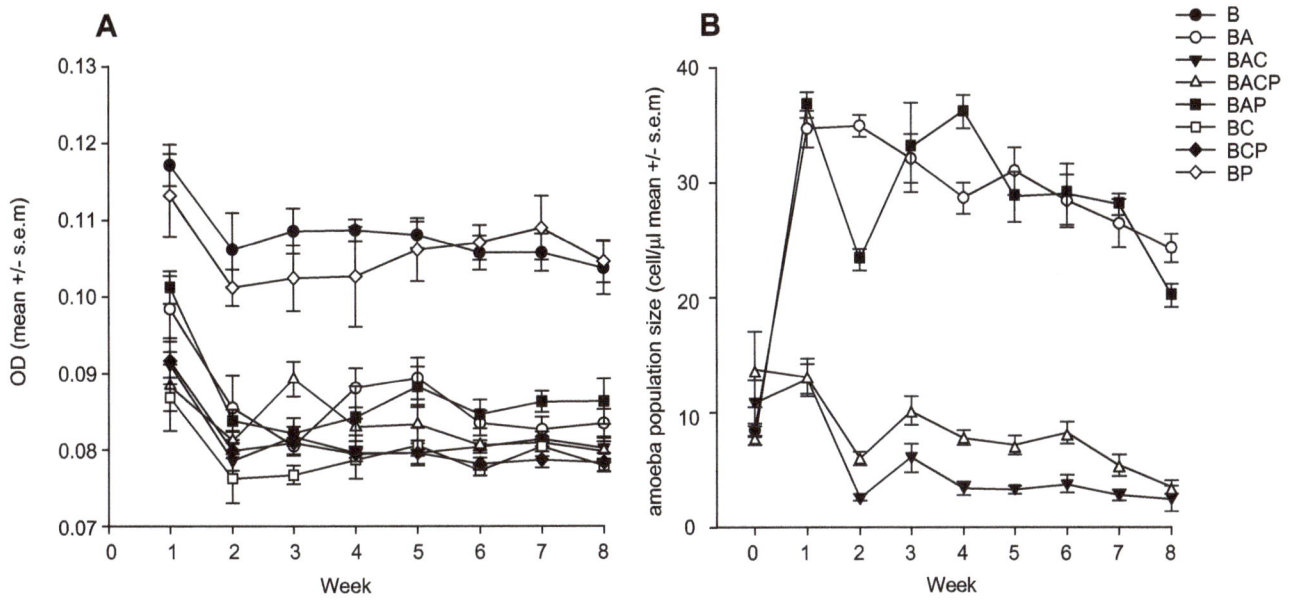

Figure 2. Bacterial biomass dynamics (A) and amoebae population dynamics (B) during the eight-week evolution experiment. The bacteria were reared alone or in several combinations of bacterial enemies (Anc.: ancestral bacterial strain DB 11; B: bacteria; A: amoebae; C: ciliate; P: phage). See Table S1 for pairwise comparisons.

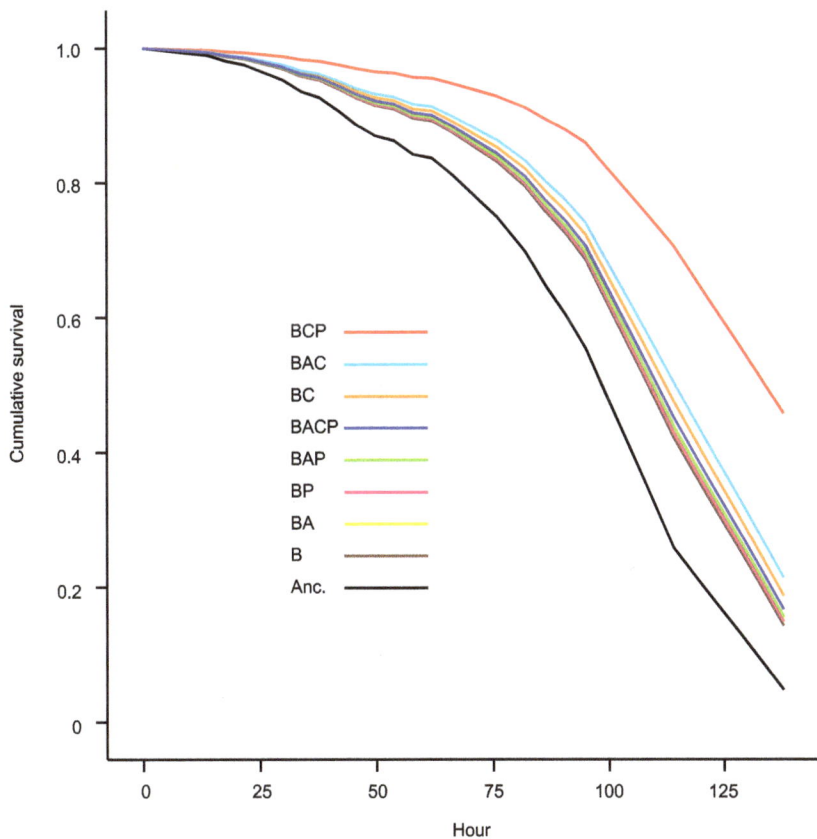

Figure 3. Cumulative survival curves of the fruit flies that were infected with evolved and ancestral bacterial clones. Anc.: ancestral bacterial strain DB 11; B: bacteria; A: amoebae; C: ciliate; P: phage. The survival curves represented the pooled survival data of the 480 fly individuals for each treatment (10 flies per vial, 6 clones per population and 8 replicates per treatment). The treatment codes are in the order of the increasing virulence. See Table S2 for pairwise comparisons.

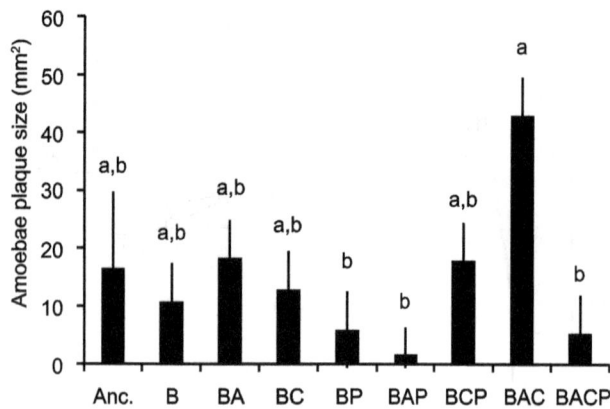

Figure 4. Sensitivity of evolved bacteria on amoebae predation measured using amoeba plaque test. Anc.: ancestral bacterial strain DB 11; B: bacteria; A: amoebae; C: ciliate; P: phage. Sensitivity is measured as a plaque size (mm²) in bacterial lawn caused by the introduced amoeba in semi-solid agar plate. Letters indicate if treatment means are statistically similar (p>0.05), after Bonferroni correction for multiple comparisons. Tests are based on the post hoc comparisons of estimated marginal means for treatments ANOVA. All bars correspond to 4 randomly picked samples from 8 replicate populations.

Figure 3; Table S2). In this group the between population variation was also very high and statistically significant. In the majority of the other groups, the between population variation was clearly non-significant (p<0.135, but in BCP, p<0.001; BAC p = 0.003). Moreover, all treatments had lower virulence than the ancestor (p<0.01 in all pairwise comparisons; Figure 3; Table S2). There was no statistical support for the difference in virulence between any other treatment groups.

Sensitivity to amoebae predation

The amoeba plaque test revealed that the sensitivity to amoeba predation (measured with the area of visible plaque formed on bacterial lawn) was highest in bacteria co-cultured with amoeba and ciliates (BAC: p<0.05 in all pairwise comparisons; Figure 4;

Table S1). The lowest amoeba sensitivity was found in the treatment where the amoeba were reared with phages (BAP; Figure 4) or with ciliates and phages (BACP; BAP vs. BACP: p = 0.66, Figure 4). The detailed result of pairwise comparisons can be found in the Table S1. Although the treatment group that had evolved with ciliates and phages (BCP) had the lower virulence than the other evolved treatment groups, its sensitivity to amoeba predation did not differ from the others in pairwise comparisons (Table S1), which suggests no clear link between amoebae predation and virulence.

Life-history traits

Treatments did not influence the maximal growth rate strongly (Table 1), however low growth rates were found in clones that had evolved with ciliates (BC; Figure 5A; Table S2). The ancestor clones had the largest yield. Evolutionary treatments did not differ greatly from each other but the clones that had evolved with phages had the lowest yield (BP; Figure 5B; Table S2). Evolutionary changes occurred most dramatically in the biofilm forming ability. The highest biofilm forming abilities were found from the clones that had evolved alone (B) or with amoeba and ciliate (BAC). Intermediate biofilm production was found in ancestral clones (Anc.) or if clones had evolved with amoebae (BA) or with all enemies (BACP). The lowest biofilm production was found if clones that had evolved with ciliate (BC), phage (BP), amoebae and phage (BAP) or with ciliate and phage (BCP) (Figure 5C; Table S1).

From the defensive traits the strongest changes were observed in growth rate and yield when the bacteria were co-cultured with amoebae. In both of the traits ancestor bacteria deviated from the clones that had undergone evolutionary treatments; ancestors had higher growth rate co-cultured with amoebae but lower yield than evolved clones (Figure 6; Table S2). Similarly, from the evolved groups the highest growth rate was found from the group that had evolved with phage and amoebae (BAP), whereas its yield with amoebae was lowest (Figure 6; Table S2). Growth rate measurements did not indicate that treatments affected the resistance of clones against ciliate predators. However, yield with ciliates was lowest if bacteria had evolved alone (B) or with amoebae and

Table 1. Estimated (ANOVA) evolutionary effects of different combinations of enemies (treatment), and population identity on bacterial virulence against *Drosophila melanogaster*, and on bacterial life-history traits, measured alone or with amoebae or ciliate.

	Treatment			Population		
	Wald	df	p	Wald	df	p
Virulence	48.6	8	<0.001	169.58	52	<0.001
	F	df1,2	p	F	df	p
Growth rate	2.465	8,32.007	0.033	0.799	52,306	0.836
Yield	2.582	8,39.627	0.023	1.349	52,306	0.066
Biofilm	12.987	8,37.001	<0.001	1.097	52,306	0.311
	F	df1,2	p	F	df	p
Growth with amoebae	9.565	8,40.485	<0.001	1.456	52,306	0.029
Yield with amoebae	3.595	8,46.032	0.003	2.885	52,306	<0.001
Growth with ciliate	1.296	8,39.223	0.274	1.304	52,306	0.091
Yield with ciliate	4.757	8,40.432	<0.001	1.449	52,306	0.031

Wald denotes Wald's test statistics, and F corresponds to F-test statistics, df denote degrees of freedom and p indicate statistical significance.

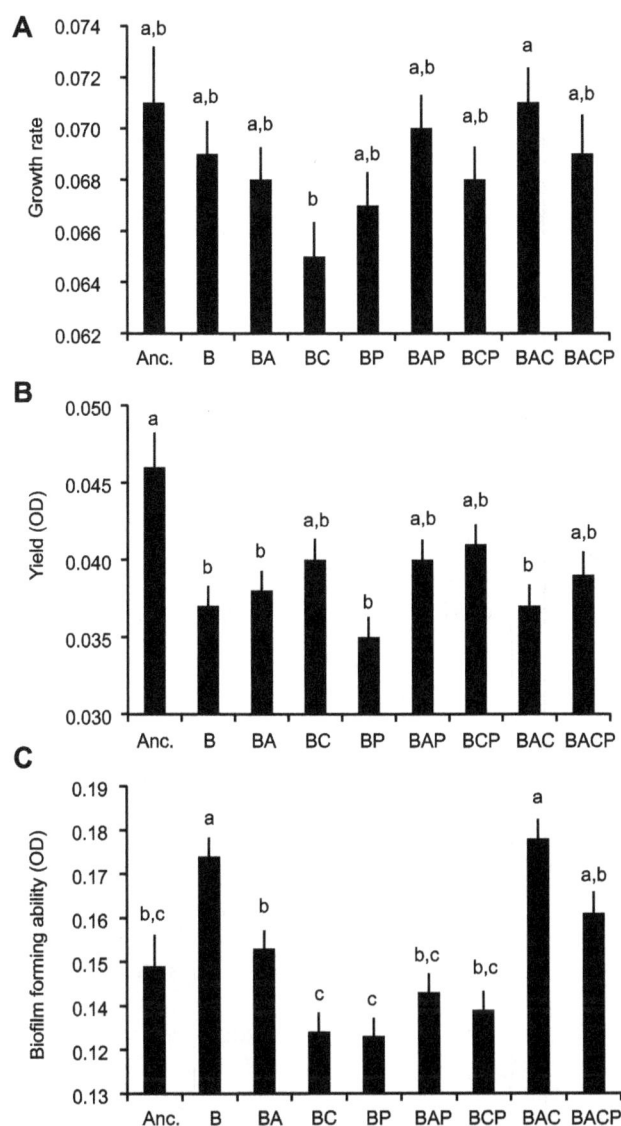

Figure 5. Growth rate (panel A), yield (panel B) and biofilm forming ability (panel C) differences between ancestral clones and clones that have evolved alone (B) or in different combinations of bacterial enemies. Anc.: ancestral bacterial strain DB 11; B: bacteria; A: amoebae; C: ciliate; P: phage. Letters indicate if treatment means are considered statistically similar (p>0.05) after Bonferroni correction for multiple comparisons. Tests are based on the post hoc comparisons of estimated marginal means for treatments of ANOVA testing the effects of treatment and population identity on these traits. Bars correspond to measurements of 6 clones from 8 replicate populations, in ancestor n = 16).

ciliate (BAC), and highest if clones had evolved with ciliate (BC) and ciliate and phage (BCP) (Figure 6C; Table 1; Table S2).

Coevolved traits

Based on the individual traits in different treatment combinations, it is difficult to get an idea of how co-ordinate traits evolved. Therefore we analysed all traits in multivariate ANOVA, followed by eigen-analysis. In this analysis we found that two dimensions dictated multivariate evolution amongst evolutionary treatments (support for two dimensions p = 0.028, for 1 dimension p<0.001). The first dimension was characterized by variation in biofilm and

ciliate defence. Those treatments that exerted strong positive selection on biofilm production had a lower yield with ciliate in free water. The second dimension of trait evolution was formed by those treatments that had increased growth rate without predators and also had a lower growth rate with ciliate. Neither of the two "major" eigen-functions linked virulence to other traits (Table 2).

When similar analysis was performed with information from the amoeba plaque test, a similar result was found, again supporting what was found for two multivariate dimensions (support for two dimensions p = 0.023, for 1 dimension p<0.001). The amoeba plaque test contributed moderately to a second eigenvector. This eigenvector had a slightly different composition than the analysis without the amoeba plaque test as the ltter contained all microcosm replicates. If anything, a higher resistance against amoebae was associated with higher biofilm forming ability and higher growth rate with ciliate. However, the amoeba plaque test clearly did not predict virulence. Since the amoeba plaque test was performed for a subset of the populations, their inclusion in more detailed measurements resulted in a smaller dataset. We base our discussion on the analysis of the larger and thus more reliable eigen-analysis without the amoeba plaque test (Table 2).

Discussion

Predators such as ciliates and amebae, and parasitic phages are expected to be the main determinants of bacterial mortality in the natural environment [8,9]. In addition to selection exerted by these predators and parasites on defensive traits, selections by bacterial natural enemies have often been suggested to lead to increased bacterial virulence [2,5,72] to such a degree that amoebae-resistance has been used as a direct proxy of strains' virulence [25,27–31,73]. However, we did not find evidence to suggest such a relationship as most of the experimental treatments had attenuated virulence. Moreover, there was no indication of coevolution of other life-history traits with virulence in evolved strains.

Contradictory to the theory that protozoan predators and phage parasites, amoebae in particular, play a strong role in the evolution of high virulence [11–13,53,72,74], we found that virulence attenuated in all of the evolved populations regardless of the presence of the enemies. This is in accordance with the previous study with *S. marcescens* that suggest that ciliates can select for attenuated virulence [39,40]. However, here we show that this is also the case with amoeba and phages [43], and under selection by multiple enemies at the same time. By far the strongest decrease in virulence was found in clones that had evolved with phages and ciliates. However, the between population variation in this group was very high and statistically significant. This was shown in separate analyses designed to test the amount of between population variance within treatments. In most of the other groups the between population variation was clearly non-significant (p<0.135, but in BCP, p<0.001; BAC p = 0.003). The between population variation is often seen as a signature of drift, mutation accumulation and lack of directional selection on traits [40,75–78]. Thus, we propose that the strong decline of virulence in ciliate-phage treatment was primarily caused by the decay of unused traits through random mutation accumulation [40,76].

To find out if evolutionary changes in virulence could be linked to changes in traits that are important for fitness outside the host, we measured growth parameters and resistance against amoeba and ciliates. However, none of the traits seemed to be determining the level of virulence amongst evolved strains. Thus, it is clear that resistance to protozoan predators or changes in other measured

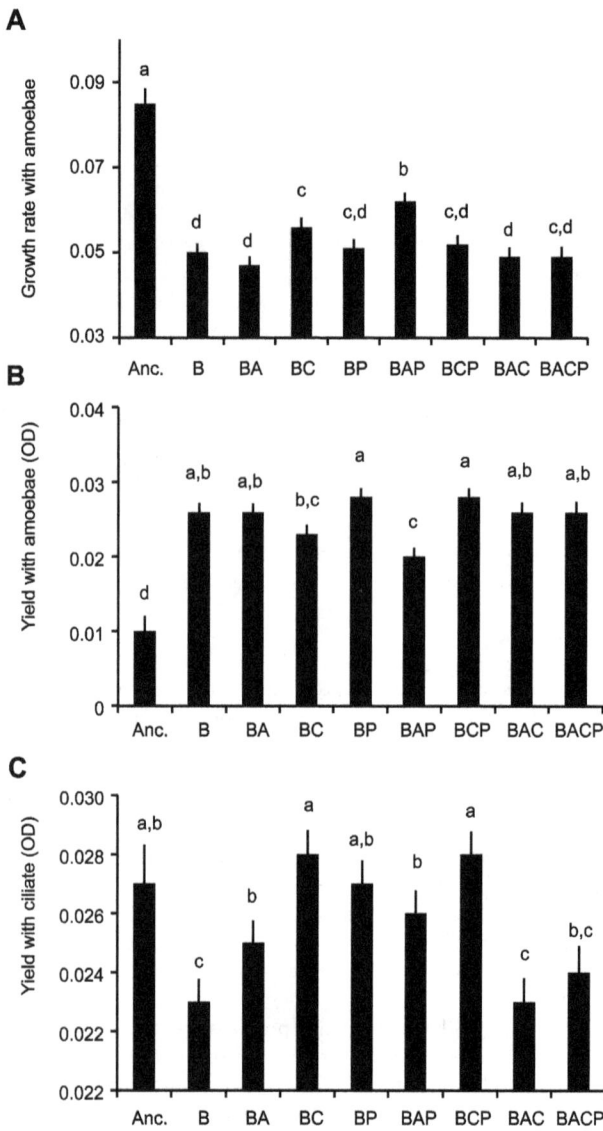

Figure 6. Growth rate with amoebae (panel A), yield with amoebae (panel B) and yield with ciliate (panel C) differences between ancestral clones and clones that have evolved alone or in different combinations of bacterial enemies. Anc.: ancestral bacterial strain DB 11; B: bacteria; A: amoebae; C: ciliate; P: phage. Letters indicate if treatment means are considered statistically similar after Bonferroni correction for multiple comparisons. Tests are based on the post hoc comparisons of estimated marginal means for treatments of ANOVA testing the effects of treatment and population identity on these traits. Bars correspond to measurements of 6 clones from 8 replicate populations, in ancestor n = 16).

Table 2. Eigenvectors describing dimensions of multivariate evolution under different kind of enemies (loadings, i.e. correlations, of original variables to new composite variable, based on MANOVA.

	Eig. 1.	Eig. 2.	Eig. 3.	Eig. 4.	Eig. 5.	Eig. 6.	Eig. 7.	Eig. 8.	Eig. 9.
Biofilm	**0.753**	0.326	−0.205	0.215	0.229	0.376	−0.343	−0.024	0.183
Yield	0.131	0.257	−0.268	0.671	−0.023	−0.155	0.150	−0.199	−0.149
Yield with amoeba	0.081	−0.223	−0.276	−0.200	0.330	0.094	0.452	−0.459	0.678
Yield with ciliate	**−0.579**	−0.253	−0.032	0.472	−0.203	0.111	−0.665	0.115	−0.026
Growth rate	0.018	**−0.518**	0.498	0.241	−0.664	−0.375	−0.106	0.245	0.022
Growth with amoeba	−0.150	0.228	0.489	−0.143	0.321	0.501	0.335	−0.357	0.511
Growth with ciliate	0.222	**0.610**	−0.483	−0.356	0.075	−0.209	0.184	0.319	0.460
Virulence	0.037	−0.149	0.080	0.171	0.176	0.236	0.202	0.666	0.063
Eigenvalues	4.692	0.9220	0.5689	0.1967	0.1039	0.0394	0.0173	<0.001	<0.001
% explained	71.74	14.10	8.70	3.01	1.59	0.60	0.27	<0.001	<0.001

Below the eigenvectors are the eigenvalues i.e. amount of variation explained by eigenvectors and the percentage of the total variation explained. First eigen function (Eig1) describes contribution of individual traits to the largest difference between the treatments. In eigenvectors that are considered significant (Eig 1 and 2, see results) the largest contributors to the evolutionary differences are highlighted with bold.

life-history traits are poor indicators for virulence in *S. marcescens*. However, in other bacterial species, virulence correlated positively with bacterial defences against predators [5,12,17,26,72] (but see [39,40,77]). In addition, several lines of research have emphasized the role of growth rate with virulence [79,80] (but see [44,77]). However, it could be that virulence might be traded off with whatever trait that is under selection in the given environment. This could have led to the clear lack of a connection between virulence and life history traits in an experiment where selection pressures are different between each treatment. This is a plausible outcome if virulence is traded off with life-history traits via finite resources. Then, any energetically costly trait under strong selection in outside-host environments could lead to virulence attenuation. Although we did not find a strong connection with virulence traits and life-history traits, we found that regardless of the evolutionary treatments, high biofilm-forming ability was closely linked with a low yield in a condition of co-culturing with ciliates in free water, and high growth rate was linked with low growth rate with ciliates. The first eigenvector (biofilm vs. yield with ciliates) could indicate that growth in biofilms does not lead to good protection against predation in free water, or that exposure to predation leads to biofilm formation only when predators are present and thus reduces cells in a free water environment. However, these findings from the eigen analysis effectively mean that there are evolutionary constraints between different life-history traits that remain unchangeable regardless of the evolutionary treatments. Moreover, since the predators (amoeba and ciliates) were effectively reducing the population densities in the long run (Figure 2A) the lack of selection on life history traits is not a plausible explanation for the obtained results.

When we compared ancestor's life history traits to evolved strains, it seemed that ancestors grew better with amoebae and were also the most virulent clone (Figure 3). However, another amoebae-resistance measurement, yield with amoebae, was actually lower for the ancestor strain than evolved strains. Moreover, in the amoebae-plaque test the ancestor strain did not excel in comparison to evolved strains. These contrasting results from measurements that should indicate the ability to resist amoebae suggest that there is a weak indication that amoebae-resistance evolved simultaneously with virulence. Therefore we suggest that the culture conditions could attenuate virulence, without any clear changes in life history traits. Interestingly, several experimental evolution studies have previously found that bacterial virulence decreases due to the exposure to the outside-host environment [39,40,81].

Alternatively, it has been suggested that if traits are not needed under particular conditions then their alleles become harmful and accumulate which leads to unused trait decay [76]. S. marcescens strain Db11 was isolated from a dead Drosophila fly over thirty years ago and has been routinely grown in highly protein-rich LB medium. Yet, the virulence of the strain has been maintained from lab to lab [55,57,82]. It is possible that the protein-enriched culture condition like LB medium (LB containing 10 g/l tryptone and 5 g/l yeast extract) was somehow needed for S. marcescens Db11 to maintain its virulence. However, a more likely scenario could be that our experimental conditions (low concentration of high C:N ratio plant detritus) selected against virulence in the 8 week evolution experiment. In addition, there might be other unknown factors that could of affected the virulence. Although predators in general effectively lowered population sizes of the bacteria (Figure 2A), spatial heterogeneity and potentially other niches created by the static cultures might lower the strength of

selection on defensive traits. For example, the biofilm could act as a protection against protozoan predation, and some bacteria might have not been under selection at all.

To summarize, we found no support for the idea that enemies outside the host could select for higher virulence, as all experimental treatments led towards lower bacterial virulence. Among evolved strains virulence was not linked to other life-history characters, suggesting that selective pressures from protozoan predators (ciliates and amoebae) and parasitic phages did not dictate virulence evolution. In conclusion, our dataset offered a case against coincidental evolution of the virulence hypothesis that expects outside-host selections, especially amoebae predation, would lead to higher bacterial virulence [2,3,5,72].

Supporting Information

Table S1   Pairwise comparisons of experimental treatment differences on amoeba population sizes, biomass in the free water phase and attached biofilm at the end of the experiment. Significant pairwise comparisons after Bonferroni correction are high-lighted with bold (critical α: $0.00178 = 0.05 \div 28$, but amoebae population size critical α: 0.008 (0.05/6)). (B = bacteria alone, BC = with ciliate; BA with amoebae; BP with phage etc.. Anc. stands for ancestor Db11 strain).

Table S2   Pairwise comparisons of experimental treatment differences on measured virulence, growth and defensive traits. Significant pairwise comparisons after Bonferroni correction are highlighted with bold (critical α: 0.00138 (0.05/36)) (B = bacteria alone, BC = with ciliate; BA with amoebae; BP with phage etc.. Anc. stands for ancestor Db11 strain).

Material S1   The macro for the Image-Pro Plus program to automatically count protozoan cells in microscopic images.

Acknowledgments

We thank Kalevi Viipale for intensive discussions on the conceptual issues. We also thank Angus Buckling, Anna-Liisa Laine, Waldron Samuel and Lauri Mikonranta for commenting on the manuscript.

Author Contributions

Conceived and designed the experiments: JZ TK AMÖ JM JL. Performed the experiments: JZ AMÖ. Analyzed the data: JZ TK AMÖ JM JL. Contributed reagents/materials/analysis tools: JZ TK AMÖ JM JL. Wrote the paper: JZ TK AMÖ JM JL.

References

1. Casadevall A, Pirofski LA (2003) The damage-response framework of microbial pathogenesis. Nat Rev Microbiol 1: 17–24.
2. Levin BR (1996) The evolution and maintenance of virulence in microparasites. Emerg Infect Dis 2: 93–102.
3. Levin BR, Svanborg Eden C (1990) Selection and evolution of virulence in bacteria: an ecumenical excursion and modest suggestion. Parasitology 100 Suppl: S103–115.
4. May RM, Anderson RM (1983) Epidemiology and Genetics in the Coevolution of Parasites and Hosts. P Roy Soc Lond B Bio 219: 281–313.
5. Adiba S, Nizak C, van Baalen M, Denamur E, Depaulis F (2010) From grazing resistance to pathogenesis: the coincidental evolution of virulence factors. PloS one 5: e11882.
6. Coombes BK, Gilmour MW, Goodman CD (2011) The evolution of virulence in non-o157 shiga toxin-producing Escherichia coli. Front Microbiol 2: 90.

7. Brown SP, Cornforth DM, Mideo N (2012) Evolution of virulence in opportunistic pathogens: generalism, plasticity, and control. Trends Microbiol 20: 336–342.
8. Jürgens K, Matz C (2002) Predation as a shaping force for the phenotypic and genotypic composition of planktonic bacteria. Antonie Van Leeuwenhoek 81: 413–434.
9. Suttle CA (2005) Viruses in the sea. Nature 437: 356–361.
10. Labrie SJ, Samson JE, Moineau S (2010) Bacteriophage resistance mechanisms. Nat Rev Microbiol 8: 317–327.
11. Matz C, Kjelleberg S (2005) Off the hook - how bacteria survive protozoan grazing. Trends Microbiol 13: 302–307.
12. Brüssow H (2007) Bacteria between protists and phages: from antipredation strategies to the evolution of pathogenicity. Mol Microbiol 65: 583–589.

13. Greub G, Raoult D (2004) Microorganisms resistant to free-living amoebae. Clin Microbiol Rev 17: 413–433.

14. Hall-Stoodley L, Costerton JW, Stoodley P (2004) Bacterial biofilms: from the natural environment to infectious diseases. Nat Rev Microbiol 2: 95–108.

15. Jousset A (2012) Ecological and evolutive implications of bacterial defences against predators. Environ Microbiol 14: 1830–1843.

16. Thurlow LR, Hanke ML, Fritz T, Angle A, Aldrich A, et al. (2011) Staphylococcus aureus biofilms prevent macrophage phagocytosis and attenuate inflammation in vivo. J Immunol 186: 6585–6596.

17. Al-Quadan T, Price CT, Abu Kwaik Y (2012) Exploitation of evolutionarily conserved amoeba and mammalian processes by Legionella. Trends Microbiol 20: 299–306.

18. Gao LY, Harb OS, AbuKwaik Y (1997) Utilization of similar mechanisms by Legionella pneumophila to parasitize two evolutionarily distant host cells, mammalian macrophages and protozoa. Infect Immun 65: 4738–4746.

19. Abu Kwaik Y, Gag LY, Stone BJ, Venkataraman C, Harb OS (1998) Invasion of protozoa by Legionella pneumophila and its role in bacterial ecology and pathogenesis. Appl Environ Microb 64: 3127–3133.

20. Ohno A, Kato N, Sakamoto R, Kimura S, Yamaguchi K (2008) Temperature-dependent parasitic relationship between Legionella pneumophila and a free-living amoeba (Acanthamoeba castellanii). Appl Environ Microb 74: 4585–4588.

21. Rowbotham TJ (1980) Preliminary report on the pathogenicity of Legionella pneumophila for freshwater and soil amoebae. J Clin Pathol 33: 1179–1183.

22. Fields BS, Benson RF, Besser RE (2002) Legionella and Legionnaires' disease: 25 years of investigation. Clin Microbiol Rev 15: 506–526.

23. Muder RR, Yu VL, Woo AH (1986) Mode of transmission of Legionella pneumophila. A critical review. Arch Intern Med 146: 1607–1612.

24. Ensminger AW, Yassin Y, Miron A, Isberg RR (2012) Experimental Evolution of Legionella pneumophila in Mouse Macrophages Leads to Strains with Altered Determinants of Environmental Survival. Plos Pathog 8.

25. Bonifait L, Charette SJ, Filion G, Gottschalk M, Grenier D (2011) Amoeba Host Model for Evaluation of Streptococcus suis Virulence. Appl Environ Microb 77: 6271–6273.

26. Cosson P, Soldati T (2008) Eat, kill or die: when amoeba meets bacteria. Curr Opin Microbiol 11: 271–276.

27. Froquet R, Lelong E, Marchetti A, Cosson P (2009) Dictyostelium discoideum: a model host to measure bacterial virulence. Nat Protoc 4: 25–30.

28. Greub G, La Scola B, Raoult D (2004) Amoebae-resisting bacteria isolated from human nasal swabs by amoebal coculture. Emerg Infect Dis 10: 470–477.

29. Hasselbring BM, Patel MK, Schell MA (2011) Dictyostelium discoideum as a Model System for Identification of Burkholderia pseudomallei Virulence Factors. Infect Immun 79: 2079–2088.

30. Lelong E, Marchetti A, Simon M, Burns JL, van Delden C, et al. (2011) Evolution of Pseudomonas aeruginosa virulence in infected patients revealed in a Dictyostelium discoideum host model. Clin Microbiol Infec 17: 1415–1420.

31. Smith MG, Gianoulis TA, Pukatzki S, Mekalanos JJ, Ornston LN, et al. (2007) New insights into Acinetobacter baumannii pathogenesis revealed by high-density pyrosequencing and transposon mutagenesis. Genes & development 21: 601–614.

32. Brüssow H, Canchaya C, Hardt WD (2004) Phages and the evolution of bacterial pathogens: from genomic rearrangements to lysogenic conversion. Microbiol Mol Biol Rev: 68: 560–602, table of contents.

33. Hacker J, Hentschel U, Dobrindt U (2003) Prokaryotic chromosomes and disease. Science 301: 790–793.

34. Hacker J, Kaper JB (2000) Pathogenicity islands and the evolution of microbes. Annu Rev Microbiol 54: 641–679.

35. Boyd EF (2012) Bacteriophage-encoded bacterial virulence factors and phage-pathogenicity island interactions. Adv Virus Res 82: 91–118.

36. Casas V, Maloy S (2011) Role of bacteriophage-encoded exotoxins in the evolution of bacterial pathogens. Future Microbiol 6: 1461–1473.

37. Bell JK, Mullen GE, Leifer CA, Mazzoni A, Davies DR, et al. (2003) Leucine-rich repeats and pathogen recognition in Toll-like receptors. Trends Immunol 24: 528–533.

38. Sahly H, Keisari Y, Crouch E, Sharon N, Ofek I (2008) Recognition of bacterial surface polysaccharides by lectins of the innate immune system and its contribution to defense against infection: the case of pulmonary pathogens. Infect Immun 76: 1322–1332.

39. Friman VP, Lindstedt C, Hiltunen T, Laakso J, Mappes J (2009) Predation on multiple trophic levels shapes the evolution of pathogen virulence. PloS one 4: e6761.

40. Mikonranta L, Friman V-P, Laakso J (2012) Life History Trade-Offs and Relaxed Selection Can Decrease Bacterial Virulence in Environmental Reservoirs. PloS one 7: e43801.

41. Josenhans C, Suerbaum S (2002) The role of motility as a virulence factor in bacteria. Int J Med Microbiol 291: 605–614.

42. Lertsethtakarn P, Ottemann KM, Hendrixson DR (2011) Motility and Chemotaxis in Campylobacter and Helicobacter. Nat Rev Microbiol, Vol 65 65: 389–410.

43. Friman VP, Hiltunen T, Jalasvuori M, Lindstedt C, Laanto E, et al. (2011) High temperature and bacteriophages can indirectly select for bacterial pathogenicity in environmental reservoirs. PloS one 6: e17651.

44. Sturm A, Heinemann M, Arnoldini M, Benecke A, Ackermann M, et al. (2011) The cost of virulence: retarded growth of Salmonella typhimurium cells expressing type III secretion system 1. Plos Pathog 7: e1002143.

45. Friedman ME, Kautter DA (1962) Effect of nutrition on the respiratory virulence of Listeria monocytogenes. J Bacteriol 83: 456–462.

46. Heckly RJ, Blank H (1980) Virulence and viability of Yersinia pestis 25 years after lyophilization. Appl Environ Microbiol 39: 541–543.

47. Midelet-Bourdin G, Leleu G, Copin S, Roche SM, Velge P, et al. (2006) Modification of a virulence-associated phenotype after growth of Listeria monocytogenes on food. J Appl Microbiol 101: 300–308.

48. Ali S, Huang Z, Ren SX (2009) Media composition influences on growth, enzyme activity, and virulence of the entomopathogen hyphomycete Isaria fumosoroseus. Entomol Exp Appl 131: 30–38.

49. Safavi SA, Shah FA, Pakdel AK, Reza Rasoulian G, Bandani AR, et al. (2007) Effect of nutrition on growth and virulence of the entomopathogenic fungus Beauveria bassiana. FEMS Microbiol Lett 270: 116–123.

50. Wu JH, Ali S, Huang Z, Ren SX, Cai SJ (2010) Media Composition Influences Growth, Enzyme Activity and Virulence of the Entomopathogen Metarhizium anisopliae (Hypocreales: Clavicipitaceae). Pak J Zool 42: 451–459.

51. Friman VP, Hiltunen T, Laakso J, Kaitala V (2008) Availability of prey resources drives evolution of predator-prey interaction. Proc Biol Sci 275: 1625–1633.

52. Hosseinidoust Z, van de Ven TG, Tufenkji N (2013) Evolution of Pseudomonas aeruginosa virulence as a result of phage predation. Appl Environ Microbiol 79: 6110–6116.

53. Molmeret M, Horn M, Wagner M, Santic M, Abu Kwaik Y (2005) Amoebae as training grounds for intracellular bacterial pathogens. Appl Environ Microbiol 71: 20–28.

54. Steinert M, Heuner K (2005) Dictyostelium as host model for pathogenesis. Cell Microbiol 7: 307–314.

55. Nehme NT, Liegeois S, Kele B, Giammarinaro P, Pradel E, et al. (2007) A model of bacterial intestinal infections in Drosophila melanogaster. Plos Pathog 3: e173.

56. Kurz CL, Chauvet S, Andres E, Aurouze M, Vallet I, et al. (2003) Virulence factors of the human opportunistic pathogen Serratia marcescens identified by in vivo screening. EMBO J 22: 1451–1460.

57. Flyg C, Kenne K, Boman HG (1980) Insect pathogenic properties of Serratia marcescens: phage-resistant mutants with a decreased resistance to Cecropia immunity and a decreased virulence to Drosophila. J Gen Microbiol 120: 173–181.

58. Grimont PA, Grimont F (1978) The genus Serratia. Annu Rev Microbiol 32: 221–248.

59. Mahlen SD (2011) Serratia infections: from military experiments to current practice. Clin Microbiol Rev 24: 755–791.

60. Sutherland KP, Porter JW, Turner JW, Thomas BJ, Looney EE, et al. (2010) Human sewage identified as likely source of white pox disease of the threatened Caribbean elkhorn coral, Acropora palmata. Environ Microbiol 12: 1122–1131.

61. Kiy T, Tiedtke A (1992) Mass Cultivation of Tetrahymena thermophila Yielding High Cell Densities and Short Generation Times. Appl Microbiol Biot 37: 576–579.

62. Kennedy GM, Morisaki JH, Champion PA (2012) Conserved mechanisms of Mycobacterium marinum pathogenesis within the environmental amoeba Acanthamoeba castellanii. Appl Environ Microbiol 78: 2049–2052.

63. Page FC (1976) An Illustrated Key to Freshwater and Soil Amoebae: With Notes on Cultivation and Ecology. Freshwater Biological Association.

64. La Scola B, Mezi L, Weiller PJ, Raoult D (2001) Isolation of Legionella anisa using an amoebic coculture procedure. J Clin Microbiol 39: 365–366.

65. Page FC (1988) A New Key to Freshwater and Soil Gymnamoebae with instructions for culture. Freshwater Biological Association.

66. Friman VP, Laakso J (2011) Pulsed-resource dynamics constrain the evolution of predator-prey interactions. Am Nat 177: 334–345.

67. Friman VP, Laakso J, Koivu-Orava M, Hiltunen T (2011) Pulsed-resource dynamics increase the asymmetry of antagonistic coevolution between a predatory protist and a prey bacterium. J Evol Biol 24: 2563–2573.

68. O'Toole GA, Kolter R (1998) Initiation of biofilm formation in Pseudomonas fluorescens WCS365 proceeds via multiple, convergent signalling pathways: a genetic analysis. Mol Microbiol 28: 449–461.

69. Laakso J, Loytynoja K, Kaitala V (2003) Environmental noise and population dynamics of the ciliated protozoa Tetrahymena thermophila in aquatic microcosms. Oikos 102: 663–671.

70. Wildschutte H, Wolfe DM, Tamewitz A, Lawrence JG (2004) Protozoan predation, diversifying selection, and the evolution of antigenic diversity in Salmonella. Proc Natl Acad Sci U S A 101: 10644–10649.

71. Potvin C (2001) ANOVA Experimental Layout and Analysis. In: Scheiner SM and Gurevitch J, editors. Design and Analysis of Ecological Experiments. New York: Oxford University Press. pp. 69–75.

72. Steinberg KM, Levin BR (2007) Grazing protozoa and the evolution of the Escherichia coli O157: H7 Shiga toxin-encoding prophage. Proc Biol Sci 274: 1921–1929.

73. Cosson P, Zulianello L, Join-Lambert O, Faurisson F, Gebbie L, et al. (2002) Pseudomonas aeruginosa virulence analyzed in a Dictyostelium discoideum host system. J Bacteriol 184: 3027–3033.

74. Casadevall A (2008) Evolution of intracellular pathogens. Nat Rev Microbiol 62: 19–33.

75. Cooper TF, Lenski RE (2010) Experimental evolution with E. coli in diverse resource environments. I. Fluctuating environments promote divergence of replicate populations. BMC Evol Biol 10: 11.

76. Hall AR, Colegrave N (2008) Decay of unused characters by selection and drift. J Evol Bio 21: 610–617.

77. Ketola T, Mikonranta L, Zhang J, Saarinen K, Ormala AM, et al. (2013) Fluctuating temperature leads to evolution of thermal generalism and preadaptation to novel environments. Evolution 67: 2936–2944.

78. Travisano M, Mongold JA, Bennett AF, Lenski RE (1995) Experimental tests of the roles of adaptation, chance, and history in evolution. Science 267: 87–90.

79. Chesbro WR, Wamola I, Bartley CH (1969) Correlation of virulence with growth rate in *Staphylococcus aureus*. Can J Microbiol 15: 723–729.

80. West SA, Buckling A (2003) Cooperation, virulence and siderophore production in bacterial parasites. Proc Biol Sci 270: 37–44.

81. Gomez P, Buckling A (2011) Bacteria-phage antagonistic coevolution in soil. Science 332: 106–109.

82. Zhang J, Friman VP, Laakso J, Mappes J (2012) Interactive effects between diet and genotypes of host and pathogen define the severity of infection. Ecology and Evolution 2: 2347–2356.

6

Identification of Novel Factors Involved in Modulating Motility of *Salmonella enterica* Serotype Typhimurium

**Lydia M. Bogomolnaya[1,2], Lindsay Aldrich[1], Yuri Ragoza[1], Marissa Talamantes[1], Katharine D. Andrews[1],
Michael McClelland[3], Helene L. Andrews-Polymenis[1]***

1 Department of Microbial Pathogenesis and Immunology, College of Medicine, Texas A&M University, Bryan, Texas, United States of America, 2 Institute of Fundamental
Medicine and Biology, Kazan Federal University, Kazan, Russia, 3 Department of Microbiology and Molecular Genetics, University of California Irvine, Irvine, California,
United States of America

Abstract

Salmonella enterica serotype Typhimurium can move through liquid using swimming motility, and across a surface by swarming motility. We generated a library of targeted deletion mutants in *Salmonella* Typhimurium strain ATCC14028, primarily in genes specific to *Salmonella*, that we have previously described. In the work presented here, we screened each individual mutant from this library for the ability to move away from the site of inoculation on swimming and swarming motility agar. Mutants in genes previously described as important for motility, such as *flgF, motA, cheY* are do not move away from the site of inoculation on plates in our screens, validating our approach. Mutants in 130 genes, not previously known to be involved in motility, had altered movement of at least one type, 9 mutants were severely impaired for both types of motility, while 33 mutants appeared defective on swimming motility plates but not swarming motility plates, and 49 mutants had reduced ability to move on swarming agar but not swimming agar. Finally, 39 mutants were determined to be hypermotile in at least one of the types of motility tested. Both mutants that appeared non-motile and hypermotile on plates were assayed for expression levels of FliC and FljB on the bacterial surface and many of them had altered levels of these proteins. The phenotypes we report are the first phenotypes ever assigned to 74 of these open reading frames, as they are annotated as 'hypothetical genes' in the Typhimurium genome.

Editor: Michael Hensel, University of Osnabrueck, Germany

Funding: This work was supported in part by NIH/NIAID grants R01AI083646, R56AI077645 awarded to HAP, R01 AI075093 awarded to MM. LMB was supported in part by a subsidy to support the Program of competitive growth of Kazan Federal University. The funders had no role in study design, data collection and analysis, decision to publish, or preparation of the manuscript.

Competing Interests: The authors have declared that no competing interests exist.

* Email: handrews@medicine.tamhsc.edu

Introduction

Infection with *Salmonella enterica* serotypes remains a serious human and animal health problem worldwide. *Salmonellae* cause an array of diseases ranging from gastroenteric disease to systemic disease including Typhoid fever and bacteremia [1]. While gastroenteritis as a result of *Salmonella* infection is common worldwide, systemic diseases caused by this organism are relatively rare in the developed world. Serotype Typhimurium is one of the two most common serotypes identified from cases of clinical disease in the United States [2]. After gaining access to a susceptible host by the oro-fecal route, *Salmonella* utilizes multiple strategies to colonize and persist. *Salmonellae* have many well studied virulence factors including Type 3 Secretion Systems (T3SS-1 and -2), lipopolysaccharides (LPS), fimbria and others, and are capable of multiple types of motility including swimming and swarming. Both types of motility require the presence of functional flagella that are composed of many proteins and consist of the following structures: the basal body, the hook and the filament. Flagellar biosynthesis is a complex, tightly regulated process where gene products are produced in the order of flagellar apparatus assembly. The current understanding of flagella structure and regulation in *Salmonella* is nicely summarized in

recent review by Chevance and Hughes [3]. Motility is linked to virulence in many pathogenic bacteria [4,5].

Swimming motility is directed movement through liquid, which is assayed using semi-solid "swimming" media, containing a low concentration of agar. Under these conditions individual bacteria swim through medium-filled spaces between agar [6]. Swimming motility is closely linked to chemotaxis [7], the ability to orient bacteria along certain chemical gradients, and is thought to allow bacteria to detect and pursue nutrients or avoid unwanted repellents [7]. Ultimately, this kind of motility allows these organisms to avoid unfavorable environments for colonization and to reach and maintain preferred niches for colonization.

Swarming motility is a multi-cellular phenomenon involving the coordinated and rapid movement of a bacterial population across a semisolid surface [8]. Factors on which swarming is known to depend include bacterial cell density, media composition and surface moistness [9–12]. Swarming is not just another form of motility but rather a part of alternative growth state and is characterized by change in gene expression of nearly one third of genome in *Salmonella* Typhimurium [4].

Recent large-scale studies have been performed to identify genetic determinants required for each type of motility in *E. coli*, an organism closely related to *Salmonella*. Both transposon mutants in *E.coli* K-12 [13], and the Keio collection, a collection of targeted deletion mutants in each non-essential open reading frame in *E. coli* K-12 [14], were screened to identify mutants with reduced motility. These comprehensive studies demonstrated that numerous genes are involved in regulation of motility in bacteria.

We have generated a library of 1023 targeted single gene deletion mutants (SGD) in virulent *Salmonella enterica* serotype Typhimurium ATCC14028s [15]. This library contains mutants in nearly all *Salmonella*-specific open reading frames as well as an additional 100 genes shared between *Salmonella* and other *Enterobacteriaceae*. We screened this collection to systematically identify those mutants in our collection that had either reduced or enhanced ability to move away from the site of inoculation on swimming and/or swarming motility agar. We identified 160 mutants with altered motility in at least one condition. Mutants with previously known motility defects (flagellar, LPS biosynthesis, chemotaxis genes) were correctly identified in our screen. We identified mutants in nine additional genes that are unable to move away from the site of inoculation on both types of agar. Furthermore, we found that the ability to swim and swarm could be uncoupled in some mutants. We also identified a significant number of mutants that had enhanced motility of one or both types ('hypermotility'). Finally, many of the motility phenotypes we identified belong to mutants with mutations in genes annotated as of unknown function (FUN, or 'orphan' genes), and are thus the first phenotypes of any kind described for these genes.

Materials and Methods

Bacterial Strains and Media

All *Salmonella enterica* serotype Typhimurium strains used in this study were derived from ATCC14028, including HA420, a spontaneous Nalidixic acid resistant isolate [16]. Targeted deletion mutants screened in this study have been described previously [15].

Salmonella strains used in this study were routinely grown on LB agar or broth, or on M9 minimal media. The growth of all mutant strains was tested on minimal M9 media [17], prior to use of these mutants in motility assays. Antibiotics were added in the following concentrations as appropriate: 50 mg/L Kanamycin sulfate, 50 mg/L of Nalidixic acid [15,16]. Media for assaying swimming and swarming motility have been described previously [11]. Swimming was assayed on plates containing 0.3% Difco Bacto Agar (LB Miller base 25 g/L), while swarming motility was assayed on 0.6% Difco Bacto Agar (LB Miller base 25 g/L, and 0.5% glucose).

Screening Individual Mutants for Swimming and Swarming Motility

Our collection of targeted deletion mutants was assayed in 96-well format for both swimming and swarming on large agar plates (15 cm diameter). Strains were inoculated into the appropriate agar with 96-pin replicator, incubated at 37°C, and closely monitored for the duration of the assay. Wild type ATCC14028s and ATCC14028r (smooth and rough LPS) were included as positive and negative controls on each motility plate. Several hours post-inoculation (3.5 hours for swimming, 5 hours for swarming), the swimming and swarming ability of each strain was evaluated by estimating the diameter of the spread of the bacteria and assigning a motility score. Individuals scoring motility were blinded to the identity of the mutants being scored, but were aware of the location of positive and negative controls inoculated on every plate. Motility was scored on a scale of 0 (completely non-motile) to 10 (extremely hypermotile) relative to wild type

ATCC14028s, which was always assigned a motility score of 5. Our scoring system allowed us to identify mutants with a range of hypo- and hyper- motility. Large-scale screening assays were performed in triplicate and were repeated on at least three separate occasions with the entire collection of mutants. The mean of the motility scores for a given assay were determined for each mutant and these are shown in Table S1. Mutants used in our large scale screening were not transduced into a new genetic background, thus motility phenotypes associated with gene loss will be confirmed in future studies.

Mutants with the most severe phenotypes (less than 25% motility of wild type or hypermotile mutants) were further evaluated by measuring the diameter of the swimming or swarming colony as compared to positive and negative controls. This step was required for identification and removal from the future study a few false positive motility candidates that during primary screening were scored as non-motile or hypermotile due to uneven transfer of cells with 96-pin replicator. Overnight cultures of each mutant were grown and normalized by OD_{600}. 3 μl was spotted on motility plates, incubated and scored as described above. The diameter of each swimming or swarming colony was measured, and compared to positive (WT ATCC14028s, smooth) and negative (ATCC14028r, rough mutant or *motA*) control isolates. Experiments were conducted in triplicate, and were repeated on three separate occasions. The identity of each mutant that displayed a statistically significant phenotype in our motility assays was verified by PCR using primers flanking the site of deletion if not previously verified. Statistical significance was determined using a Student's *t*-test and a *p*-value of <0.05.

Evaluation of Flagellin Expression

Each mutant with reduced or enhanced motility was evaluated to determine the level of FljB and FliC produced in the bacterial cell, and to determine the amount of each of these proteins that reach the bacterial surface as compared to the isogenic wild type organism. Bacterial cultures grown to stationary phase were normalized by OD_{600} and bacteria were collected by centrifugation (Eppendorf 5415D). In order to shear flagella from the bacterial surface, pelleted bacteria were resuspended in 1 ml PBS and subjected to 5 minutes of vortexing (Vortex-genie, Scientific Industries) [18]. Sheared protein was precipitated using Trichloroacetic acid (TCA) (6% final concentration) overnight at 4°C, washed twice with 300 μl of acetone and resuspended in SDS sample buffer. Precipitated protein from sheared fractions was evaluated by SDS-PAGE and Western Analysis using antibodies against FliC and FljB (Difco). The remaining bacterial pellet (without sheared flagella) was solubilized in SDS sample buffer and examined by 12% SDS-PAGE and Coomassie staining to ensure equivalent loading. Wild type organisms and *ΔfljB* and *ΔfliC* mutants were used as positive and negative controls in these assays.

Results and Discussion

Screening of a collection of targeted deletion mutants in *Salmonella* to identify determinants of motility

We have screened a library of targeted single-gene deletion mutants in many genes that are specific to *Salmonellae* and not shared with close relatives, in addition to approximately 100 mutants in shared genes that served as controls, for motility phenotypes on plates [15] (Figure 1). We identified 160 mutants with altered motility. Of these mutants, only 29 were previously connected to motility either in *Salmonella* or *E. coli*. We divided the resulting set of mutants with motility phenotypes into four

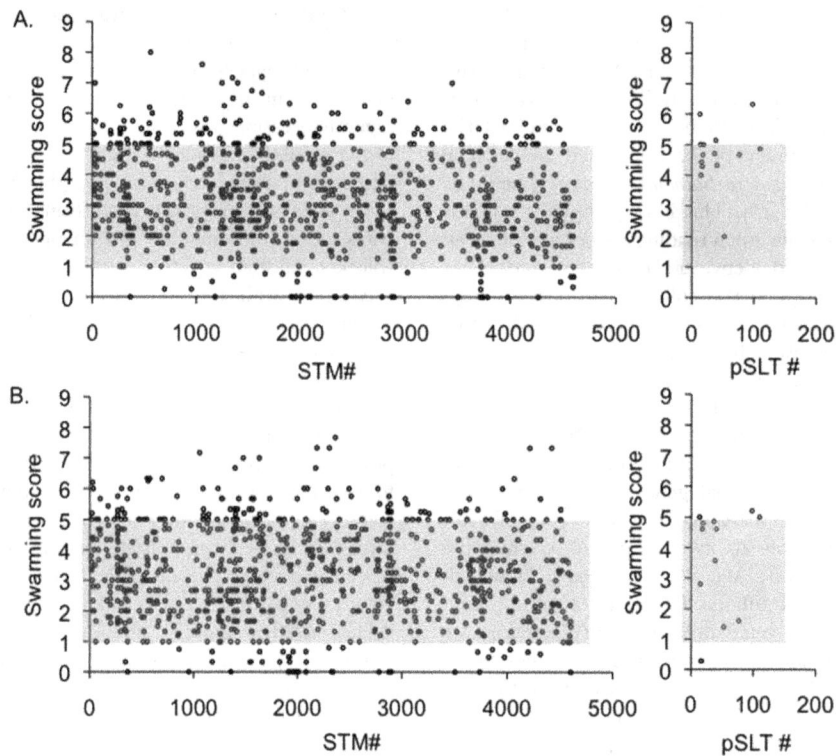

Figure 1. Motility scores of single deletion mutants of *Salmonella* Typhimurium ATCC14028 on swimming (A) and swarming (B) agar. Stationary phase cultures were transferred to motility agar, and motility was scored after 3.5 and 5 hours of incubation at 37°C on swimming or swarming plates, respectively. Motility was scored on the scale from 0 to 10 with wild type motility equal to 5. Data are presented as average swimming or swarming score from experiments with triplicate samples, performed on three independent occasions. Dots located outside of shaded area indicate scores greater than 5.25 and lower than 1.25 to define mutants with increased or reduced (<25 % of wild type) motility.

categories: (a) Mutants with defects in both types of motility; (b) Mutants with reduced swimming motility, swarming motility is unaffected; (c) Mutants reduced swarming motility only, swimming is unaffected; (d) Mutants that were hypermotile (Figure 2).

Mutants with reduced swimming and swarming motility

We identified 21 mutants that moved less than or equal to 25% of wild type from the initial site of inoculation (i.e. impaired swimming and swarming). This group contains mutations in 12 genes previously reported to be needed for swimming and swarming motility, including flagellar genes *flhA, flhB, flgF, flgG, motA, motB*, LPS biosynthesis genes *rfbN, rfbK, rfaI*, chemotaxis gene *cheY* and others (Table 1), validating our approach. Our targeted deletion collection assayed in this study contained only eight mutants in known flagellar genes, and each of the mutants in these genes had the expected non-motile phenotype.

In addition, we identified general motility phenotypes for mutants in 9 genes not previously described to be involved in motility (Table 1). Four of these mutants had deletions in genes with putative functions in transport (*aroD*), in transcription (*tctD*) and in signal transduction (*STM0343, STM0551*). The remaining five mutants with reduced movement from the site of inoculation on swimming and swarming agar had deletion mutations in fimbrial genes *sthE, STM4595* [19], *STM0699, STM2010* and *STM2880*, a gene with unknown function encoded within SPI-1 [4,20]. None of the mutants in this group had a growth defect in LB-broth (Figure S1). We note that our screening assay does not allow us to determine why mutants do not appear to spread from the site of inoculation. As such, classification of these mutants into

categories such as defects in flagellar motility, motor activity, or chemotaxis is a fascinating area of future investigation.

We hypothesized that some of the phenotypes we observed could be due to an inability to export flagellin to the bacterial surface. The external portion of the flagellum is a helical filament composed of flagellin proteins, FliC or FljB [21]. In our wild type (ATCC14028) population FliC appeared to be the major flagellin expressed (accounts for >70 % of total flagellin) and FljB expression was weak (data not shown). We tested all 22 mutants with reduced swimming and swarming motility for expression of flagellins on the bacterial surface by evaluating the amount of FliC and FljB in the sheared fraction of surface proteins by Western analysis.

Mutants in *flgF, flgG*, and *rfbK*, genes known to be important for motility [14,22], were both non-motile and displayed less flagellins on the bacterial surface in our screen. For other mutants however, (*motB, cheY*) the presence of filaments is known not to be sufficient for the ability of these mutants to move from the site of inoculation on motility plates. Our mutants in *motB* and *cheY* displayed wild type levels of flagellins on the cell surface yet were unable to spread from the site of inoculation as previously reported in the literature [23,24] (Figure 3).

Of the nine mutants with reduced ability to move from the site of inoculation on swimming and swarming agar plates in our screen that were not previously known to be needed for this ability, all had levels of FliC on the bacterial surface comparable to wild type (Figure 3). Several of the mutants we identified in this category (*STM0343, STM0551, STM0699*) expressed less of the minor flagellin FljB on the bacterial surface compared to wild type

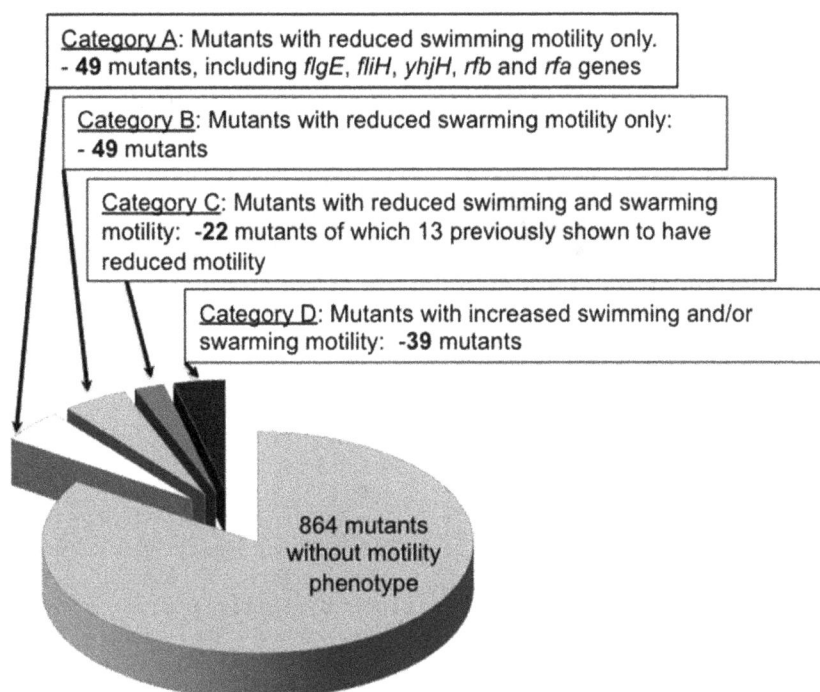

Category A: Mutants with reduced swimming motility only.
- **49** mutants, including *flgE, fliH, yhjH, rfb* and *rfa* genes

Category B: Mutants with reduced swarming motility only:
- **49** mutants

Category C: Mutants with reduced swimming and swarming motility: -**22** mutants of which 13 previously shown to have reduced motility

Category D: Mutants with increased swimming and/or swarming motility: -**39** mutants

864 mutants without motility phenotype

Figure 2. Distribution of motility phenotypes in single gene knock out collection.

Table 1. Pathway clustering of mutants with altered motility.

Functional categories*	Swimming and swarming defect	Reduced swimming only	Reduced swarming only
Motility	**flgF****, **flgG, flhA, flhB, motA, motB, fliD, fliM**	flgE, ssaV, fliH, <u>stjC</u>***	ssaU, <u>pefD</u>, <u>pefC</u>
Cell envelope biogenesis	**rfbN, rfal**	rfbP, rfbM, rfbJ, rfbC, rfbD, rfaJ, rfaG, rfaQ	
Signal transduction	STM0343, STM0551, cheY,	phoQ, **yhjH**, yjcC	yciR
Carbohydrate transport & metabolism	**rfbK**	STM0722, STM4424	STM0860, STM3780
Transport and metabolism	aroD	aroA, fur, yfeJ, pdxK, **tatC**	yliB, STM1635, sfbA, mgtB, sodA, STM0163, STM1546, STM0857
Transcription	tctD	invF, STM2912, STM3696, STM4417, **arcA**	STM0859, ydiP, torR, STM4315
Replication, recombination and repair		STM1005	STM1861
Translation, ribosomal structure and biogenesis		valS	STM1552
Posttranslational modification, protein turnover, chaperones		STM2743, sspA	
Energy production and conversion			STM0762, STM0858
Cell cycle control, cell division, chromosome partitioning			STM2594
Defense mechanism			STM4262
Not in COGs	STM0699, STM2010, STM2880, <u>sthE</u>, STM4595	STM0289, STM0295, STM0660, STM0971, STM1040, STM1331, ssaG, STM2374, sptP, sipA, <u>STM3026</u>, **rfaL**, yibR, **rfaP**, STM3783, STM4216,STM4219, sRNA candidate C1023	STM0056, STM0362, STM1131, pagC, STM1254, STM1258, STM1543, srfC, STM1632, STM1856, STM1926, STM1958, STM2303, STM2508, ygaU, STM3125, <u>lpfE</u>, STM3944, STM4030.S, STM4197, STM4204, STM4529, STM4574, STM4599, invR

*Based on COGs (Clusters of Orthologous Groups of protein).
**Mutants with previously known motility phenotype are shown in bold.
***Mutants in fimbrial genes are underlined.

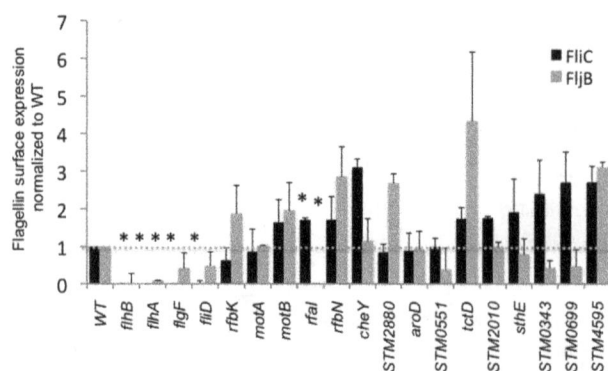

Figure 3. Reduced motility on semi-solid media in some mutants correlated with reduced expression of flagellin on the bacterial surface. Flagellins were sheared from bacterial surface, analyzed by Western blotting with antibodies to FliC (black bars) and FljB (grey bars) and blots were quantified by densitometry using Quantity One software. Surface expression of flagellins was normalized to the level on the surface of the wild type strain. Samples for each mutant were prepared in three independent experiments.

ATCC14028 (Figure 3, Figure S2). Two of these mutants (*STM0343* and *STM0551*) have deletions in genes encoding EAL-domains containing proteins, predicted to be involved in the metabolism of cyclic diguanosine monophosphate (c-di-GMP) [25,26]. Purified STM0551 possesses phosphodiesterase activity *in vitro* that is abolished by a point mutation in the EAL domain [25]. Based on the current model of c-di-GMP metabolism in bacteria, mutation in c-di-GMP phosphodiesterases results in accumulation of c-di-GMP leading to decreased expression of flagellins, and loss of motility [27]. Furthermore *STM0551* appears to be a negative regulator of type 1 fimbria. Activation of *fim* genes includes the activation of a negative regulator of motility *fimZ* [25]. Over-expression of FimZ is known to repress swimming motility in *Salmonella* [28]. In agreement with these findings, deletion of *STM0551* abrogates outward movement on both swimming and swarming agar in *Salmonella*. Thus, data from our screen is consistent with previously published work, and shows that the amount of flagellin sheared from the bacterial surface can be correlated with motility for some but not all mutants.

Mutants with reduced swimming motility

We identified 49 mutants that had reduced movement from the site of inoculation on swimming motility plates (less than 25% of wild type swimming) but appeared to behave similarly to the wild type organism on swarming motility plates (Table 1, Table S1). We were surprised that the ability to perform these two different types of motility could be uncoupled, and we note that further microscopic study of these mutants during swarming will provide clues regarding these seemingly paradoxical phenotypes. Sixteen mutants identified in this group were known to have reduced swimming motility including *flgE, fliH, arcA, yhjH* and *tatC* [4,13,14,29–33]. In addition to mutants mentioned above, we identified 33 mutants that had reduced motility on swimming agar but normal movement on swarming agar that were not previously known to be involved in swimming motility (Table S3). Some mutants in this category had deletions in genes encoding proteins involved in intracellular transport and metabolism, signal transduction and regulation of transcription based on COG predictions (Clusters of Orthologous Groups of proteins) [34] (Table 1). For example, our *Δfur* mutant was severely impaired on swimming

agar, but not on swarming agar. We found that deletion of *phoQ*, also resulted in reduced swimming but not swarming motility.

Mutants in a number of genes necessary for LPS biosynthesis/ assembly (*rfaG, rfaI, rfaL, rfaQ, rfaP, rfbC, rfbD, rfbJ, rfbM, rfbP, yibR*) had severely impaired movement on swimming motility agar. In *E. coli*, lipopolysaccharide (LPS) biosynthesis genes are not required for swimming motility [14]. Our work shows that *Salmonella* appears to require LPS biosynthesis for movement on swimming motility plates (Table 1). This finding is not unprecedented however, as others have shown that in *Salmonella rfaP and rfaJ* and in *Pseudomonas rfaL*, are required for swimming motility [22,35,36].

Mutants that had reduced outward movement from the site of inoculation on swimming agar but normal swarming movement, also included deletion mutants in genes encoding important virulence factors such as fimbria (*stjC, stdD*), Type Three Secretion System components *invF, sipA, sptP* (T3SS 1), and *ssaV* (T3SS 2) as well as Type Six Secretion System (T6SS, SPI-6 encoded) components *STM0289* and *STM0295* [37,38]. STM0289 is a member of Vrg family of proteins required for effectors delivery via T6SS. Ours is the first description of T6SS encoded on SPI-6 potential involvement in motility. Finally, fourteen mutants, the largest group of mutants that swam poorly but swarmed normally, are not annotated or do not have a previously described function or phenotype. Thus, we describe the first functional data and potential phenotypes for these genes.

Mutants with reduced swarming motility

We identified 49 mutants with reduced ability to move outward from the site of inoculation on swarming agar (less than 25% of wild type) but with normal outward migration on swimming agar (Table S4). None of the genes that we identified in this category were directly implicated previously in the ability to swarm. A limited number of reports describe genes required for swarming but not swimming motility. Those include *E.coli recA*, *S.Typhimurium flhE* and *Proteus mirabilis waaL* [39–42] and were not present in the library used for the current study. Deletion mutants in a number of transcriptional regulators, including STM1355 and *torR* (STM3824) appear to be involved in the regulation of swarming motility in *Salmonella*.

Several of these non-swarming mutants had deletions in genes involved in energy production/conversion and transport/metabolism as predicted by COG assignments (12 out of 49, Table 1). For example, mutants in *STM0762*, a gene that encodes fumarase, a TCA cycle enzyme involved in energy production, cannot swarm. Several additional mutants, *ΔSTM0849, ΔSTM0056 and ΔSTM1258*, have putative functions in transport or energy production. *STM0849* is a homolog of *yliB*, a ppGpp-dependent glutathione importer encoded on *yliABCD* operon [43–45] and *STM0056* is annotated as a putative oxalacetate decarboxylase gamma subunit [46]. Our results are consistent with previous proteomic approaches demonstrating that genes associated with energy production and *de novo* synthesis are required for swarming [47] and with the current thinking that swarming is an energetically costly process.

We identified a mutant in *srfC as* poor swarmer but normal swimmer. This gene is clustered with the flagellar class 2 genes and was determined to be under FlhDC control [48]. A polar mutation in an upstream gene in the *srfABC* operon, *srfB*, or deletion of the whole operon affected only swarming motility [49]. Similar to recent reports on swarming motility of *E. coli, Pseudomonas aeruginosa* and *Xenorhabdus nematophila* [14,50,51], mutations in fimbrial genes, *pefC* and *pefD* (encoding for usher and chaperone proteins, respectively) and in *lpfE* (long polar fimbrial minor

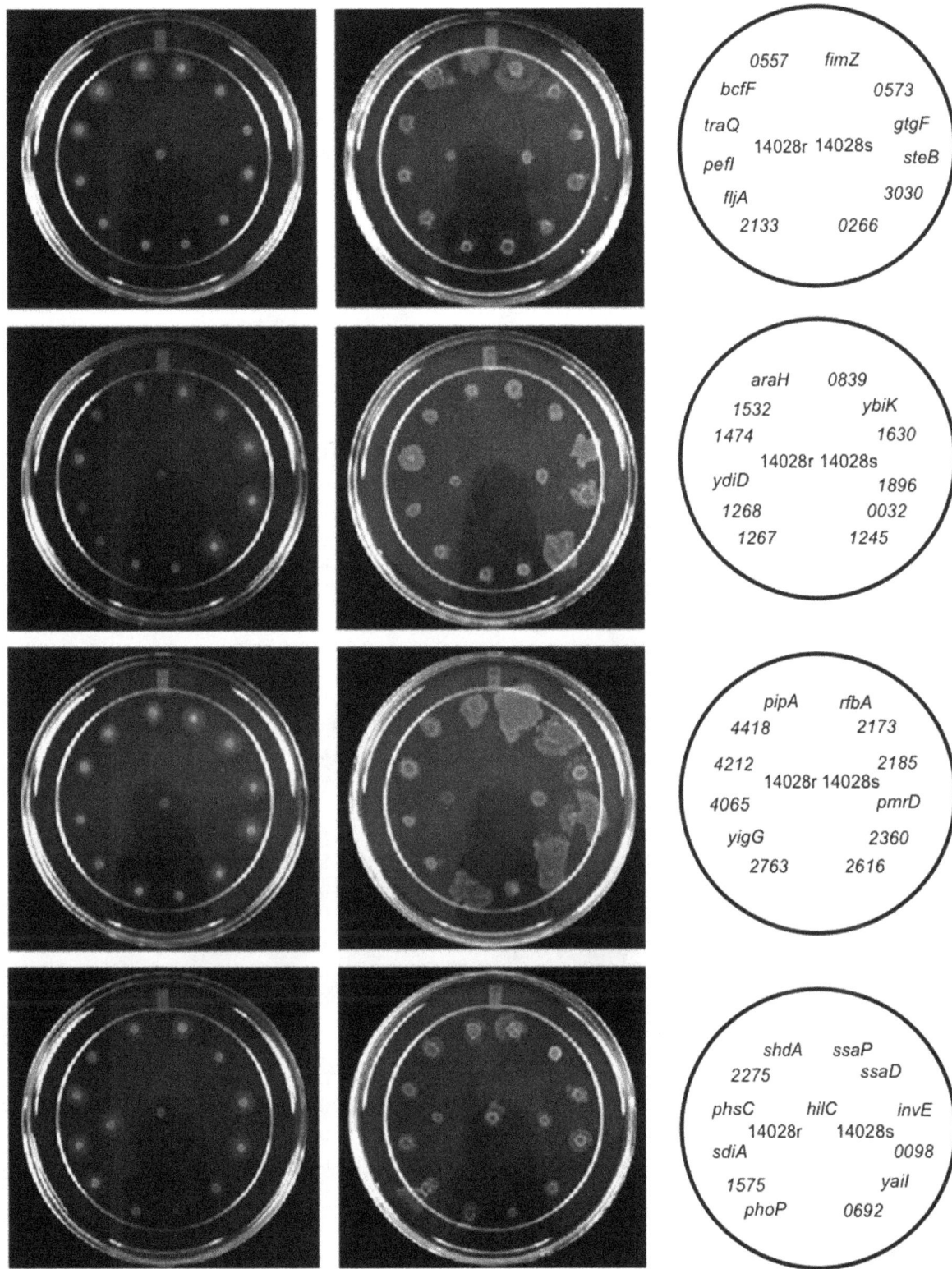

Figure 4. Confirmation of hypermotility phenotypes. Stationary cultures normalized by OD_{600} were spotted on swimming or swarming agar along with wild type strain, ATCC14028s. Rough strain, ATCC 14028r, was used as a negative control on swarming agar. After 3.5 and 5 hours of incubation at $37^{\circ}C$ on swimming or swarming plates, respectively, the diameter of each bacterial growth area was measured in three independent experiments, with each experiment performed in triplicate.

Table 2. Mutants with enhanced motility.

STM	Gene	Motility compared to wild type*	
		Swimming Mean ± SD	Swarming Mean ± SD
STM0026	bcfF	**1.32±0.26****	**2.55±0.64**
STM0032		**1.49±0.24**	1.42±0.75
STM0098		**1.65±0.37**	**1.46±0.29**
STM0266		1.13±0.35	**1.59±0.45**
STM0387	yaiI	**1.42±0.23**	1.15±0.25
STM0549	fimZ	**1.32±0.20**	1.89±1.13
STM0557	gtrC	**1.38±0.24**	**2.13±0.52**
STM0573		1.14±0.17	**1.64±0.32**
STM0839		**1.20±0.20**	**1.57±0.67**
STM0847	ybiK	**1.53±0.49**	**1.36±0.29**
STM1087	pipA	**1.30±0.22**	**1.50±0.35**
STM1231	phoP	**1.39±0.34**	**1.69±0.56**
STM1344	ydiV	**1.29±0.16**	0.93±0.18
STM1350	ydiD	**1.61±0.79**	1.20±0.63
STM1395	ssaD	**1.41±0.30**	0.93±0.24
STM1417	ssaP	**1.47±0.27**	1.25±0.64
STM1575		**1.53±0.19**	1.35±0.40
STM1629	steB	0.98±0.26	**1.71±0.33**
STM1630		**1.72±0.37**	**1.62±0.50**
STM1896		**1.50±0.21**	**1.94±1.01**
STM1950	sdiA	**1.60±0.26**	**1.49±0.45**
STM2063	phsC	**1.46±0.27**	1.23±0.30
STM2095	rfbA	**1.41±0.26**	2.08±1.11
STM2133		1.08±0.10	**1.55±0.46**
STM2173		**1.41±0.22**	**1.62±0.70**
STM2185		**1.62±0.22**	**1.49±0.56**
STM2304	pmrD	**1.49±0.23**	**1.71±0.44**
STM2360		**1.35±0.16**	1.43±0.74
STM2513	shdA	**1.37±0.23**	1.14±0.26
STM2616		**1.31±0.28**	0.90±0.09
STM2763		**1.68±0.35**	1.52±0.79
STM2770	fljA	1.05±0.08	**1.39±0.26**
STM2867	hilC	**1.67±0.26**	**1.58±0.34**
STM2897	invE	**1.56±0.30**	1.41±0.51
STM3954	yigG	**1.36±0.22**	1.19±0.33
STM4212		1.14±0.24	**1.47±0.45**
STM4418		**1.26±0.21**	1.40±0.68
PSLT013	pefI	0.98±0.19	**1.18±0.11**
PSLT098	traQ	1.04±0.22	**1.33±0.23**

* - Diameter of swimming and swarming rings were measured and compared to wild type. Results are shown as the mean of six independent experiments.
** - Bold indicates statistical significance, $p<0.05$.

protein) strongly affected swarming motility with no significant effect on swimming motility.

Interestingly, we found that a *fliB* mutant was defective in swarming, but not in swimming. FliB methylates the lysine residues of flagellin [52]. Previous studies showed that the loss of *fliB* did not affect swimming motility and it is thought that flagellin

methylation by FliB is required for *Salmonella* virulence but not for flagellin function [48]. Our data suggest that flagellin methylation is required for swarming motility.

Swarming and virulence are linked in several bacteria [53]. 21 out of 49 of our mutants with defects in swarming are deleted for genes associated with virulence. *STM1131, STM2303,*

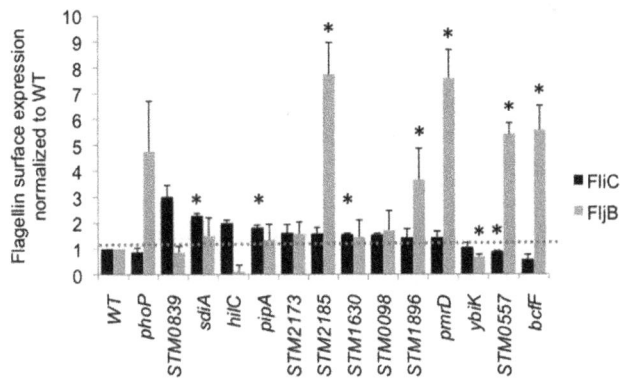

Figure 5. Elevated levels of flagellin were present on the bacterial surface of some mutants with improved swimming and swarming motility. Flagellins were sheared from bacterial surface, analyzed by Western blotting with antibodies to FliC and FljB and blots were quantified by densitometry using Quantity One software. Surface expression of flagellins was normalized to the level on the surface of the wild type strain. Samples for each mutant were prepared in three independent experiments.

STM4030.S, STM4262 (*siiF*) are reported to be needed for full virulence in mice and calves [15,54,55] and mutants in these genes have reduced swarming in our assays. *pagC*, *mgtB* and *STM0859* also had reduced motility in our assays and are part of the *phoPQ* regulon in *Salmonella* [56–58]. Finally, swarming motility was also compromised in mutants that had deletions of genes encoded on SPI-14 (*STM0859*), a region important for virulence in chickens [59]. STM0859 was recently reported to be co-regulated with the type three secretion system encoded on SPI-2 [60].

We identified a group of genes that when deleted reduce the ability to swarm to less than 25% of the ability of the wild type organism without significant reductions in the ability to move away from the site of inoculation on swimming motility plates. Some of the products of these genes may be involved in generating movement, signaling when conditions exist for swarming, surface properties or other qualities that are directly involved in the ability to swarm, while others may affect swarming indirectly. Careful quantification and microscopic examination of each of these mutants will be useful to fit the corresponding genes into an overall framework with respect to their roles in swarming motility.

The identification of hypermotile mutants

As we were using 96-well format to screen our deletion mutant library for motility phenotypes, multiple observations of each screening plate were required in order to identify phenotypes before colonies intersected obscuring individual phenotypes. Early and frequent observation allowed us to identify mutants that had increased ability to move away from the site of inoculation relative to the wild type organism. We observed thirty-nine mutants to have a larger swimming and/or swarming ring diameter as compared to the wild type observed on the same plate at the same time point post inoculation (Figure 4). Of this group, we found 14 mutants that had generalized improvement in motility, 17 mutants with increased swimming motility, and 8 mutants with increased swarming motility compared to wild type (Table 2). Movies S1, S2, and S3 show examples of mutants with increased swarming motility, able to move across swarming agar faster than wild type cells (See Supporting Information Legend).

Hypermotility phenotypes have not been well described in *Salmonella*, and there are only a limited number of reports describing hypermotility as a phenotype in bacteria, primarily in *Pseudomonas aeruginosa* and *Proteus mirabilis* [61–69] (Table S5). Thus, there is little existing data to easily validate and benchmark the hypermotility phenotypes we observed, yet a few mutations are known to promote hypermotility. First, over-expression of FimZ, a positive regulator of Type I fimbriae and negative regulator of flagellar motility, represses swimming motility in *Salmonella* [28]. A *fimZ* deletion mutant is expected to be hypermotile, and consistent with this prediction our deletion mutant in *fimZ* was indeed hypermotile (Figure 4 and Table 2).

Second, deletion of *STM1344* (*ydiV*), annotated by COG database as a gene involved in signal transduction, resulted in improved swimming motility in our assays (Table 2). YdiV negatively regulates motility via binding to the $FlhD_4C_2$ complex to prevent interaction of this complex with DNA [70]. YdiV also functions as an adaptor protein that binds FlhD and delivers $FlhD_4C_2$ to ClpXP protease for proteolytic degradation [71] and connects flagellar gene expression to nutrient starvation [72]. In agreement with previously reported role of EAL-containing protein YdiV in negative regulation of motility in uropathogenic *E. coli* [73] and *S.* Typhimurium [74], we found that *ydiV* deletion mutant in *S.* Typhimurium was more motile on swimming agar.

Third, in our screen a *ΔphoP* mutant displayed increased motility both on swimming and swarming plates (Table 2, Figure 4). PhoP, the response regulator DNA-binding protein of PhoP/PhoQ two component system [75,76], negatively regulates motility in various microorganisms including *Photorhabdus luminescens*, *Pseudomonas aeruginosa*, *Proteus mirabilis* and in uropathogenic *E. coli*, where a *phoP* nulls are hypermotile [66,77–79]. Moreover, null allele in *Proteus mirabilis phoP* homolog exhibits an elevated level of *flhDC* [66]. Finally, in uropathogenic *E. coli* inactivation of *phoP* results in the increased expression of flagellin on the bacterial surface [78]. Similar to uropathogenic *E. coli phoP* mutant in *S.* Typhimurium increased motility correlated with increased flagellin expression on the bacterial surface (Figure 5).

We hypothesized that enhanced motility of at least some mutants could be due to increased expression of flagellins on the bacterial cell surface as was previously shown for *phoP* mutant in uropathogenic *E. coli* [78] and for *ydiV* mutant in *Salmonella* and uropathogenic *E. coli* [73,80]. Therefore, we examined the amount of FliC and FljB on the bacterial surface for each mutant that displayed hypermotility both on swimming and swarming agar (Figure 5, Figure S2). We found that while our *ΔphoP* mutant had wild type level of expression for FliC, it also had more FljB on the bacterial surface than the isogenic wild type organism. Furthermore, our examination of flagellar proteins sheared from the bacterial surface showed that thirteen out of fourteen of our hypermotile mutants had more FliC or FljB on the bacterial surface (Figure 5, Figure S2) than the wild type. It seems plausible that the hypermotility phenotype we observed for this highly diverse group of mutants could be the result of up-regulation of flagella.

Interestingly, mutations in several fimbrial operons also result in hypermotility phenotypes, *ΔSTM0026* (*bcfF*) mutants were hypermotile and the *bcfABCDEFG* (bovine colonization factor) operon has previously been implicated in virulence in mice [81]. Inactivation of *bcfF* improves biofilm formation on HEp-2 tissue culture cells and chicken intestinal epithelium in comparison to wild type [82]. Mutants in FimZ (STM0549) were also hypermotile. Inactivation of FimZ reduces expression from the P_{fimA} promoter and prevents serovar Typhimurium from making type I fimbriae [83,84]. FimZ also binds to the P_{flhDC} promoter and represses the expression of *flhDC* operon.

Concluding remarks

In this study we evaluated the ability of 1023 defined non-polar single gene deletions of *Salmonella* Typhimurium to move away from the site of inoculation on swimming and swarming agar plates. Over 90% of deletions in our collection were introduced in genes present exclusively in *Salmonella* or closely related pathogenic bacteria from the *Enterobacteriaceae*. Many of the mutants used in this study did not have any functions previously assigned. We confirmed motility phenotypes associated with loss of genes involved in flagellar regulon, LPS biosynthesis and chemotaxis. We identified a number of novel contributors to bacterial motility including both known and uncharacterized genes in our pathogenicity-biased collection. Furthermore, we found that there are genes needed for both types motility, and there are genes that make a unique contribution to different kinds of motility. Finally, we identified mutations in a number of genes result in increased motility. Determination of the molecular mechanisms of the improved motility is a fascinating area of the future work.

Supporting Information

Figure S1   Mutants with severe defects in swimming and swarming motility grow indistinguishably from wild type. Overnight cultures were subcultured at 1/100 ration in LB-broth and incubated at 37°C with shaking. Bacterial growth was monitored by OD600 in three independent experiments.

Figure S2   Flagellin expression on the cell surface correlates with the ability to move on swimming and swarming agar for some, but not all mutants. Flagellins sheared from the bacterial surface from strains with decreased (A) or increased (B, C) motility grown in LB-broth were analyzed by Western blotting with μ-FliC and μ-FljB sera. The whole cell lysates for each sample were also analyzed by SDS-PAGE and stained with Coomassie as a loading control.

Table S1   Swimming and swarming motility scores from high-throughput screening.

Table S2   Confirmation of motility defects (loss of>75% motility compared to wild type) in mutants identified in primary screening.

Table S3   Mutants with defect in swimming motility only.

Table S4   Mutants with defect in swarming motility.

Table S5   Previously described hypermotility in bacteria.

Table S6   Densitometry data underlying Figures 3 and 5, and representative images of Western blots used to generate this data.

Movie S1   WT_400x_6h.mov shows examples of the swarming motility of the wild type organism.

Movie S2   fimZ_400x_6h.mov shows the swarming motility of a deletion mutant in fimZ that has increased swarming motility relative to the wild type.

Movie S3   STM1630_400x_6h.mov shows the swarming motility of a deletion mutant in STM1630 that has increased swarming motility relative to the wild type.

Author Contributions

Conceived and designed the experiments: LMB HLAP MM. Performed the experiments: LMB LA YR MT KDA. Analyzed the data: LMB KDA MM HLAP. Contributed reagents/materials/analysis tools: LMB MM HLAP. Wrote the paper: LMB MM HLAP.

References

1. Haraga A, Ohlson MB, Miller SI (2008) Salmonellae interplay with host cells. Nat Rev Microbiol 6: 53–66.
2. Rabsch W, Tschape H, Baumler AJ (2001) Non-typhoidal salmonellosis: emerging problems. Microbes Infect 3: 237–247.
3. Chevance FF, Hughes KT (2008) Coordinating assembly of a bacterial macromolecular machine. Nat Rev Microbiol 6: 455–465.
4. Wang Q, Frye JG, McClelland M, Harshey RM (2004) Gene expression patterns during swarming in Salmonella typhimurium: genes specific to surface growth and putative new motility and pathogenicity genes. Mol Microbiol 52: 169–187.
5. Ottemann KM, Miller JF (1997) Roles for motility in bacterial-host interactions. Mol Microbiol 24: 1109–1117.
6. Harshey RM (2003) Bacterial motility on a surface: many ways to a common goal. Annu Rev Microbiol 57: 249–273.
7. Adler J (1966) Chemotaxis in bacteria. Science 153: 708–716.
8. Fraser GM, Hughes C (1999) Swarming motility. Curr Opin Microbiol 2: 630–635.
9. Berg HC (2005) Swarming motility: it better be wet. Curr Biol 15: R599–600.
10. Chen BG, Turner L, Berg HC (2007) The wetting agent required for swarming in Salmonella enterica serovar typhimurium is not a surfactant. J Bacteriol 189: 8750–8753.
11. Toguchi A, Siano M, Burkart M, Harshey RM (2000) Genetics of swarming motility in Salmonella enterica serovar typhimurium: critical role for lipopolysaccharide. J Bacteriol 182: 6308–6321.
12. Wang Q, Suzuki A, Mariconda S, Porwollik S, Harshey RM (2005) Sensing wetness: a new role for the bacterial flagellum. EMBO J 24: 2034–2042.
13. Girgis HS, Liu Y, Ryu WS, Tavazoie S (2007) A comprehensive genetic characterization of bacterial motility. PLoS Genet 3: 1644–1660.

14. Inoue T, Shingaki R, Hirose S, Waki K, Mori H, et al. (2007) Genome-wide screening of genes required for swarming motility in Escherichia coli K-12. J Bacteriol 189: 950–957.
15. Santiviago CA, Reynolds MM, Porwollik S, Choi SH, Long F, et al. (2009) Analysis of pools of targeted Salmonella deletion mutants identifies novel genes affecting fitness during competitive infection in mice. PLoS Pathog 5: e1000477.
16. Bogomolnaya LM, Santiviago CA, Yang HJ, Baumler AJ, Andrews-Polymenis HL (2008) 'Form variation' of the O12 antigen is critical for persistence of Salmonella Typhimurium in the murine intestine. Mol Microbiol 70: 1105–1119.
17. Sambrook J, Fritsch EF, Maniatis T (1989) Molecular cloning: a laboratory manual. Cold Spring Harbor, N.Y.: Cold Spring Harbor Laboratory.
18. Guard-Petter J (1997) Induction of flagellation and a novel agar-penetrating flagellar structure in Salmonella enterica grown on solid media: possible consequences for serological identification. FEMS Microbiol Lett 149: 173–180.
19. McClelland M, Sanderson KE, Spieth J, Clifton SW, Latreille P, et al. (2001) Complete genome sequence of Salmonella enterica serovar Typhimurium LT2. Nature 413: 852–856.
20. Lawley TD, Chan K, Thompson LJ, Kim CC, Govoni GR, et al. (2006) Genome-wide screen for Salmonella genes required for long-term systemic infection of the mouse. PLoS Pathog 2: e11.
21. Chilcott GS, Hughes KT (2000) Coupling of flagellar gene expression to flagellar assembly in Salmonella enterica serovar typhimurium and Escherichia coli. Microbiol Mol Biol Rev 64: 694–708.
22. Nevola JJ, Stocker BA, Laux DC, Cohen PS (1985) Colonization of the mouse intestine by an avirulent Salmonella typhimurium strain and its lipopolysaccharide-defective mutants. Infect Immun 50: 152–159.
23. Mariconda S, Wang Q, Harshey RM (2006) A mechanical role for the chemotaxis system in swarming motility. Mol Microbiol 60: 1590–1602.

24. Blair DF, Kim DY, Berg HC (1991) Mutant MotB proteins in Escherichia coli. J Bacteriol 173: 4049–4055.
25. Wang KC, Hsu YH, Huang YN, Yeh KS (2012) A previously uncharacterized gene stm0551 plays a repressive role in the regulation of type 1 fimbriae in Salmonella enterica serotype Typhimurium. BMC Microbiol 12: 111.
26. Ahmad I, Lamprokostopoulou A, Le Guyon S, Streck E, Barthel M, et al. (2011) Complex c-di-GMP signaling networks mediate transition between virulence properties and biofilm formation in Salmonella enterica serovar Typhimurium. PLoS One 6: e28351.
27. Wolfe AJ, Visick KL (2008) Get the message out: cyclic-Di-GMP regulates multiple levels of flagellum-based motility. J Bacteriol 190: 463–475.
28. Clegg S, Hughes KT (2002) FimZ is a molecular link between sticking and swimming in Salmonella enterica serovar Typhimurium. J Bacteriol 184: 1209–1213.
29. Reynolds MM, Bogomolnaya L, Guo J, Aldrich L, Bokhari D, et al. (2011) Abrogation of the twin arginine transport system in Salmonella enterica serovar Typhimurium leads to colonization defects during infection. PLoS One 6: e15800.
30. Paul K, Nieto V, Carlquist WC, Blair DF, Harshey RM (2010) The c-di-GMP binding protein YcgR controls flagellar motor direction and speed to affect chemotaxis by a "backstop brake" mechanism. Mol Cell 38: 128–139.
31. Kato Y, Sugiura M, Mizuno T, Aiba H (2007) Effect of the arcA mutation on the expression of flagella genes in Escherichia coli. Biosci Biotechnol Biochem 71: 77–83.
32. Simm R, Morr M, Kader A, Nimtz M, Romling U (2004) GGDEF and EAL domains inversely regulate cyclic di-GMP levels and transition from sessility to motility. Mol Microbiol 53: 1123–1134.
33. Zhao R, Amsler CD, Matsumura P, Khan S (1996) FliG and FliM distribution in the Salmonella typhimurium cell and flagellar basal bodies. J Bacteriol 178: 258–265.
34. Tatusov RL, Natale DA, Garkavtsev IV, Tatusova TA, Shankavaram UT, et al. (2001) The COG database: new developments in phylogenetic classification of proteins from complete genomes. Nucleic Acids Res 29: 22–28.
35. Ilg K, Endt K, Misselwitz B, Stecher B, Aebi M, et al. (2009) O-antigen-negative Salmonella enterica serovar Typhimurium is attenuated in intestinal colonization but elicits colitis in streptomycin-treated mice. Infect Immun 77: 2568–2575.
36. Abeyrathne PD, Daniels C, Poon KK, Matewish MJ, Lam JS (2005) Functional characterization of WaaL, a ligase associated with linking O-antigen polysaccharide to the core of Pseudomonas aeruginosa lipopolysaccharide. J Bacteriol 187: 3002–3012.
37. Blondel CJ, Jimenez JC, Contreras I, Santiviago CA (2009) Comparative genomic analysis uncovers 3 novel loci encoding type six secretion systems differentially distributed in Salmonella serotypes. BMC Genomics 10: 354.
38. Folkesson A, Lofdahl S, Normark S (2002) The Salmonella enterica subspecies I specific centisome 7 genomic island encodes novel protein families present in bacteria living in close contact with eukaryotic cells. Res Microbiol 153: 537–545.
39. Gomez-Gomez JM, Manfredi C, Alonso JC, Blazquez J (2007) A novel role for RecA under non-stress: promotion of swarming motility in Escherichia coli K-12. BMC Biol 5: 14.
40. Stafford GP, Hughes C (2007) Salmonella typhimurium flhE, a conserved flagellar regulon gene required for swarming. Microbiology 153: 541–547.
41. Lee J, Harshey RM (2012) Loss of FlhE in the flagellar Type III secretion system allows proton influx into Salmonella and Escherichia coli. Mol Microbiol 84: 550–565.
42. Morgenstein RM, Clemmer KM, Rather PN (2010) Loss of the waaL O-antigen ligase prevents surface activation of the flagellar gene cascade in Proteus mirabilis. J Bacteriol 192: 3213–3221.
43. Haneda T, Sugimoto M, Yoshida-Ohta Y, Kodera Y, Oh-Ishi M, et al. (2010) Comparative proteomic analysis of Salmonella enterica serovar Typhimurium ppGpp-deficient mutant to identify a novel virulence protein required for intracellular survival in macrophages. BMC Microbiol 10: 324.
44. Traxler MF, Summers SM, Nguyen H-T, Zacharia VM, Hightower GA, et al. (2008) The global, ppGpp-mediated stringent response to amino acid starvation in Escherichia coli. Molecular microbiology 68: 1128–1148.
45. Suzuki H, Koyanagi T, Izuka S, Onishi A, Kumagai H (2005) The yliA, -B, -C, and -D genes of Escherichia coli K-12 encode a novel glutathione importer with an ATP-binding cassette. J Bacteriol 187: 5861–5867.
46. Gillespie JJ, Wattam AR, Cammer SA, Gabbard JL, Shukla MP, et al. (2011) PATRIC: the comprehensive bacterial bioinformatics resource with a focus on human pathogenic species. Infect Immun 79: 4286–4298.
47. Kim W, Surette MG (2004) Metabolic differentiation in actively swarming Salmonella. Mol Microbiol 54: 702–714.
48. Frye J, Karlinsey JE, Felise HR, Marzolf B, Dowidar N, et al. (2006) Identification of new flagellar genes of Salmonella enterica serovar Typhimurium. J Bacteriol 188: 2233–2243.
49. Wang Q, Mariconda S, Suzuki A, McClelland M, Harshey RM (2006) Uncovering a large set of genes that affect surface motility in Salmonella enterica serovar Typhimurium. J Bacteriol 188: 7981–7984.
50. Kohler T, Curty LK, Barja F, van Delden C, Pechere JC (2000) Swarming of Pseudomonas aeruginosa is dependent on cell-to-cell signaling and requires flagella and pili. J Bacteriol 182: 5990–5996.
51. Chandra H, Khandelwal P, Khattri A, Banerjee N (2008) Type 1 fimbriae of insecticidal bacterium Xenorhabdus nematophila is necessary for growth and colonization of its symbiotic host nematode Steinernema carpocapsiae. Environ Microbiol 10: 1285–1295.
52. Stocker BAD, McDonough MW, Ambler RP (1961) A gene determining presence or absence of epsilon-N-methyl-lysine in Salmonella flagellar protein. Nature 189: 556–558.
53. Verstraeten N, Braeken K, Debkumari B, Fauvart M, Fransaer J, et al. (2008) Living on a surface: swarming and biofilm formation. Trends Microbiol 16: 496–506.
54. Chaudhuri RR, Peters SE, Pleasance SJ, Northen H, Willers C, et al. (2009) Comprehensive identification of Salmonella enterica serovar typhimurium genes required for infection of BALB/c mice. PLoS Pathog 5: e1000529.
55. Morgan E, Campbell JD, Rowe SC, Bispham J, Stevens MP, et al. (2004) Identification of host-specific colonization factors of Salmonella enterica serovar Typhimurium. Mol Microbiol 54: 994–1010.
56. Groisman EA, Mouslim C (2006) Sensing by bacterial regulatory systems in host and non-host environments. Nat Rev Microbiol 4: 705–709.
57. Navarre WW, Halsey TA, Walthers D, Frye J, McClelland M, et al. (2005) Co-regulation of Salmonella enterica genes required for virulence and resistance to antimicrobial peptides by SlyA and PhoP/PhoQ. Mol Microbiol 56: 492–508.
58. Monsieurs P, De Keersmaecker S, Navarre WW, Bader MW, De Smet F, et al. (2005) Comparison of the PhoPQ regulon in Escherichia coli and Salmonella typhimurium. J Mol Evol 60: 462–474.
59. Shah DH, Lee MJ, Park JH, Lee JH, Eo SK, et al. (2005) Identification of Salmonella gallinarum virulence genes in a chicken infection model using PCR-based signature-tagged mutagenesis. Microbiology 151: 3957–3968.
60. Yoon H, McDermott JE, Porwollik S, McClelland M, Heffron F (2009) Coordinated regulation of virulence during systemic infection of Salmonella enterica serovar Typhimurium. PLoS Pathog 5: e1000306.
61. Waite RD, Rose RS, Rangarajan M, Aduse-Opoku J, Hashim A, et al. (2012) Pseudomonas aeruginosa possesses two putative type I signal peptidases, LepB and PA1303, each with distinct roles in physiology and virulence. J Bacteriol 194: 4521–4536.
62. Caiazza NC, Merritt JH, Brothers KM, O'Toole GA (2007) Inverse regulation of biofilm formation and swarming motility by Pseudomonas aeruginosa PA14. J Bacteriol 189: 3603–3612.
63. Kuchma SL, Ballok AE, Merritt JH, Hammond JH, Lu W, et al. (2010) Cyclic-di-GMP-mediated repression of swarming motility by Pseudomonas aeruginosa: the pilY1 gene and its impact on surface-associated behaviors. J Bacteriol 192: 2950–2964.
64. Merritt JH, Brothers KM, Kuchma SL, O'Toole GA (2007) SadC reciprocally influences biofilm formation and swarming motility via modulation of exopolysaccharide production and flagellar function. J Bacteriol 189: 8154–8164.
65. Merritt JH, Ha DG, Cowles KN, Lu W, Morales DK, et al. (2010) Specific control of Pseudomonas aeruginosa surface-associated behaviors by two c-di-GMP diguanylate cyclases. MBio 1.
66. Wang WB, Chen IC, Jiang SS, Chen HR, Hsu CY, et al. (2008) Role of RppA in the regulation of polymyxin b susceptibility, swarming, and virulence factor expression in Proteus mirabilis. Infect Immun 76: 2051–2062.
67. Clemmer KM, Rather PN (2008) The Lon protease regulates swarming motility and virulence gene expression in Proteus mirabilis. J Med Microbiol 57: 931–937.
68. Liaw SJ, Lai HC, Ho SW, Luh KT, Wang WB (2001) Characterisation of p-nitrophenylglycerol-resistant Proteus mirabilis super-swarming mutants. J Med Microbiol 50: 1039–1048.
69. Belas R, Schneider R, Melch M (1998) Characterization of Proteus mirabilis precocious swarming mutants: identification of rsbA, encoding a regulator of swarming behavior. J Bacteriol 180: 6126–6139.
70. Li B, Li N, Wang F, Guo L, Huang Y, et al. (2012) Structural insight of a concentration-dependent mechanism by which YdiV inhibits Escherichia coli flagellum biogenesis and motility. Nucleic Acids Res 40: 11073–11085.
71. Takaya A, Erhardt M, Karata K, Winterberg K, Yamamoto T, et al. (2012) YdiV: a dual function protein that targets FlhDC for ClpXP-dependent degradation by promoting release of DNA-bound FlhDC complex. Mol Microbiol 83: 1268–1284.
72. Wada T, Morizane T, Abo T, Tominaga A, Inoue-Tanaka K, et al. (2011) EAL domain protein YdiV acts as an anti-FlhD4C2 factor responsible for nutritional control of the flagellar regulon in Salmonella enterica Serovar Typhimurium. J Bacteriol 193: 1600–1611.
73. Simms AN, Mobley HL (2008) Multiple genes repress motility in uropathogenic Escherichia coli constitutively expressing type 1 fimbriae. J Bacteriol 190: 3747–3756.
74. Simm R, Remminghorst U, Ahmad I, Zakikhany K, Romling U (2009) A role for the EAL-like protein STM1344 in regulation of CsgD expression and motility in Salmonella enterica serovar Typhimurium. J Bacteriol 191: 3928–3937.
75. Miller SI, Kukral AM, Mekalanos JJ (1989) A two-component regulatory system (phoP phoQ) controls Salmonella typhimurium virulence. Proc Natl Acad Sci U S A 86: 5054–5058.
76. Groisman EA, Chiao E, Lipps CJ, Heffron F (1989) Salmonella typhimurium phoP virulence gene is a transcriptional regulator. Proc Natl Acad Sci U S A 86: 7077–7081.

77. Brinkman FS, Macfarlane EL, Warrener P, Hancock RE (2001) Evolutionary relationships among virulence-associated histidine kinases. Infect Immun 69: 5207–5211.

78. Alteri CJ, Lindner JR, Reiss DJ, Smith SN, Mobley HL (2011) The broadly conserved regulator PhoP links pathogen virulence and membrane potential in Escherichia coli. Mol Microbiol 82: 145–163.

79. Derzelle S, Turlin E, Duchaud E, Pages S, Kunst F, et al. (2004) The PhoP-PhoQ two-component regulatory system of Photorhabdus luminescens is essential for virulence in insects. J Bacteriol 186: 1270–1279.

80. Stewart MK, Cummings LA, Johnson ML, Berezow AB, Cookson BT (2011) Regulation of phenotypic heterogeneity permits Salmonella evasion of the host caspase-1 inflammatory response. Proc Natl Acad Sci U S A 108: 20742–20747.

81. Weening EH, Barker JD, Laarakker MC, Humphries AD, Tsolis RM, et al. (2005) The Salmonella enterica serotype Typhimurium lpf, bcf, stb, stc, std, and sth fimbrial operons are required for intestinal persistence in mice. Infect Immun 73: 3358–3366.

82. Ledeboer NA, Frye JG, McClelland M, Jones BD (2006) Salmonella enterica serovar Typhimurium requires the Lpf, Pef, and Tafi fimbriae for biofilm formation on HEp-2 tissue culture cells and chicken intestinal epithelium. Infect Immun 74: 3156–3169.

83. Yeh KS, Hancox LS, Clegg S (1995) Construction and characterization of a fimZ mutant of Salmonella typhimurium. J Bacteriol 177: 6861–6865.

84. Yeh KS, Tinker JK, Clegg S (2002) FimZ binds the Salmonella typhimurium fimA promoter region and may regulate its own expression with FimY. Microbiol Immunol 46: 1–10.

α-Mangostin Disrupts the Development of *Streptococcus mutans* Biofilms and Facilitates Its Mechanical Removal

Phuong Thi Mai Nguyen[1]*, **Megan L. Falsetta**[2], **Geelsu Hwang**[3], **Mireya Gonzalez-Begne**[2], **Hyun Koo**[2,3]*

1 Institute of Biotechnology, Vietnam Academy of Science and Technology, Hanoi, Vietnam, **2** Center for Oral Biology, University of Rochester Medical Center, Rochester, New York, United States of America, **3** Biofilm Research Labs, Levy Center for Oral Health, Department of Orthodontics, School of Dental Medicine, University of Pennsylvania, Philadelphia, Pennsylvania, United States of America

Abstract

α-Mangostin (αMG) has been reported to be an effective antimicrobial agent against planktonic cells of *Streptococcus mutans*, a biofilm-forming and acid-producing cariogenic organism. However, its anti-biofilm activity remains to be determined. We examined whether αMG, a xanthone purified from *Garcinia mangostana* L grown in Vietnam, disrupts the development, acidogenicity, and/or the mechanical stability of *S. mutans* biofilms. Treatment regimens simulating those experienced clinically (twice-daily, 60 s exposure each) were used to assess the bioactivity of αMG using a saliva-coated hydroxyapatite (sHA) biofilm model. Topical applications of early-formed biofilms with αMG (150 µM) effectively reduced further biomass accumulation and disrupted the 3D architecture of *S. mutans* biofilms. Biofilms treated with αMG had lower amounts of extracellular insoluble and intracellular iodophilic polysaccharides (30–45%) than those treated with vehicle control ($P<0.05$), while the number of viable bacterial counts was unaffected. Furthermore, αMG treatments significantly compromised the mechanical stability of the biofilm, facilitating its removal from the sHA surface when subjected to a constant shear stress of 0.809 N/m^2 (>3-fold biofilm detachment from sHA vs. vehicle-treated biofilms; $P<0.05$). Moreover, acid production by *S. mutans* biofilms was disrupted following αMG treatments (vs. vehicle-control, $P<0.05$). The activity of enzymes associated with glucan synthesis, acid production, and acid tolerance (glucosyltransferases B and C, phosphotransferase-PTS system, and F_1F_0-ATPase) were significantly inhibited by αMG. The expression of *manL*, encoding a key component of the mannose PTS, and *gtfB* were slightly repressed by αMG treatment ($P<0.05$), while the expression of *atpD* (encoding F-ATPase) and *gtfC* genes was unaffected. Hence, this study reveals that brief exposures to αMG can disrupt the development and structural integrity of *S. mutans* biofilms, at least in part via inhibition of key enzymatic systems associated with exopolysaccharide synthesis and acidogenicity. αMG could be an effective anti-virulence additive for the control and/or removal of cariogenic biofilms.

Editor: Jens Kreth, University of Oklahoma Health Sciences Center, United States of America

Funding: This work was supported by National Foundation for Science and Technology Development (Nafosted) grant (106.05-2011.44) to PTMN (http://www.nafosted.gov.vn); Vietnam Education Foundation (VEF) research scholar grant (VEF 2012) to PTMN (https://home.vef.gov); and National Institutes of Health (NIH) grant (DE018023) to HK. The funders had no role in study design, data collection and analysis, decision to publish, or preparation of the manuscript.

Competing Interests: The authors have declared that no competing interests exist.

* Email: koohy@dental.upenn.edu (HK); phuong_nguyen_99@yahoo.com (PTMN)

Introduction

Many infectious diseases in human are caused by virulent biofilms, including oral diseases [1]. Among them, dental caries continues to be one of the most ubiquitous and costly biofilm-dependent diseases throughout the world [2,3]. For organisms associated with caries development, the production of an extracellular polysaccharide (EPS)-rich biofilm matrix, acidification of the milieu, and the maintenance of acidic pH microenvironment in close proximity to the tooth enamel are major controlling virulence factors linked with the pathogenesis of the disease. Current therapeutic approaches to control pathogenic oral biofilms fall short; the search for new/improved agents may lead to more efficacious anti-caries therapies [4–6]. Natural products are currently regarded as potentially promising sources for new bioactive agents that may function to suppress these key virulence attributes that are associated with the establishment and maintenance of cariogenic biofilms [5].

The assembly of cariogenic biofilms results from complex interactions that occur between specific oral bacteria, the products they produce, host saliva and dietary carbohydrates, all of which occurs on pellicle-coated tooth surfaces [7,8]. *Streptococcus mutans* has been recognized as one of the key etiologic agents associated with the initiation of dental caries, although additional organisms may contribute to its pathogenesis [9]. Sucrose is considered the primary catalyst for caries development, as it serves as a substrate for the production of both EPS and acids. *S. mutans* can effectively form cariogenic biofilms when sucrose is available, because this bacterium rapidly synthesizes EPS (from sucrose) through the activity of exoenzymes (e.g. glucosyltransferases; Gtfs) [8]. At the same time, *S. mutans* produces acid and is highly aciduric, allowing it to tolerate and continue to produce acids in low pH microenvironments, while readily adapting to acidic and other environmental stresses [10–14].

EPS synthesis via *S. mutans*-derived Gtfs is critical for cariogenic biofilm formation, since the glucans produced by the secreted exoenzymes (present in the pellicle-coated tooth and on bacterial surfaces) promote local bacterial accumulation, while embedding bacteria in a diffusion-limiting matrix. These processes create highly cohesive and adhesive biofilms that are firmly attached to surfaces and are difficult to remove [15–18]. At the same time, the EPS-rich matrix shelters resident organisms from antimicrobial and other inimical influences [18–20]. In parallel, sugars (in addition to sucrose) are fermented by *S. mutans* and other acidogenic bacteria ensnared within the biofilm matrix, creating acidic microenvironments across the three-dimensional (3D) architecture and at the surface of attachment [18,21,22]. Acidification of the milieu favors growth of aciduric organisms, further enhancing EPS production and ensuring biofilm accrual and localized acid-dissolution of the enamel in areas where biofilm is present and pH is low [18,23]. Therefore, using bioactive agents that target EPS-mediated biofilm assembly and acidogenicity could disrupt the pathogenesis of dental caries in a highly effective and precise manner.

Plants are valuable sources of new bioactive compounds to combat dental caries, because they produce a wide variety of secondary metabolites, many of which have been found to have biological properties against oral pathogens *in vitro* (as reviewed in Jeon et al. [5]). *Garcinia mangostana* L. (Guttiferae) is a widely cultivated fruit tree in Southeast Asian nations, including Thailand, Sri Lanka, The Philippines, and Vietnam [24]. The pericarp of *G. mangostana* has been used in traditional medicine to treat a variety of infections. Experimental studies have demonstrated that xanthone derivatives are the major bioactive substances, exhibiting antioxidant, antitumor, anti-inflammatory, and antimicrobial activities [24–26].

Our previous work showed that αMG exhibits antimicrobial activity against planktonic *S. mutans* cells via multiple actions, particularly reducing acid production by disrupting the membrane of this organism [27]. However, the question as to whether this agent is capable of compromising the ability of *S. mutans* to develop biofilms using a clinically relevant treatment regiment (brief topical exposures) remains to be elucidated. Therefore, the aim of the present study was to investigate the potential effectiveness of topical applications of αMG and its biological actions against *S. mutans* biofilm formation on saliva-coated apatitic surfaces.

Materials and Methods

Extraction and isolation of α-mangostin

Garcinia mangostana L is a fruit plant widely available in the south of Vietnam. The dried powder of samples of *Garcinia mangostana* peels collected from Binhduong province (south of Vietnam) was used in this study. No specific permission for collection of *G. mangostana* is required for this location because it is not an endangered or protected species. Ethanolic extracts of *G. mangostana* were prepared for the initial step of αMG isolation. The dried powder of *G. mangostana* peels collected from the South of Vietnam were extracted with ethanol at room temperature, followed by an evaporation of solvent to give a dark brown gummy residue. This residue was taken up in water followed by extraction with *n*-hexane to produce the most bioactive fractions. The *n*-hexane fraction was then evaporated and dried under reduced pressure. Further separation was performed using silica gel column chromatography (Merck

Kieselgel 60, 70–230 mesh) by eluting with *n*-hexane – ethyl acetate – methanol (6:3:0.1, by volume) and 10 mL volumes of eluant were collected in test tubes. The aliquots of each fraction were subjected to thin-layer chromatography (60 F254, 1 mm plate, Merck) in a solvent system containing toluene – ethyl acetate – acetone – formic acid (5:3:1:1, by volume). Partially purified αMG was recovered from the active fractions and then further separated by silica gel column chromatography (Merck Kieselgel 60, 70–230 mesh) and eluting with *n*-hexane – chloroform – ethyl acetate – methanol (4:1:0.5:0.3, by volume), yielding a single compound, αMG, as yellow crystals. The purity of αMG was examined by high-pressure liquid chromatography connected with mass spectrometry (LCMSD- Trap-SL Mass spectra, Agilent 1100, Palo Alto, California). The chemical structure (Fig. 1) of αMG was determined using nuclear magnetic resonance (Bruker Avance 500 spectrometer, Germany).

The compound at concentration of 100, 150 and 200 μM was dissolved in 25% ethanol, which was also used as a vehicle control; treatments with 25% ethanol did not affect the viability of cells of *S. mutans* in a biofilm when compared to untreated controls. The pH of the treatment solution was maintained at 5.8 ± 0.2, based on the observation that αMG activity is best at acidic pH [27].

Preparation and treatment of the biofilm

S. mutans UA159 (ATCC 700610), a proven virulent-cariogenic strain selected for genomic sequencing, was used in this study. Biofilms of *S. mutans* were formed on saliva coated hydroxyapatite (sHA) surfaces (12.7 mm in diameter, 1 mm in thickness, Clarkson Chromatography Products Inc., South Williamsport, PA), as previously described [28]. The biofilms were grown in ultra-filtered (10 kDa MW cut-off membrane; Prep/Scale, Millipore, MA) buffered tryptone-yeast extract broth (UFTYE; 2.5% tryptone and 1.5% yeast extract with the addition of 4.35 g/L of potassium phosphate and 1 g/L of $MgSO_4 \cdot 7H_2O$, pH 7.0) with 1% sucrose at 37°C and 5% CO_2. Briefly, *S. mutans* cells in exponential growth phase were inoculated into UFTYE and applied to wells containing sHA discs placed vertically in a custom-made holder. Biofilms were allowed to form on sHA discs and were treated for the first time with the test agents or vehicle control after 6 h of development. Subsequently, the biofilms were treated at 8 am (20 h-old) and 6 pm (30 h-old), with two more additional treatments the following day (8 am; 44 h-old and 6 pm; 54 h-old). The biofilms were exposed to the treatments for 60 s, dip-washed in sterile saline solution (0.89% w/v NaCl) to remove excess agents, and then transferred to fresh culture medium [29,30]. The biofilm was analyzed after 44 h and 68 h using confocal microscopy to examine the effects on the overall 3D architecture after receiving the initial topical treatments (Figure 2). At 68 h, the biofilms were removed, homogenized and subjected to biochemical analysis as detailed previously [28]. Briefly, biomass was assessed with an aliquot of the homogenized suspension centrifuged at $10,000\,g$ for 10 min at 4°C, and the cell pellet was washed twice with water, then dried in the dry oven at 105°C for 24 h and weighed [28]. The water soluble and insoluble exopolysaccharides (EPS), and intracellular iodophilic polysaccharides (IPS) were extracted and quantified via colorimetric assays [28]. The total number of viable cells in each of the biofilms was determined by counting colony forming units (CFU), while total protein was quantified via ninhydrin assays as descrbed in Koo et al. [28]. Furthermore, the pH of the culture media of treated and untreated biofilms was monitored every 2 hours with an Orion pH electrode attached to an Orion 290 A+ pH meter (Thermo Fisher Scientific).

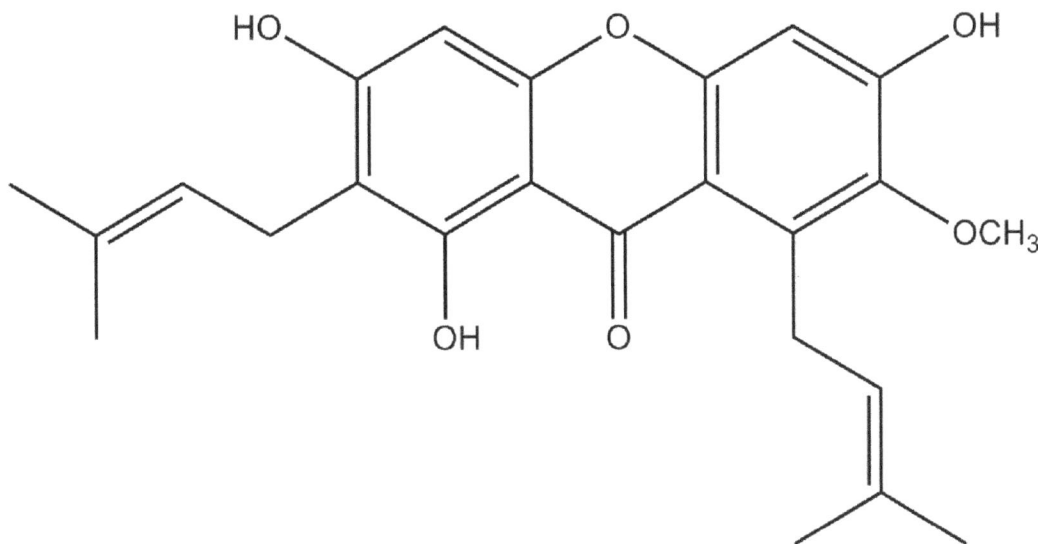

Figure 1. Chemical structure for αMG. Molecular formula: $C_{24}H_{26}O_6$. Molecular weight: 410.466.

Confocal microscopy of biofilms

The overall effect of topical applications of αMG on the 3D architecture and the spatial distribution of EPS and bacterial biomass within intact biofilms was assessed using confocal fluorescence imaging [18]. Briefly, 2.5 μM Alexa Fluor 647-labeled dextran conjugate (10,000 MW; absorbance/fluorescence emission maxima 647/668 nm; Molecular Probes Inc., Eugene, OR) was added to the culture medium during the formation and development of S. mutans biofilms. The fluorescently-labeled dextran serves as a primer for Gtf-mediated glucan synthesis and can be simultaneously incorporated during EPS matrix synthesis over the course of biofilm development, but does not stain the bacterial cells at the concentrations used in the study. The bacterial cells in the biofilms were labeled with 2.5 μM SYTO 9 green-fluorescent nucleic acid stain (480/500 nm; Molecular Probes Inc., Eugene, OR) using standard procedures [18]. Laser scanning confocal fluorescence imaging of the biofilms was performed using an Olympus FV 1000 two-photon laser scanning microscope (Olympus, Tokyo, Japan) equipped with a 10 X (0.45 numerical aperture) water immersion objective lens. Each biofilm was scanned at 5 randomly selected positions on the microscope stage and the confocal image series were generated by optical sectioning at each of these positions. Three independent experiments were conducted. The step size of z-series scanning was 2 μm. The confocal images were analyzed using software for simultaneous visualization of EPS and bacterial cells within intact biofilms [18,31,32]. Amira 5.4.1 software (Visage Imaging, San Diego, CA) was used to create 3D renderings of each structural component (EPS and bacteria) to examine the architecture of the biofilm.

Determination of mechanical stability of biofilms

The mechanical stability of the biofilms treated with or without αMG was compared using a custom built device (detailed information is in Figure S1). Biofilms were exposed to constant shear stress of 0.809 N/m^2 for 10 min, which is capable of removing S. mutans biofilm from sHA surface; such shear stress was determined as a threshold for >50% removal of untreated S. mutans biofilms from saliva-coated HA surfaces using our model. Shear stress at the biofilm surface was produced by shear flow generated via rotating paddle, and estimated based on Reynolds number of the flow (turbulent flow) and the surface friction using Blasius formula (Supplemental information). The amount of biofilm dry-weight (biomass) before and after application of shear

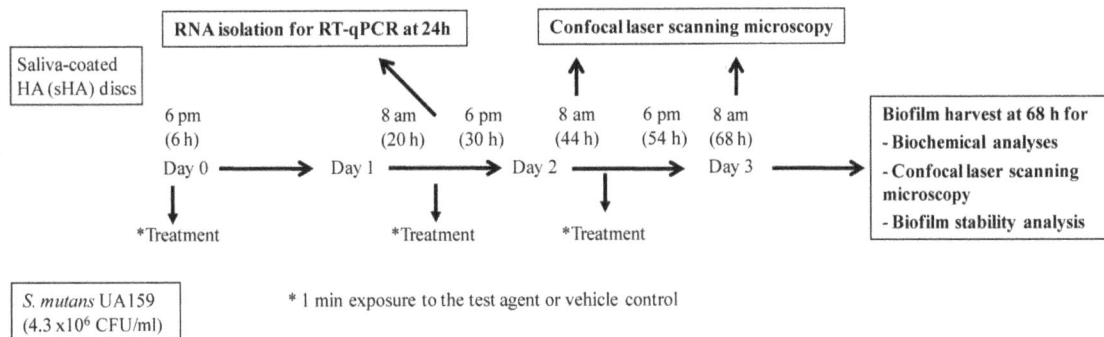

Figure 2. The experimental design for the treatment and analysis of biofilms of S. mutans. The clinical conditions of typical exposure of exogenously introduced therapeutic agents in the mouth were simulated by applying the test agent twice daily for brief exposures (60 s) at early/initial formation of the biofilm (6 h). Subsequently, the biofilms were treated twice at 8 am (20 h-old) and 6 pm (30 h-old) with two more additional treatments the following day (8 am; 44 h-old and 6 pm; 54 h-old).

Table 1. *Streptococcus mutans* UA159 biofilm composition after treatments with 150 µM αMG.

Biofilm composition	Vehicle	150 µM αMG
Dry weight (mg/biofilm)	4.73±0.41	3.00±0.45*
Protein (mg/biofilm)	2.85±0.38	1.67±0.19*
Soluble EPS (µg/biofilm)	326.6±37.2	229.7±91.0
Insoluble EPS (µg/biofilm)	1112.1±151.7	356.9±49.0*
IPS (µg/biofilm)	188.3±17.8	79.9±22.6*
CFU/biofilm	2.77E+08±5.98E+07	2.25E+08±5.50E+07

Data are expressed as the mean ± one standard deviation. For each parameter, values marked with an asterisk are significantly different from that for the vehicle control (n = 8; $P<0.05$, pair-wise comparison using Student's t test).

stress for each condition (vehicle- and αMG-treated) was determined. Then, the percentage of biofilm that remained on sHA disc surface was calculated. All experiments were performed in quadruplicates in three distinct experiments.

Gtf Docking Analyses

In the present study, different bioinformatics tools and databases were used. The crystal structure of glucosyltransferases C (GtfC) from the dental caries pathogen *Streptococcus mutans* is available in the Protein Data Bank (PDB) and was used as a receptor for docking of the αMG compound (ligand) using HEX software. Since the crystal structure of GtfB is not yet available, Phyre server [33] was used to predict ligand sites. HEX has been reported as an interactive molecular graphic program. It calculates protein-ligand docking, assuming that the ligand is rigid and then superimposes pairs of molecules using only their 3D shapes [34,35]. In addition, it uses Spherical Polar Fourier (SPF) correlations, increasing the speed of the calculations, and it also has integrated graphics software to view the final result [35–38]. PDB was used to download the crystal structure of glucansucrase from the dental caries pathogen *Streptococcus mutans* (http://www.rcsb.org/pdb/home/home.do). PubChem Compound was used for retrieving the 3D-structure of α-mangostin (http://www.ncbi.nlm.nih.gov/pccompound). MarvinSketch software was utilized for obtaining the α-mangostin structure in a PDB format (http://www.chemaxon.com/products/marvin/marvinsketch/), and the Hex-Server (HEX 6.9 software) was accessed for calculating and displaying protein-ligand docking (http://hexserver.loria.fr/). The parameters used for docking included: Correlation type (Shape only), FFT mode (3D fast life), Grid dimension (0.6), Receptor range (180), Ligand range (180), Twist range (360), and Distance range (40) were used.

Determination of Gtf activity

GtfB and GtfC were obtained from recombinant strains carrying the appropriate genes as detailed elsewhere [34]. Strain *S. milleri* KSB8 harboring the *gtfB* gene transformed from *S. mutans* GS-5 and *S. mutans* WHB 410 construct expressing *gtfC* gene only were used. The GtfB and GtfC enzymes (E.C. 2.4.1.5) were prepared from culture supernatants and purified to near homogeneity by hydroxyapatite column chromatography. The purified Gtfs (1–1.5 U) were mixed with the test compound and incubated with a [U-^{14}C-glucose]-sucrose substrate (0.2 µCi/ml; 200.0 mmol of sucrose per liter, 40 µmol of dextran 9000 per liter, and 0.02% sodium azide in adsorption buffer consisting of 50 mM KCl, 1.0 mM KPO$_4$, 1.0 mM CaCl$_2$, and 0.1 mM MgCl$_2$, pH 6.5) to a final concentration of 100 mmol of sucrose per liter (200 µl final volume) at 37°C with rocking for 4 h. For the vehicle-

control, the same reaction was carried out with 25% ethanol (v/v) replacing the test agent solutions. Glucosyltransferase activity was measured by incorporation of [U-^{14}C-glucose] from labeled sucrose into glucans [34]. The radiolabelled glucans were quantified by scintillation counting.

F-ATPase and phosphotransferase system (PTS) assays

F-ATPase and PTS activity of treated biofilm cells were determined as described by Belli and Marquis [39] and Phan et al. [40]. Biofilms were homogenized and centrifuged at 4°C, and then biofilm pellets from each sample were resuspended in 2.5 ml of 75 mM Tris-HCl buffer (pH 7.0) with 10 mM MgSO$_4$. Toluene (250 ul) was added to each biofilm cell suspension prior to vigorous vortex mixing and incubation for 5 min at 37°C. Each suspension was then subjected to two cycles of freezing in a dry ice-ethanol bath and thawing at 37°C. Permeabilized biofilm cells were harvested by centrifugation. They were then resuspended in 1.0 ml of 75 mM Tris-HCl buffer (pH 7.0) with 10 mM MgSO$_4$. The suspension was quickly frozen in a dry ice-ethanol bath and stored at −70°C for F-ATPase and PTS assays. F-ATPase activity was determined as described by Belli and Marquis [39]. The F-ATPase reaction is initiated by the addition of 30 µl of 0.5 M ATP (pH 6.0). Samples of 50 µl were removed and assayed for inorganic phosphate liberated from cleavage of ATP with reagents from American Monitor Co. (Indianapolis, IN) [39]. Phospho-transferase system (PTS) activity was assessed in terms of pyruvate production from phosphoenolpyruvate in response to glucose addition. Pyruvate was assayed by use of lactic dehydrogenase and measurements of the change in absorbance of 340 nm light associated with oxidation of NADH [39].

Reverse transcription quantitative PCR (RT-qPCR)

RT-qPCR was performed to evaluate the expression of the *gtfB*, *gtfC*, *atpD*, and *manL* genes. Biofilms were treated as described in the Figure 2. RNA was extracted and purified using standard protocols optimized for biofilms [41]. The RNA integrity numbers (RIN) of purified samples used for RT-qPCR were determined by microcapillary electrophoresis on an Agilent 2100 Bioanalyzer (Agilent Technologies, Santa Clara, CA). Purified RNA samples (RIN≥9) were stored in RNase-free water at −80°C. cDNAs were synthesized from 1 µg of purified RNA using a BioRad iScript cDNA synthesis kit (Bio-Rad Laboratories, Inc., Hercules, CA). RNA samples without reverse transcriptase were included as a negative control. The resulting cDNAs and negative controls were amplified by a MyiQ qPCR detection system with iQ SYBR Green supermix (Bio-Rad Laboratories, Inc., CA, USA) and specific primers. When Taqman probes were available, cDNAs and controls were amplified using a Bio-Rad CFX96

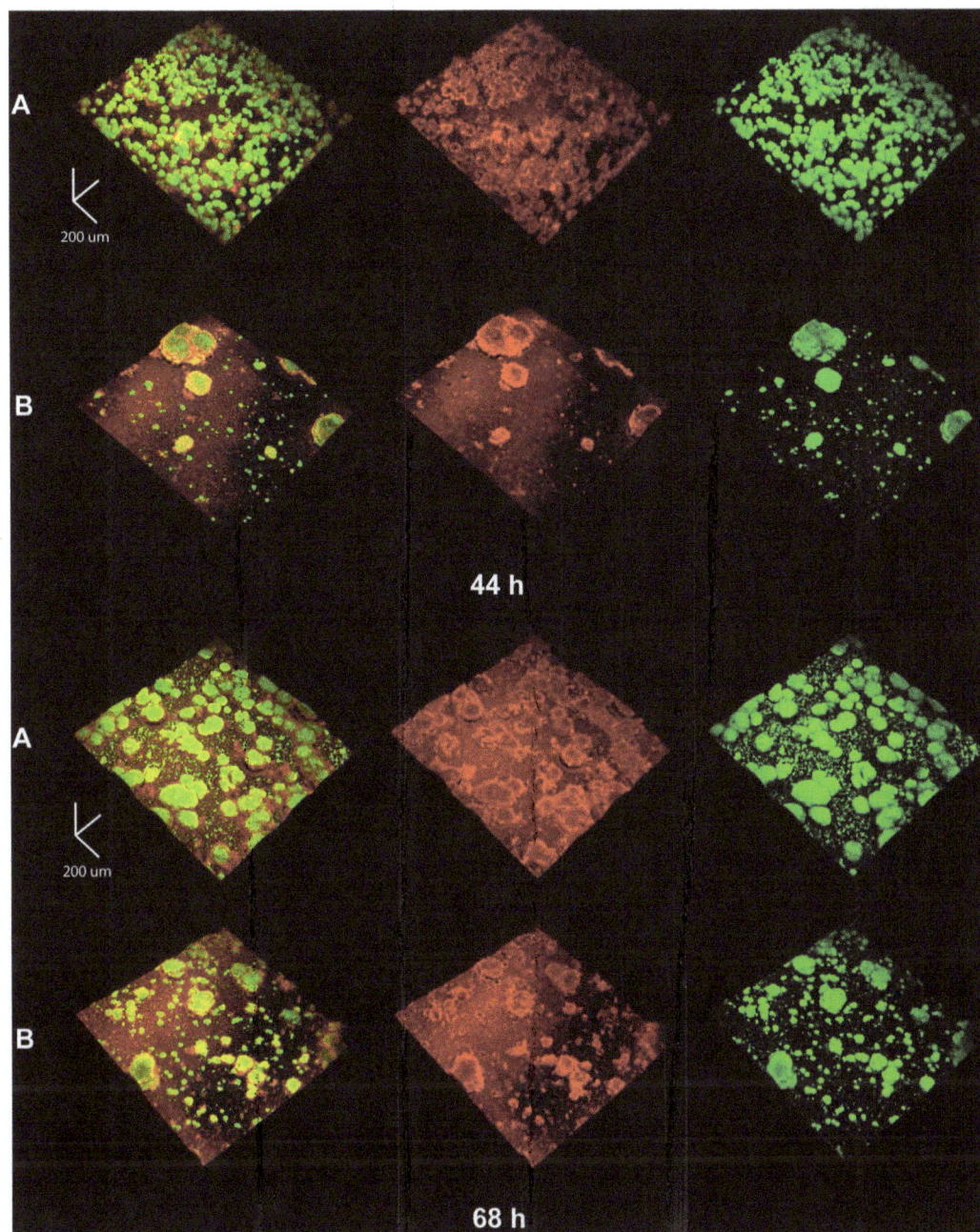

Figure 3. Representative 3D rendered images of 44 h and 68 h-old *S. mutans* biofilms following topical treatments. Biofilms were treated with the vehicle control in panel A and with 150 µM αMG in panel B. The EPS channel is in red, while bacterial cells are in green. Scale bars = 100 µm. Biofilms were formed on hydroxyapatite discs (sHA) in the presence of 1% (wt/vol) sucrose, and treated with test agents twice daily.

system (Bio-Rad Laboratories). The 16S rRNA primers/TaqMan probes were run separately, and primers/TaqMan probes for other specific targets were combined and used in a multiplex setting. For reactions with only one TaqMan probe (used for target 16S rRNA), the iQ Supermix (BioRad) was used. For multiplex reactions (*gtfB*, *gtfC*) and (*atpD*, *manL*) the iQ Multiplex Powermix (BioRad) were employed. Standard curves were used to determine the relative number of cDNA molecules, which were normalized to the relative number of 16S rRNA cDNA in each sample, as described previously [42]. 16S rRNA served as a reference gene [43]. These values were used to determine the fold-change between each treated sample and the vehicle control. The

MIQE guidelines [44] were followed for quality control of the data generated and for data analysis. The gene expression profile was determined 4 h after the topical treatment at 20 h (Figure 2), to evaluate the impact of αMG on *S. mutans* within the accumulated biofilms post-treatment. This time point represents the most active period of the biofilm development using our model, and was selected based on our biochemical data and previous studies on the dynamics of the *S. mutans* transcriptome during biofilm formation on sHA and in response to topically applied agents [43,45].

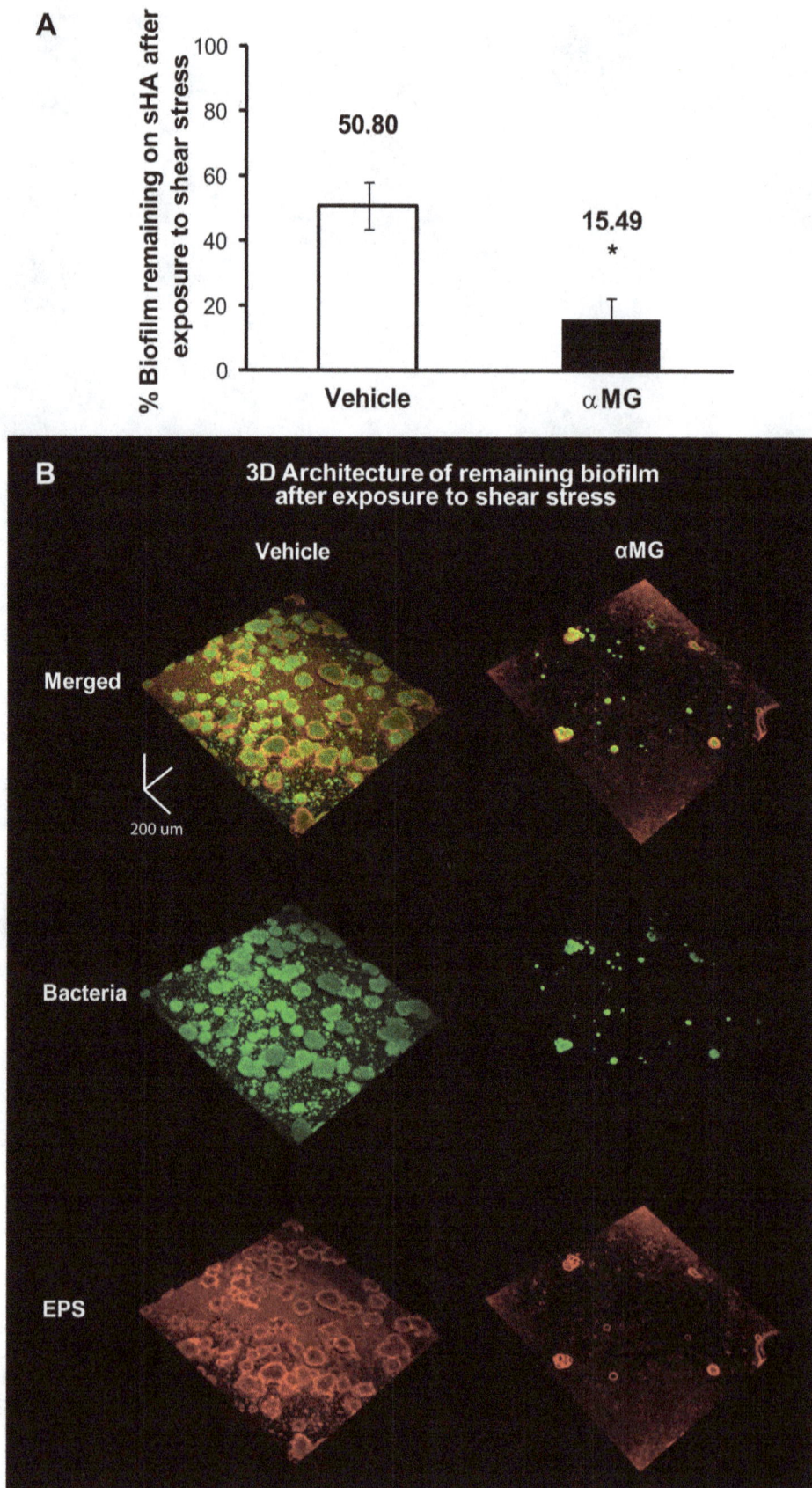

Figure 4. Biofilm mechanical stability following topical treatments. Panel A depicts the percentage of biomass of vehicle- or αMG-treated biofilms that remained on sHA after exposure to shear stress. The amount of biofilm dry-weight (biomass) before and after application of shear stress for each condition (vehicle- and αMG-treated) was determined, and the percentage of biofilm that remained on sHA disc surface was calculated. Data

are expressed as the mean ± one standard deviation. Values are significantly different from that for the vehicle control (n = 12; P<0.05, pair-wise comparison using Student's t test). Panel B shows representative 3D rendered images of treated-biofilms of *S. mutans* after shearing. The EPS channel is in red, and the bacterial cells are in green. Scale bars = 100 μm.

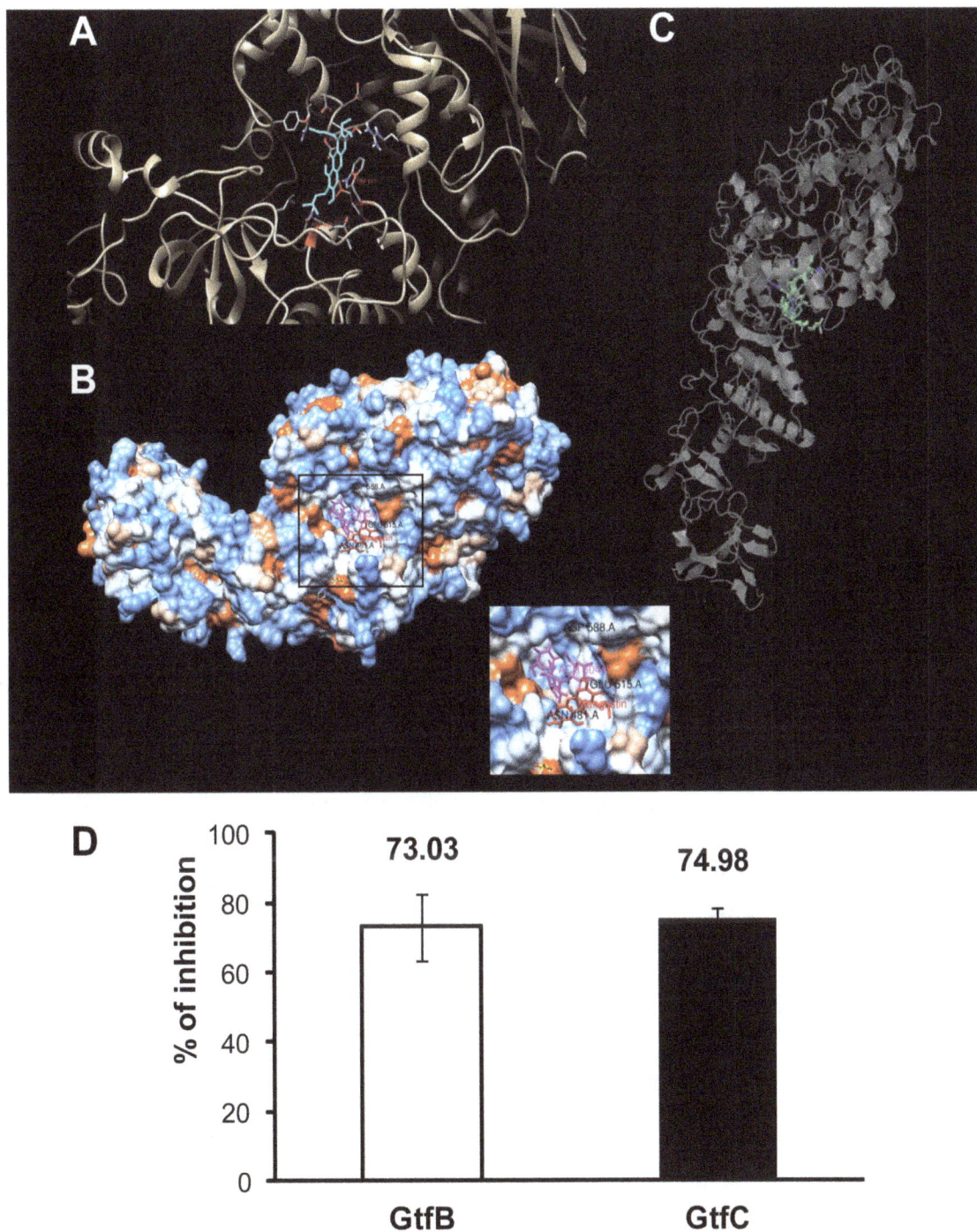

Figure 5. Snapshot of glucosyltransferase interaction with αMG compound and influence of 150 μM αMG on the activities of GtfB and GtfC. Panel A depicts the ribbon model of glucosyltransferase C (brown) docking α-mangostin (blue) using HEX-docking software Amino acids, such as Trp 517, Glu 515, Asp 588 and Asn 481 are interacting in the glucosyl binding site. Panel B depicts the surface model of glucosyltransferase C docking α-mangostin (red) and acarbose (purple) using HEX-docking software. Panel C depicts the glucosyltransferase B 3D ligand-binding site predicted model using Phyre Server. Panel D depicts Gtf activity of *S. mutans* cells when treated with αMG. The percentage of inhibition was calculated setting the vehicle control to 100% Gtf activity. Data are expressed as the mean ± one standard deviation. Values are significantly different from that for the vehicle control (n = 12; P<0.05, pair-wise comparison using Student's t test).

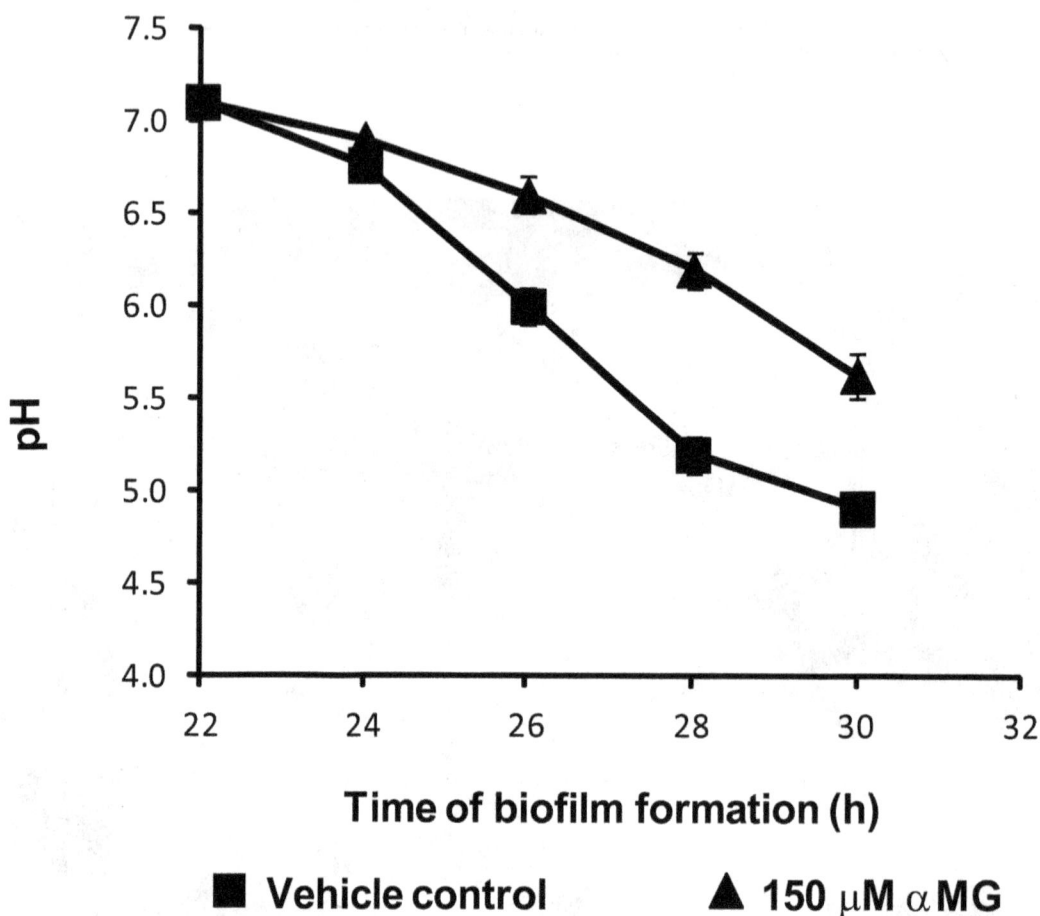

Figure 6. Effects of αMG on acid production by *S. mutans* UA159. Vehicle is represented by (■), while 150 μM αMG is represented by (▲). Data are expressed as the mean ± one standard deviation for experiments run in triplicates in at least three separate experiments.

Statistical analyses

Data are presented as the mean ± one standard deviation (SD). Pair-wise comparisons were made between test and control using Student's t-test. Statistical analysis was performed using JMP (version 3.1; SAS Institute, Cary, NC). The level of significance was set at 5%.

Results and Discussion

αMG disrupts the accumulation and acidogenicity of *S. mutans* biofilms

In our experiment, *S. mutans* biofilms were initially treated with α-mangostin (αMG) at concentrations of 100, 150, and 200 μM (Table S1) based on bioactivity against planktonic *S. mutans* cells [27] and solubility in the vehicle system. We selected a concentration of 150 μM αMG, because it was as effective as 200 μM in reducing the overall biofilm development and acid production.

The data in Table 1 indicate that treatments with 150 μM αMG significantly reduced the accumulation of *S. mutans* biofilms on saliva-coated apatitic surfaces, which resulted in less biomass (dry-weight) and less total protein compared to the vehicle control ($P<0.05$). The viability of the biofilms was not significantly impacted by the treatments. Nevertheless, short-term topical applications (one-minute exposure, twice daily) significantly reduced the amount of polysaccharides in the biofilms (Table 1).

The amount of insoluble exopolysaccharides (EPS) was drastically reduced, while the soluble EPS content was unaffected by αMG treatments. The data suggest that GtfB and GtfC, which are largely responsible for the synthesis of insoluble glucans in the biofilm matrix [8], could be targeted by αMG; while possibly having limited effects on the activity of GtfD (involved for soluble glucan synthesis). Interestingly, the amount of intracellular iodophilic polysaccharides (IPS), a glycogen-like storage polymer [46], was significantly disrupted by treatments with the agent.

Altogether, the biochemical changes inflicted by αMG may affect the matrix assembly and 3D biofilm architecture, which could disrupt the mechanical stability and adhesive strength of the treated biofilms.

αMG compromises the 3D architecture and mechanical stability of *S. mutans* biofilms

Confocal images revealed a marked impairment in the development of an insoluble EPS-matrix (in red), as well as the defective formation of bacterial clusters or microcolonies (in green) following αMG treatment, particularly at 44 h (Figure 3). The few microcolonies detected in the αMG-treated biofilms at 44 h visually appear to be larger than those treated with vehicle-control, suggesting that microcolony development was not completely inhibited. Nevertheless, the defective biofilm assembly resulted in an altered 3D architecture (at 68 h) characterized by sparsely distributed microcolonies (with many areas on the sHA surface

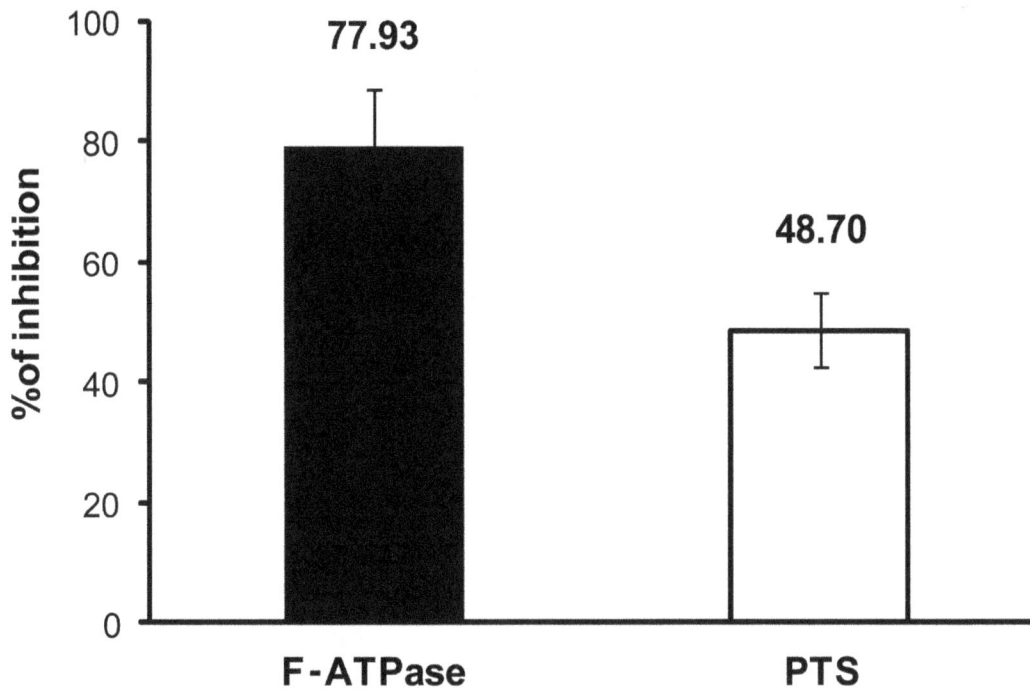

Figure 7. Effects of αMG on ATPase and PTS activities of *S. mutans* UA159. The percentage of inhibition was calculated setting the vehicle control to 100% enzymatic activity. Data are expressed as the mean ± one standard deviation. Values are significantly different from that for the vehicle control (n = 9; $P<0.05$, pair-wise comparison using Student's t test).

that were devoid of such structures), as well as a less developed EPS matrix, compared to vehicle-treated biofilms. These findings agree well with our biochemical data showing a significant reduction in the insoluble EPS content.

These structural changes may affect the stability of the biofilms treated with αMG and facilitate mechanical clearance of biofilms. The mechanical stability of biofilms appears to be dependent on the exopolysaccharide content, as EPS binds the cells together while strengthening their cohesiveness [15,47–50]. Furthermore, glucans enhance *S. mutans* adhesive strength, while the develop-

ment of multi-microcolony aggregates via EPS-cell adhesions provides structural integrity to *S. mutans* biofilms [16,18]. Thus, we hypothesized that the disruptive effects of αMG could facilitate biofilm removal and/or detachment. We investigated the impact of αMG on mechanical stability of *S. mutans* biofilms using a custom-built shear-inducing device (Figure S1).

The ability of treated-biofilms to withstand mechanical removal under shear stress was determined by measuring the amount of biofilm biomass (dry-weight) that remained on the sHA after shearing (Figure 4A). We observed that αMG-treated biofilms

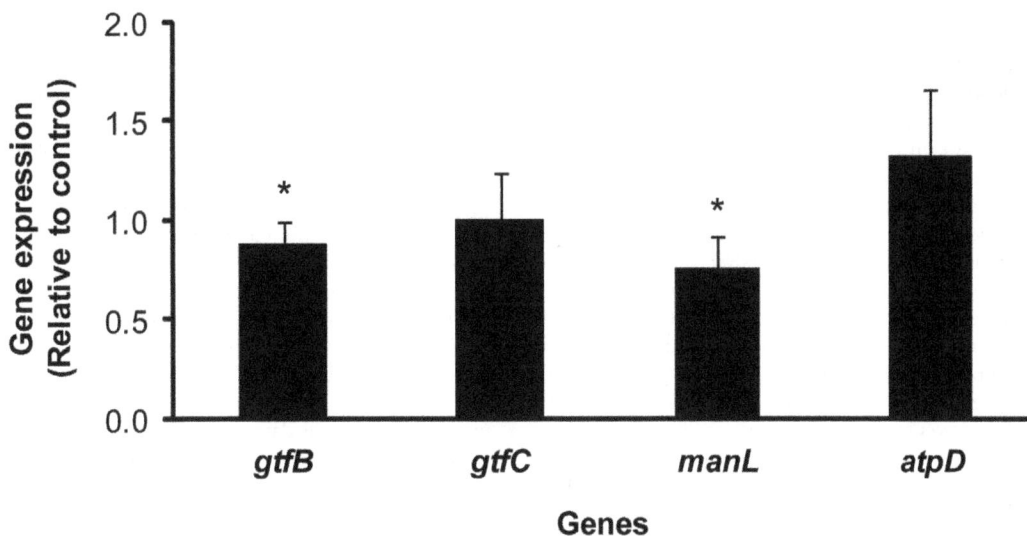

Figure 8. The expression of *S. mutans* genes *gtfB*, *gtfC*, *manL*, and *atpD* in biofilms. Fold changes ± one standard deviation. Values marked with asterisks are significantly different from that for the vehicle control (n = 8; $P<0.05$, pair-wise comparison using Student's t test).

were more effectively removed from the sHA surface (84.51% removal) than those treated with vehicle-control (49.2%; $P<0.05$) when subjected to shear stress, indicating that the mechanical stability of the biofilms was compromised by αMG. Indeed, confocal images of αMG-treated biofilms show that most of the bacterial biomass and EPS was removed, while vehicle-treated biofilms show numerous EPS-enmeshed bacterial microcolonies still attached on the sHA surface (Figure 4B). Clearly, the data demonstrate that alterations in the EPS-matrix and microcolony assembly resulted in significantly less adherent biofilms, which facilitated their mechanical clearance from the sHA surface when exposed to shear force. By reducing the production of insoluble EPS, αMG treatments could affect optimal microcolony formation and surface anchoring as well as the cell-matrix cross-linking forces and the overall viscoelasticity, which have been shown to be critical for weakening the biofilm structure [18,49,51]. Further studies shall elucidate how αMG affects the adhesion forces and rheological properties of the biofilms locally.

αMG inhibits GtfB and GtfC activity

Previous studies have shown that extracellular glucans produced by GtfB and GtfC enzymes play vital, yet distinct roles in the formation of cariogenic biofilms and are essential in the pathogenesis of dental caries (as reviewed in Bowen and Koo [8]). The glucans synthesized by GtfC assemble the initial EPS layers on the sHA surface, which provide enhanced binding sites for S. mutans colonization and accumulation [52,53]. Conversely, the highly insoluble and structurally rigid glucans formed by GtfB embed the cells, contributing to the scaffolding of the 3D EPS-rich matrix [18]. The accumulation of Gtf-derived EPS and bacteria cells mediates the construction of EPS-enmeshed microcolonies that are firmly anchored to the apatitic surface [16–18,54]. Here, we examined whether αMG is capable of inhibiting the activity of purified GtfB and GtfC enzymes, which could explain the defective assembly and attachment of the treated biofilms observed in this study. Since there is no previous data on Gtf inhibition by αMG, we initially examined the likelihood of the agent to bind Gtfs using in silico docking studies.

Docking studies support the prediction of conformation and binding affinity for selected molecules against a given target protein [55]. Therefore, docking of αMG on Gtf was carried out to explore if/how this compound might interact with the enzymes. In our study, when the GtfC enzyme was docked with αMG, the energy value obtained by HEX software was −511.36 Kcal/mol, indicating a stable and strong binding between the two molecules [55]. The best docked structure, visualized by UCSF Chimera molecular modeling system version 1.8 (http://www.cgl.ucsf.edu/chimera/download.html), showed the interaction of four amino acids (Trp 517, Glu 515, Asp 588 and Asn 481) (Figure 5A and 5B). A previous report by Ito et al. [56] indicated that binding to Glu 515 compromised the acid/base catalyst function, while interaction with Trp 517 blocked the acceptor glycosyl moiety. These observations can explain the inhibitory properties shown by acarbose when bound to Gtf-SI [56]. As displayed in figure 5B, αMG and acarbose interact with Trp 517, which provides the main frame for the glycosyl acceptor binding site. Since the crystal structure of GtfB is not yet available, Phyre server [33] was used to predict ligand sites. The obtained results highlighted the presence of hydrophobic amino acids Leu 356, Gln 35, Ala 409, Lys 408, Asn 410; Asp 878, Ser 880; Ser 884, Leu 882, Tyr 936, Phe 881; Asn 1026 and some other amino acids with electrically charged amino acids like Asp 838I (Figure 5C).

All amino acids mentioned above are found in catalytic or the glucan binding regions of both GtfB and GtfC [57–60], suggesting

that the function of these enzymes could be affected by αMG. Indeed, the enzymatic activity of purified GtfB and GtfC was impacted by αMG as shown in Figure 5D. The test agent was highly effective in reducing glucan synthesis by both enzymes, displaying more than 70% inhibition (vs. vehicle control) at 150 μM, which agrees well with the in silico analysis, as well as the biochemical (reduction of insoluble EPS content) and confocal imaging (defective assembly of EPS-matrix and impaired microcolony formation) data of the αMG-treated biofilms.

αMG affects acidogenicity of S. mutans biofilms, and disrupts F-ATPase and PTS activities

The biofilm EPS-matrix and microcolonies provide S. mutans with niches, where it survives and carries out glycolysis, even at low pH values, resulting in demineralization of the adjacent dental enamel [8,39]. In addition to the deleterious effects on biomass accumulation and structural organization, αMG also affected S. mutans biofilm acidogenicity following topical applications of the agent (Figure 6). αMG reduced both the acid production and the acid tolerance of S. mutans biofilm cells as indicated in the pH-drop profile (Figure 6). The test agent sensitized the biofilm cells to acidification to the point that the final pH value was significantly higher (~1 unit) than those treated with vehicle-control (P<0.05), suggesting that there may be disturbances in the activity of the proton-translocating membrane F-ATPase [27,39].

A previous study has shown that αMG was particularly effective in inhibiting the activity of F(H$^+$)-ATPase and PTS [27], which are critical for acid production and acid-tolerance and to ensure the optimum function of glycolysis by S. mutans within biofilms [61]. However, the assays were conducted with S. mutans grown in the planktonic phase. In this study, we examined the F-ATPase and PTS activity of biofilm cells following the treatment with αMG. The membrane-bound F-ATPase (H$^+$-translocating ATPase) is considered the primary determinant for acid tolerance [61]. During glycolysis, protons are pumped out of the cell by F-ATPase to help maintain ΔpH across the cell membrane, preventing acidification of the cytoplasm, which would typically inhibit intracellular enzymes [39]. Furthermore, under certain conditions, it also generates ATP for S. mutans growth and persistence [62]. The data in Figure 7 show that the F-ATPase activity was strongly inhibited by αMG with nearly 80% inhibition following topical treatments.

Conversely, sugar uptake by oral streptococci occurs primarily by means of the PTS system [63]. In this system, phosphoenolpyruvate (PEP), provided by glycolysis, is cleaved by Enzyme I and the phosphate group is transferred to a general phosphocarrier protein, HPr, which in turn acts as a phosphate donor to membrane-bound Enzyme II [63]. Thus, the system catalyzes the transfer of phosphate to an incoming sugar and translocation of it across the cell membrane to yield a sugar phosphate in the cytoplasm, at which point sugar is metabolized via glycolytic pathways to produce organic acids. As shown in Figure 7, the PTS activity of biofilms treated with αMG was also significantly inhibited (~50% inhibition vs. vehicle-treated biofilms, P<0.05). Although the exact nature of αMG inhibition of the F-ATPase and PTS system found in this study remains to be determined using purified enzymes, our data suggest that αMG can affect S. mutans biofilms acidogenicity by disrupting the activity of these critical membrane-associated enzymes (albeit at concentrations of 3–5 times higher than those found against planktonic cells [27].

The inhibitory effects of αMG on F-ATPase and PTS could have additional impact on biofilm composition and virulence. Cytoplasmic acidification and reduction of sugar transport not only disrupts glycolytic acid production, but also the formation and accumulation of intracellular iodophilic polysaccharides (IPS) [46], which could explain at least in part the marked reduction of IPS in the treated biofilms (Table 1). The role of IPS in *S. mutans* virulence and dental caries in general has been clearly documented [64–66]. IPS provides *S. mutans* with an endogenous source of carbohydrates that can be metabolized when exogenous fermentable substrates have been depleted within the oral cavity [67]. As a result, IPS can help to promote the formation of dental caries by prolonging the exposure of tooth surfaces to organic acids and a concomitant lower fasting pH in the matrix of the plaque [65]. Thus, the inhibition of IPS accumulation by αMG could also contribute with the overall disruptive effects of the agent on *S. mutans* biofilms acidogenicity.

αMG has limited effects on *gtfBC*, *atpD* and *manL* gene expression by *S. mutans* biofilms

Treatment of biofilms with α-mangostin could inhibit insoluble EPS synthesis and glycolytic pH drop in either of the following two ways: i) reducing enzymatic function and/or ii) affecting transcription of the genes encoding these enzymes to reduce the amount of enzyme produced. Therefore, we profiled the transcription of *gtfB*, *gtfC*, *atpD* (encoding F-ATPase), and *manL* (encoding a key component of the mannose PTS). The expression profiles of these genes are shown in Figure 8. Overall, RT-qPCR analysis showed only a slight repression of *gtfB* and *manL* after treatment with αMG ($P<0.05$), while no significant effects were observed on *gtfC* and *atpD* expression, suggesting that the reduction in EPS biomass in treated biofilms may be largely due to the impact on enzymatic function (Figure 5 and 7).

Upon biofilm establishment, the resident microorganisms, encased in an EPS-rich matrix, are difficult to remove or treat, while a highly acidogenic and aciduric biofilm environment is created [20]. In this paper, we reported that topical application of α-mangostin (αMG) can disrupt some of the major virulence properties of *S. mutans* within biofilms, impairing further biofilm accumulation and acidogenicity, while facilitating mechanical clearance. Although previous studies have shown the biological actions of αMG against planktonic cells of *S. mutans* and other organisms, this is the first study demonstrating the antibiofilm effects of this promising phytochemical agent. Analysis of our data shows that αMG could affect biofilm development by *S. mutans*

through at least three distinctive and yet interconnected ways: 1) disruption of insoluble EPS-matrix assembly at least in part by inhibiting GtfB and GtfC enzymatic activities, 2) compromising the mechanical stability, which may be linked to defective EPS production and impaired microcolony formation (thereby facilitating biofilm detachment from sHA surface), and 3) reducing acidogenicity by affecting IPS accumulation and the activities of the F-ATPase and PTS system. The results from this study indicate that GtfB and GtfC, as well as the F-ATPase and PTS enzymatic systems, are therapeutic targets of αMG.

In conclusion, our study demonstrated that the phytochemical αMG may represent a potentially useful anti-virulence additive for the control and/or removal of cariogenic biofilms. Having shown here that αMG exhibits significant bioactivity against *S. mutans* biofilms, further understanding of the molecular mechanisms of action of this agent as well as its effects on mixed-species cariogenic biofilm models are certainly warranted. Furthermore, cytotoxicity studies revealed that αMG is non-toxic and is generally regarded as safe [68–70]. Clearly, the efficacy of our treatment needs to be evaluated *in vivo* using a rodent model of dental caries.

Supporting Information

Figure S1 Biofilm mechanical strength testing device. This supplementary material shows the design of the custom-built device to evaluate biofilm mechanical strength, and the principles of shear stress calculation.

Table S1 Effects of α-mangostin on biofilm accumulation by *S. mutans* UA159.

Acknowledgments

The authors are thankful to Marlise Klein and Stacy Gregoire for technical assistance during the gene expression and Gtfs assays.

Author Contributions

Conceived and designed the experiments: HK PTMN. Performed the experiments: PTMN GH MGB. Analyzed the data: PTMN GH MGB MF. Contributed to the writing of the manuscript: HK PTMN MF MGB.

References

1. Hall-Stoodley L, Stoodley P (2009) Evolving concepts in biofilm infections. Cell Microbiol 11: 1034–1043.
2. Dye BA, Tan S, Smith V, Lewis BG, Barker LK, et al. (2007) Trends in oral health status: United States, 1988–1994 and 1999–2004. Vital Health Stat 11: 1–92.
3. Marsh PD (2003) Are dental diseases examples of ecologicalcatastrophes? Microbiology 149: 279–294.
4. Flemmig TF, Beikler T (2011) Control of oral biofilms. Periodontol 2000 55(1): 9–15.
5. Jeon JG, Rosalen PL, Falsetta ML, Koo H (2011) Natural products in caries research: current (limited) knowledge, challenges and future perspective. Caries Res 45: 243–263.
6. Marsh PD (2013) Contemporary perspective on plaque control. Br Dent J 212(12): 601–606.
7. Paes Leme AF, Koo H, Bellato CM, Bedi G, Cury JA (2006) The role of sucrose in cariogenic dental biofilm formation-new insight. J Dent Res 85: 878–887.
8. Bowen WH, Koo H (2011) Biology of *Streptococcus mutans*-derived glucosyl-transferases: role in extracellular matrix formation of cariogeneic biofilms. Caries Res 45: 69–86.
9. Nyvad B, Crielaard W, Mira A, Takahashi N, Beighton D (2013) Dental caries from a molecular microbiological perspective. Caries Res 47(2): 89–102.
10. Loesche WJ (1986) Role of *Streptococcus mutans* in human dental decay. Microbiol Rev 50: 353–380.
11. Quivey RG Jr, Kuhnert WL, Hahn K (2000) Adaptation of oral streptococci to low pH. Adv Microb Physiol 42: 239–74.
12. Burne RA, Marquis RE (2000) Alkali production by oral bacteria and protection against dental caries. FEMS Microbiol Lett 193(1): 1–6.
13. Smith EG, Spatafora GA (2012) Gene regulation in *S. mutans*: complex control in a complex environment. J Dent Res 91(2): 133–141.
14. Lemos JA, Quivey RG Jr, Koo H, Abranches J (2013) *Streptococcus mutans*: a new Gram-positive paradigm? Microbiology 159(Pt 3): 436–445.
15. Vinogradov AM, Winston M, Rupp CJ, Stoodley P (2004) Rheology of biofilms formed from the dental plaque pathogen *Streptococcus mutans*. Biofilms 1: 49–56.
16. Cross SE, Kreth J, Zhu L, Sullivan R, Shi W, et al. (2007) Nanomechanical properties of glucans and associated cell-surface adhesion of *Streptococcus mutans* probed by atomic force microscopy under in situ conditions. Microbiology 153: 3124–3132.
17. Kreth J, Zhu L, Merritt J, Shi W, Qi F (2008) Role of sucrose in the fitness of *Streptococcus mutans*. Oral Microbiol Immunol 23(Pt 12): 213–219.
18. Xiao J, Klein MI, Falsetta ML, Lu B, Delahunty CM, et al. (2012) The exopolysaccharide matrix modulates the interaction between 3D architecture and virulence of a mixed-species oral biofilm. PLoS Pathog 8: e1002623.
19. Hope CK, Wilson M (2004) Analysis of the effects of chlorhexidine on oral biofilm vitality and structure based on viability profiling and an indicator of membrane integrity. Antimicrob Agents Chemother 48: 1461–1468.
20. Koo H, Falsetta ML, Klein MI (2013) The exopolysaccharide matrix: a virulence determinant of cariogenic biofilm. J Dent Res 92(12): 1065–73.
21. Vroom JM, De Grauw KJ, Gerritsen HC, Bradshaw DJ, Marsh PD, et al. (1999) Depth penetration and detection of pH gradients in biofilms by two-photon excitation microscopy. Appl Environ Microbiol 65: 3502–3511.

22. Guo L, Hu W, He X, Lux R, McLean J, et al. (2013) Investigating acid production by *Streptococcus mutans* with a surface-displayed pH sensitive green fluorescent protein. PLoS One 8: e57182.

23. Li Y, Burne RA (2001) Regulation of the *gtfBC* and *ftf* genes of *Streptococcus mutans* in biofilms in response to pH and carbohydrate. Microbiology 147(Pt 10): 2841–2848.

24. Ee GC, Daud S, Taufiq-Yap YH, Ismail NH, Rahmani M (2006) Xanthones from *Garcinia mangostana* (Guttiferae). Nat Prod Res 20: 1067–1073.

25. Ee GC, Daud S, Izzaddin SA, Rahmani M (2008) *Garcinia mangostana*: a source of potential anti-cancer lead compounds against CEM-SS cell line. J Asian Nat Prod Res 10: 475–479.

26. Jung HA, Su BN, Keller WJ, Mehta RG, Kinghorn AD (2006) Antioxidant xanthones from the pericarp of *Garcinia mangostana* (Mangosteen). J Agric Food Chem 54: 2077–2082.

27. Nguyen PTM, Marquis RE (2011) Antimicrobial actions of alpha-mangostin against oral Streptococci. Can J Microbiol 57: 217–25.

28. Koo H, Hayacibara MF, Schobel BD, Cury JA, Rosalen PL, et al. (2003) Inhibition of *Streptococcus mutans* biofilm accumulation and polysaccharide production by apigenin and tt-farnesol. J Antimicrob Chemother 52: 782–789.

29. Koo H, Schobel B, Scott-Anne K, Watson G, Bowen WH, et al. (2005) Apigenin and tt-farnesol with fluoride effects on *Streptococcus mutans* biofilms and dental caries. J Dent Res 84(11): 1016–20.

30. Koo H, Nino de Guzman P, Schobel BD, Vacca Smith AV, Bowen WH (2006) Influence of cranberry juice on glucan-mediated processes involved in *Streptococcus mutans* biofilm development. Caries Res 40(1): 20–7.

31. Heydorn A, Nielsen AT, Hentzer M, Sternberg C, Givskov M, et al. (2000) Quantification of biofilm structures by the novel computer program COM-STAT. Microbiology 146: 2395–2407.

32. Koo H, Xiao J, Klein MI, Jeon JG (2010) Exopolysaccharides produced by *Streptococcus mutans* glucosyltransferases modulate the establishment of microcolonies within multispecies biofilms. J Bacteriol 192: 3024–3032.

33. Kelley LA, Sternberg MJ (2009) Protein structure prediction on the Web: a case study using the Phyre server. Nat Protoc 4: 363–371.

34. Koo H, Vacca Smith AM, Bowen WH, Rosalen PL, Cury JA, et al. (2000) Effects of Apis mellifera propolis on the activities of streptococcal glucosyltransferases in solution and adsorbed ontosaliva-coated hydroxyapatite. Caries Res 34(5): 418–426.

35. Venkatraman V, Ritchie DW (2012) Flexible protein docking refinement using pose-dependent normal mode analysis. Proteins 80(9): 2262–74.

36. Ritchie DW, Kemp GJ (2000) Protein docking using spherical polar fourier correlations. Proteins 39: 178–94.

37. Ritchie DW, Venkatraman V (2010) Ultra-fast FFT protein docking on graphics processors. Bioinformatics 26: 2398–405.

38. Macindoe G, Mavridis L, Venkatraman V, Devignes MD, Ritchie DW (2010) HexServer: an FFT-based protein docking server powered by graphics processors. Nucleic Acids Res. 38(Web Server issue): W445–9.

39. Belli WA, Marquis RE (1991) Adaptation of *Streptococcus mutans* and *Enterococcus hirae* to acid stress in continuous culture. Appl Environ Microbiol 57: 1134–1138.

40. Phan TN, Buckner T, Sheng J, Baldeck JD, Marquis RE (2004) Physiologic actions of zinc on inhibition of acid and alkali production by oral streptococci in suspensions and biofilms. Oral Microbiol Immunol 19(1): 31–8.

41. Cury JA, Koo H (2007) Extraction and purification of total RNA from *Streptococcus mutans* biofilms. Anal Biochem 365: 208–14.

42. Yin JL, Shackel NA, Zekry A, McGuinness PH, Richards C, et al. (2001) Real-time reverse transcriptase-polymerase chain reaction (RT-PCR) for measurement of cytokine and growth factor mRNA expression with fluorogenic probes or SYBR green. Immunol Cell Biol 79: 213–222.

43. Klein MI, DeBaz L, Agidi S, Lee H, Xie G, et al. (2010) Dynamics of *Streptococcus mutans* transcriptome in response to starch and sucrose during biofilm development. PLoS One 5: e13478.

44. Bustin SA, Benes V, Garson JA, Hellemans J, Huggett J, et al. (2009) The MIQE guidelines: minimum information for publication of quantitative real-time PCR experiments. Clin Chem 55: 611–622.

45. Falsetta ML, Klein MI, Lemos JA, Silva BB, Agidi S, et al. (2012) Novel antibiofilm chemotherapy targets exopolysaccharide synthesis and stress tolerance in *Streptococcus mutans* to modulate virulence expression in vivo. Antimicrob Agents Chemother 56: 6201–6211.

46. Hamilton IR (1990) Biochemical effects of fluoride on oral bacteria. J Dent Res 69 (spec issue): 660–667.

47. Körstgens V, Flemming HC, Wingender J, Borchard W (2001) Uniaxial compression measurement device for investigation of the mechanical stability of biofilms. J Microbiol Methods. 46(1): 9–17.

48. Cense AW, Peeters EA, Gottenbos B, Baaijens FP, Nuijs AM, et al. (2006) Mechanical properties and failure of *Streptococcus mutans* biofilms, studied using a microindentation device. J Microbiol Methods 67(3): 463–472.

49. Jones WL, Sutton MP, Mckittrick L, Stewart PS (2011) Chemical and antimicrobial treatments change the viscoelastic properties of bacterial biofilms Biofouling 27: 207–15.

50. Waters MS, Kundu S, Lin NJ, Lin-Gibson S (2014) Microstructure and mechanical properties of in situ *Streptococcus mutans* biofilms. ACS Appl Mater Interfaces 6(1): 327.

51. Simoes M, Pereira MO, Vieira MJ (2004) Effect of cationic surfactants on biofilm removal and mechanical stability. International Conference on Biofilm 2014: Structure and activity of biofilm Las Vegas, NV, USA. 171–175.

52. Schilling KM, Bowen WH (1992) Glucans synthesized in situin experimental salivary pellicle function as specific binding sites for *Streptococcus mutans*. Infect Immun 60: 284–295.

53. Venkitaraman AR, Vacca-Smith AM, Kopec LK, Bowen WH (1995) Characterization of glucosyltransferaseB, GtfC, and GtfD in solution and on the surface of hydroxyapatite. J Dent Res 74(10): 1695–1670.

54. Xiao J, Koo H (2010) Structural organization and dynamics of exopolysaccharide matrix and microcolonies formation by *Streptococcus mutans* in biofilms. J Appl Microbiol 108: 2103–2113.

55. Gundampati RK, Sahu S, Sonkar KS, Debnath M, Srivastava AK, et al. (2013) Modeling and molecular docking studies on RNase *Aspergillus niger* and *Leishmania donovani* actin: antileishmanial activity. Am J Biochem Biotechnol 9(3): 318–328.

56. Ito K, Ito S, Shimamura T, Weyand S, Kawarasaki Y, et al. (2011) Crystal structure of glucansucrase from the dental caries pathogen *Streptococcus mutans*. J Mol Biol 408(2): 177–186.

57. Monera D, Sereda TJ, Zhou NE, Kay CM, Hodges RS (1995) Relationship of side chain hydrophobicity and alpha-helical propensity on the stability of the single-stranded amphipathic alpha-helix. J Pept Sci 1: 319–329.

58. Monchois V, Willemot RM, Monsan P (1999) Glucansucrases: mechanism of action and structure-function relationships. FEMS Microbiol Rev 23: 131–151.

59. Colby SM, Russell RRB (1997) Sugar metabolism by mutans streptococci. Soc Appl Bacteriol Symp 26: 80S–88S.

60. Mooser G, Hefta SA, Paxton RJ, Shively JE, Lee TD (1991) Isolation and sequence of an active-site peptide containing a catalytic aspartic acid from two *Streptococcus sobrinus* alpha-glucosyltransferases. J Biol Chem 266: 8916–22.

61. Marquis RE, Clock SA, Mota-Meira M (2003) Fluoride and organic weak acids as modulators of microbial physiology. FEMS Microbiol Rev 26: 493–510.

62. Lemos JA, Abranches J, Burne RA (2005) Responses of cariogenic streptococci to environmental stresses. Curr Issues Mol Biol 7: 95–10.

63. Burne RA, Ahn SJ, Wen ZT, Zeng L, Lemos JA, et al. (2009) Opportunities for disrupting cariogenic biofilms. Adv Dent Res 21(1): 17–20.

64. Loesche WJ, Henry CA (1967) Intracellular microbial polysaccharide production and dental caries in a Guatemalan Indian village. Archs Oral Biol 12: 189–194.

65. Tanzer JM, Freedman ML, Woodiel FN, Eifert RL, Rinehimer LA (1976) Association of Streptococcus mutans virulence with synthesis of intracellular polysaccharide. In: Proceedings in microbiology. Aspects of dental caries. Stiles HM, Loesche WJ, O'Brien TL, editors. Special supplement to Microbiology Abstracts, vol. 3. Information Retrieval, Inc. London, 596–616.

66. Spatafora G, Rohrer K, Barnard D, Michalek SA (1995) *Streptococcus mutans* mutant that synthesizes elevated levels of intracellular polysaccharide is hypercariogenic in vivo. Infect Immun 63(7): 2556–263.

67. Hamilton IR (1976) Intracellular polysaccharide synthesis by cariogenic microorganisms. In: Proceedings in microbiology. Aspects of dental caries. Stiles HM, Loesche WJ, O'Brien TL, editors. Special supplement to Microbiology Abstracts. Vol. 3. London: Information Retrieval, Inc., 683–701.

68. Kaomongkolgit R, Jamdee K, Chaisomboon N (2009) Antifungal activity of α-mangostin against *Candida albicans*. J Oral Sci 51(3): 401–406.

69. Obolskiy D, Pischel I, Siriwatanametanon N, Heinrich M (2009) *Garcinia mangostana* L.: a phytochemical and pharmacological review. Phytother Res 23(8): 1047–1065.

70. Kosem N, Ichikawa K, Utsumi H, Moongkarndi P (2013) In vivo toxicity and antitumor activity of mangosteen extract. J Nat Med 67(2): 255–263.

8

FLS2-BAK1 Extracellular Domain Interaction Sites Required for Defense Signaling Activation

Teresa Koller, Andrew F Bent*

Department of Plant Pathology, University of Wisconsin – Madison, Madison, Wisconsin, United States of America

Abstract

Signaling initiation by receptor-like kinases (RLKs) at the plasma membrane of plant cells often requires regulatory leucine-rich repeat (LRR) RLK proteins such as SERK or BIR proteins. The present work examined how the microbe-associated molecular pattern (MAMP) receptor FLS2 builds signaling complexes with BAK1 (SERK3). We first, using in vivo methods that validate separate findings by others, demonstrated that flg22 (flagellin epitope) ligand-initiated FLS2-BAK1 extracellular domain interactions can proceed independent of intracellular domain interactions. We then explored a candidate SERK protein interaction site in the extracellular domains (ectodomains; ECDs) of the significantly different receptors FLS2, EFR (MAMP receptors), PEPR1 (damage-associated molecular pattern (DAMP) receptor), and BRI1 (hormone receptor). Repeat conservation mapping revealed a cluster of conserved solvent-exposed residues near the C-terminus of models of the folded LRR domains. However, site-directed mutagenesis of this conserved site in FLS2 did not impair FLS2-BAK1 ECD interactions, and mutations in the analogous site of EFR caused receptor maturation defects. Hence this conserved LRR C-terminal region apparently has functions other than mediating interactions with BAK1. In vivo tests of the subsequently published FLS2-flg22-BAK1 ECD co-crystal structure were then performed to functionally evaluate some of the unexpected configurations predicted by that crystal structure. In support of the crystal structure data, FLS2-BAK1 ECD interactions were no longer detected in in vivo co-immunoprecipitation experiments after site-directed mutagenesis of the FLS2 BAK1-interaction residues S554, Q530, Q627 or N674. In contrast, in vivo FLS2-mediated signaling persisted and was only minimally reduced, suggesting residual FLS2-BAK1 interaction and the limited sensitivity of co-immunoprecipitation data relative to in vivo assays for signaling outputs. However, Arabidopsis plants expressing FLS2 with the Q530A+Q627A double mutation were impaired both in detectable interaction with BAK1 and in FLS2-mediated responses, lending overall support to current models of FLS2 structure and function.

Editor: Vladimir N. Uversky, University of South Florida College of Medicine, United States of America

Funding: This research was funded by U.S. Department of Energy Basic Energy Biosciences Grant DE-FG02-02ER15342 and University of Wisconsin Hatch Funding to AFB. The funders had no role in study design, data collection and analysis, decision to publish, or preparation of the manuscript.

Competing Interests: The authors have declared that no competing interests exist.

* Email: afbent@wisc.edu

Introduction

Plants use pattern-recognition receptors (PRRs) as a first layer of defense against pathogens [1,2]. In order to engineer plants with improved pathogen recognition abilities, it is important to understand the molecular details underlying the interaction of PRRs not only with their ligands but also with their co-receptors, immediate downstream targets and other partner proteins that facilitate appropriate signaling. Several PRRs have been identified in different plant species [reviewed in 1,2]. PRRs are localized at the plasma membrane where they monitor the apoplastic space for microbe-associated molecular patterns (MAMPs), damage-associated molecular patterns (DAMPs) and apoplastic effectors. Most known PRRs are receptor-like kinases (RLKs) or receptor-like proteins (RLPs). Both receptor types consist of an extracellular domain for ligand perception and a transmembrane domain, but only the RLKs have an intracellular kinase domain. Two of the best characterized PRRs, FLS2 and EFR [3,4], carry large extracellular domains (ECDs, ectodomains) that predominantly consist of a leucine-rich repeat (LRR) domain [5,6]. The genomes of Arabidopsis and other plants each encode hundreds of LRR receptor-like kinases (LRR-RLKs) with 4 to 28 repeat units of the LRR [7].

Receptors typically exhibit high specificity for ligands with which they interact, but cells also contain co-receptors and regulatory proteins that function together with receptors and do not necessarily exhibit specificity for only a single type of ligand [8,9]. These co-receptors and regulatory proteins can be important facilitators or suppressors of signaling activation. They also allow signaling crosstalk at the plasma membrane, helping to coordinate appropriate downstream signaling in the presence of diverse endogenous and exogenous extracellular ligands. Important examples of regulatory/co-receptor RLKs include the SERK family members [8,10], BIR family members [11,12] and SOBIR1 [13].

SERK proteins have been identified in many different plant species. In Arabidopsis the family consists of five members (SERK1, SERK2, SERK3/BAK1, SERK4 and SERK5). They all have five LRRs in their ectodomain, share high overall sequence similarity and have redundant functions to various degrees. SERK proteins (mainly SERK3, also known as BAK1) have been shown to be involved in plant immunity in Arabidopsis, tomato and rice, through interactions with the receptors FLS2, EFR, PEPR1, PEPR2, Xa21, Ve1 and Eix1 [14–19]. The BAK1 co-receptor also contributes to somatic embryogenesis [20,21] and to plant development through interaction with the brassinolide hormone receptor BRI1 [22,23]. Despite impressive progress, much remains unknown about how the SERK proteins participate in all these different cell signaling tasks, and about the spatial expression of SERK proteins [24]. Studies of the SERK proteins are impeded by the redundant functions among family members and by pleiotropic effects when multiple SERK proteins are knocked out. As an example, $bak1^-$ Arabidopsis plants only have partially disrupted FLS2 signaling outputs [14,15,25,26]. A possible means of circumventing this problem of SERK functional redundancy, adopted in the present study, is to identify the specific SERK interaction site of a partner receptor and then mutate that site. If all SERKs interact with a specific receptor at similar amino acids, this approach should impair the interaction of the receptor with all SERK family members.

Recent X-ray crystallography studies provided detailed insight into the interaction of the ectodomain of BAK1 with the ectodomains of FLS2 and BRI1 [27,28], and the interaction of the ectodomains of SERK1 and BRI1 [29]. In all three cases the respective ligand promotes interaction between the ectodomains of the main receptors (FLS2 and BRI1) and the SERK co-receptors (BAK1 and SERK1). The ligand binds to the LRR domain of the main receptor, but the LRR domain of the SERK co-receptor also has multiple direct contacts with the ligand. It is surprising to see these fine-tuned co-receptor/ligand interactions, considering how many different known and potential unknown receptors and ligands BAK1 and SERK1 are able to interact with. Similar residues of the BAK1 and SERK1 ectodomains are involved in their interactions with FLS2 and BRI1. However, the residues on FLS2 and BRI1 ectodomains predicted to be used for the interactions with their SERK co-receptors are very different, not only in sequence but also in their location within the receptor LRR domain [27–29]. In BRI1 the residues interacting with co-receptors are located at the island domain, the last LRR, and the juxtamembrane domain, all close to the transmembrane domain. However, in FLS2 the BAK1-interacting residues in the crystal structure are located 108–300 amino acids from the predicted transmembrane domain, at repeats #18 to 26 of the LRR domain. This predicts a relatively recumbent orientation for the FLS2 ectodomain, bent down toward the plasma membrane (see Figure S3A).

FLS2 mediates perception of bacterial flagellin protein, an abundant MAMP, and FLS2 recognizes in particular a ~20 amino acid region that is relatively conserved across flagellins from diverse Gram-negative bacteria [1,30]. Many aspects of FLS2 structure and function have been characterized [reviewed in 31]. There is a third surprising feature of the FLS2-flg22-BAK1 ECD co-crystal structure [27]. Most research regarding FLS2 utilizes as ligand, in place of flagellin protein, a 22 amino acid "flg22" peptide whose sequence matches the recognized domain of *Pseudomonas aeruginosa* flagellin, or utilizes other small peptides based on similar sequences from various bacteria [30,32,33]. The FLS2-flg22-BAK1 ECD co-crystal structure predicts a tight pocket for the flg22 peptide, which may not be compatible with (allow sufficient space or sufficient ligand flexibility for) analogous binding of flg22 domains embedded within full-length flagellin proteins (discussed below).

In this study we first explored the possibility that a relatively universal SERK interaction site has evolved in the LRR domains

of different SERK-interacting LRR-RLKs. We also showed that flg22-dependent FLS2 interaction with BAK1 occurs via the FLS2 extracellular domains – a result subsequently shown by alternative methods by Sun *et al.* (2103). We then performed site-directed mutagenesis and functional testing of predicted LRR-RLK receptor/SERK co-receptor interaction residues, and obtained *in vivo* evidence that supports models suggested by the recently published receptor/co-receptor co-crystal structures of truncated FLS2 and BAK1. The overall goal of this study was to furnish a more clear understanding of the requirements for formation of a signaling-competent plant basal immune system MAMP receptor – an understanding that may be essential to allow future engineering of PRRs with broadened or otherwise improved performance.

Methods

Arabidopsis and *Nicotiana benthamiana* transformation

The floral dip method was used to stably transform Arabidopsis $fls2^-$ and efr^- plants. T1 seedlings were selected on 0.5x MS plates containing 25 mg/L kanamycin and 25 mg/L hygromycin. Leaves of 4-week-old *Nicotiana benthamiana* plants were infiltrated with *Agrobacterium tumefaciens* GV3101 containing the binary plasmids [34]. Proteins were harvested two days after *Agrobacterium tumefaciens* infiltration.

Co-immunoprecipitation

Transiently transformed leaf tissue from *Nicotiana benthamiana* was infiltrated with 1 μM flg22 or 1 μM elf18, or with water for mock infiltration. After 2 minutes the leaf tissues were blotted dry and frozen in liquid N_2. Then 200 mg of tissues were ground in 200 μl protein extraction buffer (50 mM Tris pH 7.5, 150 mM NaCl, 0.5% Triton X-100, 1x plant protease inhibitor cocktail (Sigma-Aldrich)). After centrifugation 300 μl supernatant was incubated with 3 μl 9E10 anti-myc antibody (Sigma-Aldrich or Covance) and rotated at 4°C for 1 h. 50 μl Protein A (Thermo Scientific) was added and the tubes were rotated at 4°C for an additional 2 h. After 3x washing with protein extraction buffer and 1x washing with ddH$_2$O the beads were resuspended in 60 μl loading buffer and boiled at 95°C for 5 min. After centrifugation the supernatant was separated on two 8% SDS-PAGE gels. For protein detection the antibodies anti-HA-HRP, anti-myc rabbit and goat-anti-rabbit-HRP (Sigma-Aldrich) were used.

Conservation mapping

Mapping of conserved regions of predicted LRR surfaces was performed using the Repeat Conservation Mapping (RCM) program at www.plantpath.wisc.edu/RCM [35], with heat map coloration range set to the minimal and maximal conservation scores of the data within each figure. The LRR domain sequences were obtained from The Arabidopsis Information Resource (TAIR) website at www.arabidopsis.org and from the National Center for Biotechnology Information (NCBI) website at www.ncbi.nlm.nih.gov. The following FLS2 non-Brassicaceae sequences were used: *Populus trichocarpa* (XP_002305701.1); *Vitis vinifera* (XP_002272319.2); *Glycine max* (XP_003532650.1); *Lotus japonicus* (AER60531.1); *Ricinus communis* (XP_002519723.1); *Sorghum bicolor* (XP_002448543.1); *Oryza sativa Japonica* (CAE02151.2); *Oryza sativa Indica* (CAH68341.1); *Hordeum vulgare* (BAJ89141.1); *Brachypodium distachyon* (XP_003581675.1).

Site-directed mutagenesis

Point mutations were generated according to the QuikChange mutagenesis kit (Agilent Technologies) on pENTR plasmids (Invitrogen) containing FLS2, FLS2-NoKinase or EFR with 35 S or native promoters [36]. Gateway LR Clonase II (Invitrogen) was used to transfer the construct into the binary plasmids pGWB13 or pGWB14 [37].

EndoH assay

Leaf tissues (60 mg) from Arabidopsis T1 plants or from transiently transformed *Nicotiana benthamiana* plants were ground in 2x SDS buffer and boiled for 5 min at 95°C. After centrifugation for 10 min at 14000 rpm at 4°C supernatants were digested with Endoglycosidase H (New England BioLabs) as per manufacturer's suggestion and separated on 8% SDS-PAGE gel. Proteins were detected using anti-HA-HRP antibody (Sigma-Aldrich).

Seedling growth inhibition

T1 Arabidopsis seedlings were grown for 6 days on 0.5x MS plates with 25 mg/L kanamycin and 25 mg/L hygromycin and 200 mg/L cefotaxime. 24 seedlings per genotype, representing 24 independent transformation events, were transferred to 24-well-plates containing 1 ml 0.5x MS liquid media per well. 12 seedlings per genotype were grown for 14 days in wells containing 1 μM flg22 and 12 seedlings per genotype were grown for 14 days in wells containing only 0.5 x MS. Seedlings were then blotted dry and weighed. The weight of each flg22-treated seedling was divided by the average weight of the mock treated seedlings of the same genotype from the same experiment, prior to determination of experiment means and standard errors.

Oxidative burst

Seven leaf discs were taken from six-week-old T1 Arabidopsis plants and incubated overnight in 1% DMSO solution. Peptide solution was added to the leaf discs and luminescence was measured by a plate reader for 0–30 min after addition of flg22 peptide. For measurement each leaf disc was in 100 μl peptide solution containing 0.5 μl 2 mg/ml horseradish peroxidase, 0.5 μl 2 mg/ml luminol in DMSO and 1 μM flg22.

Results and Discussion

Extracellular domain of FLS2 can mediate interaction with BAK1 in the presence of flg22

Full-length FLS2 and BAK1 do not detectably interact until exposure to flg22 or similar flagellin ligands, at which time interaction is immediately observed [14,15,38]. Flg22-elicited immune signaling then requires phosphorylation events among the respective kinase domains [26,38,39]. We hypothesized that the FLS2-BAK1 interaction is mediated not only intracellularly by the respective kinase domains, but also by interaction of the ectodomains. To test this we used a truncated FLS2 carrying the N-terminal ~70% of the protein including the LRR and transmembrane domains but not the predicted intracellular domains (*FLS2-NoKinase-HA*; [36]). *FLS2-NoKinase-HA* was expressed in *Nicotiana benthamiana* together with a plasmid encoding a full-length, epitope tagged *BAK1-Myc*. The transiently transformed leaves were treated with flg22 and co-immunoprecipitation experiments were performed. BAK1 and FLS2-NoKinase interact in the presence of flg22, indicating that the kinase domain of FLS2 is not needed for interaction with BAK1 *in planta* (Figure 1).

Sun *et al.* 2013 also showed ECD mediation of FLS2-BAK1 interaction [27]. Their work utilized *in vitro* mixing experiments with purified recombinant proteins, or mutated BAK1 expressed in Arabidopsis protoplasts. Our results with mutated FLS2, tested in transgenic whole plants with *FLS2* expressed under control of *FLS2* promoter sequences, are complimentary and in agreement with the results of Sun *et al.* (2103), and reveal that intracellular/kinase domain interactions of these proteins are not required for flg22-stimulated FLS2-BAK1 interaction. It is also interesting to note the previously published finding that FLS2-FLS2 interaction occurs *in planta*, with either full-length FLS2 or FLS2-NoKinase constructs [36]. At least some FLS2 exists *in planta* in FLS2-FLS2 complexes, prior to and after flagellin or flg22 exposure. FLS2-BAK1 interaction after exposure to flg22 did not appreciably deplete the overall presence of co-immunoprecipitable FLS2-FLS2 complexes [36]. Hence findings that FLS2 and BAK1 interact via LRR domains suggest either that FLS2-FLS2 interactions utilize a different side or face of the FLS2 LRR than the region that interacts with BAK1, or that different sub-pools of FLS2 are at any given moment interacting with FLS2 or BAK1. The results of Albert *et al.* (2013) and Cao *et al.* (2013) are also relevant to these updated models of PRR receptor - co-receptor structure/function [39,40]. Those studies demonstrated that *in planta* responses to flg22 are retained when hybrid FLS2 and BAK1 proteins are expressed in which the kinase domains of FLS2 and BAK1 have been reciprocally swapped [40], and that flg22-mediated FLS2-BIK1 disassociation and FLS2-BAK1 association still occur when FLS2 kinase domain mutations are present that block defense signaling. Schulze *et al.* 2010 and Schwessinger *et al.* 2011 showed that kinase-dead BAK1 still interact with FLS2, but impair FLS2 signaling [26,38]. The evidence increasingly indicates that interactions of the FLS2 and BAK1 extracellular domains are a first step in flg22 perception that can proceed relatively independent of intracellular domain structural or functional interactions.

Identification of a conserved region in the C-terminal LRRs of BAK1-interacting receptors

The SERK family members have been shown to interact with several different transmembrane LRR-RLKs involved in plant immunity and development. It is not known if the SERK interaction sites of these receptors evolved independently or originate from a common and potentially conserved SERK interaction site. We hypothesized the latter and also hypothesized that, to facilitate spatial proximity of potentially interacting extracellular domains, the relatively small ectodomains of SERK proteins would interact near the C-terminal end of the large LRR ectodomains of those partner receptors. Using Repeat Conservation Mapping [35] we searched the last seven repeats of the LRRs of the known Arabidopsis BAK1-interacting proteins FLS2 (28 total repeats in the LRR domain), EFR (21 LRRs), BRI1 (25 LRRs) and PEPR1 (26 LRRs), looking for the patch of solvent-exposed amino acids in this region that is most conserved across the four proteins. A conserved region of interest was identified (Figure 2A). Separately, we compared the solvent exposed amino acids of the whole LRR domains of eleven non-Brassicaceae FLS2s (Figure 2B). Both conservation maps revealed a conserved region at a similar location in the C-terminal LRRs. We hypothesized that this may be a somewhat universal site for interaction with SERK proteins.

FLS2-NoKinase-HA   +   +

BAK1-Myc   +   +

flg22   +   -

IP: Myc
WB: HA   FLS2-NoKinase-HA

Total
WB: HA   FLS2-NoKinase-HA

IP: Myc
WB: Myc   BAK1-Myc

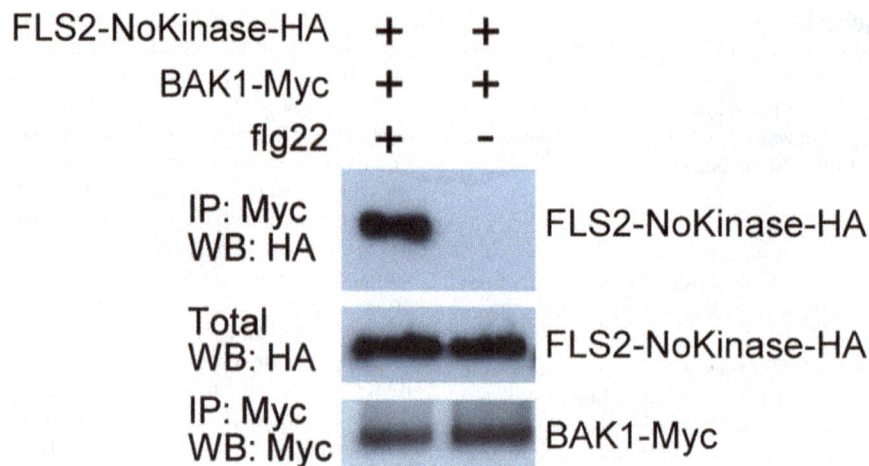

Figure 1. Extracellular domain of FLS2 can mediate interaction with BAK1. Co-immunoprecipitation experiments performed using *35S–FLS2-NoKinase-HA* (construct lacking the FLS2 intracellular domain) and *35S–BAK1-Myc* transiently expressed in *Nicotiana benthamiana* leaves by agroinfiltration. Samples were prepared for SDS-PAGE two days after agroinfiltration, two minutes after flg22 or water (mock) was infiltrated into leaves. IP: antibody used for immunoprecipitation prior to SDS-PAGE; WB: antibody used for immunodetection on protein blot; crude: SDS-PAGE and blotting of total (crude extract) protein samples. The experiment was repeated three times with similar results.

No disruption of FLS2-BAK1 interaction by mutations in the FLS2 LRR domain C-terminal region conserved among EFR, PEPR1, BRI1 and multiple FLS2s

Site-directed mutagenesis was carried out to alter residues in the identified conserved LRR C-terminal region of FLS2 and EFR (Figure 2; Figure S1A–E). D557 and S559 mutations in FLS2 were included as control mutations located in LRR sites analogous to N704/S706 and D728/S730, but outside of the conserved LRR C-terminus. The amino acids were replaced with similar yet bulkier residues in order to impair interactions. The resulting full-length receptors were expressed in *N. benthamiana* and co-immunoprecipitation experiments were then carried out, using BAK1-Myc for pull-down in the presence and absence of the corresponding ligands flg22 and elf18 in the case of FLS2 or EFR, respectively. The mutations in FLS2 did not abolish the interaction with BAK1 in the presence of flg22 (Figure 3A). As is common for agroinfiltration experiments, variable levels of expression were observed for any single transgene-encoded FLS2 protein across replicates within or between experiments, but none of the mutant proteins was reproducibly present at levels different from transgene-encoded wild-type FLS2. To ensure that interaction of the kinase domains of FLS2 and BAK1 was not masking non-interaction of mutated FLS2 and BAK1 ectodomains, the same mutations were also placed into FLS2-NoKinase constructs. In these FLS2-NoKinase variants the mutations again did not prevent interaction with BAK1 (Figure 3B).

Mutations in the conserved C-terminal LRR region of EFR cause EFR glycosylation/maturation defects

Mutations analogous to those of the preceding section were also engineered into *EFR*. These LRR domain C-terminal region mutations (Figure 2; Figure S1C, D) did cause disruption of interaction with BAK1 in the presence of elf18 (Figure 4A). This *in vivo* result could be attributable to direct impacts of the mutations on EFR-BAK1 interaction, or to defects in maturation and delivery of newly synthesized EFR out of the endoplasmic reticulum (ER) and golgi. Endoglycosidase H (EndoH) analyses were therefore conducted. EndoH cleaves incomplete glycosylation modifications present on proteins that have not successfully

passed through the ER and related endomembrane systems [41,42]. On the other hand, mature glycosylated proteins that are delivered to their functional location typically carry EndoH-resistant glycosylation [41,42]. Treatment of the EFR protein extracts with EndoH revealed defects in the mutated EFR proteins, both in *N. benthamiana* and in stable transgenic Arabidopsis *efr*- plants expressing transgene *EFR* constructs driven by native *EFR* promoter sequences (Figure 4B, C). The mutations we generated in FLS2 full-length and FLS2-NoKinase did not result in glycosylation defects (Figure S2A, B). Häweker *et al.* 2010 [42] and Sun at al. 2012 [36] showed that single amino acid changes in glycosylation sites in the EFR ectodomain result in protein degradation and several studies reported the importance of intact glycosylation enzymes for successful processing and function of EFR [43–47]. FLS2 is less sensitive to mutations in glycosylation sites [36,42]. The N590Q+S592T mutations that we placed in EFR are indeed in a Nx(S/T) predicted glycosylation site [48]. However, the EFR mutations D566E+S568T and D566F are not, yet they still disrupted correct EFR processing.

Taken together, the above results suggest that functional roles of the LRR C-terminal conserved domain of BAK1-interacting proteins (Figure 2) do not serve as a universal SERK protein interaction site. However, in EFR the integrity of this site is important for correct protein processing.

FLS2 mutations in proposed FLS2-BAK1 ECD interaction residues disrupt FLS2-BAK1 interaction in the presence of flg22

While the above work was in progress the crystal structure of FLS2-flg22-BAK1 ECD became available [27]. That important work identified in detail the interaction sites of the FLS2 and BAK1 ectodomains. Because the data are for *in vitro* crystallized protein complexes of isolated LRR domains, they may or may not capture the most functionally prominent *in vivo* configurations. Sun *et al.* [27] therefore functionally tested BAK1 mutations, and also tested FLS2 mutations in repeats #9, 11, 14 and 15 of the LRR that are predicted to mediate interaction with flg22. The sites on FLS2 predicted to mediate interaction with BAK1 did not receive mutational testing. In order to test *in vivo* the significance

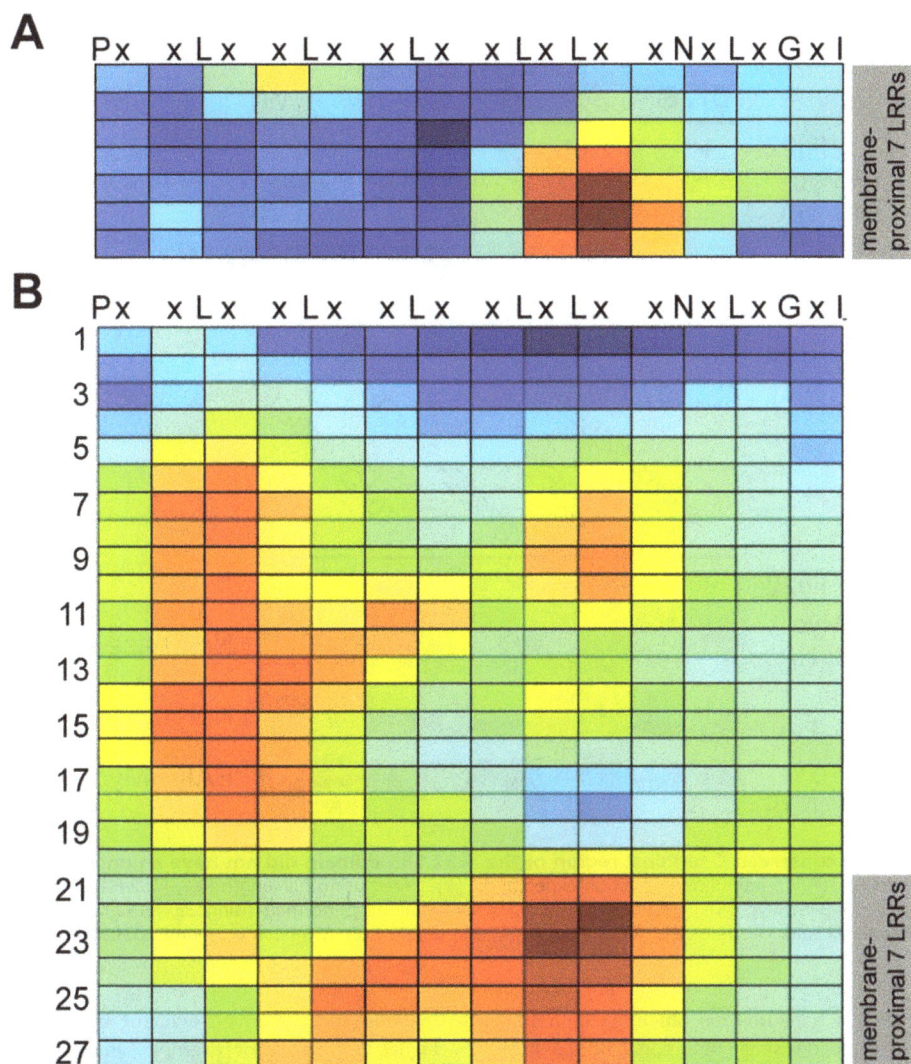

Figure 2. Repeat Conservation Mapping reveals conserved region near C-terminus of LRR domains of FLS2, EFR, BRI1 and PEPR1.
Each row represents one leucine-rich repeat (LRR) and each square represents one solvent-exposed "x" amino acid position (as per LRR consensus sequence shown at the top). Conservation score at each amino acid position is center-weighted score for the cluster of 15, 20 or 25 predicted solvent-exposed LRR amino acids surrounding that site; blue: least conserved, red: most conserved. For FLS2, the seven rows of (A) are the same repeats (same residues) as rows 21–27 of (B). (**A**) Conservation map generated by comparing the most C-terminal seven repeats of the LRR sequences of the BAK1 interacting proteins FLS2, EFR, BRI1 and PEPR1. (**B**) Conservation map generated by comparing the entire FLS2 LRR domain sequences from eleven non-Brassicaceae plant species.

of these FLS2 BAK1-interaction residues, which are likely to also mediate interaction of FLS2 with other SERK proteins, we performed site-directed mutagenesis on *FLS2-NoKinase* and full-length *FLS2*.

For FLS2 amino acids predicted in the crystal structure to form FLS2-BAK1 interaction sites [27], we changed single residues to alanine (small and relatively inactive) or to tryptophan (bulky). In addition to the single mutations we made two *FLS2-NoKinase* constructs with double mutations and one full-length *FLS2* construct with a double mutation. We had previously shown that the FLS2-NoKinase used in this work performed similarly to FLS2-full-length in flg22-dependent BAK1 co-immunoprecipitation experiments (Figure 3A, B). In *in vivo* tests of the newly predicted FLS2-BAK1 interaction sites, mutation of FLS2 residues Q530, S554, Q627 or N674 to tryptophan disrupted the flg22-stimulated interaction of FLS2-NoKinase with BAK1 (Figure 5A, B). The interaction was disrupted as well when FLS2 Q530 and

N674 were changed to alanine (Figure 5A, B). However, the FLS2 S554 and Q627 single mutations to alanine had much less impact on flg22-dependent interaction with BAK1 (Figure 5A, B), suggesting a stronger role for Q530 and N674 than S554 or Q627 in mediating FLS2-BAK1 interaction. The double alanine mutation Q530A+Q627A and the double tryptophan mutation S554W+Q627W disrupted BAK1 interaction as well (Figure 5C). The presence of abundant EndoH-insensitive bands suggested that FLS2 maturation had proceeded successfully for each of the representative FLS2 mutants S554A, Q627W and Q530A+ Q627A (Figure S2C).

Arabidopsis *fls2⁻* plants carrying FLS2-Q530A+Q627A have impaired FLS2-mediated signaling outputs

To investigate if mutations in predicted FLS2 BAK1-interaction residues not only disrupt FLS2-BAK1 interactions in co-immu-

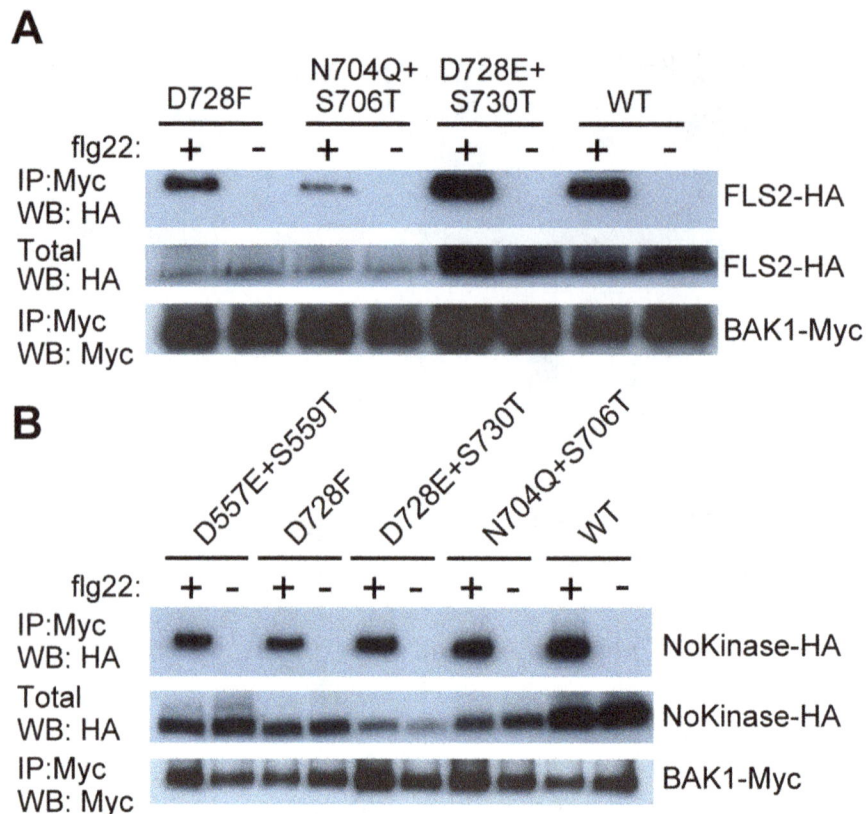

Figure 3. Mutations in the conserved C-terminal region of the FLS2 LRR domain did not have an impact on BAK1-FLS2 or BAK1-FLS2-NoKinase interaction. (**A**) Co-immunoprecipitation experiments performed using full-length P_{FLS2}-FLS2-HA, with mutations as indicated or WT (no mutations), and 35S–BAK1-Myc. (**B**) Co-immunoprecipitation experiments performed using 35S–FLS2-NoKinase-HA, with mutations as indicated or WT (no mutations), and 35S–BAK1-Myc. All samples in (A) and (B) are from *Nicotiana benthamiana*. Labeling as in Figure 1.

noprecipitation experiments but also have an impact on FLS2 signaling, we made the analogous single mutations and one of the double mutations in full-length *FLS2s*. We then tested FLS2 signaling in stably transformed *fls2⁻* Arabidopsis plants containing the mutated and HA-tagged full-length *FLS2s* under control of native *FLS2* promoter sequences. The two most widely used assays for FLS2 signaling were utilized: ROS burst assays and seedling growth inhibition assays [1]. Surprisingly, *in vivo* FLS2-mediated signaling persisted and was only minimally reduced in plants expressing most single-mutant forms of FLS2 (Figure 6A, B), including mutants that exhibited no detectable flg22-induced co-immunoprecipitation with BAK1 (Figure 5A, B). As a general trend across the multiple independent transgenic lines tested for each *FLS2* construct, mutations to alanine allowed stronger FLS2 signaling than mutations to tryptophan (Figure 6A, B). The results suggest that reduced-affinity or more transient interactions of FLS2 and BAK1 occur with many of the FLS2 mutants described in Figures 5 and 6, and that those interactions are sufficient for flg22-stimulated FLS2 signaling even if the stability of FLS2-BAK1 interactions is reduced below levels detectable in standard co-immunoprecipitation experiments. Although some FLS2 signaling capacity was still conferred by FLS2 constructs mutated at single predicted FLS2 BAK1-interaction sites, with the double mutation Q530A+Q627A FLS2-mediated signaling was significantly impaired (Figure 6C, D), supporting current models of FLS2 structure and function.

Alternative hypotheses, other than reduced-affinity or more transient interactions of FLS2 and BAK1, can be formulated

regarding the continued signaling by the FLS2 single mutants of Figure 5 and 6. For example, small sub-populations of FLS2 receptors (sufficient to initiate the levels of defense signaling observed in Figure 6) may exist in the cell that, because of different localization or post-translational modifications, continue to exhibit robust flg22-dependent interaction with BAK1 despite presence of mutations that disrupt interaction between most of the cellular FLS2 and BAK1. As another possibility, the single mutations that transition FLS2 away from high-affinity flg22-dependent binding with BAK1 may have allowed or even enhanced interaction with other SERK proteins, to an extent that allows defense signaling.

LRRs are a protein structure evolved to display widely varying surface amino acid combinations on a relatively invariant scaffold [5,6]. A previous study of over 1200 FLS2 LRR mutations of predicted LRR solvent-exposed residues at and adjacent to flg22 binding sites, carrying changes to all possible amino acids (i.e., not just to alanine), found that the vast majority of LRR surface mutations do not disrupt FLS2 function [49]. Thus the structural alterations caused by the FLS2 mutations of the present study are likely to be highly local. Their disruption of FLS2-BAK1 interactions detected via co-immunoprecipitation supports the relevance of the FLS2-flg22-BAK1 configuration in the published co-crystal structure. Mutation of FLS2 residues D557 and S559, which reside close to but outside of the BAK1-interaction residues in the solved crystal structure ([27], Figure S1), did not disrupt flg22-stimulated FLS2-BAK1 co-immunoprecipitation (Figure 3). Hence the functional disruption of signaling caused by the presumably additive effect of two alanine substitutions in FLS2

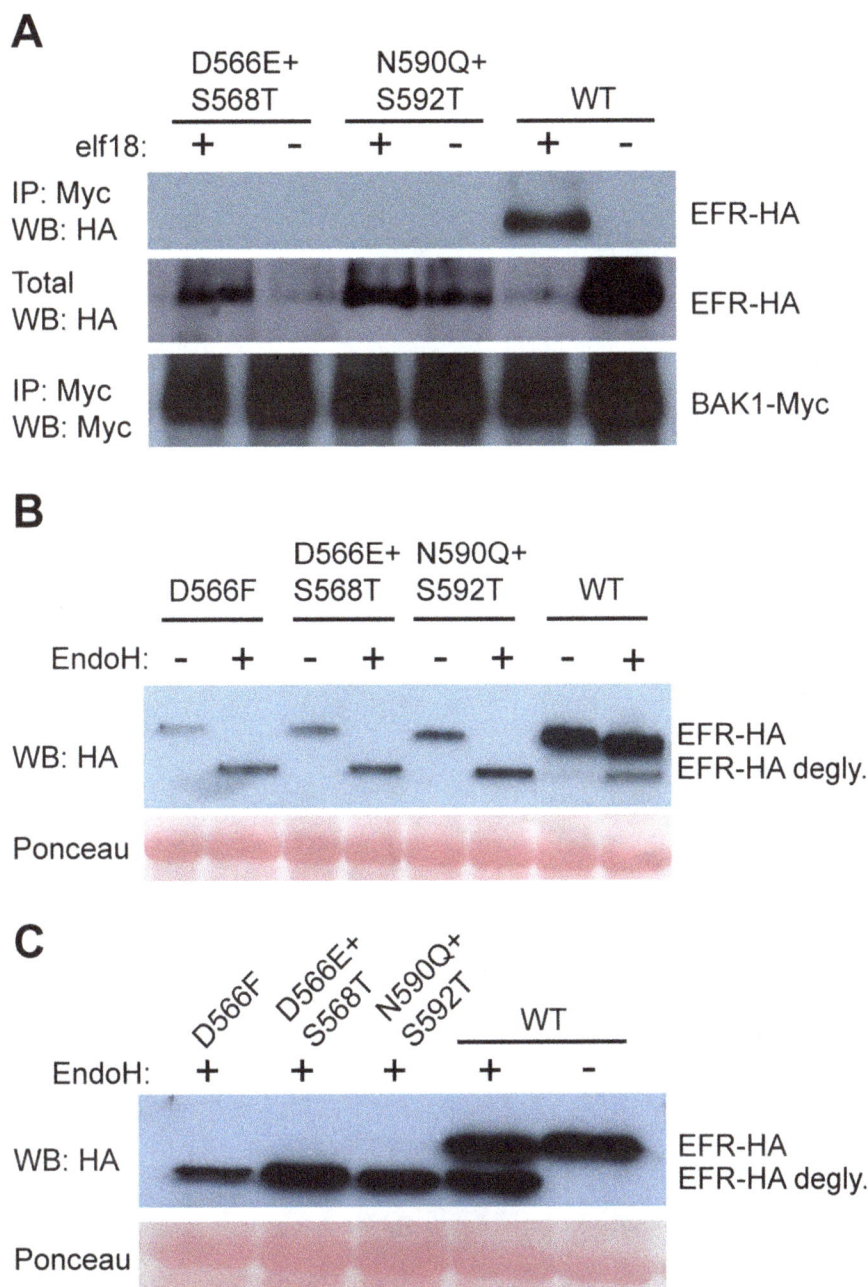

Figure 4. Mutations in the conserved C-terminal region of the EFR LRR domain disrupt EFR glycosylation and interaction with BAK1 in the presence of elf18. (**A**) Co-immunoprecipitation experiments performed using P_{EFR}-EFR-HA with mutations as indicated or WT (no mutations), and 35S–BAK1-Myc, in *Nicotiana benthamiana*. (**B, C**) Protein extracts from plants expressing P_{EFR}-EFR-HA with mutations as indicated, or WT (no mutations), not digested or digested with endoglycosidase H (EndoH). Samples in (B) are from *Nicotiana benthamiana*, samples in (C) are from stably transformed *efr* ⁻ Arabidopsis leaves. EndoH-resistant (mature) EFR is present in the EndoH-treated EFR wild type (WT) samples but is not detected for EFRs carrying the indicated mutations. Degly.: EFR pool deglycosylated by EndoH. Labeling as in Figure 1. Ponceau: blots treated with Ponceau stain to confirm even loading of total protein.

Q530A+Q627A provides further *in vivo* functional evidence indicating the requirement for this site both for FLS2-BAK1 interaction and for flg22 induction of FLS2-dependent immune signaling. Our results also indicate that, if SERK proteins other than BAK1 make residual contributions to FLS2 activation (as is suggested above and in the literature [14,15,25,26]), the FLS2 Q530A+Q627A mutations are sufficient to disrupt functional signaling mediated by those interactions as well.

Closing Observations

In this study we explored the idea of a universal SERK protein interaction site in the C-terminal repeats of the LRR ectodomains of receptors known to interact with SERK proteins. However, mutagenesis of a possible BAK1 interaction site in the ectodomains of FLS2 and EFR did not confirm this hypothesis. The subsequently available FLS2-flg22-BAK1 and BRI1-brassinolide-SERK1 extracellular domain crystal structures [27] [29], and the mutational studies in the present work, instead suggest a fine-tuned

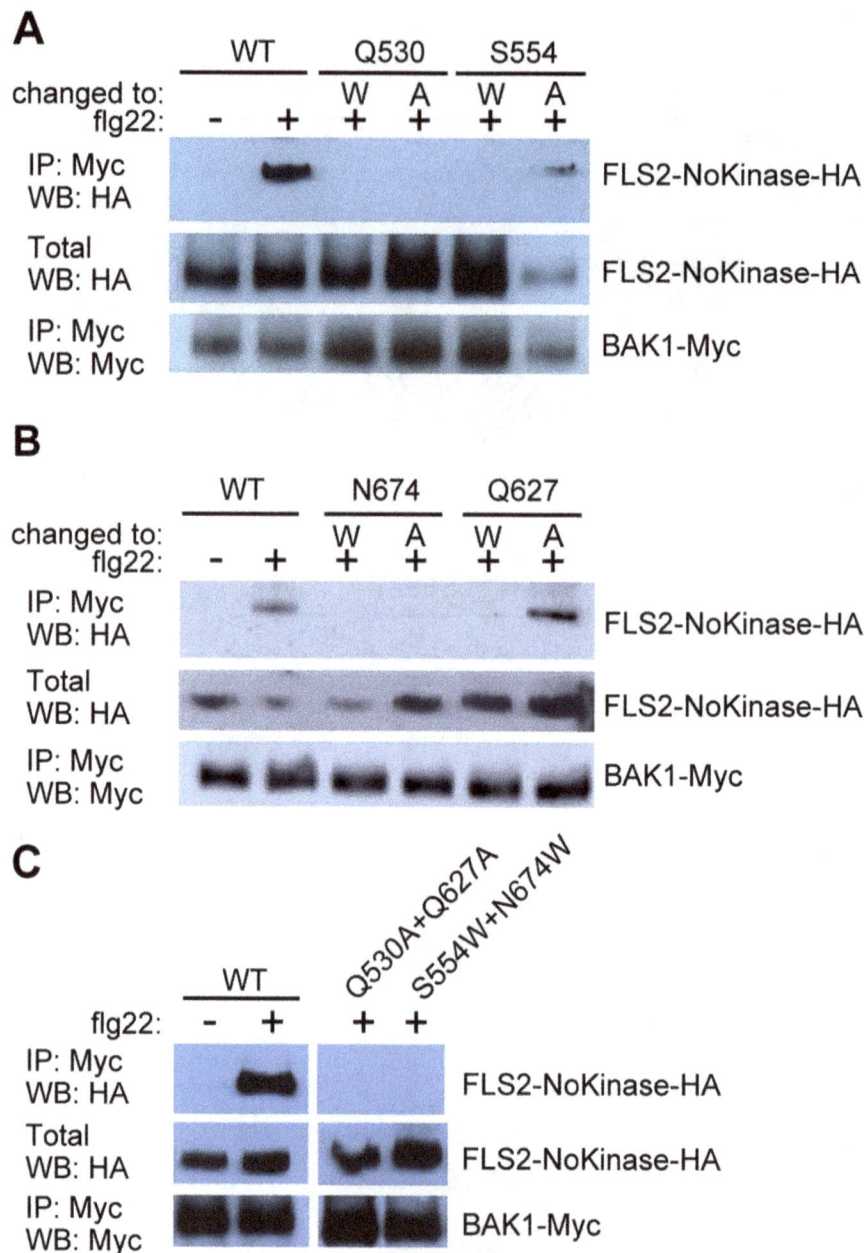

Figure 5. FLS2 residues Q530, S554, Q627 and N674 are important for FLS2-BAK1 ectodomain interaction in the presence of flg22.
Co-immunoprecipitation experiments performed in *N. benthamiana* with 35S–*FLS2-NoKinase-HA* with mutations as indicated or WT (no mutations), and with 35S–*BAK1-Myc*. Flg22-dependent interaction between FLS2-NoKinase and BAK1 not detected for (**A**) FLS2 carrying Q530A, Q530W or S554W mutations, (**B**) FLS2 carrying N674A, N674W or Q627W mutations, or (**C**) FLS2 carrying Q530A+Q627A or S554W+N674W double mutations. Labeling as in Figure 1.

interaction unique for each receptor/ligand/co-receptor complex. SERK1 and BAK1 use similar residues to interact with the BRI1 and FLS2 ectodomains, respectively. However, the SERK-interacting residues in the ectodomains of BRI1 and FLS2 are very different in terms of both the amino acid identities and their location along the large LRR macromolecule, and thus may have evolved separately.

The LRR surface region exhibiting conservation between FLS2, EFR, PEPR1 and BRI1 (Figure 2A) spans four repeats of the LRR, but overlaps with the larger region highlighted in Figure 2B that is conserved across diverse FLS2 proteins and spans the final seven repeats of the LRR. Within the larger conserved region, the

residues that are further to the left as shown in Figure 2B (or Figure S1E) encompass the BAK1 interaction site, but the residues on the right do not. The present study detected no impact of mutations in FLS2 in the Figure 2A conserved region, which in FLS2 is the same as the bottom right of the larger conserved region of Figure 2B. A previous study from our group [35] reported little or no functional impact of mutations in the upper-right area of the conserved region (the darkest red/most conserved area of Figure 2B). In that study, libraries of changes to all possible amino acids were made at the four FLS2 residues D605, S607, F633 and S634, directly above the N704/S706 and D728/S730 residues targeted in this study but in repeats #22 and 23 [35].

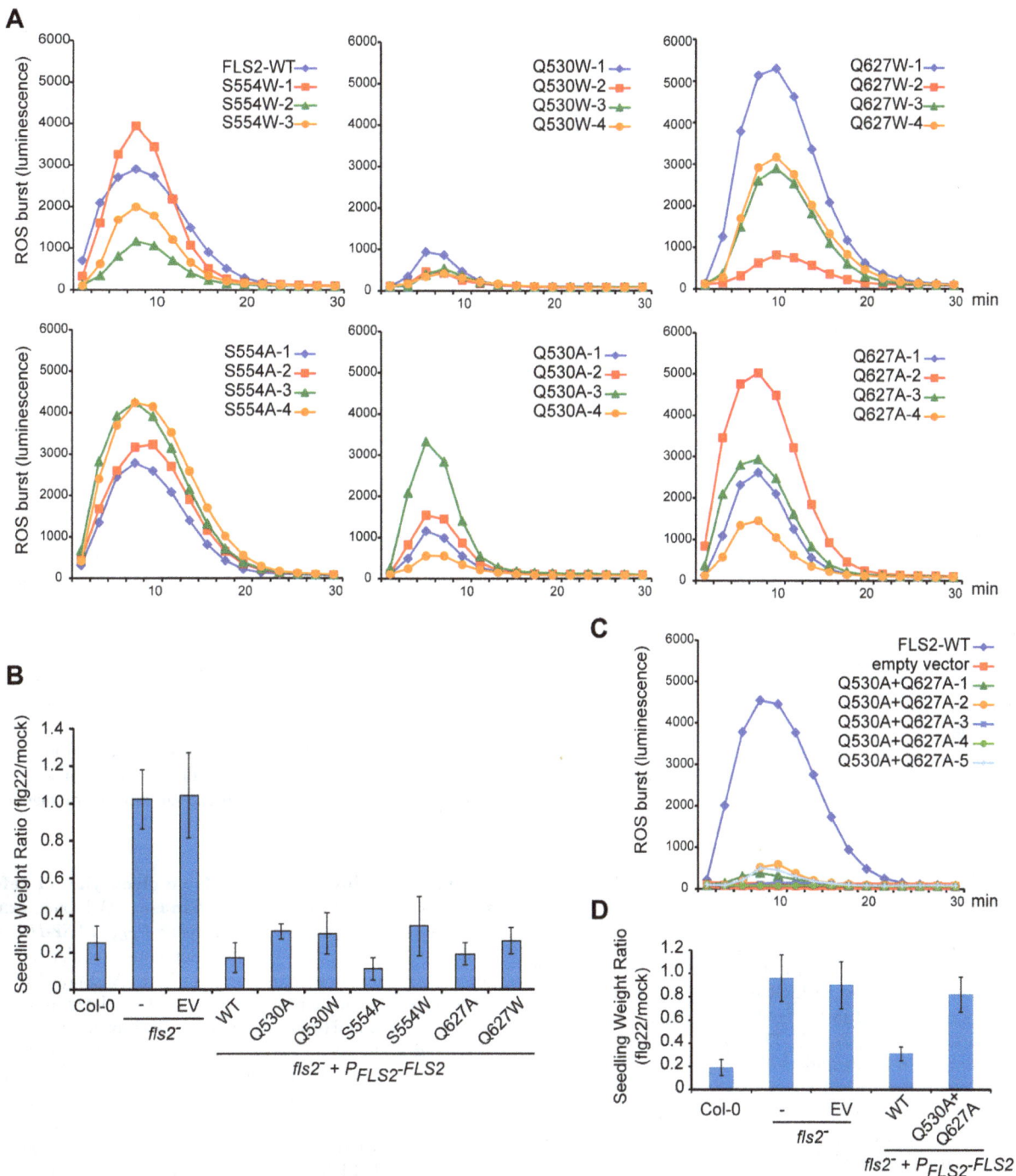

Figure 6. FLS2 signaling output impaired to various degrees in Arabidopsis *fls2⁻* plants expressing FLS2 mutations that impact FLS2-BAK1 interaction. (**A**) Reactive oxygen species (ROS) production in response to flg22 in Arabidopsis Col-0 *fls2⁻* plants stably transformed to express full-length FLS2 proteins carrying single mutations as noted, under control of *FLS2* promoter sequences. For each mutation, ROS production was recorded for 30 min. and the average for seven separately monitored leaf discs is shown for each of four independent transgenic lines (or three lines for S554W). WT: Average ROS response for six independent *fls2⁻* transformants expressing wild-type FLS2 (42 total leaf discs for WT), from same experiment. (**B**) FLS2-mediated seedling growth inhibition (SGI) in response to flg22, for plant lines as in (A). Mean and std. error of mean shown for six to eight independent transformants for each *FLS2* construct. (**C**) ROS experiment as in (A), except with five independent lines expressing FLS2 Q530A+Q627A double mutations. (**D**) Seedling growth experiment as in (B), except with twelve independent lines expressing Q530A+Q627A double mutations.

Hence it is intriguing that this right side of the region highlighted in Figure 2B, which lies along the concave β-strand surface of repeats #21–27, is highly tolerant of mutations despite being relatively conserved across FLS2 proteins from diverse plant species. It remains of interest to discover the function of this portion of the FLS2 LRR.

As a separate but related matter, it is intriguing that the set of BAK1-interacting residues of FLS2 lie not only within regions highly conserved across FLS2 proteins from diverse plant species (e.g., Q627 and N674, Figure S1E), as might be expected, but also outside of conserved regions (e.g., Q530 and S554, Figure S1E). Figure 2B and Figure S1E show regions of LRR surface residue conservation in a comparison among FLS2s from non-Brassicaceae plant species (see Methods). But even in maps of conservation among FLS2s only from Brassicaceae species (see for example [35]), the region around Arabidopsis FLS2 residues Q627 and N674 is strongly conserved while Q530, S554 and adjacent BAK1-interacting residues [27] are in an LRR surface region that is less conserved. This raises the hypothesis that there is a functionally relevant diversification of SERKs and/or this upper portion of the SERK-interaction site of FLS2, even across Brassicaceae species.

The relevance of the FLS2-flg22-BAK1 co-crystal structure to actual configurations of the protein complex within plant cells would gain stronger support if more features of the crystal structure were reconciled with other findings regarding plant FLS2s and flagellin detection. We noted in the Introduction the concern that the co-crystal, made with flg22 peptide, may not allow enough space for docking of a full-length flagellin protein at the appropriate location. Figure S3 shows hypothetical alignments of the FLS2-flg22-BAK1 ECD structure (PDB ID: 4MN8) with the structure of one Salmonella flagellin protein (PDB ID: 3A5X), placing the flg22 region of 3A5X near the apparent flg22 binding sites of FLS2 and BAK1 while attempting to minimize co-occupancy of the same space by two different molecules. The FLS2 LRR, which is notably lacking in 'loop-out' or non-LRR-consensus regions, is likely to be relatively inflexible. Flagellin monomers in solution (not polymerized with other flagellins to form flagella-like structures) are likely to be more flexible than shown, particularly in the region of the flg22 residues that form a less ordered linker between two alpha-helical regions [50,51] (see Figure S3F). Nevertheless, space-filling models (e.g., Figure S3D) demonstrate the difficulty of docking a large flagellin onto the requisite FLS2 LRR sites while also allowing space for BAK1 and not allowing co-occupancy of identical space. Importantly, even in hypothesized configurations (not shown) that might allow space for a more flexible full-length flagellin to interact with FLS2 and BAK1, the flg22 residues within a flagellin protein are apparently constrained in ways that would restrict simultaneous interaction with the majority of the FLS2 LRR surface residues that interact with the elongated flg22 in the published FLS2-flg22-BAK1 co-crystal structure (e.g., Figure S3E). FLS2, flagellins and BAK1 may associate *in vivo* in configurations that depart significantly from the co-crystal structure. However, numerous aspects of the published FLS2-flg22-BAK1 co-crystal structure are substantiated by experimental evidence ([27]; references therein; present study). Hence we consider it equally likely that the published FLS2-flg22-BAK1 co-crystal is essentially correct in representing *in vivo* configurations, and predict that flagellin proteins within plants must be fragmented rather than intact in order to form the FLS2-flagellin-BAK1 complexes that elicit plant innate immune system activation.

In the future, it also will be interesting to compare more receptor/ligand/co-receptor signaling complexes in order to learn more about the functional plasticity of co-receptors. As one example, a ligand-mediated EFR-BAK1 ectodomain complex is likely to initiate EFR signaling. Interestingly, when EFR from Arabidopsis was transferred to *Nicotiana benthamiana* or tomato (which lack an endogenous EFR) it triggered an elf18-activated immune response, indicating functional interaction of AtEFR with SERK proteins from *Nicotiana benthamiana* and tomato [52]. Thus one or more SERK proteins apparently carry sufficient

structure-function plasticity to interact with different receptors even from diverse plant species, while complying with the fine-tuned sequence constraints of the resulting receptor/ligand/co-receptor complexes. For future engineering of PRR receptors with novel ligand specificities it will be important to ensure presence of an intact SERK protein interaction site in the ectodomain of the PRR, close to or overlapping with the ligand binding site, and ensure that the co-receptors also can form PRR/ligand/co-receptor complexes with the novel ligands for which new recognition specificity is sought.

Supporting Information

Figure S1   Mutation sites in FLS2 and EFR ECDs. (A) Sites subjected to site-directed mutagenesis in the FLS2 LRR domain. Only repeats 17–28 are shown (FLS2: total 28 repeats). Green: mutation sites in the conserved LRR domain C-terminus (see also (B, D, E)). Blue: "control" mutations; sites similar to N704/D728 and S706/S730 but outside of the conserved region; blue "control" sites also adjacent to but outside of FLS2 BAK1-interaction site. Orange: mutation sites based on FLS2 BAK1-interaction sites in the FLS2-flg22-BAK1 ECD co-crystal structure. **(B)** Mutation sites as described in (A), using same color scheme as in (A). Structure is PDB ID: 4MN8 with FLS2 backbone as black ribbon, BAK1 backbone as light blue ribbon, and flg22 backbone as red ribbon. Space-filling spheres show side-chains only for mutagenized sites. **(C)** Mutation sites in the EFR LRR domain. Only repeats 17–21 are shown (EFR: total 21 repeats). Green: mutation sites in the conserved LRR domain C-terminus (see also (D)). **(D, E)** Regional LRR surface conservation maps from Arabidopsis FLS2, EFR, PEPR1 and BRI1 (D) or eleven non-Brassicaceae FLS2s (E), as shown and described in Figure 2, with x's at the FLS2 and EFR LRR domain amino acid positions described above that were subjected to site-directed mutagenesis in the present study.

Figure S2   EndoH assay reveals no glycosylation defects in mutated FLS2 and FLS2-NoKinase. (A) Protein extracts from Arabidopsis *fls2⁻* leaves carrying P_{FLS2}-FLS2-HA (with mutations as indicated, or WT = no mutations), not digested (−) or digested (+) with endoglycosidaseH (EndoH). An EndoH-resistant protein pool (characteristic of mature glycosylated proteins) is visible in all EndoH-treated samples. **(B)** Protein extracts from *Nicotiana benthamiana* carrying 35S–FLS2-NoKinase-HA (with mutations as indicated), digested with EndoH. An EndoH-resistant protein pool is visible in all EndoH-treated samples. Mutations D557E+S559T were included as control mutations located in sites of a single LRR repeat analogous to D728E+S730T, but outside of the conserved LRR C-terminus. **(C)** Protein extracts from Arabidopsis *fls2⁻* seedlings carrying P_{FLS2}-FLS2-HA (with mutations as indicated, or WT = no mutations), digested with EndoH. An EndoH-resistant protein pool is visible in all EndoH-treated samples except for the empty vector (EV) control. Ponceau stained blot shows similar loading of total protein in all lanes including EV negative control. Degly.: FLS2 pools deglycosylated by EndoH.

Figure S3   Hypothetical docking of full-length flagellin structure (PDB ID: 3A5X) to FLS2-flg22-BAK1 structure (PDB ID: 4MN8) illustrates minimal space for flagellin inside FLS2 LRR, and constraints to flg22 contact with FLS2 LRRs #3–15 if flg22 region is held within full-length flagellin. (A), (B), (C) Flagellin, hypothetically positioned so that flg22 residues within full-length flagellin are near the flg22 binding sites of FLS2 and BAK1. PDB structures 3A5X and 4MN8 superimposed at same scale; (B) and (C) are 90° rotated views of (A). Light blue: flagellin; red: flg22 residues within flagellin (3A5X). Dark blue: FLS2 LRR; green: BAK1 LRR; yellow: flg22 co-crystallized with FLS2 and BAK1 LRRs (4MN8). (D) Same view as (C), with space-filling representation of flagellin to more clearly illustrate impossible overlap of flagellin and FLS2 residues in same spatial locations in this arrangement (and other arrangements) of 3A5X and 4MN8. FLS2 and BAK1 side-chains omitted for clarity. (E) PyMol alignment of flg22 (yellow, in structure 4MN8) and flg22 region within flagellin (red, in structure 3A5X). Lower portions of flg22 in (E) (the yellow residues that are not proximal to red residues) are the N-terminal 7 residues of flg22 that associate with FLS2 LRRs #3–7 (FLS2 and BAK1 not

shown, for clarity). (F) Full length flagellin (PDB structure 3A5X) colored as in (A) but shown by itself, showing that flg22 region forms a less-ordered hinge region between flanking pairs of alpha-helical bundles.

Acknowledgments

We thank Stephen Mosher and Adam Bayless for critical reading of the manuscript.

Author Contributions

Conceived and designed the experiments: TK AFB. Performed the experiments: TK. Analyzed the data: TK AFB. Contributed to the writing of the manuscript: TK AFB.

References

1. Boller T, Felix G (2009) A renaissance of elicitors: perception of microbe-associated molecular patterns and danger signals by pattern-recognition receptors. Annu Rev Plant Biol 60: 379–406. Available: http://www.ncbi.nlm.nih.gov/entrez/query.fcgi?cmd=Retrieve&db=PubMed&dopt=Citation&list_uids=19400727.

2. Wu Y, Zhou J-M (2013) Receptor-like kinases in plant innate immunity. J Integr Plant Biol 55: 1271–1286. Available: http://www.ncbi.nlm.nih.gov/pubmed/24308571.

3. Gómez-Gómez L, Boller T (2000) FLS2: an LRR receptor-like kinase involved in the perception of the bacterial elicitor flagellin in Arabidopsis. Mol Cell 5: 1003–1011. doi:S1097-2765(00)80265-8 [pii].

4. Zipfel C, Kunze G, Chinchilla D, Caniard A, Jones JDG, et al. (2006) Perception of the Bacterial PAMP EF-Tu by the Receptor EFR Restricts Agrobacterium-Mediated Transformation. Cell 125: 749–760. doi:10.1016/j.cell.2006.03.037.

5. Kobe B, Kajava A V (2001) The leucine-rich repeat as a protein recognition motif. Curr Opin Struct Biol 11: 725–732. Available: http://www.ncbi.nlm.nih.gov/pubmed/11751054.

6. Bella J, Hindle KL, McEwan PA, Lovell SC (2008) The leucine-rich repeat structure. Cell Mol Life Sci 65: 2307–2333. Available: http://www.ncbi.nlm.nih.gov/pubmed/18408889.

7. Lehti-Shiu MD, Zou C, Hanada K, Shiu S-H (2009) Evolutionary history and stress regulation of plant receptor-like kinase/pelle genes. Plant Physiol 150: 12–26. Available: http://www.pubmedcentral.nih.gov/articlerender.fcgi?artid=2675737&tool=pmcentrez&rendertype=abstract.

8. Chinchilla D, Shan L, He P, de Vries S, Kemmerling B (2009) One for all: the receptor-associated kinase BAK1. Trends Plant Sci 14: 535–541. Available: http://www.ncbi.nlm.nih.gov/pubmed/19748302.

9. Liebrand TW, van den Burg HA, Joosten MH (2014) Two for all: receptor-associated kinases SOBIR1 and BAK1. Trends Plant Sci 19: 123–132. Available: http://www.ncbi.nlm.nih.gov/pubmed/24238702.

10. Kim BH, Kim SY, Nam KH (2013) Assessing the diverse functions of BAK1 and its homologs in arabidopsis, beyond BR signaling and PTI responses. Mol Cells 35: 7–16. Available: http://www.ncbi.nlm.nih.gov/pubmed/23269431.

11. Gao M, Wang X, Wang D, Xu F, Ding X, et al. (2009) Regulation of cell death and innate immunity by two receptor-like kinases in Arabidopsis. Cell Host Microbe 6: 34–44. Available: http://www.ncbi.nlm.nih.gov/pubmed/19616764.

12. Halter T, Imkampe J, Mazzotta S, Wierzba M, Postel S, et al. (2013) The Leucine-Rich Repeat Receptor Kinase BIR2 Is a Negative Regulator of BAK1 in Plant Immunity. Curr Biol. Available: http://www.ncbi.nlm.nih.gov/pubmed/24388849.

13. Liebrand TWH, van den Berg GCM, Zhang Z, Smit P, Cordewener JHG, et al. (2013) Receptor-like kinase SOBIR1/EVR interacts with receptor-like proteins in plant immunity against fungal infection. Proc Natl Acad Sci U S A 110: 10010–10015. Available: http://www.pubmedcentral.nih.gov/articlerender.fcgi?artid=3683720&tool=pmcentrez&rendertype=abstract.

14. Chinchilla D, Zipfel C, Robatzek S, Kemmerling B, Nürnberger T, et al. (2007) A flagellin-induced complex of the receptor FLS2 and BAK1 initiates plant defence. Nature 448: 497–500. Available: http://www.ncbi.nlm.nih.gov/pubmed/17625569.

15. Heese A, Hann DR, Gimenez-Ibanez S, Jones AME, He K, et al. (2007) The receptor-like kinase SERK3/BAK1 is a central regulator of innate immunity in plants. Proc Natl Acad Sci U S A 104: 12217–12222. Available: http://www.pubmedcentral.nih.gov/articlerender.fcgi?artid=1924592&tool=pmcentrez&rendertype=abstract.

16. Bar M, Sharfman M, Ron M, Avni A (2010) BAK1 is required for the attenuation of ethylene-inducing xylanase (Eix)-induced defense responses by the decoy receptor LeEix1. Plant J 63: 791–800. Available: http://www.ncbi.nlm.nih.gov/pubmed/20561260.

17. Fradin EF, Zhang Z, Juarez Ayala JC, Castroverde CDM, Nazar RN, et al. (2009) Genetic dissection of Verticillium wilt resistance mediated by tomato Ve1. Plant Physiol 150: 320–332. Available: http://www.pubmedcentral.nih.gov/articlerender.fcgi?artid=2675724&tool=pmcentrez&rendertype=abstract.

18. Chen X, Zuo S, Schwessinger B, Chern M, Canlas PE, et al. (2014) An XA21-Associated Kinase (OsSERK2) regulates immunity mediated by the XA21 and XA3 immune receptors. Mol Plant. Available: http://www.ncbi.nlm.nih.gov/pubmed/24482436.

19. Postel S, Küfner I, Beuter C, Mazzotta S, Schwedt A, et al. (2010) The multifunctional leucine-rich repeat receptor kinase BAK1 is implicated in Arabidopsis development and immunity. Eur J Cell Biol 89: 169–174. Available: http://www.sciencedirect.com/science/article/pii/S0171933509003306.

20. Schmidt ED, Guzzo F, Toonen MA, de Vries SC (1997) A leucine-rich repeat containing receptor-like kinase marks somatic plant cells competent to form embryos. Development 124: 2049–2062. Available: http://www.ncbi.nlm.nih.gov/pubmed/9169851.

21. Hecht V, Vielle-Calzada JP, Hartog MV, Schmidt ED, Boutilier K, et al. (2001) The Arabidopsis SOMATIC EMBRYOGENESIS RECEPTOR KINASE 1 gene is expressed in developing ovules and embryos and enhances embryogenic competence in culture. Plant Physiol 127: 803–816. Available: http://www.pubmedcentral.nih.gov/articlerender.fcgi?artid=129253&tool=pmcentrez&rendertype=abstract.

22. Li J, Wen J, Lease KA, Doke JT, Tax FE, et al. (2002) BAK1, an Arabidopsis LRR receptor-like protein kinase, interacts with BRI1 and modulates brassinosteroid signaling. Cell 110: 213–222. Available: http://www.ncbi.nlm.nih.gov/pubmed/12150929.

23. Nam KH, Li J (2002) BRI1/BAK1, a receptor kinase pair mediating brassinosteroid signaling. Cell 110: 203–212. Available: http://www.ncbi.nlm.nih.gov/pubmed/12150928.

24. Belkhadir Y, Yang L, Hetzel J, Dangl JL, Chory J (2014) The growth-defense pivot: crisis management in plants mediated by LRR-RK surface receptors. Trends Biochem Sci. Available: http://www.ncbi.nlm.nih.gov/pubmed/25089011.

25. Roux M, Schwessinger B, Albrecht C, Chinchilla D, Jones A, et al. (2011) The Arabidopsis leucine-rich repeat receptor-like kinases BAK1/SERK3 and BKK1/SERK4 are required for innate immunity to hemibiotrophic and biotrophic pathogens. Plant Cell 23: 2440–2455. Available: http://www.pubmedcentral.nih.gov/articlerender.fcgi?artid=3160018&tool=pmcentrez&rendertype=abstract.

26. Schwessinger B, Roux M, Kadota Y, Ntoukakis V, Sklenar J, et al. (2011) Phosphorylation-dependent differential regulation of plant growth, cell death, and innate immunity by the regulatory receptor-like kinase BAK1. PLoS Genet 7: e1002046. Available: http://www.pubmedcentral.nih.gov/articlerender.fcgi?artid=3085482&tool=pmcentrez&rendertype=abstract.

27. Sun Y, Li L, Macho AP, Han Z, Hu Z, et al. (2013) Structural basis for flg22-induced activation of the Arabidopsis FLS2-BAK1 immune complex. Science 342: 624–628. Available: http://www.ncbi.nlm.nih.gov/pubmed/24114786.

28. Sun Y, Han Z, Tang J, Hu Z, Chai C, et al. (2013) Structure reveals that BAK1 as a co-receptor recognizes the BRI1-bound brassinolide. Cell Res 23: 1326–1329. Available: http://www.pubmedcentral.nih.gov/articlerender.fcgi?artid=3817550&tool=pmcentrez&rendertype=abstract.

29. Santiago J, Henzler C, Hothorn M (2013) Molecular mechanism for plant steroid receptor activation by somatic embryogenesis co-receptor kinases. Science 341: 889–892. Available: http://www.ncbi.nlm.nih.gov/pubmed/23929946.

30. Felix G, Duran JD, Volko S, Boller T (1999) Plants have a sensitive perception system for the most conserved domain of bacterial flagellin. Plant J 18: 265–276. Available: http://www.ncbi.nlm.nih.gov/pubmed/10377992.

31. Robatzek S, Wirthmueller L (2012) Mapping FLS2 function to structure: LRRs, kinase and its working bits. Protoplasma 21. Available: http://www.ncbi.nlm.nih.gov/pubmed/23053766.

32. Sun W, Dunning FM, Pfund C, Weingarten R, Bent AF (2006) Within-species flagellin polymorphism in *Xanthomonas campestris pv campestris* and its impact on elicitation of Arabidopsis FLAGELLIN SENSING2-dependent defenses. Plant Cell 18: 764–779. Available: http://www.pubmedcentral.nih.gov/articlerender.fcgi?artid=1383648&tool=pmcentrez&rendertype=abstract.

33. Mueller K, Bittel P, Chinchilla D, Jehle AK, Albert M, et al. (2012) Chimeric FLS2 Receptors Reveal the Basis for Differential Flagellin Perception in Arabidopsis and Tomato. Plant Cell 24: 2213–2224. Available: http://www.ncbi.nlm.nih.gov/pubmed/22634763.

34. Tai TH, Dahlbeck D, Clark ET, Gajiwala P, Pasion R, et al. (1999) Expression of the Bs2 pepper gene confers resistance to bacterial spot disease in tomato. Proc Natl Acad Sci U S A 96: 14153–14158. Available: http://www.pubmedcentral.nih.gov/articlerender.fcgi?artid=24206&tool=pmcentrez&rendertype=abstract.

35. Helft L, Reddy V, Chen X, Koller T, Federici L, et al. (2011) LRR conservation mapping to predict functional sites within protein leucine-rich repeat domains. PLoS One 6: e21614. Available: http://www.pubmedcentral.nih.gov/articlerender.fcgi?artid=3138743&tool=pmcentrez&rendertype=abstract.

36. Sun W, Cao Y, Jansen Labby K, Bittel P, Boller T, et al. (2012) Probing the Arabidopsis flagellin receptor: FLS2-FLS2 association and the contributions of specific domains to signaling function. Plant Cell 24: 1096–1113. Available: http://www.pubmedcentral.nih.gov/articlerender.fcgi?artid=3336135&tool=pmcentrez&rendertype=abstract.

37. Nakagawa T, Kurose T, Hino T, Tanaka K, Kawamukai M, et al. (2007) Development of series of gateway binary vectors, pGWBs, for realizing efficient construction of fusion genes for plant transformation. J Biosci Bioeng 104: 34–41. Available: http://www.ncbi.nlm.nih.gov/pubmed/17697981.

38. Schulze B, Mentzel T, Jehle AK, Mueller K, Beeler S, et al. (2010) Rapid heteromerization and phosphorylation of ligand-activated plant transmembrane receptors and their associated kinase BAK1. J Biol Chem 285: 9444–9451. Available: http://www.ncbi.nlm.nih.gov/entrez/query.fcgi?cmd=Retrieve&db=PubMed&dopt=Citation&list_uids=20103591.

39. Cao Y, Aceti DJ, Sabat G, Song J, Makino S-I, et al. (2013) Mutations in FLS2 Ser-938 dissect signaling activation in FLS2-mediated Arabidopsis immunity. PLoS Pathog 9: e1003313. Available: http://www.pubmedcentral.nih.gov/articlerender.fcgi?artid=3630090&tool=pmcentrez&rendertype=abstract.

40. Albert M, Jehle AK, Fürst U, Chinchilla D, Boller T, et al. (2013) A two-hybrid-receptor assay demonstrates heteromer formation as switch-on for plant immune receptors. Plant Physiol 163: 1504–1509. Available: http://www.ncbi.nlm.nih.gov/pubmed/24130196.

41. Maley F, Trimble RB, Tarentino AL, Plummer TH (1989) Characterization of glycoproteins and their associated oligosaccharides through the use of endoglycosidases. Anal Biochem 180: 195–204. Available: http://www.ncbi.nlm.nih.gov/pubmed/2510544.

42. Häweker H, Rips S, Koiwa H, Salomon S, Saijo Y, et al. (2010) Pattern recognition receptors require N-glycosylation to mediate plant immunity. J Biol Chem 285: 4629–4636. Available: http://www.pubmedcentral.nih.gov/articlerender.fcgi?artid=2836068&tool=pmcentrez&rendertype=abstract.

43. Farid A, Malinovsky FG, Veit C, Schoberer J, Zipfel C, et al. (2013) Specialized roles of the conserved subunit OST3/6 of the oligosaccharyltransferase complex in innate immunity and tolerance to abiotic stresses. Plant Physiol 162: 24–38. Available: http://www.pubmedcentral.nih.gov/articlerender.fcgi?artid=3641206&tool=pmcentrez&rendertype=abstract.

44. Li J, Zhao-Hui C, Batoux M, Nekrasov V, Roux M, et al. (2009) Specific ER quality control components required for biogenesis of the plant innate immune receptor EFR. Proc Natl Acad Sci U S A 106: 15973–15978. Available: http://www.ncbi.nlm.nih.gov/pubmed/19717464.

45. Liu Y, Li J (2013) A conserved basic residue cluster is essential for the protein quality control function of the Arabidopsis calreticulin 3. Plant Signal Behav 8: e23864. Available: http://www.ncbi.nlm.nih.gov/pubmed/23425854.

46. Su W, Liu Y, Xia Y, Hong Z, Li J (2012) The Arabidopsis homolog of the mammalian OS-9 protein plays a key role in the endoplasmic reticulum-associated degradation of misfolded receptor-like kinases. Mol Plant 5: 929–940. Available: http://www.pubmedcentral.nih.gov/articlerender.fcgi?artid=3399701&tool=pmcentrez&rendertype=abstract.

47. Von Numers N, Survila M, Aalto M, Batoux M, Heino P, et al. (2010) Requirement of a homolog of glucosidase II beta-subunit for EFR-mediated defense signaling in Arabidopsis thaliana. Mol Plant 3: 740–750. Available: http://www.ncbi.nlm.nih.gov/pubmed/20457640.

48. Kornfeld R, Kornfeld S (1985) Assembly of asparagine-linked oligosaccharides. Annu Rev Biochem 54: 631–664. Available: http://www.ncbi.nlm.nih.gov/pubmed/3896128.

49. Dunning FM, Sun W, Jansen KL, Helft L, Bent AF (2007) Identification and mutational analysis of Arabidopsis FLS2 leucine-rich repeat domain residues that contribute to flagellin perception. Plant Cell 19: 3297–3313. Available: http://www.pubmedcentral.nih.gov/articlerender.fcgi?artid=2174712&tool=pmcentrez&rendertype=abstract.

50. Yonekura K, Maki-Yonekura S, Namba K (2003) Complete atomic model of the bacterial flagellar filament by electron cryomicroscopy. Nature 424: 643–650. Available: http://www.ncbi.nlm.nih.gov/pubmed/12904785.

51. Maki-Yonekura S, Yonekura K, Namba K (2010) Conformational change of flagellin for polymorphic supercoiling of the flagellar filament. Nat Struct Mol Biol 17: 417–422. Available: http://www.ncbi.nlm.nih.gov/pubmed/20228803.

52. Lacombe S, Rougon-Cardoso A, Sherwood E, Peeters N, Dahlbeck D, et al. (2010) Interfamily transfer of a plant pattern-recognition receptor confers broad-spectrum bacterial resistance. Nat Biotechnol 28: 365–369. Available: http://www.ncbi.nlm.nih.gov/pubmed/20231819.

9

Characterization of the YdeO Regulon in *Escherichia coli*

Yuki Yamanaka[1], Taku Oshima[3], Akira Ishihama[1,2], Kaneyoshi Yamamoto[1,2]*

1 Department of Frontier Bioscience, Hosei University, Koganei, Tokyo, Japan, **2** Micro-Nano Technology Research Center, Hosei University, Koganei, Tokyo, Japan, **3** Graduate School of Information Sciences, Nara Institute of Science and Technology, Ikoma, Nara, Japan

Abstract

Enterobacteria are able to survive under stressful conditions within animals, such as acidic conditions in the stomach, bile salts during transfer to the intestine and anaerobic conditions within the intestine. The glutamate-dependent (GAD) system plays a major role in acid resistance in *Escherichia coli*, and expression of the GAD system is controlled by the regulatory cascade consisting of EvgAS > YdeO > GadE. To understand the YdeO regulon *in vivo*, we used ChIP-chip to interrogate the *E. coli* genome for candidate YdeO binding sites. All of the seven operons identified by ChIP-chip as being potentially regulated by YdeO were confirmed as being under the direct control of YdeO using RT-qPCR, EMSA, DNaseI-footprinting and reporter assays. Within this YdeO regulon, we identified four stress-response transcription factors, DctR, NhaR, GadE, and GadW and enzymes for anaerobic respiration. Both GadE and GadW are involved in regulation of the GAD system and NhaR is an activator for the sodium/proton antiporter gene. In conjunction with co-transcribed Slp, DctR is involved in protection against metabolic endoproducts under acidic conditions. Taken all together, we suggest that YdeO is a key regulator of *E. coli* survival in both acidic and anaerobic conditions.

Editor: Dipankar Chatterji, Indian Institute of Science, India

Funding: This work is supported by MEXT-Supported Program for the Strategic Research Foundation at Private Universities and Special Coordination Funds for Promoting Science and Technology and JSPS-DST International Collaborations. The funders had no role in study design, data collection and analysis, decision to publish, or preparation of the manuscript.

Competing Interests: The authors have declared that no competing interests exist.

* Email: kanyamam@hosei.ac.jp

Introduction

Enterobacteria such as *Escherichia coli*, exist in the environment, and in the gut of warm blooded animals. To survive this switch in lifestyles, and upon ingestion by a new host, bacteria are directly exposed to various stresses and hence require sophisticated stress response systems to survive continuous changes in environment such as acidic conditions in the stomach, bile salts, and anaerobic conditions within the intestines [1]. For survival under acidic conditions, *E. coli* possesses three amino acid-dependent acid resistance systems with glutamate, arginine, and lysine [2,3,4]. The resistance mechanism involves the transient consumption of the intracellular proton by glutamate, arginine and lysine decarboxylases, and exchange of the amine products with extracellular amino acids through their respective antiporters [2,3,5,6]. The most effective system of acid resistance is the GAD (glutamic acid-dependent) system which is composed of two glutamate decarboxylase isozymes, GadA and GadB, and the cognate antiporter GadC. Expression of these components is under the control of a complex network of transcription factors, including GadE, GadX, GadW, EvgA, YdeO, and H-NS [1].

YdeO is a transcription factor, belonging to the AraC/XylS family. Knowledge about the regulatory functions of YdeO is limited except that it is known that YdeO activates transcription of the *gad* system components, *gadE*, *gadA* and *gadBC* [7,8,9]. The expression of *ydeO* is activated by the two-component system EvgSA [9,10,11], forming a regulatory cascade, EvgA > YdeO > GadE [9,12]. In this study, we performed a comprehensive interrogation of YdeO-binding sites *in vivo* on the *E. coli* genome using ChIP-chip analysis, and identified a set of YdeO-regulated genes, including four stress-response transcription factors, DctR, NhaR, GadE, and GadW, and several genes involved in respiration. Taking these observations together we propose that YdeO is the regulator which coordinates the response to acid and anaerobic conditions in *E. coli*.

Materials and Methods

E. coli strains and growth conditions

E. coli strains and plasmids used in this study are shown in Table S1. *E. coli* cells were grown at 37°C in Luria-Bertani (LB) medium. Cell growth was monitored by measuring the turbidity with a Mini photo 518R spectrophotometer (Taitec). The standard procedure for bacterial cell cultivation in this study was as follows: A single colony was isolated from an overnight culture on a LB agar plate, and inoculated into 5 ml of fresh LB medium. This liquid culture was grown overnight at 37°C, and the overnight culture was diluted 100-fold into fresh LB medium. The culture was incubated at 37°C with reciprocal shaking (160 revolutions min^{-1}) for aerobiosis or without shaking for anaerobiosis.

Introduction of a tagged gene into the *E. coli* genome

The introduction of a tagged gene into the *E. coli* genome was carried out using the method of Uzzau et al. [13]. In brief, primers were used to make PCR extensions homologous to the last portion of the targeted gene (forward primer) and to a region downstream of it (reverse primer) as follows; YDEOF-1 (forward) and YDEOR-1 (reverse) for *ydeO-3xflag*; GADE-F (forward) and GADE-R (reverse) for *gadE-3xflag*; GADW-F (forward) and GADW-R (reverse) for *gadW-3xflag* (Table S1). Amplified DNA fragments including the 3′ sequence with flag tag and a kanamycin-resistance gene were amplified by PCR using pSUB11 as a template, a pair of primers, and Ex-Taq DNA polymerase (Takara Bio). PCR products were purified using a QIAquick PCR purification kit (Qiagen), and then used directly for electro-transformation. *E. coli* carrying a lambda-Red helper plasmid, pKD46, was used to make competent cells, and were grown at 30°C in LB medium supplemented with 100 µg ml^{-1} ampicillin and 1 mM arabinose to an OD$_{600}$ of 0.4. Cells were collected by centrifugation, and washed two times with ice-cold sterile deionized water containing 10% glycerol. Aliquots (50 µl) of the bacterial suspensions in 10% glycerol were mixed with more than 1 µg of PCR product in a chilled cuvette (0.2 cm electrode gap) and subjected to a single pulse (2.5 kV) by a Gene pulser Xcell (Bio Rad). After 1 hr recovery at 37°C in 1 mL of SOC medium (2% tryptone, 0.5% yeast extract, 10 mM NaCl, 2.5 mM KCl, 10 mM MgCl$_2$, 10 mM MgSO$_4$, 20 mM glucose) containing 1 mM arabinose, half of the volume of electroporated bacteria in SOC media were spread on to LB agar plates supplemented with antibiotics for the selection of kanamycin-resistant recombinants. If none grew on the agar plate after incubation overnight at 37°C, the remainder stored was spread on to LB kan plates. The kanamycin-resistance recombinants were isolated once on LB agar at 37°C, and then examined for ampicillin sensitivity for loss of the helper plasmid.

Construction of YdeO expression plasmids

To construct pYY0401 for YdeO-3xFLAG expression, DNA fragments containing the *ydeO* coding region were amplified by PCR using *E. coli* YY5001 genomic DNA, including the *3xflag* tag at the end of *ydeO* as a template, and a pair of primers, YDEOF-2 and YDEOR-3, in which the *Bam* HI and *Eco* RI sites were included (see sequences in Table. S1). After digestion of PCR products with *Bam* HI and *Eco* RI, the PCR-amplified fragments were cloned into the pTrc99A vector containing an inducible *trc* promoter between the *Bam* HI and *Eco* RI sites. To construct pYdeO for expression of intact YdeO, DNA fragments containing the *ydeO* coding region were amplified by PCR using *E. coli* W3110 type A [14] genomic DNA as a template and the primers, YDEOF-2 and YDEOR-2 (see sequences in Table. S1). After digestion of the PCR product with *Bam* HI and *Eco* RI, the PCR-amplified fragments were ligated into the pTrc99A vector between appropriate restriction enzyme sites. To construct pYdeO-SUMO for overproduction of SUMO (Small Ubiquitin-related MOdifier) fused YdeO, DNA fragments containing the *ydeO* coding region were amplified by PCR using *E. coli* BW25113 genomic DNA as a template and the primers, YDEO-SUMO-F and YDEO-SUMO-R, in which 15-nt homologous to pE-SUMO vector (Life Sensors) digested with *Bsa* I were included (see sequences in Table. S1). The PCR-amplified fragments were cloned into the pE-SUMO vector using In-Fusion HD cloning kit (Clontech). All of the plasmids were confirmed by DNA sequencing with primers, Trc99A-F and/or Trc99A-R for pTrc99A derivatives and T7 terminator and SUMO forward for pE-SUMO derivatives.

Construction of *lacZ* and *lux* reporter plasmids

To construct a *lacZ* fusion gene, the pRS552 plasmid was used as a vector for the construction of translational fusions [15]. The promoter DNA fragment was amplified by PCR using the genome of *E. coli* W3110 type-A strain [14] as a template and a pair of

primers. The primers used were: APPC-LF and APPC-LR for pAPPC-L; YIIS-LF and YIIS-LR for pYY0503; HYAA-LF and HYAA-LR for pHYAA-L (Table S1). The PCR product was digested with *BamH* I and/or *EcoR* I and then ligated into pRS552 at the corresponding sites. A *nhaR-lux* transcription fusion was also constructed. First, DNA fragments containing the *nhaR* promoter were amplified by PCR using the primers: NHAR-lux-F and NHA-lux-R, which contained 15-nt homologous to the pLUX vector [16] digested with *Xho* I and *Bam* HI were included (see sequences in Table. S1). The PCR-amplified fragments were cloned into the pLUX vector using In-Fusion HD cloning kit (Clontech), resulting in the construction of pLUXnhaR (Table S1). All of the plasmids were confirmed by DNA sequencing using the lacZ-30R primer complementary to *lacZ* or Lux-R primer complementary to *luxC* in a vector.

ChIP-chip analysis

The ChIP-chip assay was carried out as described in previous reports [17,18,19] with a few modifications. YY0201 (Δ*ydeO*) harbouring pYY0401 (*ydeO-3xflag*) was grown to an OD$_{600}$ of 0.4 then re-incubated in LB medium containing formaldehyde (final concentration of 1%) at 37°C for 30 min. The cross-linking reaction was terminated by the addition of glycine, and cells were collected, washed, re-suspended with lysis buffer, and lysed by incubation with Lysozyme. Lysed cells were dissolved in 4 ml of IP buffer containing PMSF. The sample was then sonicated 60 times for 30 sec at 30 sec intervals on ice using a BRANSON Digital Sonifier (Branson). After centrifugation, the supernatant fraction (whole cell extract) was mixed with anti-FLAG antibody (Sigma Aldrich)-coated-protein A Dynal Dynabeads (Invitrogen) and incubated at 4°C overnight. After washing twice with IP buffer and IP salt buffer, the DNA–YdeO-3xFLAG complex bound to the beads was recovered by eluting with elution buffer (50 mM Tris–HCl pH 7.5, 10 mM EDTA, 1% SDS). YdeO-3xFLAG in whole cell extracts and in immunoprecipitated DNA fractions were digested by Pronase (Roche). DNA fragments free of cross-linked DNA–protein were purified using a QIAquick PCR purification kit (Qiagen). Recovered DNA fragments were amplified according to the random DNA amplification method using the primers, PF 43 and PF 44 described by Katou et al. [17]. PCR was performed over 30 cycles, using Phusion high-fidelity DNA polymerase (New England Biolabs). Amplified DNA fragments were terminally labeled and hybridized with the custom-designed Affymetrix oligonucleotide tiling array and raw data (CEL files) were processed using the Array edition of the In Silico Molecular Cloning (IMC) software (In Silico Biology) as previously described [18,19,20]. To detect DNA fragments by immunoprecipitation, the signal intensities of ChIP DNA were divided by those of the supernatant (Sup) fraction.

Pufirication of the YdeO protein

In a typical procedure [21], a single colony of transformed *E. coli* BL21 (DE3) was grown to OD$_{600}$ = 0.6 at 37°C with shaking in LB medium supplemented with 100 µg ml^{-1} ampicillin. The culture was then cooled on ice, induced with 4.5 mM IPTG, and incubated at 20°C overnight with shaking. Cells were isolated by centrifugation and resuspended in 400 µL of lysis buffer (100 mM NaCl, 50 mM Tris–HCl pH 8.0) containing 0.2 mM PMSF. Cells were treated with lysozyme and then subjected to sonication. Triton X-100 was added to 1% (v/v) and incubated on ice for 1 hr. The culture was centrifuged, and the supernatant was decanted and stored at 4°C. Supernatant was mixed with 2 ml of 50% Ni-nitrilotriacetic acid (NTA) agarose solution (Qiagen) and loaded onto a column. The column was washed with 10 ml of lysis

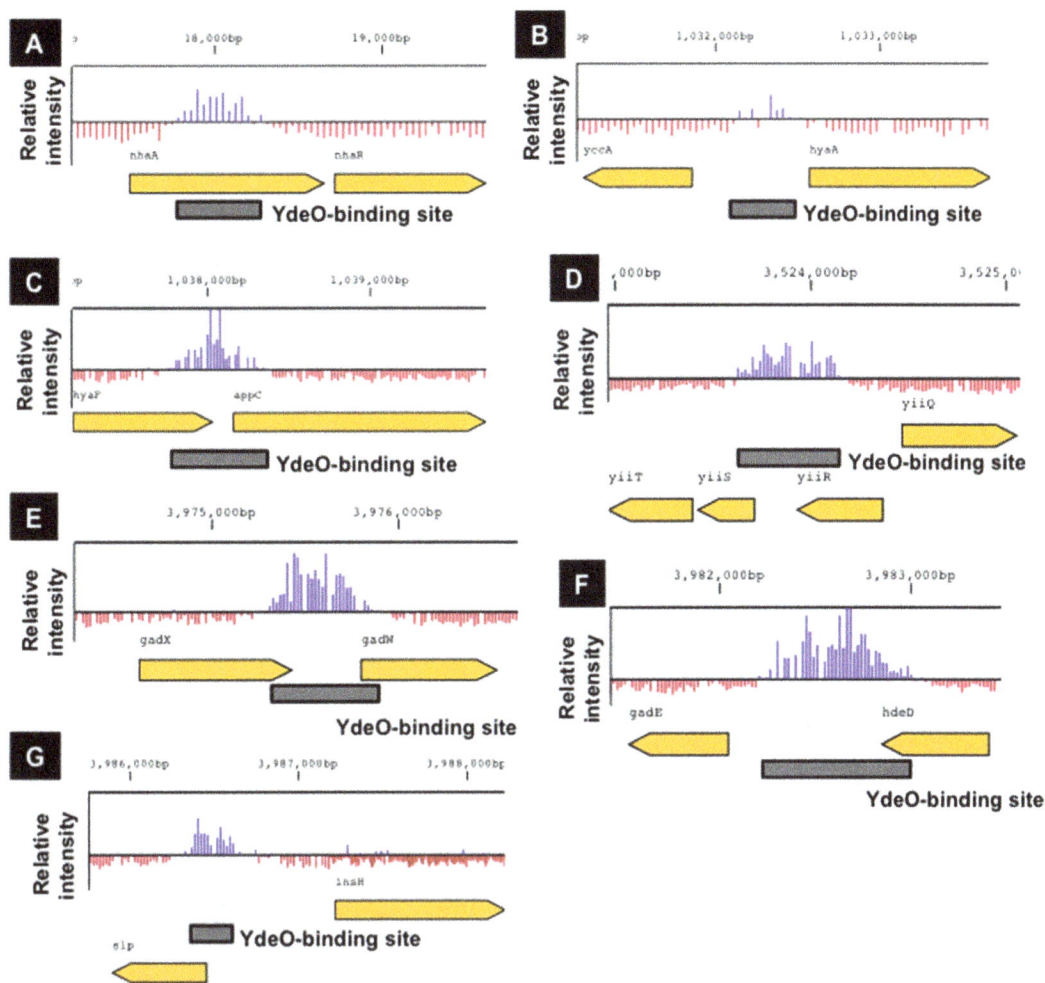

Figure 1. Genome-wide regulation of the *Escherichia coli* YdeO protein. Location of YdeO binding sites. The panel shows detailed YdeO binding data from ChIP-chip experiments at the *nhaR* (A), *hyaA* (B), *appC* (C), *yiiS* (D), *gadW* (E), *gadE* (F), and *slp* (G) genomic loci. The box indicates the YdeO-binding site.

buffer containing 1%Triton X-100, and then washed with 10 ml of lysis buffer containing 1%Triton X-100 and 25 mM imidazole. Proteins were eluted with 3 ml of each elution buffer (lysis buffer containing 1%Triton x-100 and 0.1 M, 0.2 M, 0.3 M, 0.4 M, or 0.5 M imidazole), and peak fractions of transcription factors were pooled and dialyzed against a storage buffer (50 mM Tris-HCl, pH 7.5 at 4°C, 200 mM KCl, 10 mM MgCl₂, 0.1 mM EDTA, 5 mM DTT, and 50% glycerol), and stored at –80°C until use. Protein purity was checked on SDS-PAGE.

Preparation of total RNA from *E. coli* cells

Total RNA was prepared using the as previously described [22]. A single colony of *E. coli* was grown in LB medium to $OD_{600} = 0.3$ at 37°C with shaking. Cells were harvested and total RNAs were prepared using hot phenol. In brief, total RNA was extracted with H_2O-saturated phenol and precipitated with ethanol. After digestion with RNase-free DNase I (Takara Bio), total RNA was extracted with H_2O-saturated phenol and precipitated with ethanol, and dissolved in RNase-free water. The concentration of total RNA was determined by measuring the absorbance at 260 nm. The purity of total RNA was checked by agarose gel electrophoresis.

Transcriptome analysis

To prepare fluorescently labeled cDNA, total RNA (5 μg) was used. We used the FairPlay III Microarray Labeling kit (Agilent), CyDye Cy3 mono-reactive Dye, and CyDye Cy5 mono-reactive Dye (GE Healthcare). For all experiments, two sets of RNAs from an independent colony were carried out with a pair of the fluorescence dye. The mixture containing 1 μl of Ramdom hexanucleotide primers, 5 μg of total RNA, and 12 μl of DEPC-treated water was heated at 75°C for 10 min and cooled to room temperature. After addition of 3 μl of Affinity script HC RTase (Agilent), 1X Affinity script RT buffer, 1X dNTP mixture, 75 mM DTT, and 0.5 μl of RNase block to 10 μl of RNA/primer mixture product, cDNA synthesis was carried out at 42°C for 1 hr and stopped by addition of 10 μM NaOH. The mixture was neutralized by addition of 10 μM HCl. The synthesized cDNA was purified by ethanol-precipitation and then labelled by CyDye Cy3 mono-reactive Dye or CyDye Cy5 mono-reactive Dye. The dye-coupled cDNA was purified by attached the micro spin cup.

The *E. coli* Gene Expression Microarray microarray 8×15 K (Agilent) was used. Each 300 ng of Cy3- and Cy5-labeled cDNA were mixed and added to 1X Blocking Buffer (Agilent) and 1X HI-RPM GE Hybridization Buffer (Agilent). After precipitation of impurities, 40 μl of the labelled-cDNA mixture was applied to the

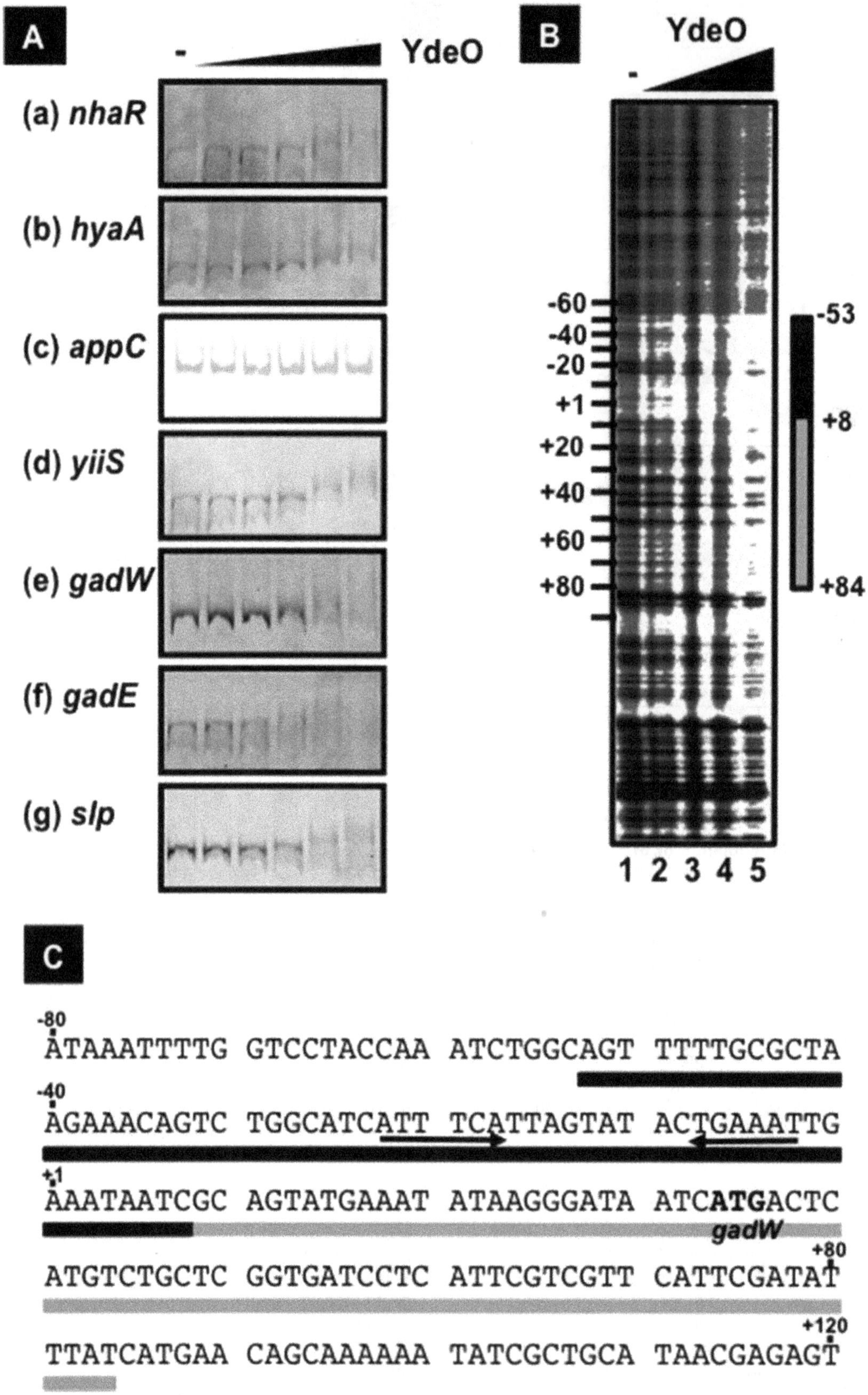

Figure 2. The binding of YdeO on target promoters. [A] The binding of YdeO to the target DNA, *nhaR* (a), *hyaA* (b), *appC* (c), *yiiS* (d), *gadW* (e), *gadE* (f), and *slp* (g). Probes were amplified by PCR using constructed reporter plasmids as templates and a pair of primers as the following; pLUXnhaR and a pair of NHAR-lux-F and Lux-R-FITC for *nhaR* probe; pHYAA-L and a pair of HYAA-LF and lacZ-30R-FITC for *hyaA* probe; pAPPC-L and a pair of APPC-LF and lacZ-30R-FITC for *appC* probe; pYY0503 and a pair of YIIS-LF and lacZ-30R-FITC for *yiiS* probe; pLUXgadWp and a pair of GADW-F-2 and

Lux-R-FITC for *gadW* probe; pLUXgadEp and a pair of GADE-SCL-F-2 and Lux-R-FITC for *gadE* probe; and pLUXslpp and a pair of SLP-F-2 and Lux-R-FITC for *slp* probe; (Table S1). Each FITC-labeled probe (1 pmol) was incubated with YdeO protein (1, 5, 10, 25, or 35 pmol) and then DNA-YdeO complex was analyzed by native PAGE. Solid and dot lines indicate the migration of free DNA probe and DNA-YdeO complex, respectively. [B] The YdeO-binding site on *gadW* promoter. FITC-labeled probe (1 pmol) was incubated with 0 (lane 1), 0.5 (lane 2), 1.0 (lane 3), 5.0 (lane 4), or 15 (lane 5) pmol YdeO protein and then digested by DNase I. Sanger ladders are synthesized using Lux-R-FITC primer and pLUXgadWp plasmid as a template. A bar indicates the major region protected from DNase I digestion. The numbers represent the position from the transcription start site of *gadW*p1 promoter. [C] The sequence of YdeO-binding on *gadW* promoter. Black and gray bars indicate the major and minor YdeO-binding regions, respectively, as shown in [B]. The initiation codon of *gadW* coding is represented as a bold triplet. The numbers represent the position from the transcription start site of *gadW*p1 promoter.

DNA chip, and the hybridization was carried out at 65°C for 17 hr. The DNA chip was washed at room temperature with Agilent Gene Expression Wash Buffer 1 (Agilent) and at 37°C with Agilent Gene Expression Wash Buffer 2 (Agilent). The DNA chip was scanned with an Agilent G2565CA microarray scanner Ver. 8.1, and the intensities of both Cy3 and Cy5 were quantified by Feature Extraction Ver. 8.1. And then, the Cy5/Cy3 ratios were calculated from the normalized values.

RT-qPCR

Total RNAs were transcribed to cDNA with random primers using Primer Script 1st strand cDNA synthesis Kit (Takara Bio). Quantitative PCR (qPCR) was conducted using SYBR Green PCR Master Mix (Applied Biosystems). Pairs of primers used are described in Table S1. The cDNA templates were twofold serially diluted and used in the qPCR assays. The qPCR reaction mixtures, each containing 12.5 µl of 2X Power SYBR Green PCR Master Mix (Applied Biosystems), 0.225 µl of each primer (10 µM stock), 9.55 µl of water, and 2.5 µl of cDNA, were amplified under the following thermal cycle conditions of: 50°C for 2 min and 95°C for 10 min followed by 40 cycles of 15 sec at 95°C and then 60 sec at 60°C. The expression levels of the 16 S rRNA gene were used for normalization of data, and the relative expression levels were quantified using 'Delta–delta method' presented by PE Applied Biosystems (Perkin Elmer) as described in previous reports [23,24]. The results presented are averages of the results from the replicate experiments ± standard errors of the means (SEM).

EMSA

Probes were amplified by PCR using the previously constructed reporter plasmids as templates, with a pair of primers: a specific primer and an FITC-labeled primer. PCR products with FITC at their termini were purified using the QIAquick PCR purification kit (Qiagen). For gel shift assays, mixtures of the FITC-labeled probes and purified SUMO-YdeO were incubated at 37°C for 30 min in gel shift buffer (50 mM Tris-HCl, pH 7.8 at 37°C, 50 mM NaCl, 3 mM Mg acetate, 0.1 mM EDTA, 0.1 mM DTT, and 0.37 µM BSA) containing 0.2 µg ml^{-1} salmon sperm DNA. After addition of a DNA dye solution, the mixture was directly subjected to 4% or 7% PAGE. Fluorescent-labeled DNA in gels was detected using Typhoon 9410 (Amersham Biosciences).

DNase I footprinting analysis

The probe was amplified by PCR using a pLUXgadWp as a template, primer pairs GADW-F-2 and Lux-R-FITC, and Ex Taq DNA polymerase (Takara). 1.0 pmol of a FITC-labeled probe was incubated at 37°C for 30 min with purified SUMO-YdeO (0.5 to 15 pmol) in 25 µl of gel shift buffer (50 mM Tris-HCl, pH 7.8 at 37°C, 50 mM NaCl, 3 mM Mg acetate, 0.1 mM EDTA, 0.1 mM DTT, and 0.37 µM BSA). After incubation for 30 min, DNA was digested by DNase I (Takara Bio) for 30 s at 25°C, and then the reaction was terminated by addition of phenol. DNA was precipitated by ethanol, dissolved in formamide dye solution,

and analyzed by electrophoresis on a DNA analyzer DSQ-2000L (Shimadu).

Measurement of luciferase activity in *E. coli*

A single colony of a strain freshly transformed with one of the luciferase reporter plasmids (Table S1) was grown in LB medium supplemented with 50 µg ml^{-1} kanamycin to $OD_{600} = 0.3$ at 37°C with shaking. At this point, the culture was transferred to a microtiter plate (96-well micro-titer) to start monitoring reporter activity measurement in an automated plate reader MTP-880 (Corona). The Lux (luciferase activity) reads were then divided by the equivalent OD reads (Lux/OD) to approximate Lux activity unit per cell mass for each well. The Lux/OD values of the three technical replicate wells of each culture were averaged.

Measurement of β-galactosidase activity in *E. coli*

E. coli cells were grown in LB medium and subjected to measurement of β-galactosidase activity with *o*-nitrophenyl-D-galactopyranoside as described in the previous report [11].

Western blotting analysis.

E. coli cells grown in LB medium were harvested by centrifugation and re-suspended in 0.4 ml lysis buffer containing 8 M urea and sonicated. After centrifugation, the same volume of supernatant was subjected to 15% SDS-PAGE and blotted on to PVDF membranes using an iBlot semi-dry transfer apparatus (Invitrogen). Membranes were first immuno-detected with anti-FLAG (Sigma), anti-NhaR serum (Lab stock), or anti-α (Neoclone) and HRP-conjugated anti-mouse IgG (Nacalai Tesque) antibodies and then developed with a chemiluminescence kit (Nacalai Tesque). The image was analyzed with a LAS-4000 IR multi colour imager (Fuji Film).

Results

Identification of YdeO associated sites *in vivo* within the *E. coli* genome

To identify the genes directly regulated by YdeO, we first determined the genome-wide distribution of YdeO-binding sites by ChIP-chip (Chromatin ImmunoPrecipitation-DNA chip) analysis. For this purpose, we inserted a *3xflag* tail into the 3' end of the *ydeO* gene in the genome and tried to prepare YdeO-DNA complexes for ChIP-chip analysis from the YY5001 strain harbouring *ydeO-3xflag* grown in LB medium at 37°C with shaking. The level of YdeO-3xFLAG expression was, however, not enough to isolate YdeO-DNA complexes using the anti-FLAG antibody. We then constructed plasmid pYY0401 for the expression of YdeO-3xFLAG and transformed it into the *ydeO*-deficient mutant. The *ydeO*-deficient mutant transformed with pYY0401 was grown until it reached log phase and was then treated with formaldehyde for DNA-protein cross-linking. The *E. coli* cells were disrupted with sonication to prepare a whole cell extract from which YdeO-DNA complexes were isolated, sonicat-

Table 1. Genes up-regulated by YdeO expression.

	Gene name	Transcriptoin units[a]	Wild type (Log$_{10}$ ratio[b])		$\Delta gadE$ (Log$_{10}$ ratio[b])	
			1st[c]	2nd[c]	1st[c]	2nd[c]
In both wild type and $\Delta gadE$	adiC	adiC	1.68	1.68	1.65	1.42
	appC	appCBA	1.66	1.85	1.73	1.64
	appB	appCBA	1.71	1.69	1.71	1.60
	appA	appCBA	1.34	1.33	1.04	1.12
	dnaK	dnaK-tpke11-dnaJ	0.74	0.57	−0.01	0.46
	dnaJ	dnaK-tpke12-dnaJ	0.66	0.57	0.04	0.75
	hyaA	hyaABCDEF	1.46	1.54	1.87	1.17
	hyaB	hyaABCDEF	1.61	1.50	2.16	1.11
	hyaC	hyaABCDEF	1.46	1.53	1.84	1.20
	hyaD	hyaABCDEF	1.22	1.29	1.57	1.03
	hyaE	hyaABCDEF	1.19	1.02	1.45	0.92
	hyaF	hyaABCDEF	1.15	1.09	1.4\5	1.09
	ibpA	ibpAB	0.65	0.50	0.09	0.64
	katE	katE	0.78	0.72	0.08	0.46
	metK	metK	0.64	0.63	0.03	0.60
	nhaA	nhaAR	0.53	0.50	0.54	0.40
	yehX	osmF-yehYXW	0.67	0.63	0.22	0.46
	slp	slp-dctR	1.99	2.15	2.41	1.79
	dctR	slp-dctR	1.85	1.16	2.08	1.82
	thrA	thrLABC	0.50	0.59	0.04	0.49
	ybaS	ybaST	1.55	1.52	1.38	0.87
	ybaT	ybaST	1.31	1.22	1.29	0.73
	ynaI	ynaI	0.77	0.74	0.70	0.63
In wild type but not $\Delta gadE$	aidB	aidB	1.11	1.13	0.03	−0.31
	blc	blc	0.61	0.64	0.35	−0.27
	cbpA	cbpAM	0.57	0.53	−0.15	−0.50
	cbpM	cbpAM	0.62	0.53	−0.04	−0.47
	dps	dps	0.56	0.55	0.41	−0.18
	elaB	elaB	0.60	0.53	0.10	0.08
	gabT	gabDTP	0.52	0.52	0.18	0.11
	gadA	gadAX	2.03	2.23	−0.09	0.04
	gadX	gadAX	1.03	0.87	0.27	−0.05
	gadC	gadCB	2.18	2.13	0.05	−0.10
	gadB	gadCB	2.23	2.22	0.09	0.05
	gadW	gadW	0.51	0.50	0.15	0.02
	hdeA	hdeAB-yhiD	1.89	2.06	0.81	0.75
	hdeB	hdeAB-yhiD	1.85	2.10	1.04	0.75
	hdeD	hdeD	1.72	1.84	0.60	0.27
	mdtE	mdtEF	1.89	1.94	−0.40	−0.49
	mdtF	mdtEF	1.93	1.93	0.50	0.18
	osmF	osmF-yehYXW	0.61	0.60	0.11	0.17
	yehY	osmF-yehYXW	0.59	0.65	0.25	0.29
	pagP	pagP	0.92	0.95	−0.38	0.21
	sufA	sufABCDSE	0.93	0.77	−0.18	−0.20
	sufB	sufABCDSE	0.56	0.50	−0.25	−0.31
	wrbA	wrbA-yccJ	0.76	0.70	0.40	0.13
	yccJ	wrbA-yccJ	0.73	0.64	0.55	0.14
	ycaC	ycaC	0.59	0.58	0.28	0.12

Table 1. Cont.

| Gene name | Transcriptoin units[a] | Wild type (Log$_{10}$ ratio[b]) | | $\Delta gadE$ (Log$_{10}$ ratio[b]) | |
		1st[c]	2nd[c]	1st[c]	2nd[c]
yfcG	yfcG	0.68	0.61	0.18	0.05
ygaM	ygaM	0.63	0.62	0.12	−0.03
yhiM	yhiM	1.98	1.75	0.33	0.23
yjjU	yjjUV	0.58	0.63	−0.39	−0.41
yjjV	yjjUV	0.58	0.62	−0.11	−0.28

[a]Transcriptional unit is represented according to the Regulon DB (http://regulondb.ccg.unam.mx/).
[b]The processed intensity was calculated by Agilent Future Extraction. More than 0.5 of log ratio in WT.
[c]Experiment was independently performed twice (each ratio is shown as 1st and 2nd).

ed and subjected to immune-precipitation using anti-FLAG antibody. After the pronase treatment, ChIP DNA fragments were isolated from the YdeO-DNA complexes for mapping on the genome. As an internal reference for the specific binding of YdeO with its targets, we interrogated the association of YdeO with the gadE promoter, the only known target of YdeO. After PCR amplification from the ChIP DNA samples using specific primers, the gadE promoter could be specifically amplified (data not shown).

To identify the genome-wide YdeO-binding sites on the entire E. coli genome, Sup (the whole extract DNA) and ChIP samples were each labelled and subjected to hybridization on a tiling array. Seven chromosomal regions were determined with high-level signal peaks indicating YdeO-binding, which were distinguishable from the background intensities (Fig. 1), including the gadEp2p3 promoters (Fig. 1F), the only known direct target of YdeO [9]. Six additional YdeO-binding sites were identified by ChIP-chip and were located within intergenic chromosomal regions. These included the intergenic spacer between yccA (an inner membrane protein) and hyaA (hydrogenase I) (Fig. 1B); the intergenic spacer upstream of appC (cytochrome bd-II oxidase) (Fig. 1C); the intergenic spacer upstream of the yiiS gene (a conserved protein)

(Fig. 1D); the intergenic spacer upstream of the gadW gene (the gad operon regulator) (Fig. 1E); the intergenic spacer upstream of the gadE gene (the gad operon regulator) (Fig. 1F); and the intergenic spacer upstream of the slp gene (an outer membrane lipoprotein) (Fig. 1G). Although one YdeO-binding site was located inside the nhaA ORF another binding site was identified upstream of nhaR (Fig. 1A), in which the nhaR promoter has previously been identified [25]. Thus, all of YdeO binding sites were found in the vicinity of possible promoters (see below).

Identification of YdeO-binding in vitro to the seven targets

In order to confirm the direct interaction of YdeO to the seven target sequences determined by ChIP-chip, we performed the EMSA assay. Firstly we failed to purify the YdeO protein using the pET system, because the over-expressed YdeO proteins formed inclusion bodies in E. coli cells. Next YdeO was over-expressed as a His-SUMO fusion, and the His-SUMO-tagged YdeO protein could be purified in soluble forms by affinity chromatography with Ni-NTA agarose (data not shown). After treatment with SUMO protease to remove the His-SUMO tag, the intact YdeO protein, however, became insoluble. Then we used this His-SUMO-tagged

Table 2. Induction of gene expression by YdeO.

Gene name	Log$_{10}$ ratio
nhaA	0.62±0.02
nhaR	0.31±0.05
yccA	−0.04±0.04
hyaA	0.72±0.20
hyaF	1.66±0.04
appC	2.64±0.04
appA	1.72±0.03
yiiT/uspD	0.70±0.03
yiiS	0.72±0.02
gadW	0.39±0.04
gadE	2.91±0.06
mdtF	2.36±0.07
slp	2.86±0.04
dctR	1.47±0.25

Ratio (ydeO$^+$/vector) ± SEM is determined by RT-qPCR as described in Materials and methods.

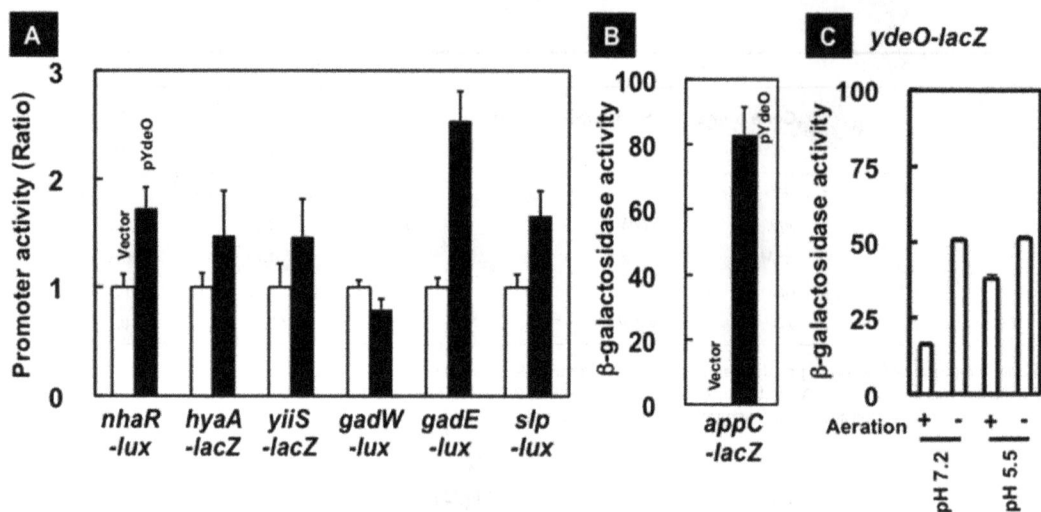

Figure 3. Reporter assays for transcriptional regulation by YdeO. [A] YdeO-expression induces the expression of target promoters. YY0201/pLUXnhaR (nhaR-lux), HYAA-JL (hyaA-lacZ), YY1101 (yiiS-lacZ), YY0201/pLUXgadWp (gadW-lux), YY0201/pLUXgadE (gadE-lux), and YY0201/pLUXslpp (slp-lux), and were transformed with pTrc99A (vector, white bar) and pYdeO (ydeO, black bar). Transformants grew until logarithmic phase and β-galactosidase and luciferase activities of cultures were measured as described in Materials and methods. The data show the average of independent eight experiments with standard deviation as the ratio of a vector-transformant. [B] APPC-JL (appC-lacZ) was transformed with pTrc99A (vector) or pYdeO (ydeO). Transformants grew until logarithmic phase and β-galactosidase was measured as decribed in [A]. The data show the average of independent eight experiments with standard deviation as the Miller unit. [C] The ydeO expression induced under anaerobic conditions. The activity of ydeO promoter was measured in YY0101 growing in LB medium with pH 7.2 and 5.5 under aerobic (+) and anaerobic (−) conditions at logarithmic phase.

YdeO as the test protein. The purified His-SUMO-YdeO protein bound to the gadEp2p3 promoters, the only known target of YdeO (Fig. 2A-f), in good agreement with the previous report [9]. Besides the gadE promoter, His-SUMO-YdeO formed complexes with the nhaR (Fig. 2A-a), hyaA (Fig. 2A-b), yiiS (Fig. 2A-d), gadW (Fig. 2A-e), and slp (Fig. 2A-g) promoters, which were observed as a smeared band, in the presence of 10-fold molar excess of YdeO over the DNA probes. A detectable level of the YdeO-probe complex was not formed with the appC promoter even in the presence of 35-fold molar excess of YdeO (Fig. 2A-c). These results indicate that YdeO directly binds to at least these six sites. YdeO-DNA was detected as a smeared band in several cases, implying the cooperative binding of YdeO at the higher concentration. Since the association of YdeO with the appC promoter was observed only in vivo (see Fig. 1), this association might require another factor(s) for effective binding.

Regulation in vivo of the predicted targets by YdeO: Transcriptome and RT-qPCR assays

We analyzed the alteration in the E. coli K-12 transcriptome caused by the over-expression of YdeO from a plasmid. E. coli KP7600 harboring pYY0401 (ydeO-3xflag) or the empty expression vector, pTrc99A, were grown until log phase under the same conditions used for ChIP-chip analysis, and total RNAs from these cultures were subjected to transcriptome analysis. Amongst genes downstream of a YdeO-binding site, 19 genes, (nhaA, nhaR, hyaA, hyaB, hyaC, hyaD, hyaE, hyaF, appC, appB, appA, yiiS, yiiT/uspD, slp, dctR, gadE, mdtE, mdtF, and gadW) were induced more than 3-fold by the over-expression of YdeO; while three genes, yccA, yiiR, and yhiS, were not affected in both duplicate experiments. (Table S2 and see also Table 1). These 19 genes induced by YdeO constitute a total of 7 transcriptional units, nhaAR, hyaABCDEF, appCBA, yiiS-yiiT/uspD, slp-dctR, gadE-mdtEF, and gadW. All 7 of these operons carry promoters containing YdeO-binding sites (see Fig. 2), and thus should be

under the direct control of YdeO. We also examined the induction of these transcriptional units by the expression of YdeO by RT-qPCR after expression of YdeO. Transcripts of some representative genes from each operon were measured using specific pairs of the respective primers (Table S1). Transcripts were found to increase for all seven operons, nhaAR, hyaABCDEF, appCBA, yiiS-uspD(yiiT), gadW, gadE-mdtEF, and slp-dctR, in the ydeO-expressing cells (Table 2).

We also measured the level of mRNAs in the ydeO-deficeint mutant, but detectable differences were not found for the mRNA from YdeO-target genes between the wild-type and the ydeO mutant. Transcript of yccA, an opposite direction gene from hyaA, was also not affected in the presence and absence of the YdeO-expressing plasmid (Table 2). Although the yiiS and uspD genes, encoding conserved proteins with unidentified function, were expressed even without the over-expression of YdeO, their expressions were further increased after YdeO expression. These results altogether indicate that YdeO plays a role as a positive regulator for expression of all seven operons, nhaAR, hyaABC-DEF, appCBA, yiiS-yiiT/uspD, gadW, gadE-mdtEF, and slp-dctR.

Regulation in vivo of the predicted targets by YdeO: Reporter assay

To confirm the positive role of YdeO on expression of the newly identified target promoters, we performed the reporter assay using the lacZ reporter [15] and lux reporter [16] systems. The translation fusions, hyaA-lacZ, appC-lacZ, and yiiS-lacZ, on the pRS552 derivative plasmids were introduced at the attachment (att) site of the E. coli YY0201 chromosome using the λRS45 phage, resulting in isolation of HYAA-JL (hyaA-lacZ), APPC-JL (appC-lacZ), and YY1101 (yiiS-lacZ). Three E. coli lysogens containing hyaA, appC, and yiiS translational lac fusions in their chromosomes were transformed with either the YdeO-expression plasmid or the vector plasmid. The β-galactosidase activities in these transformants were measured in log-phase (Fig. 3A). YdeO-

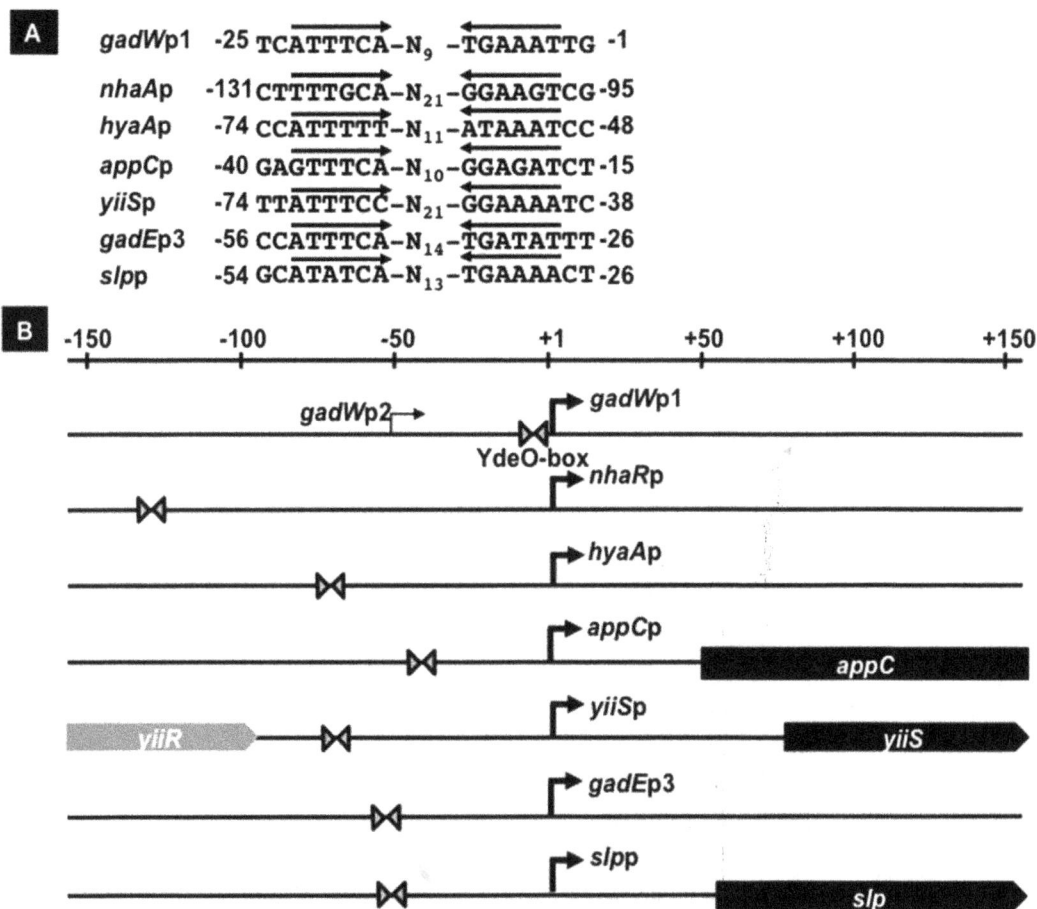

Figure 4. The characterization and location of YdeO-box on target promoters. We examined the conservation of the inverted repeat across seven YdeO-binding regions detected *in vivo* by ChIP-chip analysis, 131-bp on *nhaR* promoter, 216-bp on *hyaA* promoter, 139-bp on *appC* promoter, 217-bp on *yiiS* promoter, 181-bp on *gadW* promoter, 241-bp on *gadE* promoter, and 145-bp on *slp* promoter. [A] The panel shows the DNA sequence, containing the identified hexa-mer repeat (YdeO-box). The YdeO-box identified in all of promoters located on seven binding sites of YdeO. The number indicates the distance from each transcription start site (RegulonDB [http://regulondb.ccg.unam.mx]). [B] Organization of the promoters controlled by YdeO is shown. The locations of a hexamer of YdeO-binding sites (triangle) at relative positions from the transcription initiation site (solid arrow) are shown for the promoters. The filled bars represent open reading frames of the target genes.

expression was found to induce the expression of all these test promoters, *hyaA-lacZ*, *yiiS-lacZ*, and *appC-lacZ* (Fig. 3A and B).

In the cases of *hyaA-lacZ* and *yiiS-lacZ*, the promoter activity increased approximately 1.5 fold upon expression of YdeO. The

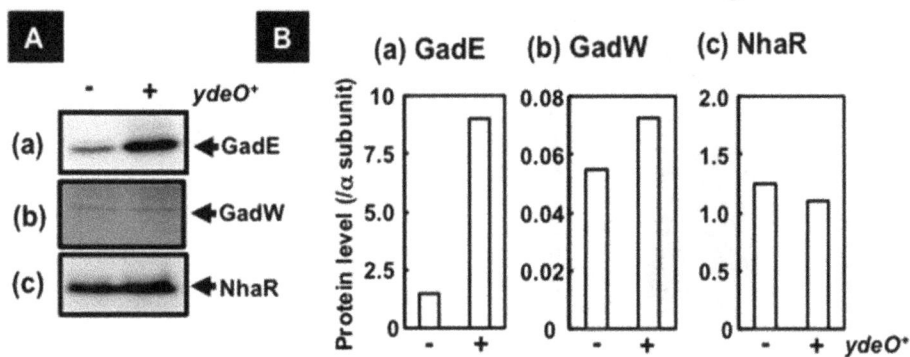

Figure 5. Induction of NhaR, GadE, and GadW by YdeO in *E. coli*. YY5002 (*gadE-3xflag*) (a), YY5003 (*gadW-3xflag*) (b), and BW25113, (c) harboring pTrc99A (−) or pYdeO (+), were grown in LB medium until logarithmic phase. After centrifugation the lysate solution was prepared in lysis buffer containing 8 M urea by sonication. The lysates were subjected to western blotting as described in Materials and Methods. Anti-FLAG (SIGMA) and anti-NhaR (Lab preparation) were used for detection of GadE-3xFLAG/GadW-3xFLAG, and NhaR, respectively [A]. The amounts of GadE-3xFLAG, GadW-3xFLAg, and NhaR were represented as the ratio of level of RNA polymerase-α subunit, detected by anti-α (Neoclone) [B].

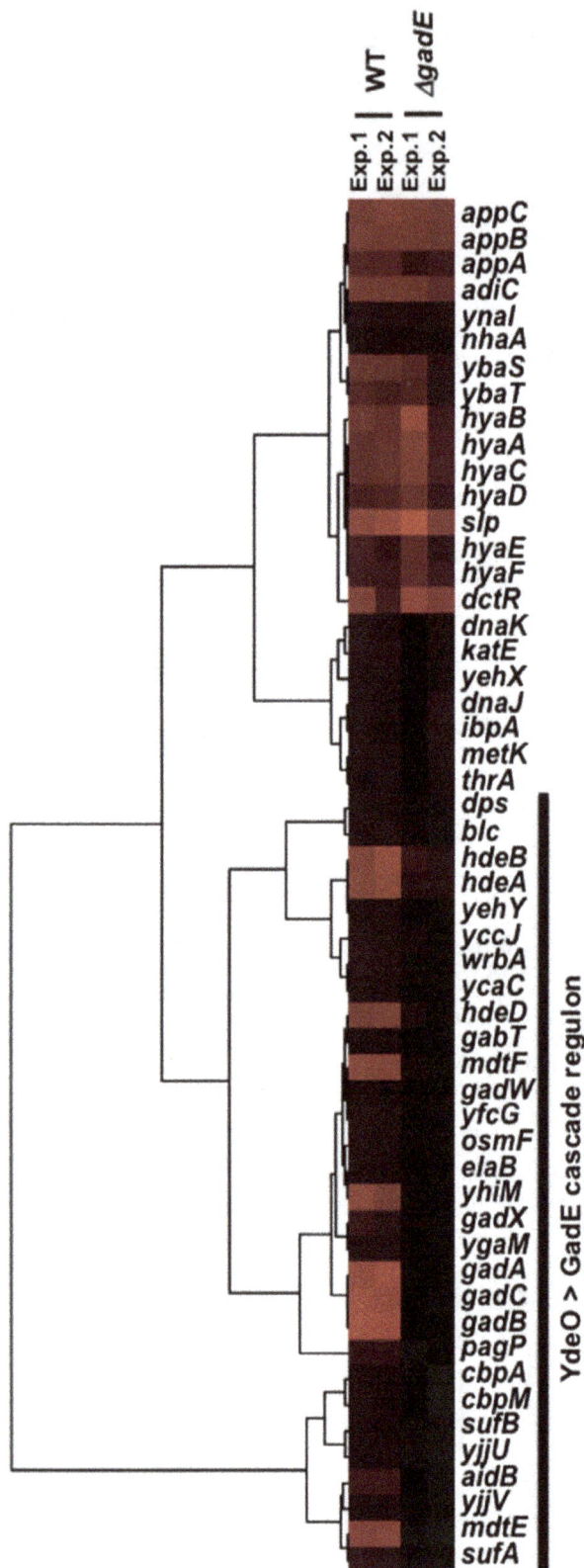

Figure 6. YdeO > GadE cascade regulation in *E. coli*. This shows the clustering pattern of expression of genes induced by the *ydeO*-expression in the parent (KP7600) and *gadE*-deficient mutant (JD25278) by Cluster 3.0 (http://bonsai.hgc.jp/~mdehoon/software/cluster/software.htm).

detectable level of expression was not observed for *appC-lacZ* in the absence of YdeO expression but a high-level of *appC-lacZ* activity was detected upon expression of YdeO (Fig. 3B). The result indicates YdeO has a positive role in activation of the *appC*, *hyaA*, and *yiiS* promoters, in agreement with the observation by transcriptome and RT-qPCR (see above).

The *nhaR*, *slp*, *gadE* and *gadW* promoters were too weak for quantitation by the LacZ reporter system, so we then employed the more sensitive Lux reporter system. The *lux* reporter plasmids of four transcription fusions, *slp-lux*, *gadE-lux* and *gadW-lux* (kindly provide by Peter Lund [16]) and the *nhaR-lux* plasmid [constructed in this study], were introduced into YY0201 *E. coli* carrying either the vector plasmid or the YdeO-expression plasmid. The expression of *nhaR-lux*, *slp-lux*, and *gadE-lux* was found to be activated in the presence of the YdeO-expressing plasmid (Fig. 3A), indicating that YdeO is also a positive regulator for these promoters. Recently RNA-seq analysis indicated the presence of a novel *nhaR* promoter inside the coding region of *nhaA* [25]. The binding site of YdeO is located upstream of this putative promoter (see above). Accordingly the constructed *nhaR-lux* reporter plasmid containing this novel *nhaR* promoter was also activated in the presence of YdeO expression (Fig. 3A).

The expression level of *gadW-lux* stayed unaltered with and without the YdeO-expression plasmid. It is inconsistent with the RT-qPCR result that the mRNA level of *gadW* increased in the presence of YdeO expression as detected by RT-qPCR (Table 2). This apparent disagreement might be due to translational inhibition of *gadW-lux* by the anti-sense RNA of *gadW*, named *gadY*, encoded in the *gadW*-lux plasmid.

Recognition sequence of YdeO transcription factor

To identify the YdeO-binding sequence, we performed DNase I footprinting of the *gadW* promoter with increasing concentrations of YdeO. At low protein levels, YdeO protected the region from −53 to +8 of the *gadW* promoter (Fig. 2B, lanes 2–4). In the presence of 15-fold molar excess of YdeO, the protected region by YdeO expanded from −53 to +84 of the *gadW* promoter possibly due to protein-protein interaction (Fig. 2B, lane 5) in agreement with the smeared band formation observed by EMSA (see above). Within the core YdeO-binding region, the inverted repeat of hexa-nucleotides, 5′-ATTTCA-3′, was identified (see Figs. 2C and 4A).

Using this YdeO-box sequence, we searched for this inverted repeat within the seven YdeO-binding regions detected by *in vivo* by ChIP-chip analysis, and identified this inverted repeat sequence of all the YdeO-binding regions at various positions between −131 to −1 with respect to the transcription start site (Fig. 4). The length of spacer between the 5′-ATTTCA-3′ hexa-nucleotide sequence ranges from 9 to 21 nucleotides (Fig. 4A). Recent studies show that YpdB and YehT bind to the direct repeat of their specific sequence separated by a 9- and 13-bp spacer, respectively, in *E. coli* [26,27]. Previous work shows that the spacer length of the specific DNA binding region is diverse for the *E. coli* transcription factor CpxR [28]. Therefore, we have denoted the inverted repeat as the YdeO-box (Fig. 4).

Induction of NhaR, GadE, and GadW by YdeO

Four transcription factors, the LysR-type NhaR, the LuxR-type GadE, and the AraC-type GadW, were found to be under the direct control of YdeO (see Figs. 1–4). NhaR is an activator of a sodium/proton antiporter gene [29] and both GadE and GadW are involved in regulation of the genes for glutamate-dependent acid resistance system [8,9]. In addition to these three transcription factors, the gene encoding the CadC-like transcription factor DctR is located downstream of the *slp* gene which codes for a

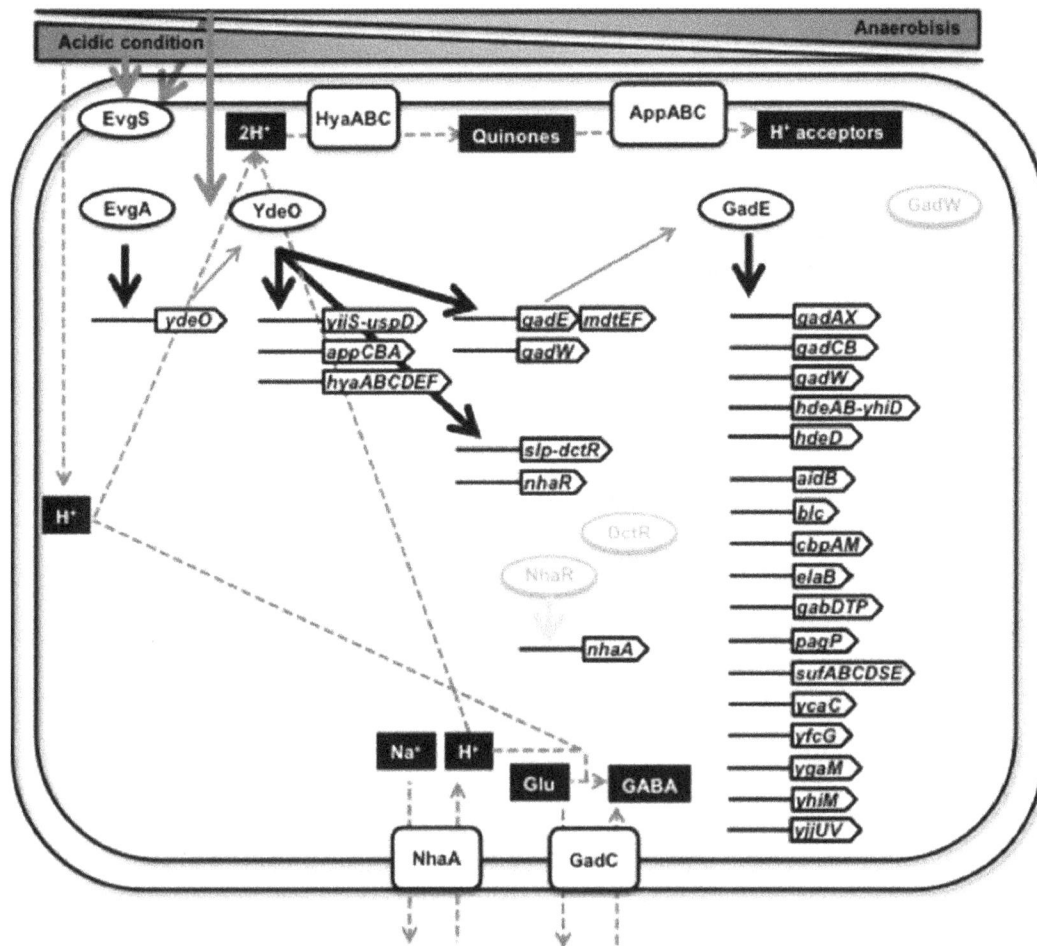

Figure 7. EvgAS > YdeO > DctR/NhaR/GadE/GadW regulatory network in *E. coli*.

starvation lipoprotein, and is considered to be co-transcribed with the *slp* gene. DctR is involved in protection against metabolic endproducts under acidic conditions [30]. To examine the involvement of YdeO in control of expression of the three transcription factors, the cellular level of these proteins in *E. coli*, with or without the YdeO-expressing plasmid, were measured by Western blotting assay.

To perform the Western blotting assay of GadE and GadW using the anti-FLAG antibody, we constructed *E. coli* strains YY5002 and YY5003 including *3xflag* tag at the 3′-terminal end of *gadE* and *gadW*, respectively, on the *E. coli* chromosome. The YdeO-expression plasmid, pYdeO (*ydeO*), and the empty expression vector pTrc99A were transformed into these *E. coli* strains and the transformants were grown in LB medium until log phase. The whole-cell lysates were prepared, and subjected to Western blotting assay by using anti-NhaR, anti-FLAG, and anti-RpoA for detection of NhaR, GadE-3xFLAG and GadW-3xFLAG, and RNA polymerase α subunit, respectively. All transformants with or without the YdeO-expressing plasmid retain approximately a constant amount of the α subunit of RNA polymerase (data not shown). The level of GadE increased in the YY5002 harboring the YdeO-expression plasmid, supporting the prediction that the *gadE* gene is under the direct and positive control of YdeO. However, we failed to detect NhaR and GadW even in the presence of YdeO expression (Fig. 5).

Search for the whole set of genes regulated by YdeO > GadE

To obtain the gene expression profile of the YdeO > GadE cascade, we performed a transcriptome assay. *E. coli* wild-type KP7600 and *gadE*-deficient JD25278 harbouring pTrc99A and pYY0401 (*ydeO-3xflag*) were incubated in LB medium at 37°C with shaking until log phase and total RNA from these cultures was subjected to transcriptome analysis under standard experimental conditions as described in Materials and methods. The results revealed that a total of 106 genes were markedly affected by YdeO expression in the wild-type and included 53 up- and the same number of down-regulated genes (Tables S2 and S3). Among the 53 genes up-regulated by YdeO expression, clustering analysis showed 23 genes were induced in both the parent strain and the *gadE*-deficient mutant and 30 genes induced in the wild-type but not the *gadE*-deficient mutant (Fig. 6). The observed alteration of the transcriptome profile caused by deletion of the *gadE* gene was similar to that reported by Masuda and Church [31]. Genes induced in both strains are organized into a total of 12 transcriptional units (Table 1), including five transcription units, *hyaABCDEF, appCBA, slp-dctR,* and *nhaAR,* that are under the direct control of YdeO (see above). On the other hand, the rest of the 30 up-regulated genes forming 21 transcription units were induced in the wild-type but not in the *gadE*-deficient mutant (Table 1), indicating that these 21 transcription units are under the direct control of GadE but the indirect control of YdeO. This set of 21 transcription units includes the hitherto identified GadE targets, *gadA*, *gadB*, and *gadC* [9]. On the other hand, detectable change was not observed in the transcription pattern between the parent strain and the *gadW*-deficient mutant, consistent with the lack of YdeO-dependent GadW expression under the conditions herein employed (Figs. 3 and 5). The *yehX* gene was induced by the YdeO-expression plasmid in both the parent strain and the *gadE* mutant but the *osmF* and *yehY* genes, and parts of *osmF-yehYXW* transcription unit, were not induced in the *gadE* mutant (Fig. 6 and Table 1), implying that GadE activates the known promoters located at the upstream of *yehX* which is possibly activated by YdeO.

Physiological roles of YdeO in response to environmental stresses

The level of translational control of the YdeO regulator itself was analyzed using a reporter assay with the *ydeO-lacZ* fusion. In *E. coli* YY0101 (*ydeO-lacZ*) grown under aerobic conditions, β-galactosidase activity from *ydeO-lacZ* increased two-fold under the acidic condition of pH 5.5 compared with pH 7.0 (Fig. 3C). Interestingly the high level of *ydeO-lacZ* was detected in both pH 5.5 and 7.0 when *E. coli* were grown under anaerobic conditions (Fig. 3C), implying that YdeO plays a role in *E. coli* respiration under anaerobic conditions, such as in the animal intestine. Previously, we identified that the transcription of *ydeO* is induced by exposure to ultraviolet light via the two-component system EvgSA two-component system [11]. In agreement with this finding, *ydeO* expression was not induced in the *evgA*-defective mutant under both acidic and anaerobic conditions (data not shown).

Discussion

The YdeO regulon

Here we have identified a total of seven YdeO-binding sites on the *E. coli* genome using ChIP-chip and transcription analyses *in vivo*. The EMSA experiments showed that purified YdeO also binds *in vitro* to these six sites (see Fig. 2). The reporter and RT-qPCR assays indicated that all of the promoters located downstream of these YdeO-binding sites are activated by YdeO (see Table 1 and Fig. 3). The hexa-nucleotide repeat 5′-ATTTCA-3′, which we have named the YdeO box, is conserved in all of YdeO-binding sites we identified experimentally (see Fig. 4). Even though this YdeO-box like sequence exists within the *appC* promoter, which is located immediately downstream of a YdeO-binding site (see Fig. 1), the binding *in vitro* of YdeO to the *appC* promoter probe was not high (see Fig. 2), implying that an as yet unidentified additional transcription factor or DNA secondary structure is needed for efficient binding of YdeO to the target promoter. Since the *appC* promoter is transcribed *in vivo* by RNA polymerase containing the RpoS sigma factor and is induced by AppY [32,33], one possibility is that AppY and/or RpoS sigma are required for the efficient binding of YdeO to the *appC* promoter. Thus, we conclude that YdeO is a positive regulator for transcription of operons controlled by seven promoters, the *nhaR* promoter, *hyaA* promoter, *appC* promoter, *yiiS* promoter, *slp* promoter, *gadE* promoter, and *gadW* promoter (Fig. 7).

Transcription cascade: EvgSA > YdeO > NhaR, GadE, GadW

E. coli responds to temporary low pH using the glutamate-dependent acid resistant system, which involves two complex regulatory systems: EvgAS > YdeO > GadE; and Crp> RpoS >

GadX > GadW [9,12]. In this study, we showed that YdeO directly regulates the expression of three transcription factor genes, *nhaR*, *gadE,* and *gadW* (see Fig. 7), proving the novel transcription cascade: EvgAS > YdeO > NhaR/GadE/GadW.

YdeO not only plays a regulatory role in positive feedback loop of EvgAS > YdeO > GadE pathway, but also a positive role in the GadXW pathway, thereby linking the GadE- and GadXW-pathways for acid resistance. The GadXW circuit is believed to function during stationary phase. YdeO-overexpression induced GadE-dependent transcription of the *gadW* gene but GadW protein was not detected in growing *E. coli* cell (Fig. 6), suggesting that stationary phase specific factors are required for GadW.

Transcriptome analysis identified the set of genes directly regulated by YdeO or indirectly through the YdeO > GadE cascade (Table 2; see Fig. 7 for the summary model). GadE induced by YdeO stimulated the transcription of *hdeAB-yhiD*, *hdeD*, *gadAX*, *gadCB*, *mdtEF*, *gadW*, and *yhiM* as well as those previously reported promoters [9,34,35]. The GAD cluster including *hdeAB-yhiD*, *hdeD*, *gadAX*, *gadCB*, *gadE*, *mdtEF*, and *gadW*, is necessary for glutamine-dependent acid resistance [9,34]. Recently the *yhiM* gene was reported to be essential for growth at pH 2.5 and is necessary for glutamine- and lysine-dependent acid resistance, but is not required for arginine-dependent acid resistance [36]. In addition of these operons, the YdeO > GadE cascade induced a total of 19 operons including *aidB*, *blc*, *cbpAM*, *elaB*, *gabDTP*, *pagP*, *sufABCDSE*, *ycaC*, *yfcG*, *ygaM*, and *yjjUV* (see Table 1), of which the *yfcG* gene encodes a disulfide reductase [37] and the *sufABCDSE* operon encodes the complex biosynthetic machinery for iron-sulfur clusters in several enzymes which have critical cysteine residues [38], suggesting a relationship between the function of YdeO and cysteine metabolism.

The physiological role of the YdeO regulon

In addition of the hitherto-identified target *gadE*, we have identified a total of seven operons belonging to the YdeO regulon. The expression of *ydeO* is induced under acidic conditions (see Fig. 3C). In good agreement, the *gadE* operon encodes the master activator for expression of *gadA* and *gadBC*, which are involved in the glutamate-dependent acid resistance system which works for consumption of intracellular protons by glutamate decarboxylation. In addition to acid conditions, the expression of *ydeO* is also induced under anaerobic growth in both neutral and acid conditions. Two YdeO-regulated targets, *hyaABCDEF* and *appCBA*, encode a hydrogenase and a quinone oxidase, respectively, both being involved in bacterial respiration. The HyaABC complex oxidizes dihydrogen to two protons, following release of them to the outside of the membrane, and donation of the electrons to the quinone pool. The AppBC complex donates electrons taken by a quinone to intracellular oxygen, consuming an intracellular proton per electron (Fig. 7), resulting in H_2O production via oxygen [39]. Thus, the *hyaABCDEF* operon contributes to the consumption of the intracellular proton while the *appCBA* operon contributes to the utilization of reduced quinone. Taken together, these physiological systems activated by YdeO stimulate stress response and respiration. These findings also suggest that YdeO activated genes play an important role in primary adaptation, which enables the cell to colonize animal intestines by contributing to adaptation to acidic conditions in the stomach and to anaerobic conditions in the intestine.

Supporting Information

Table S1 Bacterial strains, phage, plasmids, and oligo-nucleotides used in this study. *E. coli* K-12 derivatives used in this study were indicated with characterizations. The used bacteriophage and plasmids were also shown. Oligonucleotides were represented with DNA sequences.

Table S2 The genes affected by expression of *ydeO* gene in *E. coli* KP7600. Transcriptome analysis was performed using total RNAs from KP7600 harboring pTrc99A (vector) and pYY0401 (*ydeO-3xflag*) as described in Materials and methods. The *E. coli* Gene Expression Microarray microarray 8×15 K (Agilent) hybridized by the fluorescent cDNAs was scanned with an Agilent G2565CA microarray scanner Ver. 8.1, the intensities of both Cy3 and Cy5 were quantified by Feature Extraction Ver. 8.1, and then, the Cy5/Cy3 ratios were calculated from the normalized values.

Table S3 The genes affected by expression of *ydeO* gene in the *gadE*-deficient *E. coli* mutant. Transcriptome analysis

was performed using total RNAs from JD25278 (*gadE*::mini-Tn10) harboring pTrc99A (vector) and pYY0401 (*ydeO-3xflag*) as described in Table S2. The intensities quantified by Feature Extraction Ver. 8.1 and the Cy5/Cy3 ratios were represented.

Acknowledgments

We gratefully acknowledge Jon Hobman (Nottingham University) for valuable comments and proofreading of the manuscript.

This work is supported by MEXT-Supported Program for the Strategic Research Foundation at Private Universities and Special Coordination Funds for Promoting Science and Technology and JSPS-DST International Collaborations. We are grateful to Peter Lund for providing lux plasmids. We also thank the National BioResource Project (NBRP) of Japan for providing *E. coli* strains.

Author Contributions

Conceived and designed the experiments: YY TO KY. Performed the experiments: YY TO. Analyzed the data: YY TO AI KY. Contributed reagents/materials/analysis tools: TO AI KY. Contributed to the writing of the manuscript: TO AI KY.

References

1. Foster JW (2004) *Escherichia coli* acid resistance: tales of an amateur acidophile. Nature Rev Microbiol 2: 898–907.
2. Castanie-Cornet MP, Penfound TA, Smith D, Elliott JF, Foster JW (1999) Control of acid resistance in *Escherichia coli*. J Bacteriol 181: 3525–3535.
3. Iyer R, Williams C, Miller C (2003) Arginine-agmatine antiporter in extreme acid resistance in *Escherichia coli*. J Bacteriol 185: 6556–6561.
4. Lin J, Lee IS, Frey J, Slonczewski JL, Foster JW (1995) Comparative analysis of extreme acid survival in *Salmonella typhimurium*, *Shigella flexneri*, and *Escherichia coli*. J Bacteriol 177: 4097–4104.
5. de Biase D, Tramonti A, Bossa F, Visca P (1999) The response to stationary-phase stress conditions in *Escherichia coli*: role and regulation of the glutamic acid decarboxylase system. Mol Microbiol 32: 1198–1211.
6. Hersh BM, Farooq FT, Barstad DN, Blankenshorn DL, Slonczewski JL (1996) A glutamate-dependent acid resistance gene in *Escherichia coli*. J Bacteriol 178: 3978–3981.
7. Hommais F, Krin E, Coppee JY, Lacroix C, Yeramian E, et al. (2004) GadE (YhiE): a novel activator involved in the response to acid environment in *Escherichia coli*. Microbiology 150: 61–72.
8. Ma Z, Gong S, Richard H, Tucker DL, Conway T, et al. (2003) GadE (YhiE) activates glutamate decarboxylase-dependent acid resistance in *Escherichia coli* K-12. Mol Microbiol 49: 1309–1320.
9. Ma Z, Masuda N, Foster JW (2004) Characterization of EvgAS-YdeO-GadE branched regulatory circuit governing glutamate-dependent acid resistance in *Escherichia coli*. J Bacteriol 186: 7378–7389.
10. Itou J, Eguchi Y, Utsumi R (2009) Molecular mechanism of transcriptional cascade initiated by the EvgS/EvgA system in *Escherichia coli* K-12. Biosci Biotechnol Biochem 73: 870–878.
11. Yamanaka Y, Ishihama A, Yamamoto K (2012) Induction of YdeO, a regulator for acid resistance genes, by ultraviolet irradiation in *Escherichia coli*. Biosci Biotechnol Biochem 76: 1236–1238.
12. Yamamoto K (2014) The hierarchic network of metal-response transcription factors in *Escherichia coli*. Biosci Biotechnol Biochem 78: 737–747.
13. Uzzau S, Figueroa-Bossi N, Rubino S, Bossi L (2001) Epitope tagging of chromosomal genes in Salmonella. Proc Natl Acad Sci USA 94: 13997–14001.
14. Jishage M, Ishihama A (1997) Variation in RNA polymerase sigma subunit composition within different stocks of *Escherichia coli* strain W3110. J Bacteriol 179: 959–963.
15. Simon RW, Hausman F, Kleckner N (1987) Improved single and multicopy lac-based cloning vectors for protein and operon fusions. Gene 53: 85–96.
16. Burton NA, Johnson MD, Antczak P, Robinson A, Lund PA (2010) Novel aspects of the acid response network of *E. coli* K-12 are revealed by a study of transcriptional dynamics. J Mol Biol 401: 726–742.
17. Katou Y, Kaneshiro K, Aburatani H, Shirahige K (2006) Genomic approach for the understanding of dynamic aspect of chromosome behavior, Methods Enzymol 409: 389–410.
18. Uyar E, Kurokawa K, Yoshimura M, Ishikawa S, Ogasawara N, et al. (2009) Differential binding profiles of StpA in wild-type and *h-ns* mutant cells: a comparative analysis of cooperative partners by chromatin immunoprecipitation-microarray analysis. J Bacteriol 191: 2388–2391.
19. Chumsakul O, Takahashi H, Oshima T, Hishimoto T, Kanaya S, et al. (2011) Genome-wide binding profiles of the *Bacillus subtilis* transition state regulator

AbrB and its homolog Abh reveals their interactive role in transcriptional regulation, Nucleic Acids Res 39: 414–428.
20. Ueda T, Takahashi H, Uyar E, Ishikawa S, Ogasawara N, et al. (2013) Functions of the Hha and YdgT proteins in transcriptional silencing by the nucleoid proteins, H-NS and StpA, in *Escherichia coli*. DNA Res 20: 263–271.
21. Yamamoto K, Hirao K, Oshima T, Aiba H, Utsumi R, et al. (2005) Functional characterization *in vitro* of all two-component signal transduction systems from *Escherichia coli*. J Biol Chem 280: 1448–1456.
22. Yamamoto K, Ishihama A (2005) Transcriptional response of *Escherichia coli* to external copper. Mol Microbiol 56: 215–227.
23. Pfaffl MW (2001) A new mathematical model for relative quantification in real-time RT-PCR. Nucleic Acids Res 29: e45.
24. Kailasan Vanaja S, Bergholz TM, Whittam TS (2009) Characterization of the *Escherichia coli* O157:H7 Sakai GadE regulon. J Bacteriol 191: 1868–1877.
25. Salgado H, Peralta-Gil M, Gama-Castro S, Santos-Zavaleta A, Muñiz-Rascado L, et al. (2013) RegulonDB v8.0: omics data sets, evolutionary conservation, regulatory phrases, cross-validated gold standards and more. Nucleic Acids Res 41: D203–13.
26. Fried L, Behr S, Jung K (2013) Identification of a target gene and activating stimulus for the YpdA/YpdB histidine kinase/response regulator system in *Escherichia coli*. J Bacteriol 195: 807–815. J Bacteriol 194: 4272–4284.
27. Kraxenberger T, Fried L, Behr S, Jung K (2012) First insights into the unexplored two-component system YehU/YehT in Escherichia coli.
28. Yamamoto K, Ishihama A (2006) Characterization of copper-inducible promoters regulated by CpxA/CpxR in *Escherichia coli*. Biosci Biotechnol Biochem 70: 1688–1695.
29. Rahav-Manor O, Carmel O, Karpel R, Taglicht D, Glaser G, et al. (1992) NhaR, a protein homologous to a family of bacterial regulatory proteins (LysR), regulates *nhaA*, the sodium proton antiporter gene in *Escherichia coli*. J Biol Chem 267: 10433–10438.
30. Mates AK, Sayad AK, Foster JW (2007) Products of the *Escherichia coli* acid fitness island attenuate metabolic stress at extreme low pH and mediate a cell density-dependent acid resistance. J Bacteriol 189: 2759–2768.
31. Masuda N, Church GM (2003) Regulatory network of acid resistance genes in *Escherichia coli*. Mol Microbiol 48: 699–712.
32. Atlung T, Brøndsted L (1994) Role of the transcriptional activator AppY in regulation of the *cyx-appA* operon of *Escherichia coli* by anaerobiosis, phosphate starvation, and growth phase. J Bacteriol 176: 5414–5422.
33. Brøndsted L, Atlung T (1996) Effect of growth conditions on expression of the acid phosphatase (*cyx-appA*) operon and the *appY* gene, which encodes a transcriptional activator of *Escherichia coli*. J Bacteriol 178: 1556–1564.
34. Masuda N, Church GM (2002) *Escherichia coli* gene expression responsive to levels of the response regulator EvgA. J Bacteriol 184: 6225–6234.
35. Krin E, Danchin A, Soutourina O (2010) RcsB plays a central role in H-NS-dependent regulation of motility and acid stress resistance in *Escherichia coli*. Res Microbiol 161: 363–371.
36. Nguyen TM, Sparks-Thissen RL (2012) The inner membrane protein, YhiM, is necessary for *Escherichia coli* (*E. coli*) survival in acidic conditions. Arch Microbiol 194: 637–641.
37. Wadington MC, Ladner JE, Stourman NV, Harp JM, Armstrong RN (2009) Analysis of the structure and function of YfcG from *Escherichia coli* reveals an efficient and unique disulfide bond reductase. Biochemistry 48: 6559–6561.

38. Vinella D, Brochier-Armanet C, Loiseau L, Talla E, Barras F (2009) Iron-sulfur (Fe/S) protein biogenesis: phylogenomic and genetic studies of A-type carriers. PLoS Genet 5: e1000497.

39. Borisov VB, Gennis RB, Hemp J, Verkhovsky MI (2011) The cytochrome bd respiratory oxygen reductases. Biochim Biophys Acta 1807: 1398–1413.

10

Characterisation of Pellicles Formed by *Acinetobacter baumannii* at the Air-Liquid Interface

Yassine Nait Chabane[1], Sara Marti[1]*, Christophe Rihouey[1], Stéphane Alexandre[1], Julie Hardouin[1], Olivier Lesouhaitier[2], Jordi Vila[3], Jeffrey B. Kaplan[4], Thierry Jouenne[1], Emmanuelle Dé[1]*

1 Unité Mixte de Recherche 6270 CNRS - Laboratory "Polymères, Biopolymères, Surfaces", University of Rouen, Mont-Saint-Aignan, France, **2** Laboratory of "Microbiologie Signaux et Micro-Environnement" - Equipe d'Accueil 4312, University of Rouen, Evreux, France, **3** Department of Microbiology, Hospital Clinic, Barcelona, Spain, **4** Department of Biology, American University, Washington, District of Columbia, United States of America

Abstract

The clinical importance of *Acinetobacter baumannii* is partly due to its natural ability to survive in the hospital environment. This persistence may be explained by its capacity to form biofilms and, interestingly, *A. baumannii* can form pellicles at the air-liquid interface more readily than other less pathogenic *Acinetobacter* species. Pellicles from twenty-six strains were morphologically classified into three groups: I) egg-shaped (27%); II) ball-shaped (50%); and III) irregular pellicles (23%). One strain representative of each group was further analysed by Brewster's Angle Microscopy to follow pellicle development, demonstrating that their formation did not require anchoring to a solid surface. Total carbohydrate analysis of the matrix showed three main components: Glucose, GlcNAc and Kdo. Dispersin B, an enzyme that hydrolyzes poly-*N*-acetylglucosamine (PNAG) polysaccharide, inhibited *A. baumannii* pellicle formation, suggesting that this exopolysaccharide contributes to pellicle formation. Also associated with the pellicle matrix were three subunits of pili assembled by chaperon-usher systems: the major CsuA/B, A1S_1510 (presented 45% of identity with the main pilin F17-A from enterotoxigenic *Escherichia coli* pili) and A1S_2091. The presence of both PNAG polysaccharide and pili systems in matrix of pellicles might contribute to the virulence of this emerging pathogen.

Editor: Adam Driks, Loyola University Medical Center, United States of America

Funding: SM has a post-doctoral fellowship from the Sociedad Espanola de Enfermedades Infecciosas y Microbiología Clinica (SEIMC) and YNC has a doctoral fellowship from the Haute-Normandie region. The funders had no role in study design, data collection and analysis, decision to publish, or preparation of the manuscript.

Competing Interests: The authors have declared that no competing interests exist.

* Email: saramarti2@yahoo.es (SM); emmanuelle.de@univ-rouen.fr (ED)

Introduction

A biofilm is an organized community of bacterial cells surrounded by a protective self-secreted matrix of extracellular polymeric substances (EPS) [1,2]. Biofilms attached to biotic or abiotic surfaces have been extensively studied. Nevertheless, bacterial aggregation can also take place at the air-liquid interface and in suspensions [3]. The biofilm formed at the air-liquid interface, generally referred to as "pellicle", is a floating structure that requires a high organization due to the lack of a solid surface for initial attachment [4,5].

An important component of the biofilm is the EPS matrix, a protective cover that maintains a cohesive structure and interacts with the external environment to allow the entrance of specific substances. It can act as a recycling centre to keep lysed cells and nutrients available for the bacterial community [6,7]. The EPS matrix is mainly composed of polysaccharides, proteins, nucleic acids and lipids [6,7]. Some of these molecules such as cell motility-associated appendages, fimbriae or pili, contribute to the initial stages of biofilm formation [7,8]. The matrix is highly hydrated preventing biofilm desiccation and it may also contribute to antimicrobial resistance by decreasing the transport of these substances into the biofilm [1]. These characteristics are very important especially in nosocomial pathogens such as *Acinetobacter baumannii* because the biofilm gives them a protection from the hospital environment.

Over the last two decades, *A. baumannii* has emerged as a problematic opportunistic pathogen associated with nosocomial infections, such as pneumonia, bacteraemia or meningitis [9–12]. This species has been considered the paradigm of multiresistant bacteria due to its remarkable capacity to acquire mechanisms of resistance to antimicrobial agents. Moreover, its ability to persist in the hospital environment accounts for its emergence. This persistence and resistance to desiccation could be directly associated to biofilm formation [13,14]. Indeed, *A. baumannii* can attach to biotic and abiotic surfaces in a process that has been associated with the presence of several factors: the pili assembly systems, the production of the Bap (Biofilm associated protein) surface-adhesion protein and the autotransporter Ata [15–17]. OmpA, the major outer membrane protein is also required for attachment to epithelial cells [18] and type IV fimbriae promote bacterial motility, enhancing bacterial adhesion [19]. Although most attention has been focused on the biofilm formed on solid surfaces, *A. baumannii* also forms thick pellicles at the air-liquid interface [20–22], a favourable niche because bacteria can obtain nutrients from the liquid media and oxygen from the air [5]. Note that this type of biofilm has been mostly associated to the more pathogenic *Acinetobacter* spp. [21] and as such its characterisation, especially the EPS matrix, is important to understand the interactions between the pellicle and the external environment.

This study aimed to explore and characterize *A. baumannii* pellicles and their EPS matrix. These structures were morphologically examined using different microscopy approaches and clustered into three different groups. A representative sample from each morphological group was studied in depth to determine the principal components of the EPS matrix *i.e.* the polysaccharide and proteins secreted to form this protective cover.

Materials and Methods

Bacterial strains & growth conditions

Eighty-six epidemiologically unrelated *A. baumannii* clinical isolates (Table S1) were screened in this study: 81 isolates collected in Spain during the GEIH-Ab2000 project [23,24]; 2 isolates from the ICU in Hospital Charles Nicolle (Rouen, France); 3 isolates from the Hospital Clinic (Barcelona, Spain).

Pellicles were grown at 25°C in Mueller Hinton Broth (MHB) (Oxoid, France) or in T-broth medium (10 g/L bacto tryptone, 5 g/L NaCl) supplemented with 20 μg/ml congo red (CR-TB) to examine the production of cellulose [25] using initial inocula equivalent to an OD_{600} value of 0.01.

Pellicle formation assay

Standing 2 mL cultures in MHB were grown for 72 h in polystyrene tubes (Ø 13 mm × H 75 mm). Pellicle formation was identified visually (Figure 1); isolates were considered positive when the surface of the culture was covered with an opaque layer.

Pellicle cohesion was examined by manually inverting the tubes (Figure 1C). The experiments were performed in duplicate at three independent time-points.

Microscopy analyses

Inversed Optical Microscopy. Standing 2 mL cultures in CR-TB were grown for 24 h in 24-well polystyrene plates and examined using a LEICA DM LM (Wetzlar, Germany) at ×5 amplification in the inverse mode. Photographs were taken with a Sony CMA-D2 camera, (Zaventem, Belgium) and the software Archimed (Microvision Instruments, Evry, France).

Brewster Angle Microscopy (BAM). Brewster angle microscopy is an optical microscopy technique used to visualize materials located at an interface without the use of optical probes. It is based on the effect that no reflection occurs when a clean surface is lighted with p-polarized incoming light at the Brewster angle. The presence of materials at the interface leads to a partial light reflection allowing their visualization. Standing 15 mL cultures in MHB were grown for 24 h in 6-well polystyrene plates. The microscope used was a BAM 2 plus (NFT, Germany) with 2 μm resolution. Images were taken at different growth times to monitor pellicle evolution and are presented after a deinterlacing process to remove artefacts. Bacterial clusters were counted manually by examining the digital images superimposed by a micrometer grid. This allowed us to evaluate the number and size of the bacterial clusters.

Scanning Electron Microscopy (SEM). Standing 7 mL cultures in MHB were grown for 24 h in 6-well polystyrene plates. To visualize the liquid-facing interface, pellicles formed on the surface of the liquid medium were recovered with collodion-coated cover slides as previously described by Henk *et al.* [26]. To

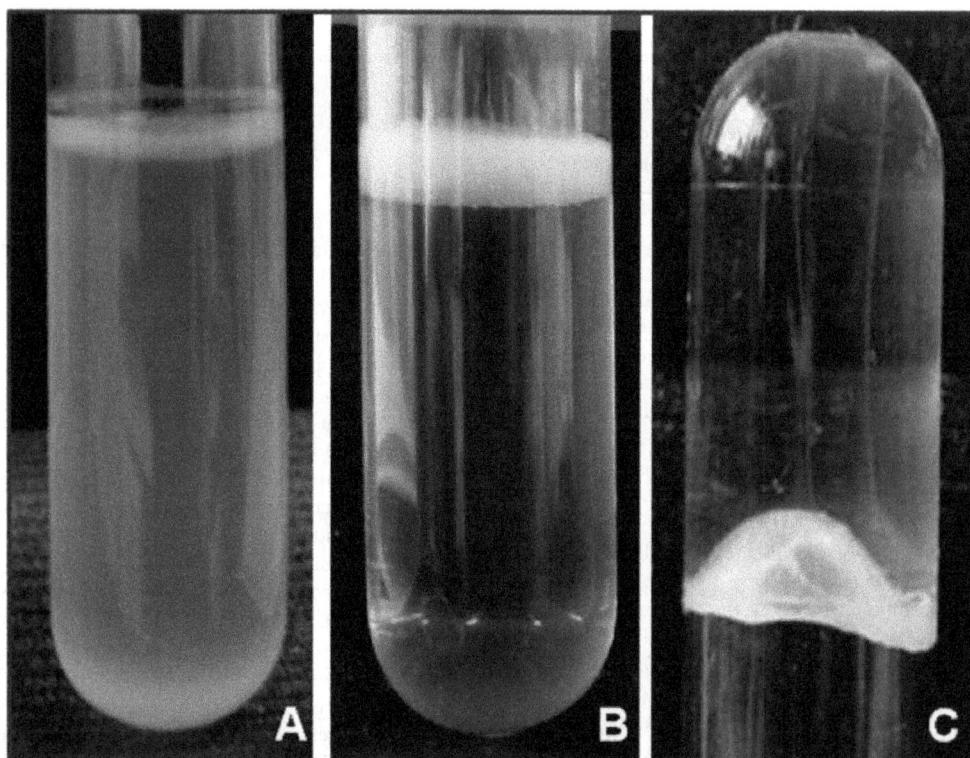

Figure 1. Biofilm formation by *A. baumannii*. A) Strain negative for pellicle formation (turbid culture); **B)** Strain forming a pellicle on the top of the liquid media (culture broth transparent); **C)** Inverted tube to examine pellicle strength.

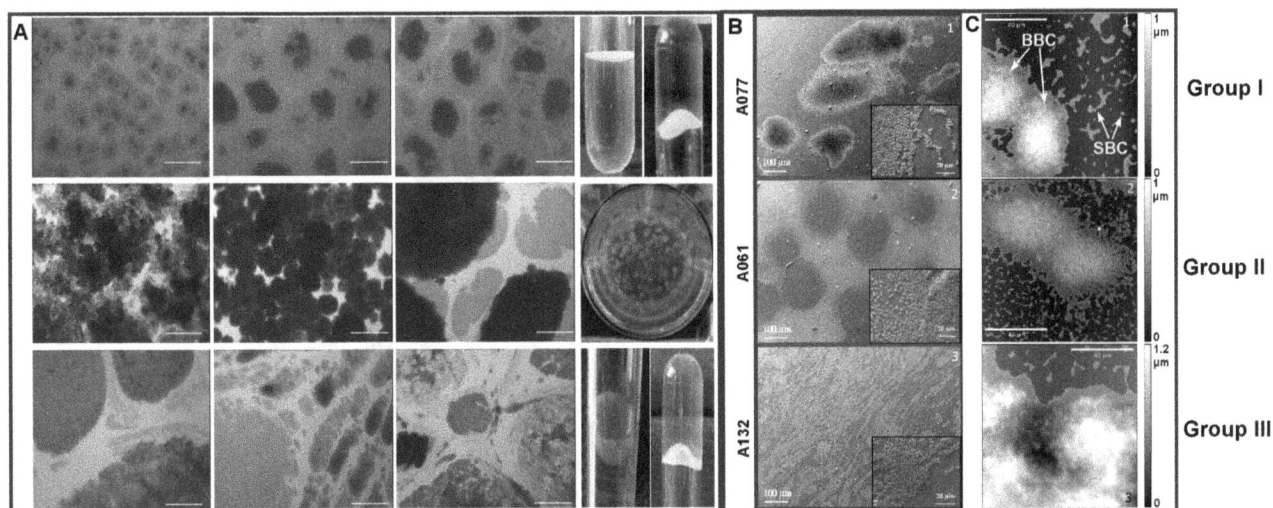

Figure 2. Different morphogroups of pellicles formed by *A. baumannii:* **Group I (egg-shaped); Group II (ball-shaped); Group III (irregular pellicles).** Section **A**: Pellicles were grown for 24 h in CR-TB and examined by inverse optical microscopy (three examples of strains from each morphogroup). White scale bar = 200 µm. Pellicles formed by strains from Group I and Group III were strong enough to support the weight of the medium; Section **B**: SEM, insets with enlarged scale; Section **C**: AFM; 100 µm×100 µm images taken after 48 h (A132) and 24 h (A061; A077) growth to be in a similar development stages of pellicles. Big bacterial clusters (BBC) and small bacterial clusters (SBC) are shown by arrows.

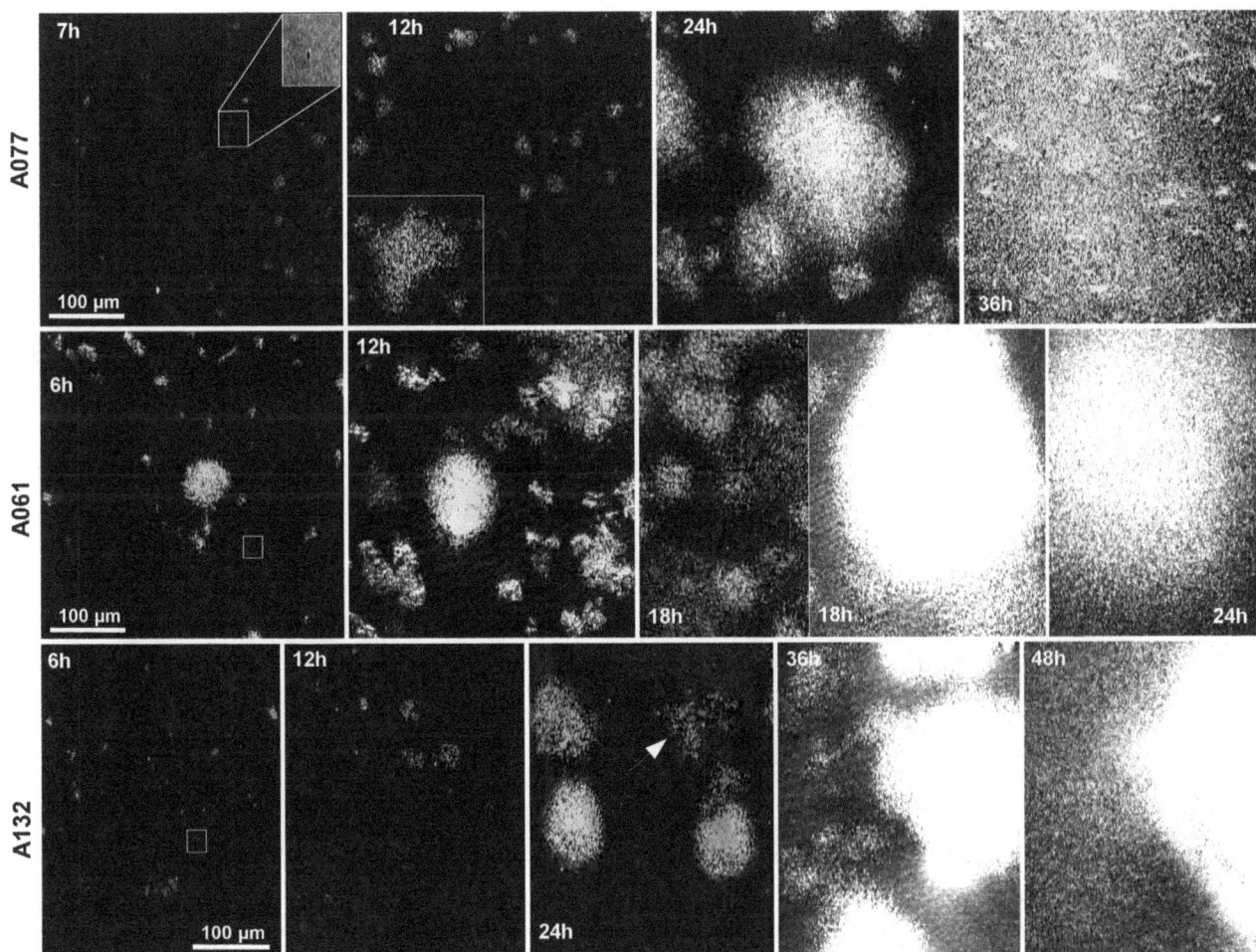

Figure 3. Brewster Angle Microscopy to follow *A. baumannii* **pellicle development.** Squared boxes showed initial dark clusters (for A077, the insert box is obtained with a two-fold magnification and an enhanced contrast). White arrow points large dark one-bacteria-in-height clusters.

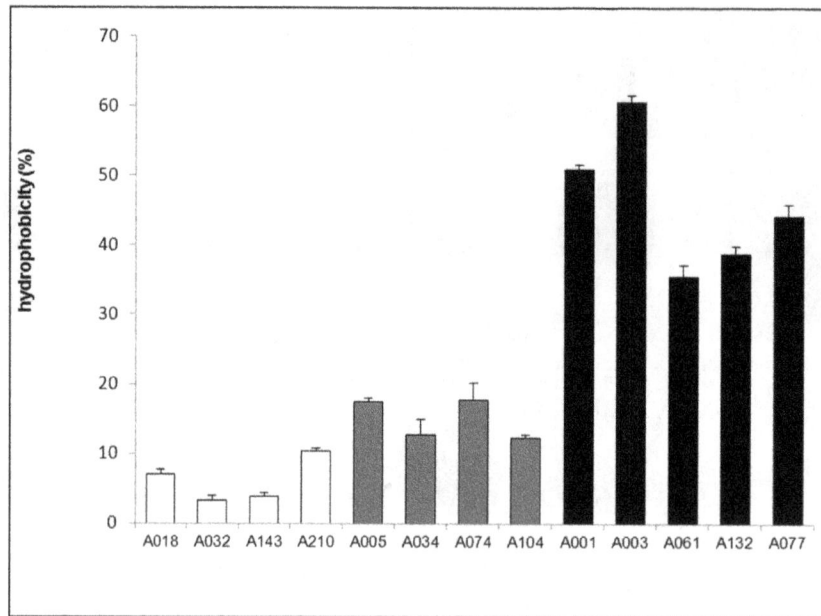

Figure 4. Bacterial hydrophobicity. Cellular hydrophobicity was measured and compared between non-biofilm forming (Pellicle – Biofilm -), pellicle-forming (Pellicle +) and solid biofilm-forming (Biofilm +) *A. baumannii* strains. Results are presented as means of at least 3 measurements for each strain. Error bars represent standard error with a 95% confidence interval. Results of ANOVA test, comparing hydrophobicity between the three groups, give p-values <0.001.

visualize the biofilm air side, a non-coated cover slide was introduced, allowing the pellicle to sit on it. Samples were examined by SEM as previously described by Collet *et al.* [27].

Atomic Force Microscopy (AFM). Water and air-facing sides of 24-h grown pellicles were visualized by AFM as described by Marti *et al.* [28]. AFM imaging was performed by using a Nanoscope III Multimode microscope (Veeco instrument, Santa Barbara, Ca, USA) with a 100 μm piezoelectric scanner. Imaging was achieved in the air in the contact mode. The cantilevers used were characterized by a low spring constant of about 0.06 N/m and were equipped with sharpened Si-tips. All the measurements were performed with the feedback loop on (constant force from

10^{-9} to 10^{-8} N). All images are presented in the height mode (black and white palette for height: dark for low zones, light for high zones) and are top-view images. A flatten operation was usually done on all images. This was achieved either with the nanoscope software (flatten order 2 or 3) or using the free Gwiddion AFM software downloadable at http://gwyddion.net/ (three points levelling, polynomial background order 2).

Bacterial hydrophobicity

Bacterial adherence to hexadecane was adapted from Rosenberg *et al.* [29]. After an overnight planktonic culture (10 mL) at 25°C, bacteria were harvested by centrifugation (1500×g for

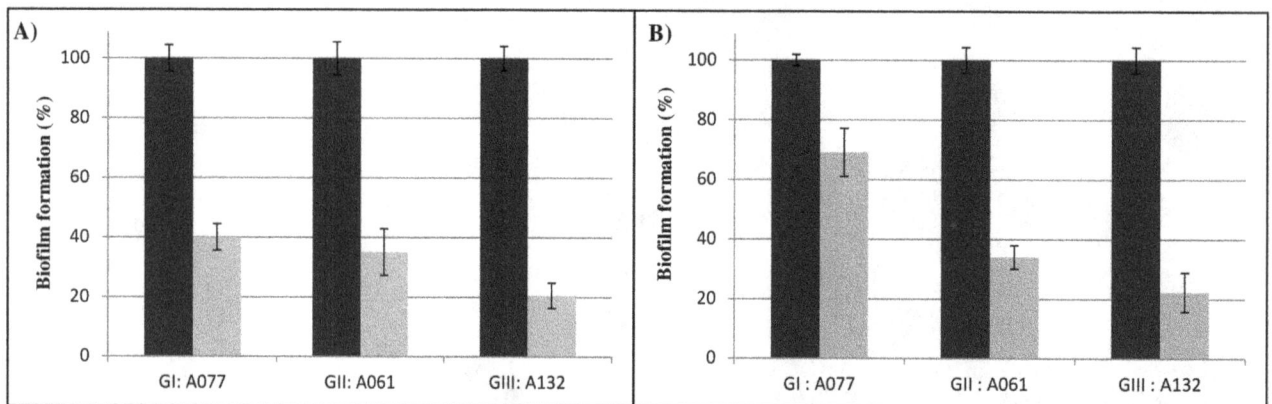

Figure 5. Effect of Dispersin B on pellicle formation for each morphotype. A077 strain from morphotype I, A061 strain from morphotype II and A132 strain from morphotype III in presence of 50 μg/mL Dispersin B (grey) or without Dispersin B (black). **A)** Dispersin B was added at the beginning of the pellicle growth and biofilm formation was quantified at 24 h by cristal violet assay; **B)** Dispersin B was added on a 24 h-pellicle and biofilm formation was quantified after additional 24 h. Results are presented as means (from at least 3 replicates for each condition). Error bars represent standard error with a 95% confidence interval. Results of ANOVA test, comparing the pellicle formation with and without Dispersin B for each strain, give p-values <0.001.

Table 1. Types and characteristics of pellicles produced by A. baumannii.

Groups	Strain	Isolate source	MDR*	Hydrophobicity (%)	Monosaccharide composition (% mol)					
					Glc	Kdo	GlcNAc	GalA	GalNAc	ManAc
I – Egg-shaped	A077	Urine	S	44±2	48.6±4.0	29.3±2.3	18.6±4.0	2.2±0.2	0.5±0.6	0.9±0.8
II – Ball-shaped	A061	Urine	MDR	35±2	28.1±2.5	11.0±1.3	51.5±3.6	–	9.4±1.0	–
III – Irregular	A132	Urine	S	38±1	43.3±5.0	27.6±1.9	25.0±4.9	1.9±0.3	1.5±1.3	0.6±0.5

(*) S for sensitive, MDR for Multidrug-resistant as defined by Magiorakos et al. [52].

15 min) and the resulting pellet was washed and resuspended in sterile phosphate buffered saline solution (PBS) to a final absorbance value, A_{400} between 1.5 and 1.6. To 1.2 mL of bacterial suspension, 0.2 mL of Hexadecane (Acros Organics, France) were added and vortexed vigorously for 2 min. After 15 min standing at room temperature to allow phase separation, the absorbance was measured again. Bacterial hydrophobicity (%) was calculated as follows:

$$Hydrophobicity = \left[\left(A_{400\,initial} - A_{400\,final} \right) / A_{400\,initial} \right] \times 100$$

Pellicle and crude matrix isolation

Biofilm matrix isolation and purification were adapted from Friedman et al. [30] with some modifications. Standing 1 L bacterial cultures on MHB in 2 L glass Erlenmeyers were inoculated to an initial OD_{600} value of 0.01 and left at 25°C without shaking for 5 days. Pellicles were recovered from the top of the culture with a 10 mL pipette, avoiding the culture medium, and washed once with dH_2O. Pellets were resuspended with 10 mL of 1M NaOH, vortexed 30 s every 2 min for 15 min and centrifuged in a Beckman ultracentrifuge (Optima L-90 K Ultracentrifuge, CA, USA) at $19,000 \times g$ (70 Ti Rotor) for 1 h at 4°C. Supernatants were filtered through a 0.45 μm cellulose filter and neutralized with concentrated HCl. The matrix was precipitated at 4°C overnight, after addition of 3 volumes of cold 95% ethanol, centrifuged to recover the precipitated matrix, washed with 70% ethanol and then dried. Purified matrix was resuspended in dH_2O and dialyzed overnight against water.

Protein and polysaccharide quantification

Protein concentrations were determined using either the DC-protein assay (Bio-Rad, France) or the micro-BCA method from Pierce (Fisher, France). Quantification of neutral sugars was estimated by the modified phenol-sulfuric acid method [31] using Glucose (Glc), Galactose (Gal) and 3-Deoxy-manno-oct-2-ulopyranosonic acid (Kdo) standard curves. Briefly, 30 μL of 5% phenol aqueous solution (w/v) were automatically added (PerkinElmer: JANUS automated workstation, Courtaboeuf, France) to 30 μL of diluted purified pellicles (in triplicate) or standards solutions in a quartz 96-well microplate (Hellma, 730.009-QG). The plate was stirred for 1 min at 500 rpm on shaker (IKA, TTS 3 shaker with microtiter attachment) placed in controlled mechanical ventilation area. Then, 180 μL concentrated sulphuric acid (95%) were added and the plate was incubated in the oven for 20 minutes at 110°C. Finally, the plate was stirred for 5 min at 200 rpm and cooled to room temperature before reading the absorbance at 490 nm (A_{490}) (Victor 3 spectrophotometer, PerkinElmer, Courtaboeuf, France). According to the monosaccharide composition of the EPS matrices and the linear responses of the Dubois assay for the standard sugars solutions, the quantity of detectable unmodified sugars (Q in mg) in the tested solution was calculated according to the following equation:

$$Q_{detectable\,sugar} = A_{490} / \left[L \times \sum_i \left(molar\%_i \times MM_i \times \varepsilon_i \right) \right]$$

where i is one detectable sugar and $molar\%_i$, MM_i and ε_i are respectively, the molar percentage of sugar as determined by GC experiment, the molar mass and the extinction coefficient of the phenol-sugar condensed obtained from standard sugars calibration

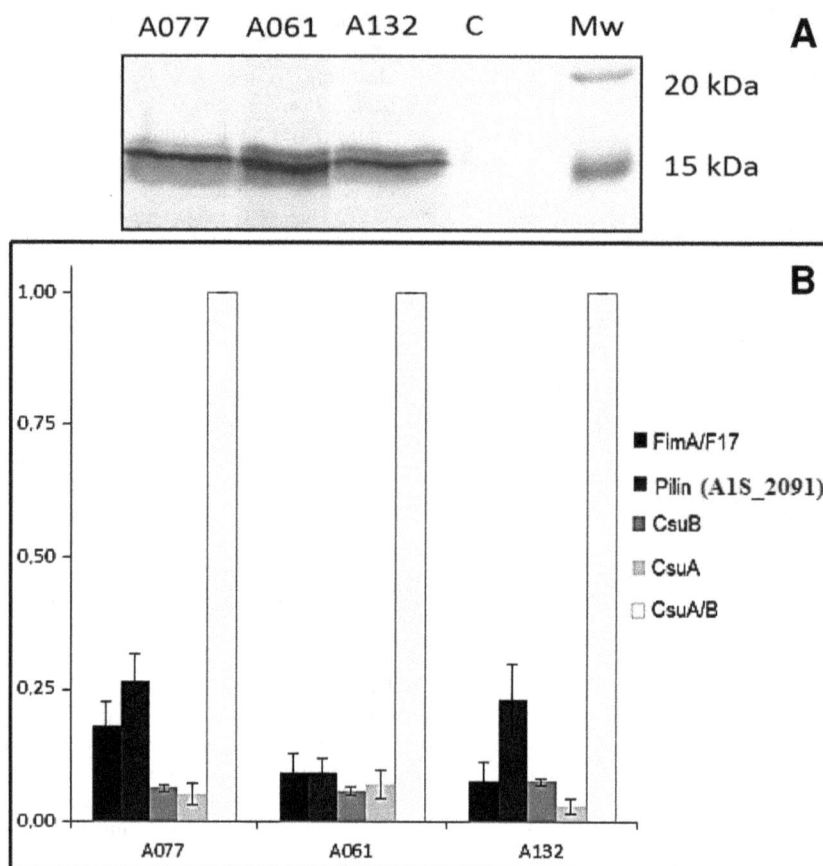

Figure 6. Matrix associated proteins of the pellicles formed by each representative strain of the three morphotypes. A) separated and visualized by SDS-PAGE and silver nitrate staining respectively, after treatment with 0.2% cellulase. **B)** Histogram giving relative abundances of proteins calculated according to the minimum-maximum normalization method from 3 biological replicates and 2 technical replicates. Error bars represent standard error with a 95% confidence interval.

curves. L is constant in the reading process and the expression can be simplified to:

$$Q_{detectable\,sugar} = A_{490} / \left[\sum_i \left(molar\%_i \times MM_i \times \varepsilon_i \right) \right]$$

Monosaccharide composition

25 μL of 4 mM inositol solution were added as internal standard to 500 μL of purified pellicles. Samples were freeze-dried and submitted to 16 h methanolysis at 80°C with 200 μL of 1 M methanolic-HCl. After evaporation of methanol, samples were re-acetylated by addition of 20 μL anhydrous acetic anhydride and 20 μL pyridine. The resulting N-acetyl methyl glycosides (methyl ester) were dried, converted into their trimethylsilyl derivatives and separated by gas chromatography (GC). The gas chromatograph (Varian GC3800, Les Ullis, France) was equipped with a flame ionization detector, a WCOT fused silica capillary column as stationary phase (Varian CP-Sil 5 CB length 25 m, i.d. 0.25 mm, film thickness 0.25 μm) and helium as gas vector (constant pressure 20 psi). The oven temperature program was: 2 min at 120°C, 10°C/min to 160°C, and 1.5°C/min to 220°C and then 20°C/min to 280°C. Sugar quantification

was done by integration of peaks and determination of the corresponding molar values using response factors established with standard monosaccharides. Quantity of each carbohydrate is expressed in molar percentage from derivatization of 1 mg of total sugars and results from the analysis of three pellicles from each strain and a technical duplicate.

Dispersin B effect on pellicle formation

Pellicles from each morphogroup (*i.e.*, A061, A077 and A132 strains) were grown on 24-wells plates in MHB with or without 50 μg/mL Dispersin B for 24 h at 25°C without shaking. Pellicle formation was quantified according to the protocol described by O'Toole and Kolter [32]. For testing the activity of Dispersin B on 24-h pellicles, 50 μg/mL of the enzyme were added and pellicle formation was similarly quantified after 24 h of additional growth. All experiments were performed at least in triplicate. Statistical analyses were performed using Prism Graph Pad 4 software and significant differences between pellicles growth with or without Dispersin B were assessed by *P* values.

Secreted protein identification

Purified matrixes were treated with 0.02% cellulase (Sigma-Aldrich, Lyon, France) at 37°C overnight with shaking. Samples were separated by discontinuous SDS-polyacrylamide gel electrophoresis (4.5% stacking, 15% separating gel) and gels were stained with 0.1% (wt/vol) Coomassie brilliant blue G 250 solution.

Excised bands were washed several times with water and dried for 2 hours. Trypsin digestion was performed as described by Marti *et al.* [28]. Identification of secreted proteins was performed by LC-MS/MS. All experiments were done on a LTQ-Orbitrap Elite (Thermo Scientific) coupled with an Easy nano-Liquid Chromatography II system (Thermo Scientific). Tryptic peptides were eluted from 2 to 55% of B solution (0.1% formic acid in acetonitrile) over 14 min, from 55 to 100% of B solution in 1 min and 10 min at 100% of B solution. The samples were analysed using the CID (Top20) method. Raw data files were processed using Proteome Discoverer 1.3 software (Thermo Scientific). Peak lists were searched using the MASCOT search engine (Matrix Science) against the database *A. baumannii* ATCC17978 protein sequences at http://www.ncbi.nlm.nih.gov. Database searches were performed with the following parameters: 2 missed trypsin cleavage sites allowed; variable modifications: carbamidomethylation of cystein, oxidation of methionine. Quantitation was based on label-free approach using spectral counting (peptide spectral match – PSM). Protein relative abundance was expressed after minimum-maximum normalization method.

Results

Pellicle formation

Culture medium (2 mL) was inoculated with a low bacterial cell density (OD$_{600}$ = 0.01) and grown at 25°C without shaking. After 24 hours, a thin pellicle started to form at the surface of the liquid media; this pellicle grew in thickness and by the end of the third day, an opaque and solid structure covered the whole liquid surface. Eighty-six *A. baumannii* clinical isolates were visually analysed for pellicle formation and twenty-six (30%) pellicle-forming isolates were separated for further analysis (Table S1). Pellicles were diverse depending on the isolates, with no growth difference between MHB or CR-TB, and were characterized as thick pellicles at the air-liquid interface attached to the wall of the tube, with transparent liquid media underneath, in contrast to negative samples that grew as turbid cultures with no pellicle attached to the tube (Figure 1A&B). Some pellicles were so strong that they could support, for several minutes, the weight of the liquid medium (2 g) when the tube was inverted (Figure 1C). Bacterial growth on congo-red containing agar plates was not useful to differentiate between pellicle-forming and non-forming strains (data not shown).

Pellicle morphology

Twenty-six non related pellicle-forming *A. baumannii* isolates were classified by inverse optical microscopy into three different morphologic groups named from their appearance (Figure 2A, Table S1): 'egg-shaped' for its similarity with fried eggs observed by inverse microscopy, 'ball-shaped' for the presence of ball-like aggregations in liquid media, and 'irregular' pellicles for the differences in size and shape of the bacterial aggregates. In the ball-shaped group, non-associated bacterial aggregates can be observed floating on the surface of the liquid medium. These balls remained attached to a thin bacterial layer that covered the whole surface. It could be observed after removing the liquid medium underneath. The ball-shaped morphology was the most common, with 50% of the strains; the remaining strains were equally divided into the other two groups (7 and 6 strains, respectively).

A representative strain from each morphologic group was selected for further analyses by SEM and AFM (Figure 2B&C). SEM images of A077 (Group I: egg-shaped); A061 (Group II: ball-shaped); A132 (Group III: irregular pellicle) confirmed the morphologies already observed. AFM images allowed to assess the topography of the different morphotypes and showed big bacterial clusters surrounded by smaller ones (<50 bacteria) with a single layer in height (Figure 2C). A 3D comparison showed that A061 formed the most homogeneous pellicle with fairly round and uniform hills, A077 pellicle was less uniform and A132 pellicle was completely heterogeneous in height. For all the strains, the big bacterial clusters were surrounded by a layer with a height corresponding to a single bacterium. The width of this layer reached 4 to 9 μm for A061 and A077, when it could spread up to 20 μm for A132.

Pellicle growth by BAM

Images taken few minutes after placing the bacterial suspension in the plate showed that the light was slightly reflected because of the organic monolayer (lipids and amphiphile proteins from the growth medium) already formed at the air-liquid interface. This monolayer remained stable and static over the first hour. After two hours, bright and dark objects started to appear on the surface. Actually, bacteria clusters organized in the plane of the surface lead locally to a decrease in light reflection and then correspond to the dark objects. When bacteria start to assemble as three dimensional clusters, the incoming laser light is locally no longer at the Brewster angle leading to a high light reflection. This is responsible for the observed bright objects. The area of these initial clusters was around 25 μm^2 (20 to 30 pixels). Then, these clusters were composed of 20 to 30 bacteria. Bright and dark clusters grew in size during the next hours with a widespread conversion from dark into bright (Figure 3, see squared boxes).

Isolate A061 (Group II). For 6 h, cluster growth (size and reflective intensity) was continuous, growing as independent units and also merging with nearby clusters. The total number of clusters was stable for at least the next 6 hours, with new clusters still formed at this stage of pellicle formation (see Figure 3, A061-6 h & 12 h). The overall surface was covered by the pellicle after 36 h.

Isolates A077 (Group I) and A132 (Group III). Pellicle formation was more heterogeneous up to the 18 h of growth for A077 and 36 h for A132 (Figure 3, A077- 12 h & A132- 24 h) and clusters were characterized by variable area and height. This phenomenon was even more evident for the A132 strain, which at 24 h growth, still presented large dark one bacterium in height clusters (see arrow in Figure 3, A132- 24 h) as well as large bright clusters, showing a certain level of competition between the building of the biofilm in the plane and in height. This could be responsible for the slower rate of pellicle formation.

Bacterial hydrophobicity

Cellular hydrophobicity was studied and compared between non-biofilm forming, pellicle-forming, and solid biofilm-forming (formed on the wall of the tube) *A. baumannii* strains by an extraction method in organic solvent (Figure 4). Pellicle-forming strains were markedly more hydrophobic (35–60% of the bacterial extraction by hexadecane) than solid biofilm-forming strains (12–17%) or non-biofilm forming counterparts (3–10%).

Monosaccharide composition

Gas chromatography analysis of the purified matrix polysaccharide was performed on 1 mg sample from each morphogroup representative strain (Table 1). Total carbohydrate analysis confirmed that pellicle matrices contained a carbohydrate-rich material; the main monosaccharide components were glucose and N-acetyl-glucosamine, with differences among groups (Table 1). Isolates from groups I and III showed a high conserved carbohydrate composition with a matrix composed mainly of glucose residues while the isolate A061 from group II contained above all N-acetyl glucosamine (Table 1). The activity of Dispersin B, which is an endoglycosidase specific to β-1-6-linked poly-glucosamine molecules [33] was significant on pellicle formation when it was added at the beginning of the culture (60 to 80% of decrease for all morphotypes, Figure 5A). The activity of the enzyme remained effective on 24-h pellicles (Figure 5B).

Identification of matrix-associated proteins

Proteins embedded within the pellicle matrix were difficult to analyse due to the high amount of polysaccharides interfering with migration. Consequently, purified matrices were treated with cellulase previous to SDS-PAGE separation. No high molecular weight proteins were detected associated to the pellicle matrix but different low molecular weight proteins were identified in the samples from each morphogroup (Figure 6 & Table S2). Three proteic bands (located at about 15 kDa) were cut then digested and analysed by LC-MS/MS. The most abundant proteins were identified as pilins. As showed in the Figure 6 (and emphasized by psm values, see Table S2), CsuA/B (A1S_2218) was the most abundant protein in all matrices. This protein as well as CsuA (A1S_2217) and CsuB (A1S_2216), are secreted subunits of the CsuA/BABCDE chaperon-usher pili assembly system, which has been described to be involved in the attachment and biofilm formation of A. baumannii on abiotic surfaces [16]. Another identified pilin was the protein A1S_2091, which could be the major subunit of a chaperone–usher pili assembly (CU) system consisting in 4 proteins (A1S_2088-2091, reversed ORFs) in which A1S_2090 would be the assembly chaperone and A1S_2089 the fimbrial usher. After a sequence analysis and a BLASTP in the data bank, we found this CU system extremely well conserved in A. baumannii (>90% identity except in ACICU strain in which it was absent), and in the other Acinetobacter species, i.e. A. pittii, A. nosocomialis and A. calcoaceticus (>80% identity, see Table S3). Finally, the subunit coded by A1S_1507 gene was also identified as part of an operon coding for a last A1S_1510-1507 CU system, (again reversed ORFs in ATCC 17978). This CU system is constituted by the pilin FimA (A1S_1510 sometimes annotated F17-A protein in A. baumannii genomes), then a PapD chaperone (A1S_1509), a PapC porin (A1S_1508) and the tip adhesion pilin could be the A1S_1507 protein. This operon has already been shown to be involved in biofilm formation on solid support [34]. As shown by the Table S4, this operon is also well conserved in A. baumannii (except in AYE strain where the operon showed the insertion of an ISAbaI in A1S_1508 usher protein), and was found in ACB complex as well as in A. junii.

Discussion

As for many bacterial species, the biofilm formation by A. baumannii has been linked to device-associated infections, antibiotic resistance and survival on dry surfaces [13,35,36]. Recent studies have also confirmed the capacity of this species to form pellicles at the air-liquid interface [20–22], reporting a higher frequency in A. baumannii and A. nosocomialis (formerly Acinetobacter genospecies 13TU), than in other less pathogenic Acinetobacter spp. [21].

Previous studies on Pseudomonas and Salmonella spp. described distinctive colony morphologies on congo red agar plates, identifying pellicle forming bacteria that produced dry and wrinkled colonies, defined as rdar in Salmonella spp. or wrinkly spreaders for Pseudomonas fluorescens [37–39]. This approach was not suitable with A. baumannii pellicle-forming isolates, since their growth, colour and morphology on congo red agar were

highly variable. Consequently, pellicles were studied in liquid media to observe not only morphological differences among the strains, but also pellicle cohesion and matrix composition. The screening performed by optical microscopy on 26 pellicle-forming strains (30% of our collection), allowed their clustering into three morphological groups (Table 1, Figure 2). By contrast Koza et al. [5] classified P. fluorescens pellicles into four classes on basis of phenotype and physical robustness, considering their strength, sugar component and ability to attach at the meniscus. In our study, all the morphological groups were attached to a certain extent to the walls of the tube. Consequently, BAM microscopy examinations performed during the initial stages of pellicle formation aimed to elucidate whether this structure could develop without a solid anchor or bacteria attached to the wall were expanding and covering the liquid media as already described for P. fluorescens wrinkly spreaders [39]. Observations (Figure 3) pointed out that, contrary to P. fluorescens [39], Acinetobacter bacteria emerged anywhere on the liquid interface, formed aggregations that grew covering the whole surface and attached afterwards to the tube.

In wrinkly spreader P. fluorescens, bacterial recruitment to the air-liquid interface was associated to cell hydrophobicity combined with surface-interactions which would lead bacteria to be at the surface rather than submerged [5,39]. Our results are consistent with these data; indeed, the bacterial pellicle forming ability was associated with high strain hydrophobicity (Figure 4). These results disagree with those of McQueary et al. who reported no correlation between cell hydrophobicity and pellicle formation in A. baumannii [22]. This discrepancy may be attributed to the temperature used in their assays (37°C), as pellicles in A. baumannii are more commonly observed at 25°C [20,21]. This hydrophobicity that may be associated with an obligate aerobic status could explain why some A. baumannii strains were able to grow at the surface of the liquid media.

The microscopy analysis of the A. baumannii pellicles was followed by a chemical characterization of the matrix components. EPS matrix purification is a complex procedure since bacterial cells have to be separated from this carbohydrate rich substance and aggressive methodologies may damage bacterial cells leading to the release of cytoplasmic material [7]. In the present study, high molecular weight proteins were not detected and the identified matrix associated proteins were mostly pilin-like proteins. Although not directly considered as matrix components, these extracellular appendages are known to stabilize the structure, especially to maintain pellicles at the air-liquid interface, and have been widely related to the matrix [7,8]. We confirmed that, at least, three pili systems are present in matrix pellicle and could be involved in the cohesion of the matrix structure. The first one is the Csu operon which is required for biofilm formation on solid surface [16] and would contribute to A. baumannii pellicle building, the CsuA/B pilin being the most abundant protein detected in matrix. The second pili system is a CU system from which we identified the pilin A1S_1510, annotated as FimA or F17aA protein, which presents 46% identity with the F17A pilin, member of F17A-G pili system expressed in E. coli. In enterotoxigenic E. coli, these pili mediate attachment to intestinal microvilli, leading to diarrhea or septicemia in ruminants [40,41]. Disruption of the tip adhesin (A1S_1507) gene was shown to induce a severe decrease in biofilm formation on polypropylene plates [34]. Of note, the homologue of this operon in A. baylyi, the AcuADCG operon, was described to be involved in formation of thin pili required for adhesion to polystyrene and erythrocytes [42]. Finally, the identified pilin A1S_2091 is part of the 4-proteins CU system that has not been described before, which is extremely well conserved in the ACB complex but did not present any significant identity with some known systems.

Matrix exopolysaccharide composition has been shown to be complex (*e.g.*, *P. aeruginosa* biofilms matrix containing among others Pel, Psl and alginate polysaccharides [43,44]) but also variable in monomer units and linkages used to form the polysaccharide [8]. Cellulose-like polysaccharides, such as Pel in *P. aeruginosa* PA14 [30], rich in glucose subunits, are usually associated to solid biofilm structures and often to pellicle formation. Partially acetylated cellulose has also proved to be important for the formation of strong pellicles by wrinkly spreader *P. fluorescens* and *Salmonella* spp. [8,25,43]. In the present study, carbohydrate analysis corroborated the similarities of robustness observed between the morphogroups I and III, presenting a matrix composed mainly of glucose. In contrast, the isolate from the morphogroup II possess a matrix mainly composed of GlcNAc, an amino sugar with important structural roles in cell surface [45]. GlcNAc also forms part of a different type of surface polymer, PNAG, that plays an important role as virulence factor in staphylococcal biofilm formation [46] and has also been identified in numerous Gram-negative pathogenic bacteria. It has been recently reported in *A. baumannii* to be essential for maintaining the integrity of biofilm formed in dynamic environments [47]. In our case, the reactivity of Dispersin B (that specifically cleaves the β-(1,6)-linked N-acetylglucosamine polymer) on pellicle formation of each morphogroup suggests that their pellicle matrix may contain PNAG. Finally, the identification of a component as Kdo purified matrix revealed the presence of lipopolysaccharide (LPS), which instead of being a cellular contamination, could have been released into the external surrounding by the bacteria. In *P. fluorescens* pellicles, the interaction between LPS and cellulose gives strength and integrity, playing an important role in the relative hydrophobicity of the cell and bacterial attachment to surfaces [39].

In conclusion, as already observed in *P. aeruginosa* [48], the matrix of *A. baumannii* pellicles is complex, containing different extracellular polymeric substances, *i.e.* several exopolysaccharides, lipopolysaccharide, cell surface appendages (and probably eDNA, [49]), that may be contributing factors for the persistence and epidemicity of *A. baumannii* in the clinical environment [13,35,50]. In the hospital environment, pellicles could be formed in small droplets and therefore, might be involved in bacterial colonization of humidifiers, respiratory systems and any moist surfaces or apparatus already described as potential sources for *A. baumannii* isolation in hospital outbreaks [51]. Bacterial clusters may eventually be detached from the pellicle and move freely to cause infection if transfer into the patient occurs, *i.e.* through a respiratory system. As pellicles are more readily formed at 25°C than 37°C, bacteria in the pellicle could revert to the planktonic state and cause infection. For this reason, hospital antiseptics and disinfectants should target this biofilm structure, basically the matrix components that give stability and protect bacteria from external hazards.

Supporting Information

Table S1   A. baumannii clinical isolates used in this study.

Table S2   Matrix associated proteins identified in the three morphotypes of pellicles produced by A. baumannii.

Table S3   Percentages of identity of A1S_2091–2088 proteins with homologues in related organisms.

Table S4   Percentages of identity of A1S_1510–1507 proteins with homologues in related organisms.

Acknowledgments

We thank the Grupo de Estudio de la Infección Hospitalaria (GEIH) from the Sociedad Española de Enfermedades Infecciosas y Microbiología Clínica (SEIMC) for supplying *A. baumannii* strains. We thank Dr. I. Zimmerlin for SEM experiments and Dr. Laurent Coquet for the protein identifications. We thank Kane Biotech Inc. for providing us Dispersin B.

Author Contributions

Conceived and designed the experiments: OL TJ JV ED. Performed the experiments: YNC SM CR SA JH OL JK. Analyzed the data: YNC SM CR SA JH OL TJ ED. Contributed reagents/materials/analysis tools: JV JK. Contributed to the writing of the manuscript: TJ JK ED.

References

1. Donlan RM (2002) Biofilms: microbial life on surfaces. Emerg. Infect Dis. 8: 881–890.
2. Hall-Stoodley L, Stoodley P (2009) Evolving concepts in biofilm infections. Cell Microbiol 11: 1034–1043.
3. Davey ME, O'Toole GA (2000) Microbial biofilms: from ecology to molecular genetics. Microbiol Mol Biol Rev 64: 847–867.
4. Branda SS, Vik S, Friedman L, Kolter R (2005) Biofilms: the matrix revisited. Trends Microbiol 13: 20–26.
5. Koza A, Hallett PD, Moon CD, Spiers AJ (2009) Characterization of a novel air-liquid interface biofilm of *Pseudomonas fluorescens* SBW25. Microbiology 155: 1397–1406.
6. Flemming HC, Neu TR, Wozniak DJ (2007) The EPS matrix: the "house of biofilm cells". J Bacteriol 189: 7945–7947.
7. Flemming HC, Wingender J (2010) The biofilm matrix. Nat. Rev Microbiol 8: 623–633.
8. Pamp SJ, Gjermansen M, Tolker-Nielsen T (2007) The biofilm matrix: a sticky framework, p 37–69. In Kjelleberg S, Givskov M (eds.), The biofilm mode of life: mechanisms and adaptations, Horizon Bioscience, Norfolk, UK.
9. Bergogne-Berezin E, Towner KJ (1996) *Acinetobacter* spp. as nosocomial pathogens: microbiological, clinical, and epidemiological features. Clin Microbiol Rev 9: 148–165.
10. Dijkshoorn L, Nemec A, Seifert H (2007) An increasing threat in hospitals: multidrug-resistant *Acinetobacter baumannii*. Nat. Rev Microbiol 5: 939–951.
11. Howard A, O'Donoghue M, Feeney A, Sleator RD (2012) *Acinetobacter baumannii*: an emerging opportunistic pathogen. Virulence. 3: 243–250.
12. Visca P, Seifert H, Towner KJ (2011) *Acinetobacter* infection-an emerging threat to human health. IUBMB. Life 63: 1048–1054.
13. Espinal P, Marti S, Vila J (2012) Effect of biofilm formation on the survival of Acinetobacter baumannii on dry surfaces. J Hosp. Infect 80: 56–60.
14. Vila J, Marti S, Sanchez-Cespedes J (2007) Porins, efflux pumps and multidrug resistance in *Acinetobacter baumannii*. J Antimicrob Chemother 59: 1210–1215.
15. Loehfelm TW, Luke NR, Campagnari AA (2008) Identification and characterization of an *Acinetobacter baumannii* biofilm-associated protein. J Bacteriol 190: 1036–1044.
16. Tomaras AP, Dorsey CW, Edelmann RE, Actis LA (2003) Attachment to and biofilm formation on abiotic surfaces by *Acinetobacter baumannii*: involvement of a novel chaperone-usher pili assembly system. Microbiology 149: 3473–3484.
17. Bentancor LV, Camacho-Peiro A, Bozkurt-Guzel C, Pier GB, Maira-Litran T (2012) Identification of Ata, a multifunctional trimeric autotransporter of *Acinetobacter baumannii*. J Bacteriol 194: 3950–3960.
18. Gaddy JA, Tomaras AP, Actis LA (2009) The *Acinetobacter baumannii* 19606 OmpA protein plays a role in biofilm formation on abiotic surfaces and in the interaction of this pathogen with eukaryotic cells. Infect Immun 77: 3150–3160.
19. Eijkelkamp BA, Stroeher UH, Hassan KA, Papadimitrious MS, Paulsen IT, et al. (2011) Adherence and motility characteristics of clinical *Acinetobacter baumannii* isolates. FEMS Microbiol Lett 323: 44–51.
20. Eijkelkamp BA, Stroeher UH, Hassan KA, Elbourne LD, Paulsen IT, et al. (2013) H-NS Plays a Role in Expression of *Acinetobacter baumannii* Virulence Features. Infect Immun 81: 2574–2583.

21. Marti S, Rodriguez-Bano J, Catel-Ferreira M, Jouenne T, Vila J, et al. (2011) Biofilm formation at the solid-liquid and air-liquid interfaces by *Acinetobacter* species. BMC Res Notes 4: 5.
22. McQueary CN, Actis LA (2011) *Acinetobacter baumannii* biofilms: variations among strains and correlations with other cell properties. J Microbiol 49: 243–250.
23. Rodriguez-Bano J, Cisneros JM, Fernandez-Cuenca F, Ribera A, Vila J, et al. (2004) Clinical features and epidemiology of *Acinetobacter baumannii* colonization and infection in Spanish hospitals. Infect Control Hosp Epidemiol 25: 819–824.
24. Vila J, Riberaa A, Marcoa F, Ruiza J, Mensaa J, et al. (2002) Activity of clinafloxacin, compared with six other quinolones, against *Acinetobacter baumannii* clinical isolates. J Antimicrob Chemother 49: 471–477.
25. Spiers AJ, Bohannon J, Gehrig SM, Rainey PB (2003) Biofilm formation at the air-liquid interface by the *Pseudomonas fluorescens* SBW25 wrinkly spreader requires an acetylated form of cellulose. Mol Microbiol 50: 15–27.
26. Henk MC (2004) Method for collecting air-water interface microbes suitable for subsequent microscopy and molecular analysis in both research and teaching laboratories. Appl Environ Microbiol 70: 2486–2493.
27. Collet A, Cosette P, Beloin C, Ghigo JM, Rihouey C, et al. (2008) Impact of rpoS deletion on the proteome of *Escherichia coli* grown planktonically and as biofilm. J Proteome Res 7: 4659–4669.
28. Marti S, Nait CY, Alexandre S, Coquet L, Vila J, et al. (2011) Growth of *Acinetobacter baumannii* in pellicle enhanced the expression of potential virulence factors. PLoS One 6: e26030.
29. Rosenberg M, Perry A, Bayer EA, Gutnick DL, Rosenberg E, et al. (1981) Adherence of *Acinetobacter calcoaceticus* RAG-1 to human epithelial cells and to hexadecane. Infect Immun 33: 29–33.
30. Friedman L, Kolter R (2004) Genes involved in matrix formation in *Pseudomonas aeruginosa* PA14 biofilms. Mol Microbiol 51: 675–690.
31. Dubois M, Gilles KA, Hamilton JK, Rebers PA, Smith F (1956) Colorimetric method for determination of sugars and related substances. Anal Chem 28: 350–356.
32. O'Toole GA, Kolter R (1998) Initiation of biofilm formation in *Pseudomonas fluorescens* WCS365 proceeds via multiple, convergent signalling pathways: a genetic analysis. Mol Microbiol 28: 449–461.
33. Kaplan JB, Velliyagounder K, Ragunath C, Rohde H, Mack D, et al. (2004) Genes involved in the synthesis and degradation of matrix polysaccharide in *Actinobacillus actinomycetemcomitans* and *Actinobacillus pleuropneumoniae* biofilms. J Bacteriol 186: 8213–8220.
34. Rumbo-Feal S, Gómez MJ, Gayoso C, Álvarez-Fraga L, Cabral MP, et al. (2013) Whole transcriptome analysis of *Acinetobacter baumannii* assessed by RNA-sequencing reveals different mRNA expression profiles in biofilm compared to planktonic cells. PLoS One. 8: e72968.
35. McConnell MJ, Actis L, Pachon J (2013) *Acinetobacter baumannii*: human infections, factors contributing to pathogenesis and animal models. FEMS Microbiol Rev 37: 130–155.
36. Rodriguez-Bano J, Marti S, Soto S, Fernandez-Cuenca F, Cisneros JM, et al. (2008) Biofilm formation in *Acinetobacter baumannii*: associated features and clinical implications. Clin Microbiol Infect 14: 276–278.
37. Romling U, Rohde M (1999) Flagella modulate the multicellular behavior of Salmonella typhimurium on the community level. FEMS Microbiol Lett. 180: 91–102.
38. Solano C, Garcia B, Valle J, Berasain C, Ghigo JM, et al. (2002) Genetic analysis of *Salmonella enteritidis* biofilm formation: critical role of cellulose. Mol Microbiol 43: 793–808.
39. Spiers AJ, Rainey PB (2005) The *Pseudomonas fluorescens* SBW25 wrinkly spreader biofilm requires attachment factor, cellulose fibre and LPS interactions to maintain strength and integrity. Microbiology 151: 2829–2839.
40. Buts L, Bouckaert J, De GE, Loris R, Oscarson S, et al. (2003) The fimbrial adhesin F17-G of enterotoxigenic *Escherichia coli* has an immunoglobulin-like lectin domain that binds N-acetylglucosamine. Mol Microbiol 49: 705–715.
41. Girardeau JP (1980) A new in vitro technique for attachment to intestinal villi using enteropathogenic *Escherichia coli*. Ann Microbiol (Paris) 131B: 31–37.
42. Gohl O, Friedrich A, Hoppert M, Averhoff B (2006) The thin pili of *Acinetobacter* sp. strain BD413 mediate adhesion to biotic and abiotic surfaces. Appl Environ Microbiol 72: 1394–1401.
43. Mann EE, Wozniak DJ (2012) Pseudomonas biofilm matrix composition and niche biology. FEMS Microbiol Rev 36: 893–916.
44. Ryder C, Byrd M, Wozniak DJ (2007) Role of polysaccharides in *Pseudomonas aeruginosa* biofilm development. Curr Opin Microbiol 10: 644–648.
45. Konopka JB (2012) N-acetylglucosamine (GlcNAc) functions in cell signaling. Scientifica (Cairo.) 2012.
46. O'Gara JP (2007) ica and beyond: biofilm mechanisms and regulation in *Staphylococcus epidermidis* and *Staphylococcus aureus*. FEMS Microbiol Lett 270: 179–188.
47. Choi AH, Slamti L, Avci FY, Pier GB, Maira-Litran T (2009) The *pgaABCD* locus of *Acinetobacter baumannii* encodes the production of poly-beta-1-6-N-acetylglucosamine, which is critical for biofilm formation. J Bacteriol 191: 5953–5963.
48. Yang L, Hu Y, Liu Y, Zhang J, Ulstrup J, et al. (2011) Distinct roles of extracellular polymeric substances in *Pseudomonas aeruginosa* biofilm development. Environ. Microbiol 13: 1705–1717.
49. Tetz GV, Artemenko NK, Tetz VV (2009) Effect of DNase and antibiotics on biofilm characteristics. Antimicrob. Agents Chemother. 53: 1204–1209.
50. Peleg AY, Seifert H, Paterson DL (2008) *Acinetobacter baumannii*: emergence of a successful pathogen. Clin Microbiol Rev 21: 538–582.
51. Towner KJ (2009). *Acinetobacter*: an old friend, but a new enemy. J Hosp. Infect 73: 355–363.
52. Magiorakos AP, Srinivasan A, Carey RB, Carmeli Y, Falagas ME, et al. (2012) Multidrug-resistant, extensively drug-resistant and pandrug-resistant bacteria: an international expert proposal for interim standard definitions for acquired resistance. Clin Microbiol Infec 18: 268–281.

11

Erythrocytic Mobilization Enhanced by the Granulocyte Colony-Stimulating Factor Is Associated with Reduced Anthrax-Lethal-Toxin-Induced Mortality in Mice

Hsin-Hou Chang[1,2], Ya-Wen Chiang[1], Ting-Kai Lin[1], Guan-Ling Lin[2], You-Yen Lin[2], Jyh-Hwa Kau[3], Hsin-Hsien Huang[4], Hui-Ling Hsu[4], Jen-Hung Wang[5], Der-Shan Sun[1,2]*

1 Department of Molecular Biology and Human Genetics, Tzu-Chi University, Hualien, Taiwan, 2 Institute of Medical Sciences, Tzu-Chi University, Hualien, Taiwan, 3 Department of Microbiology and Immunology, National Defense Medical Center, Taipei, Taiwan, 4 Institute of Preventive Medicine, National Defense Medical Center, Taipei, Taiwan, 5 Department of Medical Research, Tzu Chi General Hospital, Hualien, Taiwan

Abstract

Anthrax lethal toxin (LT), one of the primary virulence factors of *Bacillus anthracis*, causes anthrax-like symptoms and death in animals. Experiments have indicated that levels of erythrocytopenia and hypoxic stress are associated with disease severity after administering LT. In this study, the granulocyte colony-stimulating factor (G-CSF) was used as a therapeutic agent to ameliorate anthrax-LT- and spore-induced mortality in C57BL/6J mice. We demonstrated that G-CSF promoted the mobilization of mature erythrocytes to peripheral blood, resulting in a significantly faster recovery from erythrocytopenia. In addition, combined treatment using G-CSF and erythropoietin tended to ameliorate *B. anthracis*-spore-elicited mortality in mice. Although specific treatments against LT-mediated pathogenesis remain elusive, these results may be useful in developing feasible strategies to treat anthrax.

Editor: Nupur Gangopadhyay, University of Pittsburgh, United States of America

Funding: This work was supported by grants of National Science Council http://web1.nsc.gov.tw/mp.aspx (NSC 96-2311-B-320-005-MY3 and NSC 99-2311-B-320-003-MY3) and Tzu-Chi University http://www.tcu.edu.tw/ (610400130). The funders had no role in study design, data collection and analysis, decision to publish, or preparation of the manuscript.

Competing Interests: The authors have declared that no competing interests exist.

* Email: dssun@mail.tcu.edu.tw

Introduction

Infection with *Bacillus anthracis*, a gram-positive spore-forming bacterium, can lead to life-threatening anthrax [1]. Anthrax lethal toxin (LT) is comprised of a protective antigen (PA, 83 kDa) and lethal factor (LF, 90 kDa) [2–4], and is one of the primary virulence factors of *B. anthracis*. LF is a specific metalloprotease for mitogen-activated protein kinase (MAPK) kinases (MKKs/MEKs) [5], and can thus disrupt MAPK signaling cascades including p38 MAPK, p42/44 extracellular signal-regulated kinase (ERK), and c-Jun N-terminal kinase (JNK) [6,7]. All of these 3 MAPK pathways are critical in maintaining fundamental cellular homeostasis, including cell proliferation, differentiation, and apoptosis [8]. LF can be delivered into cells by PA, a cell-receptor binding component [3,9]. Although experimental LT treatments may not reproduce the full pathogenesis of anthrax, LT studies in cell or animal models have revealed certain pathogenic progressions. Various cell types, which include macrophages [10,11], dendritic cells [12], lymphocytes [13,14], erythrocytes [15], and megakaryocytes [16], cardiomyocytes [17], and smooth muscle [17], are sensitive to LT treatment. LT has been shown to suppress the differentiation and maturation of the progenitors of macrophages, megakaryocytes, and erythrocytes [15,16,18]. In addition, blood cell count analyses have indicated that LT treatment significantly reduced levels of circulating red blood cells (RBCs) and platelets in mice [15,16], suggesting multiple targets of

LT in hematopoietic lineage cells. Deficiencies of platelets and RBCs may lead to hemorrhage and lethal hypoxic damage [19,20]. Because high levels of LT accumulate in the body when anthrax enters the bacteremia stage, death is typically inevitable even after aggressive antibiotic treatments. This suggests that a specific treatment to overcome the toxic effect is crucial in controlling the disease [21]. Unfortunately, an effective therapeutic approach against LT remains elusive.

Cytokine treatments, particularly hematopoietic growth factors, have been used in various clinical settings to rescue pathological defects [22]. Our previous demonstration was the first to indicate that thrombopoietin (TPO), a megakaryopoiesis-enhancing cytokine [23], can ameliorate LT-induced thrombopoiesis suppression, thrombocytopenia, and likely reduce the mortality in mice [16]. Our data also revealed that erythropoietin (EPO), a potent erythropoiesis-stimulating cytokine [24], ameliorated LT-induced erythropoiesis suppression (particularly those precursors in early erythropoiesis stages), erythrocytopenia, and reduced mortality rates from 100% to 50% after lethal-dose LT challenges in experimental mice [15]. Bone marrow is the primary stem niche supporting erythropoiesis, displaying technically-divided 4-differentiation stages of erythroblasts based on the expression levels of surface markers CD71 and TER-119 [25]. The transferrin receptor (CD71) is first expressed on early erythroblasts, such as erythroid burst-forming units (BFU-Es) and erythroid colony-forming units (CFU-Es) cells. Erythrocytic CD71 is downregulated by more mature erythroblasts [26]. By contrast, TER-119 is primarily expressed on relatively mature erythroblasts, reticulocytes, and mature erythrocytes [27]. Accordingly, 4 cell populations (CD71^{high}TER-119^{med}, CD71^{high}TER-119^{high}, CD71^{med}-TER-119^{high}, and CD71^{low}TER-119^{high}) can be defined, which are morphologically equivalent to proerythroblasts (flow cytometry-gated region 1; R1), basophilic erythroblasts (R2), late basophilic and polychromatophilic erythroblasts (R3), and orthochromatophilic erythroblasts (R4), from the early to late stages of erythroid differentiation, respectively [25]. Following these approaches, we are thus able to characterize the mechanism to use hematopoietic cytokines/growth factors as ameliorative agents to rescue anthrax LT-induced mortality [15,16].

The granulocyte colony-stimulating factor (G-CSF) has been found to regulate granulopoiesis [28], and is a multifunctional cytokine. For example, it has been found to stimulate cell proliferation, differentiation, enhance hematopoiesis, mobilize hematopoietic stem cells, and induce anti-apoptotic and anti-inflammatory effects [29–31]. Both G-CSF and EPO are U.S. FDA-approved drugs. Although the mechanism remains unclear, combined treatments using G-CSF and EPO were shown to ameliorate aplastic anemia in patients with myelodysplastic syndrome [32–36]. Given that EPO treatments are beneficial for LT-challenged mice [15], we hypothesized that combining G-CSF and EPO may be useful in treating anthrax. Consequently, we used mouse models to discuss the combined treatments of G-CSF and EPO on reduced anthrax LT and spore-induced mortality. In addition, we also discussed the differential erythropoietic regulation in response to G-CSF and EPO treatments.

Materials and Methods

Ethics Statement

Our research approaches involving experimental mice were approved by the Institutional Animal Care and Use Committee of Tzu Chi University (Approval ID: 98104) and the National Defense Medical Center (Approval ID: AN-100-04).

Toxins and spores

B. anthracis-derived LT was purified according to previously described procedures [49]. LT was delivered in a 1:5 ratio of LF and PA [16]. Spores derived from the *B. anthracis*-nonencapsulated mutant strain (pXO1$^+$, pXO2$^-$) were purchased from the American Type Culture Collection (Manassas, VA, USA) (ATCC 14186).

Erythroid colony-forming cell assay

The erythroid colony-forming cell assay was conducted according to the manufacturer's instructions (MethoCult M3334, StemCell Technologies). For the *in vitro* erythroid colony-forming cell assay, bone marrow cells were collected from the femurs and tibiae of C57BL/6J mice. C57BL/6J mice (males, 8–10 wk of age) were obtained from the National Laboratory Animal Center (Taipei, Taiwan) and kept in a specific pathogen-free (SPF) environment in the experimental animal center of Tzu Chi University. For the *ex vivo* erythroid colony-forming cell assay, C57BL/6J mice were retro-orbitally injected with 55 μg/kg/d of recombinant human G-CSF (Filgrastim, Kirin, Tokyo, Japan) in 250 μl saline, once daily for 5 d, initiated 5 d before the challenges of a lethal dose of LT (1.5 mg/kg in 250 μl saline, retro-orbitally injected). Treatments using saline, G-CSF, and LT alone were used as comparison controls. Bone marrow cells were collected at 69 h after LT treatment and flushed with Roswell Park Memorial Institute medium (RPMI)-1640 containing 20% anticoagulant acid

citrate dextrose formula A (ACD-A: 38 mM citric acid, 75 mM trisodium citrate, 139 mM D-glucose, 12.5 mM EDTA [15]). After depleting RBCs by adding a hypotonic buffer (153 mM NH$_4$Cl and 17 mM Tris-HCl) at room temperature for 10 min, 100 μl of remaining cells (9×10^5/ml) were resuspended in Iscove's Modified Dulbecco's Medium (IMDM) (StemCell Technologies) and mixed with 1 ml of semisolid methylcellulose-based medium containing 3 units of EPO. Finally, each 1.1 ml of methylcellulose-cell suspension was mixed with or without a dose of G-CSF (20 ng/ml or 764 ng/ml) and duplicate seeded in 35-mm dishes. A G-CSF dose of 20 ng/ml was used in the colony-forming cell assay for hematopoietic cells [50]. Because the volume of mice blood is 70–80 ml/kg [51], a dose of 764 ng/ml approximated the dose used to ameliorate LT-induced mortality in the experiments. Two doses of G-CSF (20 ng/ml and 764 ng/ml) were added to the medium supplement of the erythroid colony-forming cell assay. The cultures were incubated at 37°C for 14 d. Dynamic changes in colony number were measured on Days 3, 7, and 14 after initiating the colony assay. The erythroid colonies were separated into 3 groups by size: small (8–50 cells), medium (more than 50, but less than 200 cells), and large (more than 200 cells).

Analysis of erythropoiesis in bone marrow

Bone marrow cells were purified and blocked with 5% bovine serum albumin in RPMI medium at 37°C for 1 h and incubated in 500 μl RPMI-1640 medium with 1 μl of fluorescein (FITC)-conjugated rat anti-mouse CD71 antibody (BioLegend) and 3 μl of R-Phycoerythrin (R-PE)-conjugated rat anti-mouse TER-119 antibody (BD Immunocytometry System) at 37°C for 1 h. After washing with phosphate-buffered saline (PBS), the cells were measured and analyzed using a FACSCalibur flow cytometer and the CellQuest^{TM} Pro program (Becton-Dickinson).

Flow cytometry analysis of peripheral blood cells of G-CSF-treated EGFP mice

EGFP mice [C57BL/6J-Tg (Pgk1-EGFP) 03Narl, males, 10–12 wk of age] were obtained from the National Laboratory Animal Center (Taipei, Taiwan) and maintained in the aforementioned SPF environments. EGFP mice were retro-orbitally injected with G-CSF (55 μg/kg/d in 250 μl saline) once daily for 4 d. To detect the erythrocytes' specific surface markers, 50 μl of retro-orbital blood samples were obtained 22, 44, 66, and 94 h after the initial G-CSF injection, and subsequently mixed with 450 μl of anticoagulant ACD-A (1:9). Cells were incubated in 300 μl of RPMI-1640 medium with 3 μl of the R-Phycoerythrin (R-PE)-conjugated rat anti-mouse TER-119 antibody (BD Immunocytometry System) at 37°C for 1 h. After washing with PBS, the cells were analyzed using a FACSCalibur flow cytometer and the CellQuest^{TM} Pro program.

G-CSF treatment to reduce LT-induced mortality

C57BL/6J mice (males, 8–10 wk of age) were retro-orbitally injected with 55 μg/kg/d of recombinant human G-CSF in 250 μl saline, once daily for 5 d, initiated 5 d before or 1 d after the challenges of a lethal dose of LT (1.5 mg/kg in 250 μl saline, retro-orbitally injected). Treatments using saline, G-CSF, and LT alone were used as comparison controls. Because no suitable potential predictor of death/survival exists for LT-challenged mice, we used death as an endpoint for the survival experiment. The survival time and mortality of mice were recorded after the LT challenges. LT treatment in mice did not induce obvious discomfort and body weight loss, except for reducing activities. The experimental mice were continually monitored up to 250 h

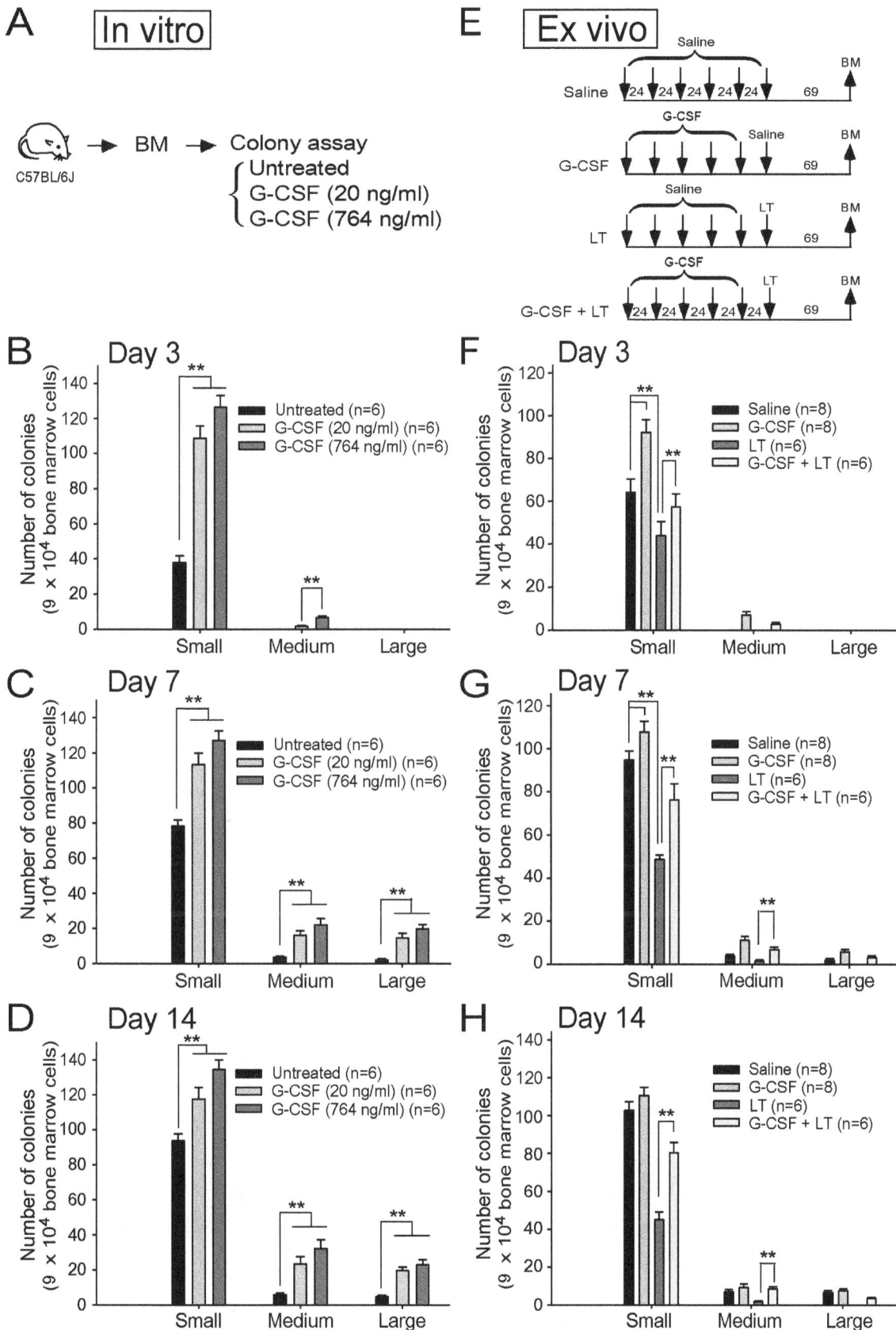

Figure 1. Erythroid colony-forming cell assays to measure the effect of G-CSF on erythropoiesis. The experimental outlines of *in vitro* (A) and *ex vivo* (E) analyses are shown. An *in vitro* assay was performed using murine bone marrow (BM) cells that were incubated with [20 ng/ml (n = 6) or 764 ng/ml (n = 6)], or without G-CSF (n = 6). The colonies were quantified on Days 3 (B), 7 (C), and 14 (D). Untreated bone marrow cells were used

as a control. Colony numbers of bone marrow cells from mice, which were treated with G-CSF (n = 8), LT (n = 6), or G-CSF and LT (n = 6), were measured on Days 3 (F), 7 (G), and 14 (H) following the *ex vivo* colony assay. Bone marrow cells from saline treated mice (n = 8) served as controls. **$P<0.01$ was compared between the indicated groups. Data are shown as mean ± standard deviation (SD) and represent results from 2 independent experiments. The mouse drawing used in this and all following figures was originally published in *Blood*. Huang, H. S., Sun, D. S., Lien, T. S. and Chang, H. H. Dendritic cells modulate platelet activity in IVIg-mediated amelioration of ITP in mice. *Blood*. 2010; 116: 5002–5009. © the American Society of Hematology.

for every 4–6 h. All surviving mice were monitored each day for 2 subsequent mo. For hematopoietic parameters, 50 μl of retro-orbital blood samples were collected at 22, 44, and 66 h after LT challenges and analyzed by an automated hematology analyzer (KX-21, Sysmex Corporation).

Combined treatments with G-CSF and EPO to reduce anthrax-spore-induced mortality

C57BL/6J mice were retro-orbitally injected with G-CSF (55 μg/kg/d in 250 μl saline) daily for 5 consecutive d or injected

with a combination of recombinant human EPO (rhEPO, Neorecormon, Roche, Mannheim, Germany) (2 IU/g, in 250 μl saline) twice at 24 and 48 h after injecting spores (1×10^7 in 1 ml saline, intraperitoneal injection). The survival times and mortality of mice were recorded. Because no suitable potential predictor of death/survival exists for spore-challenged mice, we used death as an endpoint for the survival experiment. Spore treatment in mice did not induce obvious discomfort and body weight loss, except for reducing activities. The experimental mice were continually

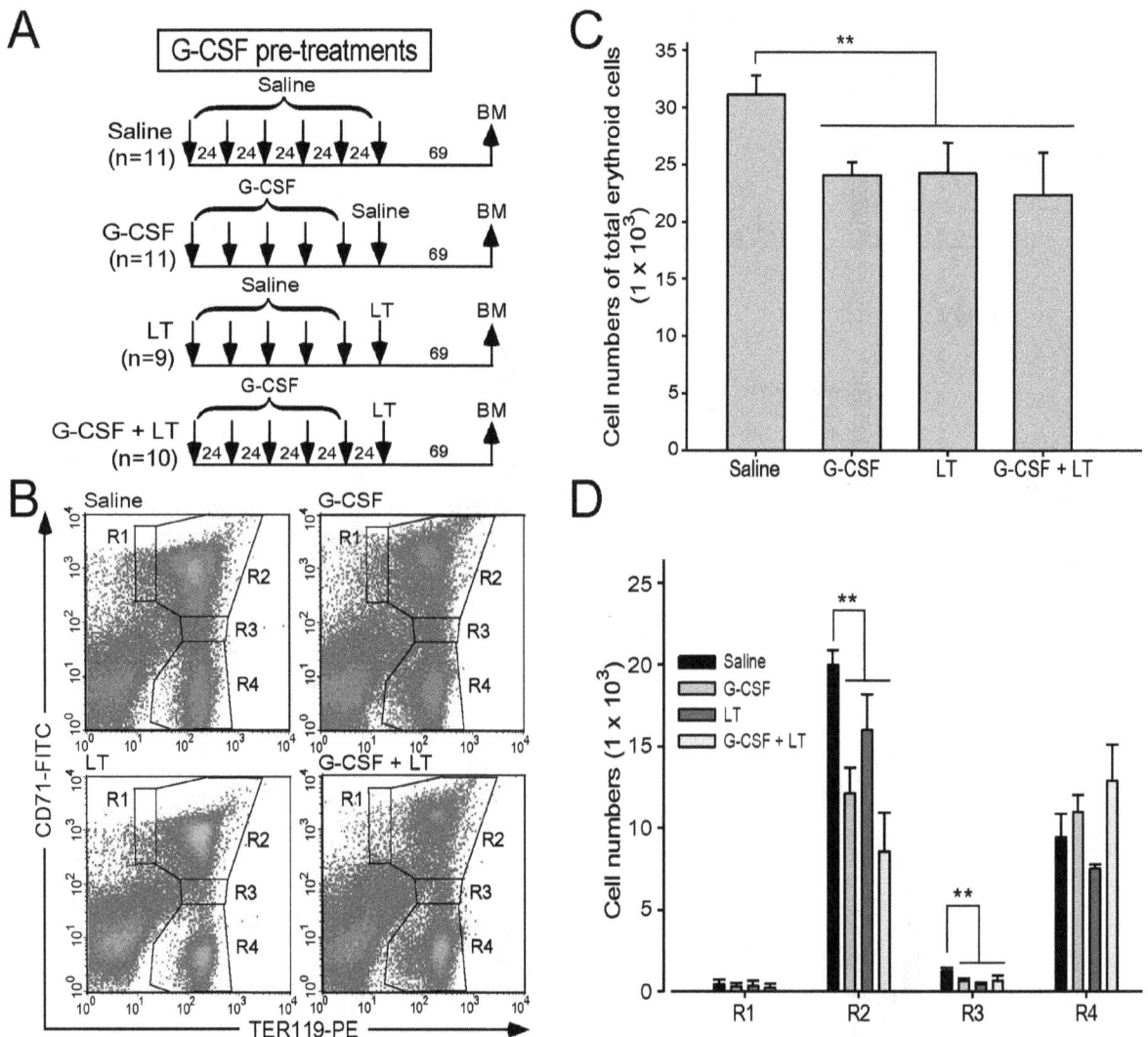

Figure 2. Regulation of G-CSF on bone marrow erythroblast populations. The experimental outlines are illustrated (A). Mice were treated with saline (n = 11), G-CSF (n = 11), LT (n = 9), or G-CSF plus LT (n = 10). Flow cytometry analysis analyzed erythroblast populations of BM cells at 69 h after LT challenges. The erythroblast cells were gated as R1 (CD71[high], TER-119[med]), R2 (CD71[high], TER-119[high]), R3 (CD71[med], TER-119[high]), and R4 (CD71[low], TER-119[high]) in all groups (B) as described [25]. The cell numbers of all erythroblast cells (sum of R1 to R4) (C) and individual erythroblast (R1, R2, R3, and R4) (D) in each group were quantified. **$P<0.01$ was compared between indicated groups. Data are showed as mean ± SD and represent the results from 2 independent experiments.

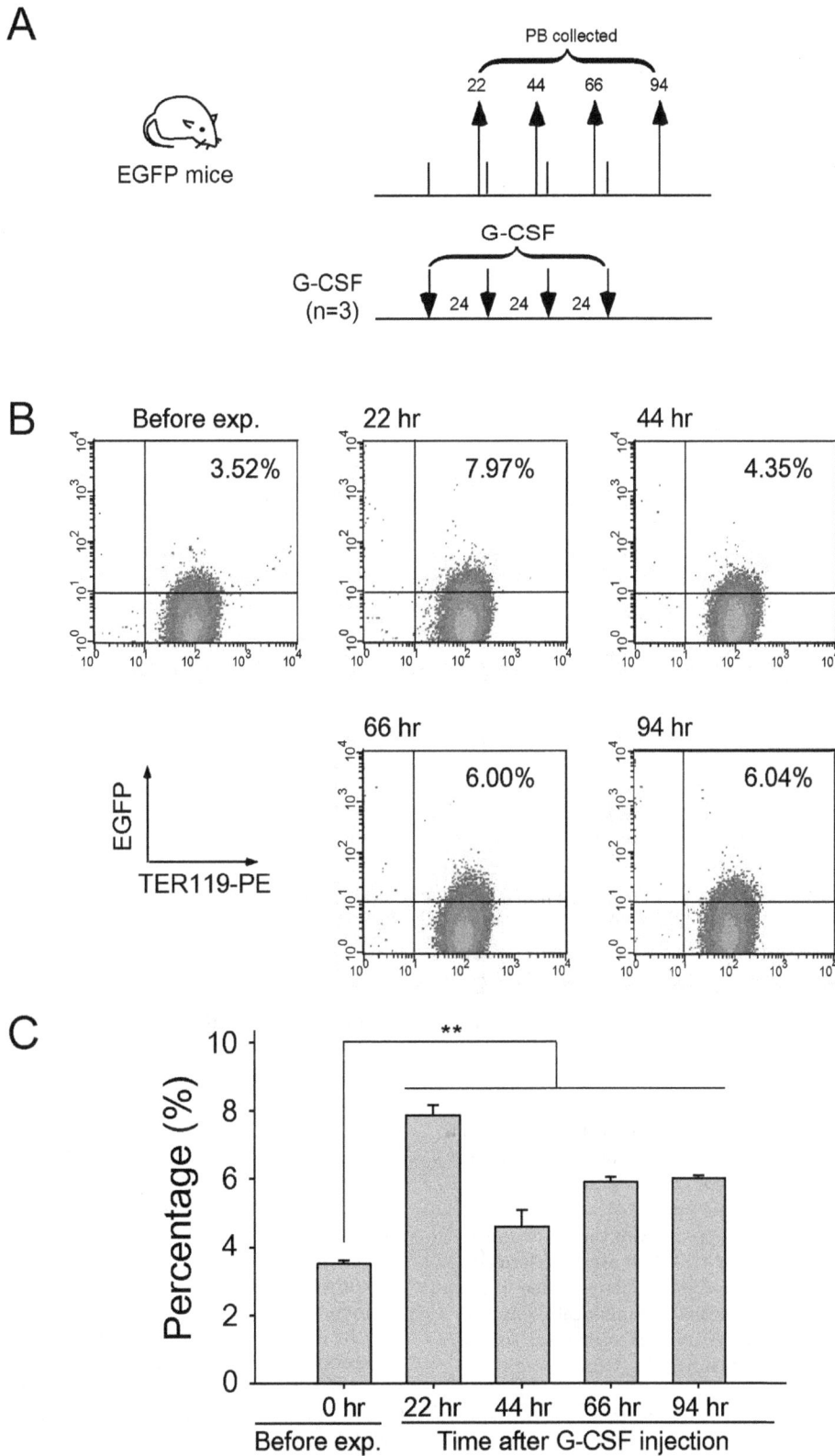

Figure 3. Mobilization of newly synthesized erythrocytes into peripheral blood by G-CSF. The experimental outline is illustrated (A). The EGFP mice were injected with G-CSF (n = 3). The percentage of EGFP$^+$/TER119$^+$ cells in peripheral blood (PB) was analyzed by flow cytometry (B) and quantified (C) at 22, 44, 66, and 94 h after G-CSF injection. PB collected from mice before G-CSF injection served as the negative control. $**P<0.01$ was compared to the negative control. Data are shown as mean $\pm$ SD.

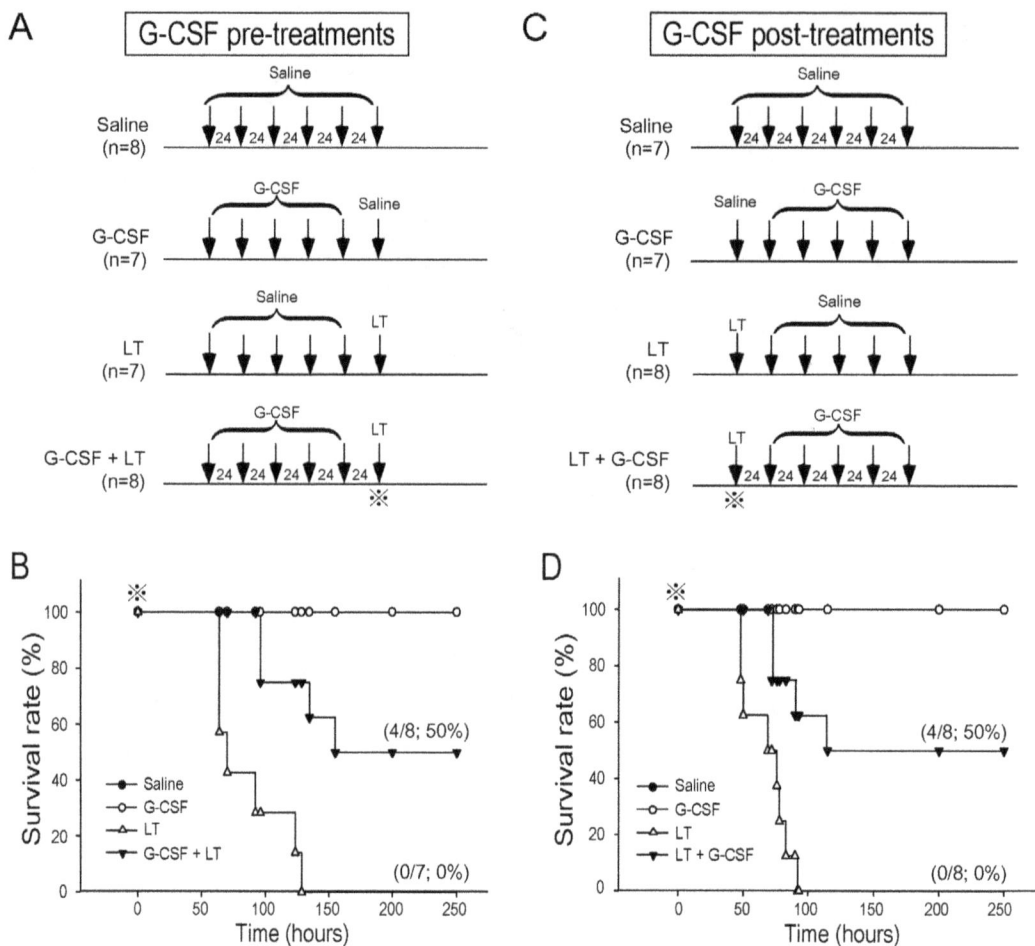

Figure 4. G-CSF treatments ameliorated LT-elicited mortality in mice. The experimental timetable (A), (C), and the survival rates of mice pre-treated (B) and post-treated (D) with G-CSF, LT, or G-CSF and LT are indicated. Saline treated mice served as negative controls. The symbol (※) in (A) to (D) indicates the onset time point for recording survival rates.

monitored up to 15 d for every 4–6 h. All surviving mice were monitored each day for 2 subsequent mo.

Statistics

All results are presented as the mean ± SD (standard deviation) for each group. Data significance was examined by one-way ANOVA followed by the post-hoc Bonferroni-corrected t-test. Univariate Kaplan-Meier analysis was used to compare the difference in survival rate between groups with various treatments. P-values were calculated and log-rank tests were performed to determine statistical significance. A probability of type 1 error $\alpha = 0.05$ was recognized as the threshold of statistical significance. Statistical analysis was conducted using the statistical software SPSS, version 17.0 (SPSS Inc., Chicago, IL, USA).

Results

G-CSF treatment promoted erythrocytic differentiation and proliferation *in vitro* and *ex vivo*

To elucidate the role of G-CSF on erythrocytic differentiation and proliferation, an *in vitro* erythroid colony-forming cell assay was performed to quantify BFU-Es and CFU-Es. Control groups without using filgrastim G-CSF supplements formed only medium-sized colonies by Day 7 (Figure 1C). By contrast, G-CSF

treatments accelerated the formation of medium-sized colonies, which appeared earlier by Day 3 (Figure 1B). Following G-CSF treatment, the numbers of colonies in all colony sizes were greater than those in the untreated control groups (Figure 1B–1D). Based on the traditional concept that G-CSF primarily regulates granulopoiesis [28], those colonies to be affected by G-CSF treatment may not be exclusively erythroid-origin cells. Consequently, 3, 3'-diaminobenzidine tetrahydrochloride (DAB) [37] was used to identify erythroid colonies, and the pseudoperoxidase activity of erythroid cells was stained on Day 14. Compared with the untreated groups, the number of DAB$^+$ colonies was greater in G-CSF-treated groups (Figure S1). This data indicated that G-CSF treatment enhances the proliferation and differentiation of erythrocytes *in vitro*. Further experiments were performed to investigate the effect of G-CSF and LT treatments *ex vivo* (Figure 1E). The number of erythroid colonies sharply decreased with LT treatment on Days 3, 7 and 14 (Figure 1F–1H). Medium-sized colonies first appeared on Day 3 in G-CSF treated groups (Figure 1F), compared with saline and LT-treated groups, whereas medium-sized colonies only appeared on Day 7 (Figure 1G). In addition, G-CSF treatments ameliorated LT-induced suppression on erythropoiesis in the *ex vivo* erythroid colony-forming cell assay (Figure 1F–1H). These results indicated that G-CSF promoted erythrocytic proliferation and differentiation *in vitro* and *ex vivo*

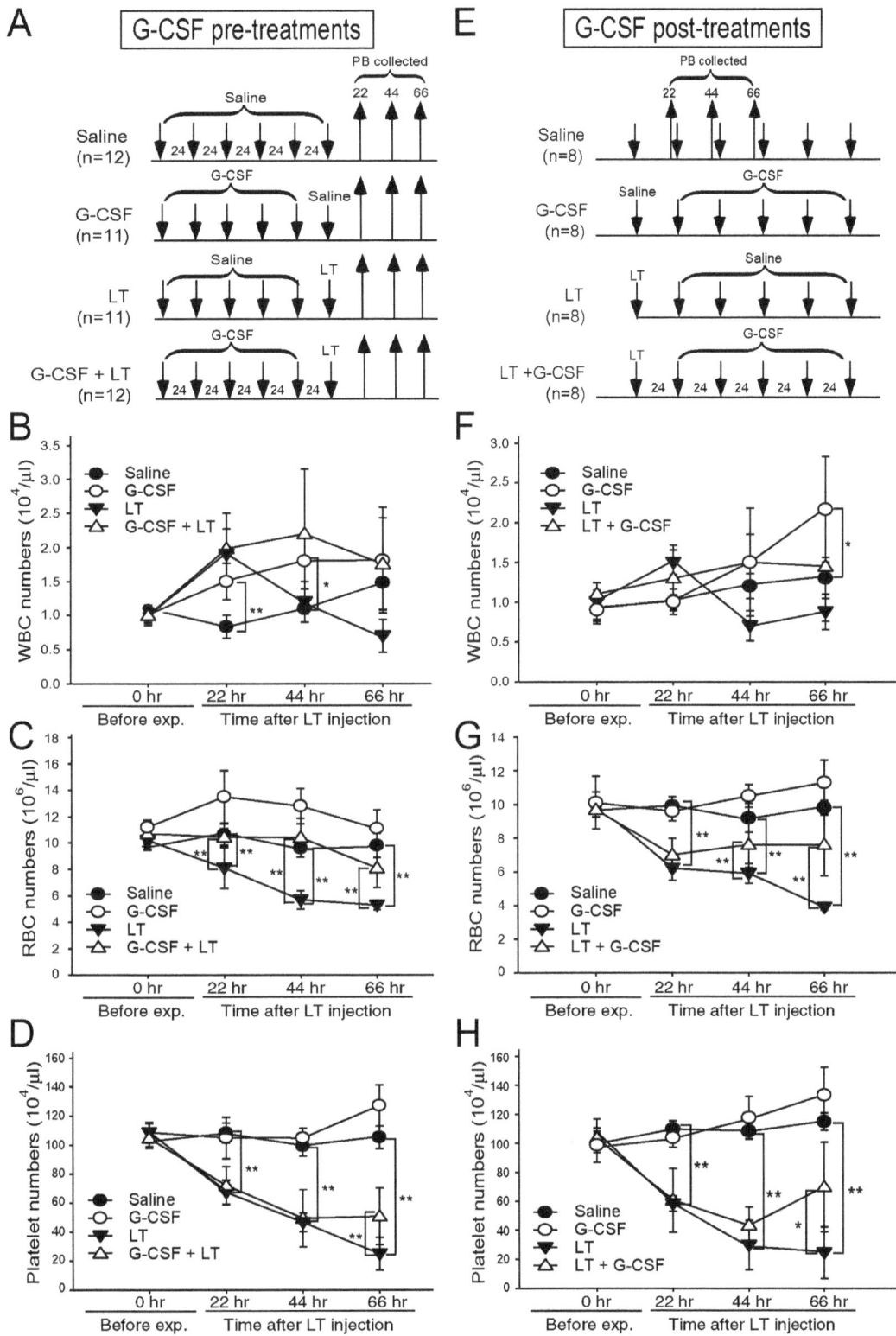

Figure 5. Amelioration of LT-induced erythrocytopenic response by G-CSF. The experimental outlines are indicated (A), (E). Mice were treated with saline (n = 12), G-CSF (n = 11), LT (n = 11), and G-CSF plus LT (n = 12) before (A) or saline (n = 8), G-CSF (n = 8), LT (n = 8), and LT plus G-CSF (n = 8) after (E) the LT challenges; their WBC, RBC, and platelet counts were subsequently analyzed at 22, 44, and 66 h after the LT challenges. Saline-treated mice were used as negative controls. *$P < 0.05$, **$P < 0.01$ were compared between the indicated groups. Data are shown as mean ± SD and represent the results from 2 independent experiments.

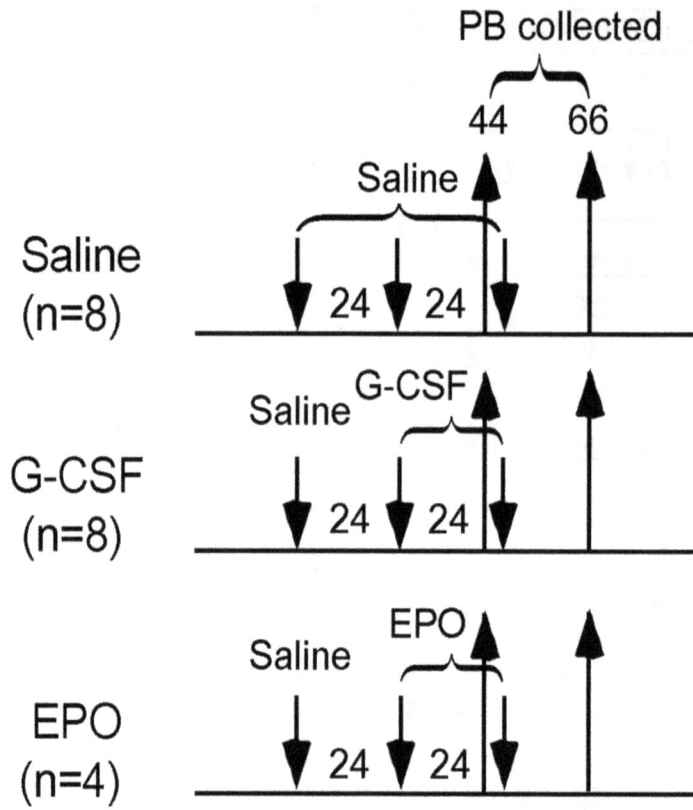

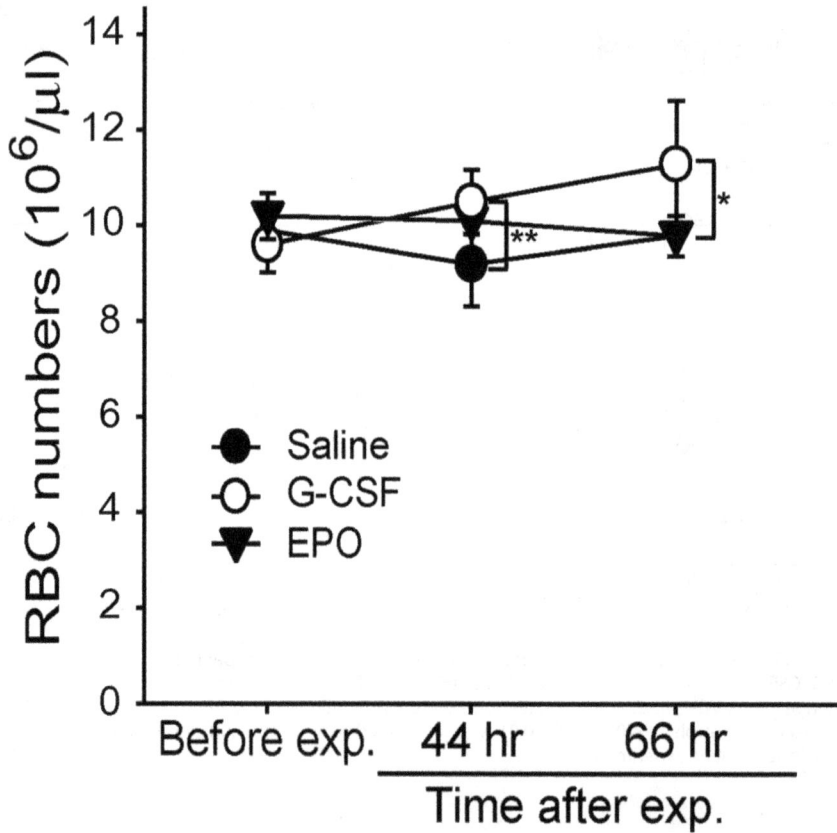

Figure 6. Fast mobilization of erythrocytes into peripheral blood by G-CSF versus EPO treatments. Experimental outline for measuring the PB-RBC counts of mice treated with G-CSF (n = 8) or EPO (n = 4) for 2 consecutive d (A). The PB RBC counts were measured before the experiments and at 44 and 66 h after the first saline injection (B). Data are shown as mean ± SD and represent the results from 2 independent experiments. Saline treated groups (n = 8) were used as the negative control. *P<0.05, **P<0.01 were compared to the negative control.

and ameliorated LT-induced erythropoiesis suppression in the *ex vivo* erythroid colony-forming cell assay.

G-CSF treatment promoted mobilization of newly synthesized RBC to peripheral blood

After the promising analyses *in vitro* and *ex vivo*, this study investigated the erythropoietic progression *in vivo*. Our previous report revealed that LT suppressed erythropoiesis in bone marrow [15]. Following similar approaches [15,25], we used surface expression of CD71 and TER-119 to verify the maturation status of various bone marrow erythroblasts under the G-CSF treatments with or without anthrax LT challenges. Although we found that G-CSF treatment rescued LT-induced erythrocytopenia (please see the following section), G-CSF pre-treatments could not overcome LT-mediated suppression on the cell numbers of both total erythroblast and individual subpopulations of erythroblast (R1 to R4 populations) (Figure 2C and 2D). This prompted us to verify whether G-CSF could mobilize mature erythrocytes into peripheral blood; we employed C57BL/6J mice with the whole-body-expressing enhanced-green-fluorescence-protein (EGFP) transgene. Prior to G-CSF analyses, we found that only a small fraction of EGFP+ RBCs was detectable in the peripheral blood of normal control groups (Figure 3, 3.5% cells, before exp. groups). This is likely because mature erythrocytes do not have a nucleus, and that newly differentiated RBCs, rather than aged RBCs, express detectable EGFP. We employed an acute hemorrhage model, in which 35% of total blood was removed, to provoke the natural induction of erythropoiesis to investigate whether newly synthesized erythrocytes contain additional fluorescence. Our data revealed that EGFP+/TER-119+ erythrocytes increased consistently by Days 2, 4, and 6 after acute anemia (Figure S2). This suggested that EGFP+/TER-119+ cells are newly synthesized erythrocytes. Using the same strategy, analyses revealed that G-CSF treatment can mobilize newly synthesized erythrocytes to peripheral blood in mice (Figure 3, beginning at 22 h after G-CSF injections).

G-CSF treatment reduced LT-mediated mortality, erythrocytopenia, and thrombocytopenia

To investigate the ameliorative effect of G-CSF on LT, C57BL/6J mice were treated with G-CSF according to the manufacturer's instructions (once daily for 5 d). Treatments of G-CSF were initiated 5 d before (Figure 4A) or 1 d after the challenges of a lethal dose of LT (Figure 4C). LT initiated mortality within 48 to 129 h (Figure 4B and 4D, LT groups). Administration of 5 doses of G-CSF before (Figure 4A) and after (Figure 4C) the LT challenges significantly improved survival rates (Figure 4B and 4D) (P<0.01). Treatments using saline, G-CSF, and LT alone served as the controls (Figure 4B and 4D). The peripheral white blood cell (WBC) counts of the G-CSF-treated groups increased approximately 2-fold at 22 and 44 h (Figure 5A and 5B), as well as at 66 h (Figure 5E and 5F) following the implementation of differing G-CSF regimens; this is in consistent with a previous G-CSF report [38]. Notably, both G-CSF treatments significantly ameliorated LT-induced erythrocytopenia (Figure 5C, G-CSF + LT vs. LT; Figure 5G, LT + G-CSF vs. LT). Compared with RBC counts, the ameliorative effect of G-CSF on LT-induced thrombocytopenia was somewhat later and was observed at

approximately 66 h after LT treatment (Figure 5D and 5H). These results indicated that G-CSF positively regulated both RBC and platelet counts.

G-CSF treatment induced erythrocytes to mobilize into peripheral blood faster than EPO

Our previous study showed that EPO up-regulated RBC counts in peripheral blood [15]. To compare the EPO and G-CSF treatments in their efficiency at increasing RBC counts, mice were injected with 2 doses of either G-CSF or EPO. The circulating RBC counts were measured (Figure 6). Our data revealed that G-CSF induced a faster increase of RBC counts, within 20 h of the first G-CSF administration, than the EPO treatment, in which no increased RBC counts were observed (Figure 6B, G-CSF vs. EPO). These results suggested that G-CSF induced a faster mobilization of erythrocytes into peripheral blood than that of EPO.

Combined G-CSF and EPO had an ameliorative effect on anthrax-spore-induced mortality in C57BL/6J mice

Our previous report suggested that EPO ameliorates LT-mediated erythrocytopenia by enhancing erythropoiesis [15]. Because G-CSF increases the erythrocyte supply through a diverse mechanism by enhancing the mobilization of erythrocytes into peripheral blood, these results prompted us to investigate whether combined treatments using G-CSF and EPO may be more effective than respective single treatments alone. The analysis indicated that EPO treatment did not exert a protective effect on anthrax-spore-challenged mice (Figure 7B, Spore + EPO vs. Spore only). This is consistent with another line of evidence; the survival rates of anthrax LT-challenged mice increased only 25% following EPO post-treatment (Figure S3, LT + EPO vs. LT, P = 0.101). By contrast, G-CSF treatments with or without EPO effectively increased the survival rate of anthrax spore-challenged mice from 18.75% to 37.5% (Figure 7, Spore + G-CSF vs. Spore only). Post-treatments combining G-CSF and EPO prolonged the survival period of anthrax-spore-challenged mice (Figure 7, Spore + G-CSF + EPO vs. Spore + G-CSF). Statistical analysis revealed that the P value is marginally significant (Figure 7, P = 0.094, Spore + G-CSF + EPO vs. Spore; P = 0.088, Spore + G-CSF + EPO vs. Spore + EPO). These results suggested that combined treatment using G-CSF and EPO tended to ameliorate anthrax-spore-induced mortality in mice.

Discussion

This study demonstrated that G-CSF, a stimulating factor for granulopoiesis, enhanced erythrocytic mobilization, by which it enhanced RBC counts in peripheral blood. This likely therefore rescued anthrax LT-induced anemia and mortality in mice.

Our *in vitro* (Figure 1A–1D) and *ex vivo* (Figure 1E–1H) evidence suggested that G-CSF may promote erythropoiesis. The *in vivo* analyses revealed that G-CSF ameliorated LT-induced erythrocytopenia in peripheral blood (Figure 5C and 5G), but did not increase erythroblast cell numbers in bone marrow (Figure 2). Therefore, the effects of G-CSF on erythropoiesis *in vivo* require clarification. One study demonstrated that G-CSF had a negative effect on bone marrow erythropoiesis in mice [39].

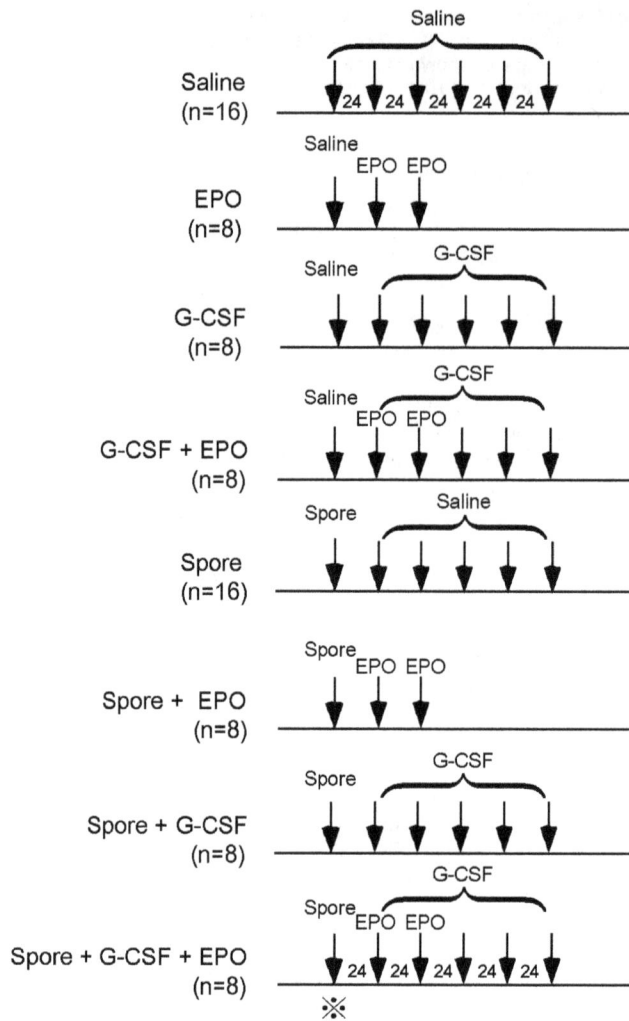

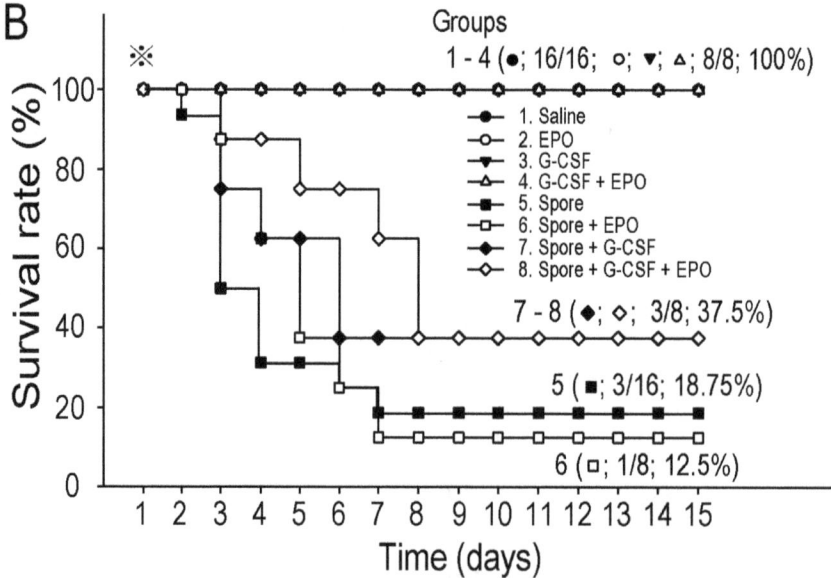

Figure 7. Post-treatments of G-CSF and EPO tended to ameliorate anthrax spore elicited-lethality in mice. Experimental outlines (A) and the survival rates of mice treated with G-CSF (n = 8), EPO (n = 8), and G-CSF combined with EPO (n = 8) after anthrax spore injection (B). Saline (n = 16), EPO (n = 8), G-CSF (n = 8), G-CSF and EPO (n = 8), and anthrax spore (n = 16) injected groups were used for comparisons. The symbol (※) in (A) and (B) indicates the onset time point for recording survival rates. Data represent the results from 2 independent experiments.

However, another study demonstrated that G-CSF treatments in humans increased immature reticulocytes in peripheral blood [40]. Clinical observations also found that the reticulocyte fraction, an assessment of immature erythroid cells in peripheral blood, was an early surrogate marker for the rise of CD34$^+$ hematopoietic stem cells during G-CSF mobilization [41]. This evidence suggests that G-CSF is involved in regulating erythroid precursor cells in humans.

G-CSF induced fast mobilization of RBCs to peripheral blood within 20 h of the first G-CSF administration, compared with the EPO treatment (Figure 6). This data is consistent with the EGFP mice experiment, regarding the time in which G-CSF treatment increased the percentage of newly synthesized erythrocytes in peripheral blood (Figure 3). Thus, a single post-treatment dose of G-CSF was sufficient to save approximately 40% of mice within 24 h (Figure 4D, 100% and 60% survival rates in LT + G-CSF and LT, 48 h groups, respectively). The advantage of fast mobilization of RBCs of G-CSF can also be confirmed by the higher survival rate of G-CSF compared with EPO in LT-induced (Figure 4D vs. Figure S3B) and LT-spore-induced (Figure 7B) mortality in mice.

G-CSF is a multi-function cytokine that stimulates the proliferation and differentiation of myeloid precursors and modulates mature cells [29]. G-CSF is also used to prevent or shorten neutropenia in chemotherapy-induced or primary congenital neutropenia [42], and mobilize hematopoietic stem cells to peripheral blood for transplantation [43]. Probably because of the anti-apoptotic and anti-inflammatory effects, G-CSF has also been used to treat nonhematopoietic targets including cerebral ischemia [44], spinal cord ischemia [45], infarct heart [46], and end stage liver disease [47]. Consequently, the effects of G-CSF on other cell types (e.g., macrophage, lymphocyte, endothelial, dendritic cells, cardiomyocytes, and smooth muscle) may not be completely excluded from the rescue mechanism of G-CSF. Although the mechanism is unclear, combination therapy using G-CSF and EPO has been used to treat myelodysplastic anemia in clinical settings [32–36,48]. This suggests that the medical community has empirically recognized the RBC-enhancing effect of the combination treatment using G-CSF and EPO. One critical aspect of G-CSF is the rapid induction of peripheral RBCs, a property superior to EPO, which may be applied to anthrax or other diseases with urgent RBC and oxygen demands. Further detailed and well-designed clinical studies are required to explore the therapeutic potential of combination treatment using G-CSF and EPO.

In this study, we demonstrated that G-CSF mobilized erythrocytes into peripheral blood. In addition, combined treatments of G-CSF and EPO tended to ameliorate anthrax spore-induced mortality. An optimized rescue protocol may provide new perspectives and assist the development of a feasible therapeutic strategy against anthrax.

Supporting Information

Figure S1 G-CSF treatments enhanced bone marrow erythroid colony numbers. An *in vitro* erythroid colony-forming cell assay was performed using murine bone marrow (BM) cells incubated with [20 ng/ml (n = 6) or 764 ng/ml (n = 6)] or without G-CSF (n = 6). The erythroid colonies were confirmed by 3, 3′-diaminobenzidine tetrahydrochloride (DAB) staining (A) and quantified on Day 14 (B). Untreated BM cells were used as the control. **$P<0.01$ was compared to the untreated groups. Scale bar: 500 μm. Data are shown as mean ± SD.

Figure S2 Mobilization of newly synthesized erythrocytes (EGFP$^+$/TER-119$^+$) into peripheral blood by acute anemia in EGFP transgenic mice. Experimental timetable used in acute anima assay (A). During acute anima (after aspirating 35% of total blood), the population of EGFP$^+$/TER-119$^+$ cells in peripheral blood (PB) of EGFP mice was gated as R1 and quantified. The total cell number was defined as 100%. The percentage of R1 was analyzed by flow cytometry (B) on Day 2, 4, and 6 after the removal of 35% of total blood. PB samples from mice before acute anemia were used as negative controls. Data were collected from 2 representative EGFP mice.

Figure S3 Post-treatments of EPO increased the survival rates of LT-challenged mice. Experimental timetable (A). The survival rates of mice challenged with EPO (n = 4), LT (n = 4), or LT plus EPO (n = 4) are shown (B). Saline-treated mice were used as controls (n = 4). C57BL/6J mice were retro-orbitally injected with recombinant human EPO (rhEPO, Neorecormon, Roche, Mannheim, Germany) (2 IU/g, in 250 μl saline) twice every 24 h after the challenges of a lethal dose of LT (1.5 mg/kg in 250 μl saline, retro-orbitally injected). The symbol (✳) in (A) and (B) indicates the onset time point for recording the survival rates of mice.

Acknowledgments

We wish to thank Professor Yu MS for providing 3, 3′-diaminobenzidine tetrahydrochloride (DAB) tablets. We also wish to thank Professor Wang MH and his team, the Experimental Animal Center of Tzu-Chi University for animal care.

Author Contributions

Conceived and designed the experiments: HHC DSS. Performed the experiments: YWC TKL GLL YYL JHK HHH HLH. Analyzed the data: HHC JHW DSS. Contributed reagents/materials/analysis tools: JHK HHH HLH JHW. Wrote the paper: HHC DSS.

References

1. Mock M, Fouet A (2001) Anthrax. Annu Rev Microbiol 55: 647–671.
2. Brossier F, Mock M (2001) Toxins of Bacillus anthracis. Toxicon 39: 1747–1755.
3. Collier RJ, Young JA (2003) Anthrax toxin. Annu Rev Cell Dev Biol 19: 45–70.
4. Mourez M (2004) Anthrax toxins. Rev Physiol Biochem Pharmacol 152: 135–164.
5. Bardwell AJ, Abdollahi M, Bardwell L (2004) Anthrax lethal factor-cleavage products of MAPK (mitogen-activated protein kinase) kinases exhibit reduced binding to their cognate MAPKs. Biochem J 378: 569–577.
6. Hagemann C, Blank JL (2001) The ups and downs of MEK kinase interactions. Cell Signal 13: 863–875.
7. Wada T, Penninger JM (2004) Mitogen-activated protein kinases in apoptosis regulation. Oncogene 23: 2838–2849.
8. Raman M, Chen W, Cobb MH (2007) Differential regulation and properties of MAPKs. Oncogene 26: 3100–3112.
9. Moayeri M, Leppla SH (2004) The roles of anthrax toxin in pathogenesis. Curr Opin Microbiol 7: 19–24.
10. Kau JH, Sun DS, Huang HS, Lien TS, Huang HH, et al. (2010) Sublethal doses of anthrax lethal toxin on the suppression of macrophage phagocytosis. PLoS One 5: e14289.
11. Muehlbauer SM, Evering TH, Bonuccelli G, Squires RC, Ashton AW, et al. (2007) Anthrax lethal toxin kills macrophages in a strain-specific manner by apoptosis or caspase-1-mediated necrosis. Cell Cycle 6: 758–766.
12. Alileche A, Serfass ER, Muehlbauer SM, Porcelli SA, Brojatsch J (2005) Anthrax lethal toxin-mediated killing of human and murine dendritic cells impairs the adaptive immune response. PLoS Pathog 1: e19.

13. Comer JE, Chopra AK, Peterson JW, Konig R (2005) Direct inhibition of T-lymphocyte activation by anthrax toxins in vivo. Infect Immun 73: 8275–8281.

14. Fang H, Xu L, Chen TY, Cyr JM, Frucht DM (2006) Anthrax lethal toxin has direct and potent inhibitory effects on B cell proliferation and immunoglobulin production. J Immunol 176: 6155–6161.

15. Chang HH, Wang TP, Chen PK, Lin YY, Liao CH, et al. (2013) Erythropoiesis suppression is associated with anthrax lethal toxin-mediated pathogenic progression. PLoS One 8: e71718.

16. Chen PK, Chang HH, Lin GL, Wang TP, Lai YL, et al. (2013) Suppressive effects of anthrax lethal toxin on megakaryopoiesis. PLoS One 8: e59512.

17. Liu S, Zhang Y, Moayeri M, Liu J, Crown D, et al. (2013) Key tissue targets responsible for anthrax-toxin-induced lethality. Nature 501: 63–68.

18. Kassam A, Der SD, Mogridge J (2005) Differentiation of human monocytic cell lines confers susceptibility to Bacillus anthracis lethal toxin. Cell Microbiol 7: 281–292.

19. Kau JH, Sun DS, Tsai WJ, Shyu HF, Huang HH, et al. (2005) Antiplatelet activities of anthrax lethal toxin are associated with suppressed p42/44 and p38 mitogen-activated protein kinase pathways in the platelets. J Infect Dis 192: 1465–1474.

20. Moayeri M, Haines D, Young HA, Leppla SH (2003) Bacillus anthracis lethal toxin induces TNF-alpha-independent hypoxia-mediated toxicity in mice. J Clin Invest 112: 670–682.

21. Rainey GJ, Young JA (2004) Antitoxins: novel strategies to target agents of bioterrorism. Nat Rev Microbiol 2: 721–726.

22. Wadhwa M, Thorpe R (2008) Haematopoietic growth factors and their therapeutic use. Thromb Haemost 99: 863–873.

23. Debili N, Wendling F, Katz A, Guichard J, Breton-Gorius J, et al. (1995) The Mpl-ligand or thrombopoietin or megakaryocyte growth and differentiative factor has both direct proliferative and differentiative activities on human megakaryocyte progenitors. Blood 86: 2516–2525.

24. Fisher JW (2003) Erythropoietin: physiology and pharmacology update. Exp Biol Med (Maywood) 228: 1–14.

25. Socolovsky M, Nam H, Fleming MD, Haase VH, Brugnara C, et al. (2001) Ineffective erythropoiesis in Stat5a(−/−)5b(−/−) mice due to decreased survival of early erythroblasts. Blood 98: 3261–3273.

26. Trowbridge IS, Lesley J, Schulte R (1982) Murine cell surface transferrin receptor: studies with an anti-receptor monoclonal antibody. J Cell Physiol 112: 403–410.

27. Kina T, Ikuta K, Takayama E, Wada K, Majumdar AS, et al. (2000) The monoclonal antibody TER-119 recognizes a molecule associated with glycophorin A and specifically marks the late stages of murine erythroid lineage pathways. Br J Haematol 109: 280–287.

28. Hubel K, Dale DC, Liles WC (2002) Therapeutic use of cytokines to modulate phagocyte function for the treatment of infectious diseases: current status of granulocyte colony-stimulating factor, granulocyte-macrophage colony-stimulating factor, macrophage colony-stimulating factor, and interferon-gamma. J Infect Dis 185: 1490–1501.

29. Morstyn G, Burgess AW (1988) Hemopoietic growth factors: a review. Cancer Res 48: 5624–5637.

30. Metcalf D (2008) Hematopoietic cytokines. Blood 111: 485–491.

31. Xiao BG, Lu CZ, Link H (2007) Cell biology and clinical promise of G-CSF: immunomodulation and neuroprotection. J Cell Mol Med 11: 1272–1290.

32. Bessho M, Hirashima K, Asano S, Ikeda Y, Ogawa N, et al. (1997) Treatment of the anemia of aplastic anemia patients with recombinant human erythropoietin in combination with granulocyte colony-stimulating factor: a multicenter randomized controlled study. Multicenter Study Group. Eur J Haematol 58: 265–272.

33. Greenberg PL, Sun Z, Miller KB, Bennett JM, Tallman MS, et al. (2009) Treatment of myelodysplastic syndrome patients with erythropoietin with or without granulocyte colony-stimulating factor: results of a prospective randomized phase 3 trial by the Eastern Cooperative Oncology Group (E1996). Blood 114: 2393–2400.

34. Hellstrom-Lindberg E, Gulbrandsen N, Lindberg G, Ahlgren T, Dahl IM, et al. (2003) A validated decision model for treating the anaemia of myelodysplastic syndromes with erythropoietin + granulocyte colony-stimulating factor: significant effects on quality of life. Br J Haematol 120: 1037–1046.

35. Hellstrom-Lindberg E, Negrin R, Stein R, Krantz S, Lindberg G, et al. (1997) Erythroid response to treatment with G-CSF plus erythropoietin for the anaemia of patients with myelodysplastic syndromes: proposal for a predictive model. Br J Haematol 99: 344–351.

36. Negrin RS, Stein R, Doherty K, Cornwell J, Vardiman J, et al. (1996) Maintenance treatment of the anemia of myelodysplastic syndromes with recombinant human granulocyte colony-stimulating factor and erythropoietin: evidence for in vivo synergy. Blood 87: 4076–4081.

37. Ogawa M, Parmley RT, Bank HL, Spicer SS (1976) Human marrow erythropoiesis in culture. I. Characterization of methylcellulose colony assay. Blood 48: 407–417.

38. Kikuta T, Shimazaki C, Ashihara E, Sudo Y, Hirai H, et al. (2000) Mobilization of hematopoietic primitive and committed progenitor cells into blood in mice by anti-vascular adhesion molecule-1 antibody alone or in combination with granulocyte colony-stimulating factor. Exp Hematol 28: 311–317.

39. Nijhof W, De Haan G, Dontje B, Loeffler M (1994) Effects of G-CSF on erythropoiesis. Ann N Y Acad Sci 718: 312–324; discussion 324–315.

40. Park K, Im T, Sasaki A, Yamane T, Nakao Y, et al. (1991) Positive effect of granulocyte-colony stimulating factor on erythropoiesis in humans. Osaka City Med J 37: 123–132.

41. Remacha AF, Martino R, Sureda A, Sarda MP, Sola C, et al. (1996) Changes in reticulocyte fractions during peripheral stem cell harvesting: role in monitoring stem cell collection. Bone Marrow Transplant 17: 163–168.

42. Page AV, Liles WC (2011) Colony-stimulating factors in the prevention and management of infectious diseases. Infect Dis Clin North Am 25: 803–817.

43. Rankin SM (2012) Chemokines and adult bone marrow stem cells. Immunol Lett 145: 47–54.

44. Abe K, Yamashita T, Takizawa S, Kuroda S, Kinouchi H, et al. (2012) Stem cell therapy for cerebral ischemia: from basic science to clinical applications. J Cereb Blood Flow Metab 32: 1317–1331.

45. Chen WF, Jean YH, Sung CS, Wu GJ, Huang SY, et al. (2008) Intrathecally injected granulocyte colony-stimulating factor produced neuroprotective effects in spinal cord ischemia via the mitogen-activated protein kinase and Akt pathways. Neuroscience 153: 31–43.

46. Baldo MP, Davel AP, Damas-Souza DM, Nicoletti-Carvalho JE, Bordin S, et al. (2011) The antiapoptotic effect of granulocyte colony-stimulating factor reduces infarct size and prevents heart failure development in rats. Cell Physiol Biochem 28: 33–40.

47. Gaia S, Smedile A, Omede P, Olivero A, Sanavio F, et al. (2006) Feasibility and safety of G-CSF administration to induce bone marrow-derived cells mobilization in patients with end stage liver disease. J Hepatol 45: 13–19.

48. Jadersten M, Montgomery SM, Dybedal I, Porwit-MacDonald A, Hellstrom-Lindberg E (2005) Long-term outcome of treatment of anemia in MDS with erythropoietin and G-CSF. Blood 106: 803–811.

49. Kau JH, Lin CG, Huang HH, Hsu HL, Chen KC, et al. (2002) Calyculin A sensitive protein phosphatase is required for Bacillus anthracis lethal toxin induced cytotoxicity. Curr Microbiol 44: 106–111.

50. Sarma NJ, Takeda A, Yaseen NR (2010) Colony forming cell (CFC) assay for human hematopoietic cells. J Vis Exp.

51. Harkness JE, Wagner JE (1995) Clinical procedures: Biology and Medicine of Rabbits and Rodents 93.

12

Sulfonamide-Resistant Bacteria and Their Resistance Genes in Soils Fertilized with Manures from Jiangsu Province, Southeastern China

Na Wang[1,2⅋], **Xiaohong Yang**[3⅋], **Shaojun Jiao**[2], **Jun Zhang**[3], **Boping Ye**[3*¶], **Shixiang Gao**[1*¶]

1 State Key Laboratory of Pollution Control and Resource Reuse, School of the Environment, Nanjing University, Nanjing, 210093, China, **2** Nanjing Institute of Environmental Science, Ministry of Environmental Protection of China, Nanjing, 210042, China, **3** School of Life Science and Technology, China Pharmaceutical University, Nanjing, 210009, China

Abstract

Antibiotic-resistant bacteria and genes are recognized as new environmental pollutants that warrant special concern. There were few reports on veterinary antibiotic-resistant bacteria and genes in China. This work systematically analyzed the prevalence and distribution of sulfonamide resistance genes in soils from the environments around poultry and livestock farms in Jiangsu Province, Southeastern China. The results showed that the animal manure application made the spread and abundance of antibiotic resistance genes (ARGs) increasingly in the soil. The frequency of sulfonamide resistance genes was *sul*1 > *sul*2 > *sul*3 in pig-manured soil DNA and *sul*2 > *sul*1 > *sul*3 in chicken-manured soil DNA. Further analysis suggested that the frequency distribution of the *sul* genes in the genomic DNA and plasmids of the SR isolates from manured soil was *sul*2 > *sul*1 > *sul*3 overall ($p<0.05$). The combination of *sul*1 and *sul*2 was the most frequent, and the co-existence of *sul*1 and *sul*3 was not found either in the genomic DNA or plasmids. The sample type, animal type and sampling time can influence the prevalence and distribution pattern of sulfonamide resistance genes. The present study also indicated that *Bacillus, Pseudomonas* and *Shigella* were the most prevalent *sul*-positive genera in the soil, suggesting a potential human health risk. The above results could be important in the evaluation of antibiotic-resistant bacteria and genes from manure as sources of agricultural soil pollution; the results also demonstrate the necessity and urgency of the regulation and supervision of veterinary antibiotics in China.

Editor: Jose Luis Balcazar, Catalan Institute for Water Research (ICRA), Spain

Funding: The funding source of this study are the Commonwealth and Environmental Protection project granted by the MEP "The health risk assessment and management technology of veterinary medicine" (201109038), "Study of determination method, pollution level and pollution control strategy of antibiotic-resistant gene in China" (201309031). The funders had no role in study design, data collection and analysis, decision to publish, or preparation of the manuscript.

Competing Interests: The authors have declared that no competing interests exist.

* Email: yebp@cpu.edu.cn (BY); ecsxg@nju.edu.cn (SG)

⅋ These authors contributed equally to this work.

¶ These authors also contributed equally to this work.

Introduction

In the past few decades, veterinary antibiotics have been widely used in many countries to treat disease and promote animal growth. However, this release together with antibiotic-resistant bacteria (ARB) is a great concern recently [1], primarily because the land application of antibiotic-polluted manure in agricultural practice not only introduced bacteria carrying antibiotic resistance genes (ARGs) into the soil but also had a significant effect on the ARB promotion and selection. In the soil, antibiotics provide a positive selective pressure for these bacteria [2]. The horizontal transfer of ARGs between bacteria is an important factor in resistance dissemination [3]. It is worth noting that some ARB in soil and manure are phylogenetically close to human pathogens, making genetic exchange more likely [3]. Evidence from the last 35 years demonstrates that there was consistent correlation between the use of antibiotic-contaminated manure on farms and the transfer of ARGs in human pathogens, as well as the direct

shift of ARB from animals to humans [4]. Therefore, ARGs are recognized as new environmental pollutants, and special concern is warranted due to their potential environmental and human health risks.

The used amount of veterinary medicines in China is more than that of other countries. According to a 2007 survey, the usage of antibiotics in livestock was almost half of the total antibiotics produced in China, which was 210,000 tons [5]. It was approximately 10-fold higher than in the USA and approximately 300-fold higher than in the UK [6]. It would be a good chance to analyze the impact of livestock practices on ARGs in the environment in China, where the animal farm was large-scale and the antibiotics usage was great [7]. However, there are few reports on veterinary ARGs in China.

Sulfonamides are synthetic veterinary antibiotics that are the most widely used veterinary antibiotics in China, the European Union and some developing countries due to their low costs [8,9].

However, sulfonamides were ranged as "High priority" of veterinary medicines, due to the high potential to reach the environment [10]. Sulfonamide resistance is primarily mediated by the *sul*1, *sul*2 and *sul*3 genes encoding dihydropteroate synthetase (DHPS) with a low affinity for sulfonamides [11–13]. A wide range of bacterial species harbor these genes, which are located in transposons and in self-transferable or mobilizable plasmids with a broad host range; these genes manifest multiple antibiotic resistance that is co-selected by sulfonamides [14–16].

Numerous recent studies have focused solely on the prevalence of sulfonamide resistance genes in bacterial isolates from manured agricultural soils or on the quantification of the total ARGs from environmental soil media to reflect the resistance reservoir. Few studies have systematically covered the identity of sulfonamide-resistant (SR) bacteria and the distribution patterns of sulfonamide ARGs in the total soil DNA and in sulfonamide-resistant bacteria.

The objectives of this study were (i) to determine the influence of the fertilization with antibiotic-polluted manure on the selection of sulfonamide ARB and ARGs and (ii) to investigate the distribution pattern of the *sul*1, *sul*2 and *sul*3 genes in the total soil DNA and the identified SR bacteria. Furthermore, (iii) the identification of the SR bacteria genera and description of the genotypes in each genus were also conducted to identify resistant opportunistic pathogens that increased the risk of ARGs affecting public health. To the best of our knowledge, this is the first comprehensive study of sulfonamide ARB and ARGs in livestock and poultry farms in China. The present study could be important in the evaluation of the pollution of soils used for agriculture by ARB and ARGs from manure; this study also demonstrates the necessity and urgency for the regulation and supervision of veterinary antibiotics in China.

Materials and Methods

Sampling

Soil samples from 10 sites were studied, including four pig farms, four chicken farms, one non-arable agricultural area and one mountain forest. The animal feeding farms of different sizes and scales were selected (detailed information about the sampling sites and the person in charge of sampling are given in Table S1 in File S1). The study was permitted and approved by the Ministry of Environmental Protection, China. The land accessed was not privately owned or protected. No protected species were sampled. There were vegetable cultivation area and grain planting area, which were all fertilized with animal manure, in each animal feeding farm. Therefore, two replicates of 1 kg soil samples for each type in every animal feeding farm were collected from depth of 10 to 15 cm, loaded into sterile glass flasks. The soil samples of the same type in different animal feeding farms were mixed (50 g from each source) to processed within 1 to 2 days after collection. The following description was the name rule of samples: (i) samples from the vegetable region of pig farms collected in the winter, the mixture of which was marked as PVW; (ii) samples from the agricultural region of pig farms collected in winter, the mixture of which was marked as PAW; (iii) samples from the vegetable region of pig farms collected in the summer, the mixture of which was marked as PVS; (iv) samples from the agricultural region of pig farms collected in the summer, the mixture of which was marked as PAS; (v) samples from the vegetable region of chicken farms, the mixture of which was marked as CV; (vi) samples from the agricultural region of chicken farms, the mixture of which was marked as CA; (vii) non-arable soils (marked as NA) where manure was not used for a few years near a Nanjing chicken farm; and (viii) forest soil collected from the Fangshan mountain in the Jiangning district of Nanjing (manure and/or antibiotics were not used),

which was marked as F. Soil P represents the mixture of soil samples from a pig farm in winter, and soil C is the mixture of soil samples from a chicken farm. The manure (M) was obtained from chickens that were treated with sulfonamides.

For each sample, 100 g was taken for the isolation of SR bacteria and the measurement of sulfonamide residues, and the remainder was stored at 4°C for DNA extraction. Meanwhile, the concentration of sulfonamides in the samples was analyzed in this study using a previously published method [17].

Viable plate counts

The isolation of SR bacteria from the soil or manure was performed by cultivating bacteria on nutrient broth agar plates containing 60 µg/ml sulfadiazine (SDZ) [15] followed by the spread plate technique [17]. Total bacteria from samples M, F, NA, P and C were cultivated on nutrient broth agar plates without SDZ. In brief, 1.0 ml of each soil sample solution, which was prepared by dissolving 5 g of soil in 45 ml of sterile physiological saline (0.9% NaCl), was mixed with 9 ml of sterile physiological saline. The process was repeated to make additional serial 10-fold dilutions, i.e., 10^{-3}, 10^{-4}, 10^{-5} and 10^{-6}. After 2–5 days of incubation at 37°C, the number of resistant bacteria on the agar plates were counted to calculate the colony-forming units (CFUs) per gram of soil with the following formula: CFU/g soil = 45 × average colony number × dilution factor. For subsequent analyses, SR isolates were randomly picked from the plates of each soil sample, with a total of 237 SR bacterial isolates, including 6 isolates from M; 1 isolate from F; 2 isolates from NA; 65, 57, 25 and 25 isolates from PVW, PAW, PVS, and PAS, respectively; and 20 and 36 isolates from CV and CA, respectively. All bacterial strains were stored at −80°C in nutrient broth medium containing 15% glycerol.

DNA extraction

Total soil DNA was extracted from 0.5 g of soil using a PowerSoil® DNA Isolation Kit (MoBio Laboratories, Carlsbad, California, USA) following the manufacturer's instructions. SR isolates were cultured at 37°C overnight with constant shaking at 200 rpm/min in 5 ml of LB supplemented with 60 µg/ml SDZ. DNA extraction was performed with 3.0 ml of cultured SR isolates using the TIANamp bacteria DNA kit (Tiangen, Beijing, China). The plasmids were extracted with the Biomiga EZgene™ Plasmid Miniprep kit (Biomiga, USA) following the manufacturer's protocol. The genomic DNA and plasmids were examined by 1% and 1.5% agarose gel electrophoresis, respectively. Moreover, the λDNA and DNA5000 were used as the marker of genomic DNA and pasmid, respectively. Usually, the molecular weight of genomic DNA was greater than that of the plasmid.

The detection of the *sul*1, *sul*2, and *sul*3 genes in the SR isolates

The prevalence of the *sul*1, *sul*2, and *sul*3 genes in the genomic DNA and plasmids of the isolates was examined via PCR with gene-specific primers (Table S2 in File S1). The amplification conditions for the *sul*1 and *sul*2 genes were as follows: 94°C for 5 min; 30 cycles of 94°C for 30 s, 69°C for 30 s and 72°C for 45 s; and one cycle of 72°C for 7 min. The amplification conditions for the *sul*3 gene were 94°C for 5 min, 30 cycles of 94°C for 30 s, 52°C for 30 s and 72°C for 60 s, and one cycle of 72°C for 7 min. Gel electrophoresis was performed on 1.5% agarose gels. The CA01 (a bacteria from soil CA) plasmid containing the *sul*1 gene was used as the positive control for the detection of the *sul*1 gene; the M01 (bacteria from chicken manure) plasmid containing the *sul*2 and *sul*3 genes was used as the positive control for the detection of the *sul*2 or *sul*3 genes. *E. coli* DH5α cells were used as the negative control. When the PCR product appeared as a single clear band with the same migration profile as the corresponding gene control, the isolate was counted as positive for that gene.

Quantitative PCR

The relative abundances of the *sul*1, *sul*2, and *sul*3 genes in the soil DNA were determined in triplicate via SYBR Green-based real-time PCR on a CFX96 Touch Real-Time PCR Detection System. The primer sequences are listed in Table S3 in File S1. Each 10-μl reaction mixture contained 5 μl of SYBR Premix (Cwbio, China), 1 μl of 2 μM forward and reverse primer mix, 1 μl of template, and 3 μl of ddH$_2$O. The PCR conditions were 95°C for 10 min, followed by 39 cycles of 95°C for 15 s and 60°C for 60 s. The samples were assessed via $2^{-\Delta\Delta Ct}$ relative quantitative analysis to compare the relative abundance of the *sul* genes among samples. All samples were analyzed in triplicate. The CA01 (a bacteria from soil CA) plasmid containing the *sul*1 gene was used as the positive control for the detection of the *sul*1 gene; the M01 (bacteria from chicken manure) plasmid containing the *sul*2 and *sul*3 genes was used as the positive control for the detection of the *sul*2 or *sul*3 genes. *E. coli* DH5α cells were used as the negative control.

16S rRNA sequencing of SR isolates

The complete 16S rRNA gene was used to identify the genera present in the bacterial isolates. Genomic DNA was used as the template for the PCR amplification of the 16S rRNA gene using the universal bacterial 16S rRNA primers 27F and 1492R (Table S2 in File S1). Each 50-μl reaction mixture consisted of 1 to 4 μl of genomic DNA, Taq plus polymerase buffer containing 1.5 mM MgCl$_2$, 0.2 mM each of the 4 deoxynucleoside triphosphates (dNTPs), 1 mM each of the 27F and 1492R primers, and 1 U of Taq plus polymerase (Tiangen). PCR was performed using a Bio-Rad thermal cycler under the following conditions: 94°C for 5 min, followed by 30 cycles of 94°C for 30 s, 58°C for 30 s, and 72°C for 1.5 min, and 1 cycle of 72°C for 10 min. The PCR products were separated via electrophoresis on 1.0% agarose gels. The PCR amplicons were sequenced by Sangon (Shanghai, China). A pair-wise 16S rRNA gene sequence similarity was performed using the EzTaxon server (http://www.eztaxon.org/) [18] and NCBI BLAST (http://blast.ncbi.nlm.nih.gov/blast.cgi). A bacterial genus was considered present when a sample 16S rRNA gene sequence was ≥97% identical to the reference sequence of the bacteria in that genus.

Statistical analysis

The statistical analysis was performed using SAS 9.1. The group mean levels were analyzed via a one-way Analysis of Variance (ANOVA). Statistical significance was defined as a *p*-value≤0.05. This *p*-value was chosen because the standard error associated with CFU plating and qPCR results are generally approximately 5% of the mean. The mean and standard error (SE) displayed in the figures were generated using the means procedure without transformation.

Results and Discussion

Enumeration of the total culturable microbial populations and SR Bacteria in the soil

The number of total culturable microbial populations on the nutrient agar ranged from 1.96×10^7 to 9.75×10^7 CFU/g soil and that of the SR isolates on the nutrient agar ranged from 4.5×10^5 to 9.0×10^7 CFU/g soil (Figure 1), which were higher than those of the reported aquaculture-agriculture ponds (3.0×10^4 to 1.6×10^6 and 3.0×10^2 to 4.1×10^4, respectively) [19]. The higher numbers of total bacteria and SR isolates were found in chicken manure (9.75×10^7 and 9.00×10^7, respectively), which was most likely due to the amount of easily accessible nutrients in the manure that stimulated the growth of bacteria [20]. The number of SR bacteria from the soils affected by pig or chicken manure (3.02×10^6 to 9.40×10^6 CFU/g soil) was higher than that from non-arable soil (1.96×10^6 CFU/g soil) or forest soil (4.5×10^5 CFU/g soil). This difference was most likely due to the application of manure to the soil. Previous studies reported that manure from treated pigs was rich in antibiotics and bacteria carrying ARGs, which were both transferred to the soil via fertilization [3,10]. Furthermore, the number of SR isolates from the vegetable soils was significantly higher than that from the agricultural soils (5.96×10^6 and 3.02×10^6 CFU/g soil for PVW and PAW, respectively; 9.40×10^6 and 4.98×10^6 CFU/g soil for PVS and PAS, respectively; 7.50×10^6 and 4.11×10^6 CFU/g soil for CV and CA, respectively). Because liquid manure or wastewater was frequently used to irrigate the vegetable region, manure was more frequently applied to the vegetable soils than to the agricultural soils, and the repeated application of manure to the vegetable soils may have increased bacterial resistance. Additionally, the mean number of SR isolates from the winter soils (4.49×10^6 CFU/g soil for PW) was lower than that from the summer soils (7.19×10^6 CFU/g soil for PS). This difference most likely occurred because the temperature in the summer is more suitable for the growth of bacteria than that in the winter.

The concentration sums of sulfadiazine, sulfamerazine, sulfathiazole, sulfamethazine, sulfadimethazine and sulfamethoxazole were 4503, 0, 0.536, 35.6, 25.9, 15.8, 12.6, 239 and 193 μg/kg in the mixed samples of M, F, NA, PVW, PAW, PVS, PAS, CV and CA, respectively. The number of cultivable bacteria was not consistent with the concentration of antibiotic sulfonamides in the soil. The pollution level of sulfonamides was found to be significantly higher in chicken farms than in pig farms, but there was no significant difference among the numbers of cultivable bacteria.

Characterization of SR bacteria

All 237 SR isolates that were identified via 16S r RNA belonged to 26 typical soil bacteria genera, including *Achromobacter*, *Arthrobacter*, *Bacillus*, *Brevibacterium*, *Chryseobacterium*, *Citrobacter*, *Cupriavidus*, *Escherichia*, *Flavobacterium*, *Hydrogenophaga*, *Klebsiella*, *Lysinibacillus*, *Massilia*, *Microbacterium*, *Microvirga*, *Pseudomonas*, *Pseudoxanthomonas*, *Rhizobium*, *Rhodococcus*, *Shigella*, *Sphingobacterium*, *Sphingopyxis*, *Staphylococcus*, *Stenotrophomonas*, *Streptococcus*, and *Streptomyces*. *Bacillus* was the most prevalent genus in all 9 environmental samples with a frequency of 43.88%, followed by *Pseudomonas* and *Shigella* (11.39% and 8.02%, respectively; Figure 2). However, it is reported that *Acinetobacter* was abundant in pig wastewater in Vietnam [21]. Both pig- and chicken-manured soil samples were rich in bacteria species; for example, 12 genera were found in PVW and CA (see Figure S1).

Relative abundance of the *sul* genes in the soils

A qPCR analysis of sulfonamide resistance genes was performed on the total DNA extracted directly from the soil. There was significant variation in the relative quantities of the *sul*1, *sul*2, and *sul*3 genes in the DNA extracted from the eight types of soils (see Figure 3). The DNA from the pig-manured soils (PVW, PAW, PVS and PAS) contained relatively higher copy numbers of *sul*1

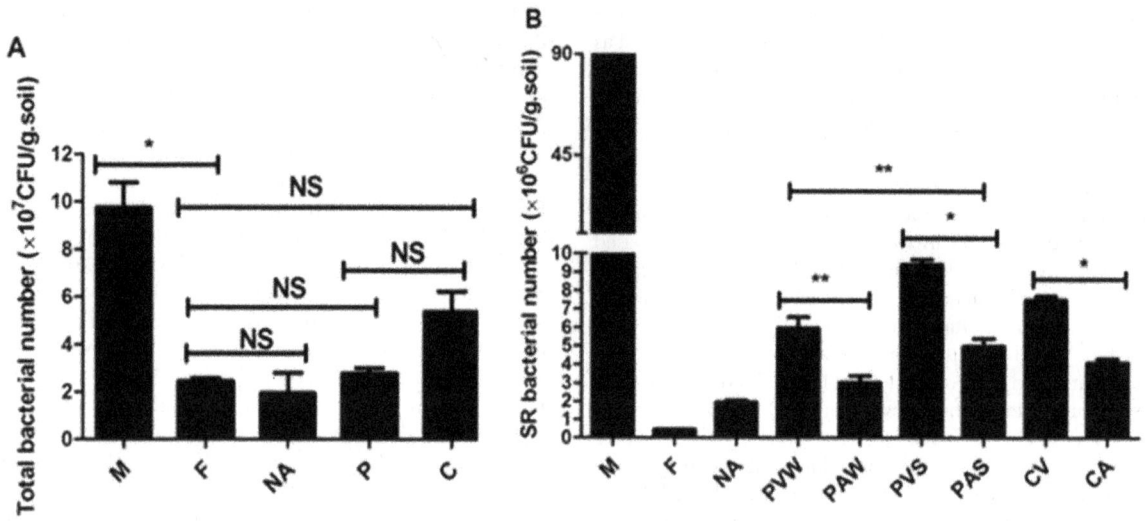

Figure 1. Numbers of cultivable bacteria. (M = Manure, F = Forest, NA = non-arable fied, P = Pig, C = Chicken, W = winter, V = vegetable garden soil, A = agricultural soil; *p≤0.05, **p≤0.01, n = 3; NS, not significant).

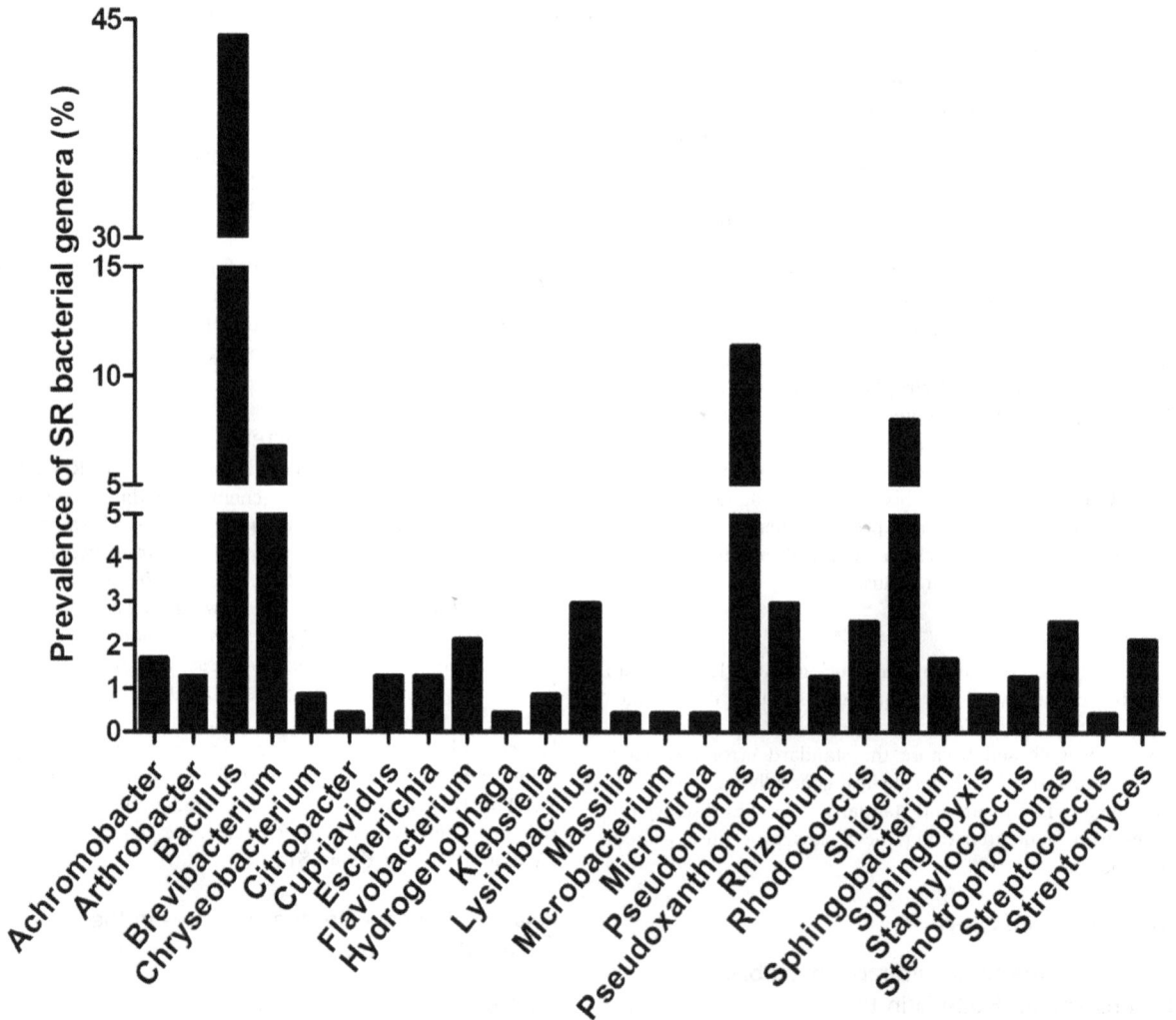

Figure 2. The genera of SR bacteria and their detected frequency in all sampling sites.

than *sul2*. Comparatively, the relative quantity of the *sul1* gene in the chicken-manured soils was lower than that of the *sul2* gene. Additionally, the *sul3* genes were detected at low relative quantities in the DNA extracted from the eight soils but were not detected via PCR in bacteria isolated from forest and pig-manured agricultural (summer) soils. The results of our study were consistent with other reports that demonstrated that the repeated application of manure from pigs or chickens treated with SDZ increased the transfer and abundance of ARGs in the soil [3,10,20]. Furthermore, good positive linear correlations were observed between the relative abundance of the *sul2* genes and the number of culturable SR isolates in the soil. For the *sul2* gene and sum of the three *sul* genes, the correlation coefficients (R^2) were 0.95 and 0.65, respectively ($p<0.05$). However, the abundance of *sul1* and *sul3* showed no significant correlation with the numbers of culturable SR isolates in the soil ($R^2 = 0.44$, $p>0.05$ for *sul1* and $R^2 = 0.39$, $p>0.05$ for *sul3*). This lack of a correlation could be attributed to the fact that the viable plate counts method only sampled microbes that were culturable and expressed their ARGs under those conditions, so most of the microbes carrying *sul1* and *sul3* genes may not be culturable. The other probable reason was that some "silent" or unexpressed *sul1* and *sul3* genes may be existed in the isolates of soils, which could be horizontally transferred or expressed under other conditions.

In brief, the number of culturable SR isolates in the soil can reflect the total relative abundance of the three *sul* genes, showing that the plate count method was effective in assessing the antibiotic resistance risk of the soil. Therefore, the diversity of ARGs enriched at the farm level should be the focus of more attention.

Distribution of *sul* genes in SR isolates

The number and percentage of isolates carrying the *sul* genes in their genomic DNA and plasmids are summarized in Table 1 and Table 2. The distribution and spread of SR genes in the soil microbes are sufficiently frequent to warrant special concern. The *sul1*, *sul2*, and *sul3* genes were all detected at a frequency of 100% in the genomic DNA and plasmids of the SR isolates from the manure sample, indicating that ARGs were extensively harbored in the chromosome and mobile genetic elements of the bacteria in manure, leading to the high potential of horizontal gene transfer of ARGs in soil. Interestingly, the *sul2* genes were only present in the genomic DNA of the isolates collected from forest soil and non-arable soil, which had no history of manure application. This finding may be attributed to the notion that the *sul1* and *sul3* genotype in genomic DNA maybe associate with the amended manure. However, the *sul1*, *sul2* and *sul3* genes were all located in the plasmids of the isolates from non-arable soil but were absent from the plasmids of the isolates from the forest soil; a potential explanation for this difference could be that the bacteria carrying *sul* genes in the manured soil may transfer to the nearby region by aerosolization or runoff, then horizontal transfer occurred in close bacteria via plasmids.

For the manured soil, the frequency distribution of the *sul* genes in the genomic DNA and plasmids of the SR isolates investigated overall followed a trend of *sul2* > *sul1* > *sul3* ($p<0.05$). This result was in contrast to several previous studies showing that the *sul1* gene was more prevalent than the *sul2* gene in the DNA from manure and manured soils [10,15] due to different conditions in various countries. The *sul3* gene was found at low frequencies in

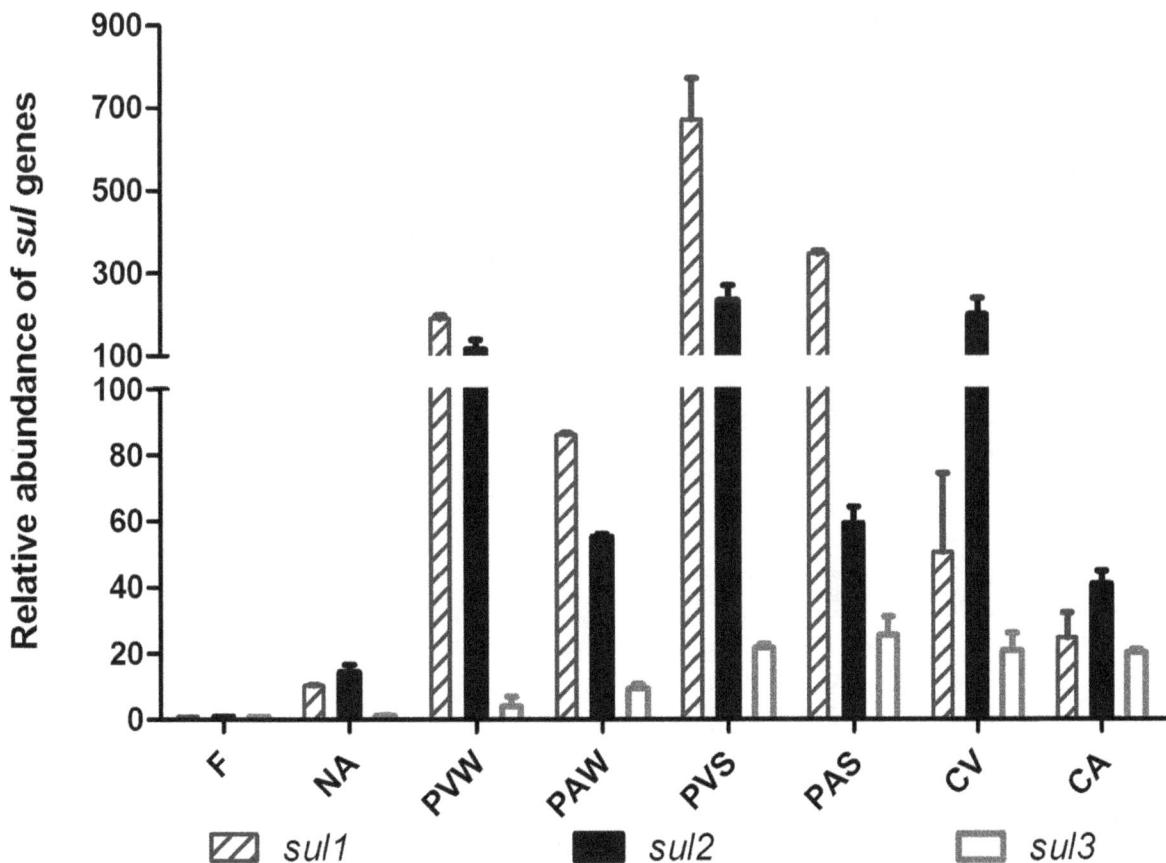

Figure 3. Relative quantity of sulfonamides resistant genes in soils with and without manure treatment.

Table 1. Distribution of *sul1*, *sul2* and *sul3* genes in genomic DNA and plasmid of SR isolates (in samples M, F, NA, CV and CA).

sul gene combination		M (n = 6a/6b) NO. of isolates (%)		F (n = 1/0) NO. of isolates (%)		NA (n = 2/2) NO. of isolates (%)		CV (n = 20/20) NO. of isolates (%)		CA (n = 36/36) NO. of isolates (%)	
		Genomic DNA	Plasmid DNA	Genomic DNA	Plasmid DNA	Genomic DNA	Plasmid DNA	Genomic DNA	Plasmid DNA	Genomic DNA	Plasmid DNA
Single genes	*sul1*	0 (0.0)	0 (0.0)	0 (0.0)	0 (0.0)	0 (0.0)	0 (0.0)	0 (0.0)	1 (5.0)	2 (5.6)	0 (0.0)
	sul2	0 (0.0)	0 (0.0)	1 (100.0)	0 (0.0)	2 (100.0)	0 (0.0)	2 (10.0)	3 (15.0)	3 (8.3)	2 (5.6)
	sul3	0 (0.0)	0 (0.0)	0 (0.0)	0 (0.0)	0 (0.0)	0 (0.0)	0 (0.0)	0 (0.0)	1 (2.8)	0 (0.0)
Two genes	*sul1+sul2*	0 (0.0)	0 (0.0)	0 (0.0)	0 (0.0)	0 (0.0)	0 (0.0)	4 (20.0)	7 (35.0)	10 (27.8)	2 (5.6)
	sul1+sul3	0 (0.0)	0 (0.0)	0 (0.0)	0 (0.0)	0 (0.0)	0 (0.0)	0 (0.0)	0 (0.0)	0 (0.0)	0 (0.0)
	sul2+sul3	0 (0.0)	0 (0.0)	0 (0.0)	0 (0.0)	0 (0.0)	1 (50.0)	0 (0.0)	0 (0.0)	1 (2.8)	9 (25.0)
Three genes	*sul1+sul2 +sul3*	6 (100.0)	6 (100.0)	0 (0.0)	0 (0.0)	0 (0.0)	1 (50.0)	12 (60.0)	9 (45.0)	19 (52.8)	23 (63.9)
None		0 (0.0)	0 (0.0)	0 (0.0)	0 (0.0)	0 (0.0)	0 (0.0)	0 (0.0)	0 (0.0)	0 (0.0)	0 (0.0)
Total	*sul1*	6 (100.0)	6 (100.0)	0 (0.0)	0 (0.0)	0 (0.0)	1 (50.0)	16 (80.00)	17 (85.0)	27 (75.0)	25 (69.4)
	sul2	6 (100.0)	6 (100.0)	1 (100.0)	0 (0.0)	2 (100)	2 (100.0)	19 (95.0)	19 (95.0)	29 (80.6)	36 (100.0)
	sul3	6 (100.0)	6 (100.0)	0 (0.0)	0 (0.0)	0 (0.0)	2 (100.0)	13 (65.0)	9 (45.0)	21 (58.3)	32 (88.9)
Total of SR isolate positive for *sul* genes		6 (100.0)	6 (100.0)	1 (100.0)	0 (0.0)	2 (100.0)	2 (100.0)	19 (95.0)	20 (100.0)	36 (100.0)	36 (100.0)

a = genomic DNA, b = plasmid.

Table 2. Distribution of sul1, sul2 and sul3 genes in genomic DNA and plasmid of SR isolates (in samples PVW, PAW, PVS and PAS).

sul gene combination		PVW (n=65/47)		PAW (n=57/43)		PVS (n=25/22)		PAS (n=25/22)	
		NO. of isolates (%)		NO. of isolates (%)		NO. of isolates (%)		NO. of isolates (%)	
		Genomic DNA	Plasmid DNA	Genomic DNA	Plasmid DNA	Genomic DNA	Plasmid DNA	Genomic DNA	Plasmid DNA
Single genes	sul1	0 (0.0)	23 (48.9)	0 (0.0)	19 (44.2)	15 (60.0)	0 (0.0)	24 (96.0)	0 (0.0)
	sul2	28 (43.1)	3 (6.4)	15 (26.3)	8 (18.6)	1 (4.0)	15 (68.2)	0 (0.0)	16 (72.7)
	sul3	0 (0.0)	1 (2.1)	0 (0.0)	2 (4.7)	0 (0.0)	0 (0.0)	0 (0.0)	0 (0.0)
Two genes	sul1+sul2	34 (52.3)	11 (23.4)	41 (71.9)	9 (20.9)	1 (4.0)	6 (27.3)	1 (4.0)	6 (27.3)
	sul1+sul3	0 (0.0)	2 (4.3)	0 (0.0)	0 (0.0)	0 (0.0)	0 (0.0)	0 (0.0)	0 (0.0)
	sul2+sul3	0 (0.0)	0 (0.0)	1 (1.8)	0 (0.0)	0 (0.0)	1 (4.5)	0 (0.0)	0 (0.0)
Three genes	sul1+sul2+sul3	1 (1.5)	0 (0.0)	0 (0.0)	0 (0.0)	0 (0.0)	0 (0.0)	0 (0.0)	0 (0.0)
None		2 (3.1)	7 (14.9)	0 (0.0)	5 (11.6)	8 (32.0)	0 (0.0)	0 (0.0)	0 (0.0)
Total	sul1	35 (53.8)	36 (76.6)	41 (71.9)	24 (55.8)	16 (64.0)	6 (27.3)	25 (100.0)	6 (27.3)
	sul2	63 (96.9)	14 (29.8)	57 (100.0)	17 (39.5)	2 (8.0)	21 (95.5)	1 (4.0)	22 (100.0)
	sul3	1 (1.5)	3 (6.4)	1 (1.8)	2 (4.7)	0 (0.0)	1 (4.5)	0 (0.0)	0 (0.0)
Total of SR isolate positive for sul genes		63 (96.9)	40 (85.1)	57 (100.0)	38 (88.4)	17 (68.0)	22 (100.0)	25 (100.0)	22 (100.0)

a = genomic DNA, b = plasmid.

our samples, whereas recently, Suzuki et al showed that sul3 was major sul in seawater [22]. Hoa et al. suggested that most of the *sul* genes are located on the chromosome [15]. However, there was no significant difference between the overall percentage of the isolates carrying the *sul* genes located on the genomic DNA and those on the plasmids in our study. It was interesting to note that the frequency order of the *sul1* and *sul2* genes from the isolates of the pig-manured soils for the genomic DNA was opposite that for the plasmids. In the isolates collected from the pig-manured soils in winter, *sul2* was the most prevalent gene located within the genomic DNA (96.9% and 100.0% in PVW and PAW, respectively) followed by *sul1* (53.8% and 71.9% in PVW and PAW, respectively); *sul1* was the most prevalent gene located on plasmids (76.6% and 55.8% in PVW and PAW, respectively) followed by *sul2* (29.8% and 39.5% in PVW and PAW, respectively). However, in the isolates collected from pig-manured soil in summer, the order of *sul1* (64.0% and 100.0% in PVW and PAW, respectively) > *sul2* (8.0% and 4.0% in PVW and PAW, respectively) in the genomic DNA and *sul2* (95.0% and 100.0% in PVW and PAW, respectively) > *sul1* (27.3% and 27.3% in PVW and PAW, respectively) in the plasmids was determined. We concluded that in most isolates, *sul1* and *sul2* were located in the different mobile elements and transferred at different rates.

Furthermore, the animal type was a significant factor influencing the expression frequency of *sul* genes, which showed that the frequency in the chicken-manured soil was higher than that in the pig-manured soil ($p < 0.05$), which was consistent with the data of the concentration of sulfonamides in the soil.

We also determined the co-presence of any two different *sul* genes on the chromosome and plasmids in a single isolate. The combination of *sul1* and *sul2* on the chromosome was the most frequent and was present in PVW, PAW, PVS, PAS, CV and CA (52.3%, 71.9%, 4.0%, 4.0%, 20.0% and 27.8%, respectively), and the *sul1*, *sul2* and *sul3* genes were highly co-present on the chromosomes of M, CV and CA (100%, 60.0% and 52.8%, respectively). The co-presence of *sul2* and *sul3* was only detected in two isolates from PAW and CA, respectively, and the co-existence of *sul1* and *sul3* was not detected in any SR isolates. The *sul1* and *sul2* genes were also frequently detected together in the plasmids (23.4%, 20.9%, 27.3%, 27.3%, 35.0% and 5.6% in PVW, PAW, PVS, PAS, CV and CA, respectively). In contrast, the co-presence of *sul2* and *sul3* was only detected in NA (50.0%), PVS (4.5%) and CA (25.0%), and the co-presence of *sul1* and *sul3* was not found in any plasmids. Furthermore, the three *sul* genes were co-present in the plasmids of M (100%), NA (50.0%), CV (45.0%), and CA (63.9%). We concluded that the combination of *sul1* and *sul2* was the most frequent and that the co-existence of *sul1* and *sul3* was not found in the genomic DNA or plasmids. Based on these results, the co-presence of the three *sul* genes was only in the isolates from manure and soil from chicken farms, suggesting that there was a positive correlation between the frequency of the co-presence of the three *sul* genes and the time and amount of repeated manure applications.

In summary, the *sul* genes, either individually or in combinations of two or three, were present in the SR isolates at high frequencies. Nearly all plasmids from the SR isolates contained the *sul* genes (with the exception of F). This observation suggests that the resistance that we observed in most cases was linked to plasmids or other mobile genetic elements, which theoretically have transfer potential. The SR isolates could possibly carry these *sul* genes through gene transfer under selection conditions, leading to an increase in antibiotic resistance among bacteria.

SR bacterial and *sul* genes

The distribution of *sul* genes in bacteria species is listed in Table 3. *Bacillus* was the most prevalent *sul*-positive genus in the soil samples of this study, carrying the *sul* genes in 43.88% of the total isolates; thus, this genus could be the main reservoir of the *sul* genes. This finding was not consistent with other studies that showed that *Acinetobacter* was the dominant genus in aquatic environments (wastewater and shrimp ponds of north Vietnam) and manured agricultural clay soils and slurry samples in the United Kingdom [15,16]. Except for different environments, what makes the difference of genus may be the different condition of culture, such as 28 or 30°C incubation in these two references, not 37°C. It was reported that *Bacillus* spp. have developed resistance to most antibiotic groups, but only a few species of *Bacillus* have been reported to be sensitive to sulfonamides [23]. *Pseudomonas* and *Shigella* were the second and third most prevalent, carrying the *sul* genes in 11.39% and 8.02% of all isolates, respectively. Ventilator-acquired pneumonia, respiratory tract infections in immunocompromised patients and chronic respiratory infections in cystic fibrosis patients were associated with the *Pseudomonas* species (especially *P. aeruginosa*) [24]. *Enterobacteriaceae* species including *Shigella*, *Klebsiella*, and *Escherichia* have represented some of the most dominant bacterial infections over the last 30 years [24]. In the Henan Province of China, 72.6% of infections were caused by *Shigella* strains in 2006 [25].

To the best of our knowledge, this report is the first on *sul* genes in *Chryseobacterium*, *Cupriavidus*, *Flavobacterium*, *Hydrogenophaga*, *Lysinibacillus*, *Massilia*, *Microbacterium*, *Microvirga*, *Pseudoxanthomonas*, *Rhizibium*, *Rhodococcus*, *Sphingopyxis*, *Staphylococcus*, *Streptococcus*, and *Streptomyces* from soils and the first that indicates the widespread presence of ARB in the arable soils of China. Previous studies demonstrated the co-presence of *sul1*, *sul2* and *sul3* in a single cell; this was detected in *Acinetobacter*, *Bacillus*, *Psychrobacter*, *Escherichia coli*, and *Salmonella* [15,16,26,27]. In our study, these three *sul* genes were simultaneously found in *Arthrobacter*, *Brevibacterium*, *Citrobacter*, *Cupriavidus*, *Flavobacterium*, *Lysinibacillus*, *Pseudomonas*, *Pseudoxanthomonas*, *Rhizibium*, *Sphingobacterium*, *Staphylococcus*, *Stenotrophomonas*, *Streptococcus*, and *Streptomyces*, with the exception of three genera (*Bacillus*, *Escherichia*, and *Shigella*). This result indicates that the three *sul* genes are common and widely distributed in ARB in soil. Additionally, the *sul3* gene was detected for the first time in *Achromobacter*, *Chryseobacterium*, *Citrobacter*, *Cupriavidus*, *Flavobacterium*, *Lysinibacillus*, *Pseudoxanthomonas*, *Rhizibium*, *Sphingobacterium*, *Staphylococcus*, *Streptococcus*, and *Streptomyces* from arable soils.

It was revealed that the manured soils could be a reservoir of sulfonamide ARBs and ARGs, according to the observation of high frequency of various combinations of the *sul* genes in bacteria of manured agricultural soils, which may bring potential hazards to human and ecosystem health. Therefore, the diversity of ARGs and ARB enriched at the farm level should be the focus of more attention.

Conclusion

A comprehensive study of sulfonamide ARB and ARGs in livestock and poultry farms in Jiangsu Province of China revealed that the fertilization with antibiotic-polluted manure had a significant influence on the selection of sulfonamide ARB and ARGs. The sample type, animal type and sampling time may affect the prevalence and distribution rule of SR genes. The results from the identification of the SR bacteria genus and the description of the genotypes in the genus revealed that resistant

Table 3. Summary of *sul* genotype of sul-positive bacterial species isolated.

Genus	No. of total sul-positive isolates (%)	Source of isolates	*sul* genotype	No. of sul-positive isolates
Achromobacter	4 (1.69)	NA, PVW, CV	*sul2*	2
			sul1 sul2	1
			sul2 sul3	1
Arthrobacter	3 (1.27)	CV, CA	*sul1 sul2*	1
			sul1 sul2 sul3	2
Bacillus	104 (43.88)	F, PVW, PAW, PVS, PAS, CV, CA	*sul1*	2
			sul2	23
			sul1 sul2	66
			sul2 sul3	1
			sul1 sul2 sul3	12
Brevibacterium	16 (6.75)	PVW, PAW, PVS, PAS	*sul2*	4
			sul1 sul2	11
			sul1 sul2 sul3	1
Chryseobacterium	2 (0.84)	PVW, PVS	*sul1 sul2*	2
Citrobacter	1 (0.42)	CA	*sul1 sul2 sul3*	1
Cupriavidus	3 (1.27)	CA	*sul1 sul2 sul3*	3
Escherichia	3 (1.27)	PVW, CA	*sul1 sul2*	1
			sul1 sul2 sul3	2
Flavobacterium	5 (2.11)	CV, CA	*sul2*	1
			sul1 sul2	1
			sul1 sul2 sul3	3
Hydrogenophaga	1 (0.42)	PVS	*sul1 sul2*	1
Klebsiella	2 (0.84)	PAS	*sul1 sul2*	2
Lysinibacillus	7 (2.95)	PVW, PAW, PAS	*sul1*	1
			sul1 sul2	4
			sul1 sul2 sul3	2
Massilia	1 (0.42)	PVW	*sul2*	1
Microbacterium	1 (0.42)	PAW	*sul1 sul2*	1
Microvirga	1 (0.42)	PAS	*sul1 sul2*	1
Pseudomonas	27 (11.39)	PVW, PAW, CV	*sul2*	1
			sul1 sul2	23
			sul1 sul2 sul3	3
Pseudoxanthomonas	7 (2.95)	PVW, PVS	*sul2*	2
			sul1 sul2	4
			sul1 sul2 sul3	1
Rhizobium	3 (1.27)	PVS, CV	*sul1 sul2*	2
			sul1 sul2 sul3	1
Rhodococcus	6 (2.53)	PVW, PAW, PVS, PAS	*sul1 sul2*	6
Shigella	19 (8.02)	CV, CA, M	*sul1 sul2 sul3*	19

Table 3. Cont.

Genus	No. of total sul-positive isolates (%)	Source of isolates	*sul* genotype	No. of sul-positive isolates
Sphingobacterium	4 (1.69)	VA	*sul*1 *sul*2	1
			*sul*1 *sul*2 *sul*3	3
Sphingopyxis	2 (0.84)	PVW, PAW	*sul*2	1
			*sul*1 *sul*2	1
Staphylococcus	3 (1.27)	PAW, CA	*sul*1 *sul*2 *sul*3	3
Stenotrophomonas	6 (2.53)	CV, CA, NA	*sul*2 *sul*3	1
			*sul*1 *sul*2 *sul*3	5
Streptococcus	1 (0.42)	CV	*sul*1 *sul*2 *sul*3	1
Streptomyces	5 (2.11)	PVW, PAW, CA	*sul*1 *sul*2	1
			*sul*1 *sul*2 *sul*3	4

opportunistic pathogens increased the risk of ARGs affecting public health. Overall, the high frequency of various combinations of the *sul* genes in manured agricultural soil samples of Southeastern China should be the focus of more attention, and the regulation and supervision of veterinary antibiotics are urgently needed in China.

Supporting Information

Figure S1 Prevalences of SR bacteria belonging to different genera identified in the studied soils.

File S1 Contains the following files: **Table S1**. Detailed information on sampling sites in present study. **Table S2**. Primers for PCR in this Study. **Table S3**. Primers for quantitative PCR in this Study.

Author Contributions

Conceived and designed the experiments: NW. Performed the experiments: XHY SJJ JZ. Analyzed the data: XHY NW. Contributed reagents/materials/analysis tools: NW BPY XHY. Contributed to the writing of the manuscript: NW XHY. Guided the experiment: BPY SXG.

References

1. Ghosh S, LaPara TM (2007) The effects of subtherapeutic antibiotic use in farm animals on the proliferation and persistence of antibiotic resistance among soil bacteria. ISME J 1: 191–203.
2. Popowska M, Rzeczycka M, Miernik A, Krawczyk-Balska A, Walsh F, et al. (2012) Influence of soil use on prevalence of tetracycline, streptomycin, and erythromycin resistance and associated resistance genes. Antimicrob Agents Chemother 56: 1434–1443.
3. Heuer H, Schmitt H, Smalla K (2011) Antibiotic resistance gene spread due to manure application on agricultural fields. Curr Opin Microbiol 14: 236–243.
4. Marshall BM, Levy SB (2011) Food animals and antimicrobials: impacts on human health. Clinical microbiology reviews 24: 718–733.
5. Hvistendahl M (2012) Public Health China Takes Aim at Rampant Antibiotic Resistance. Science 336: 795–795.
6. Kim K-R, Owens G, Kwon S-I, So K-H, Lee D-B, et al. (2011) Occurrence and environmental fate of veterinary antibiotics in the terrestrial environment. Water Air Soil Poll 214: 163–174.
7. Zhu YG, Johnson TA, Su JQ, Qiao M, Guo GX, et al. (2013) Diverse and abundant antibiotic resistance genes in Chinese swine farms. Proc Natl Acad Sci U S A 110: 3435–3440.
8. Ungemach F (1999) Figures on quantities of antibacterials used for different purposes in the EU countries and interpretation. Acta Vet Scand Suppl 93: 89–97; discussion 97–88, 111–117.
9. Kools SA, Moltmann JF, Knacker T (2008) Estimating the use of veterinary medicines in the European Union. Regul Toxicol Pharm 50: 59–65.
10. Heuer H, Smalla K (2007) Manure and sulfadiazine synergistically increased bacterial antibiotic resistance in soil over at least two months. Environ Microbiol 9: 657–666.
11. Skold O (2000) Sulfonamide resistance: mechanisms and trends. Drug Resist Updat 3: 155–160.
12. Perreten V, Boerlin P (2003) A new sulfonamide resistance gene (sul3) in Escherichia coli is widespread in the pig population of Switzerland. Antimicrob Agents Chemother 47: 1169–1172.
13. Yun MK, Wu Y, Li Z, Zhao Y, Waddell MB, et al. (2012) Catalysis and sulfa drug resistance in dihydropteroate synthase. Science 335: 1110–1114.
14. Heuer H, Szczepanowski R, Schneiker S, Pühler A, Top E, et al. (2004) The complete sequences of plasmids pB2 and pB3 provide evidence for a recent ancestor of the IncP-1β group without any accessory genes. Microbiology 150: 3591–3599.
15. Hoa PTP, Nonaka L, Hung Viet P, Suzuki S (2008) Detection of the sul1, sul2, and sul3 genes in sulfonamide-resistant bacteria from wastewater and shrimp ponds of north Vietnam. Sci Total Environ 405: 377–384.
16. Byrne-Bailey KG, Gaze WH, Kay P, Boxall AB, Hawkey PM, et al. (2009) Prevalence of sulfonamide resistance genes in bacterial isolates from manured agricultural soils and pig slurry in the United Kingdom. Antimicrob Agents Chemother 53: 696–702.
17. Sengeløv G, Agersø Y, Halling-Sørensen B, Baloda SB, Andersen JS, et al. (2003) Bacterial antibiotic resistance levels in Danish farmland as a result of treatment with pig manure slurry. Environ Int 28: 587–595.
18. Chun J, Lee JH, Jung Y, Kim M, Kim S, et al. (2007) EzTaxon: a web-based tool for the identification of prokaryotes based on 16S ribosomal RNA gene sequences. Int J Syst Evol Microbiol 57: 2259–2261.
19. Hoa PTP, Managaki S, Nakada N, Takada H, Anh DH, et al. (2010) Abundance of sulfonamide-resistant bacteria and their resistance genes in integrated aquaculture-agriculture ponds, North Vietnam. Interdisciplinary studies on environmental chemistry - biological responses to contaminants. Tokyo: TERRAPUB. 15–22.
20. Jechalke S, Kopmann C, Rosendahl I, Groeneweg J, Weichelt V, et al. (2013) Increased abundance and transferability of resistance genes after field application of manure from sulfadiazine-treated pigs. Appl Environ Microbiol 79: 1704–1711.
21. Hoa PTP, Managaki S, Nakada N, Takada H, Shimizu A, et al. (2011) Antibiotic contamination and occurrence of antibiotic-resistant bacteria in aquatic environments of northern Vietnam. Science of the Total Environment 409: 2894–2901.

22. Suzuki S, Ogo M, Miller TW, Shimizu A, Takada H, et al. (2013) Who possesses drug resistance genes in the aquatic environment sulfamethoxazole (SMX) resistance genes among the bacterial community in water environment of Metro-Manila, Philippines. Frontiers in Microbiology 4.

23. Valderas MW, Bourne PC, Barrow WW (2007) Genetic basis for sulfonamide resistance in Bacillus anthracis. Microb Drug Resist 13: 11–20.

24. Diene SM, Rolain J-M (2013) Investigation of antibiotic resistance in the genomic era of multidrug-resistant Gram-negative bacilli, especially Enterobacteriaceae, Pseudomonas and Acinetobacter. Expert Rev Anti-Infe 11: 277–296.

25. Xia S, Xu B, Huang L, Zhao JY, Ran L, et al. (2011) Prevalence and characterization of human Shigella infections in Henan Province, China, in 2006. J Clin Microbiol 49: 232–242.

26. Antunes P, Machado J, Sousa JC, Peixe L (2005) Dissemination of sulfonamide resistance genes (sul1, sul2, and sul3) in Portuguese Salmonella enterica strains and relation with integrons. Antimicrob Agents Chemother 49: 836–839.

27. Hammerum AM, Sandvang D, Andersen SR, Seyfarth AM, Porsbo LJ, et al. (2006) Detection of sul1, sul2 and sul3 in sulphonamide resistant Escherichia coli isolates obtained from healthy humans, pork and pigs in Denmark. Int J Food Microbiol 106: 235–237.

13

Replication Rates of *Mycobacterium tuberculosis* in Human Macrophages Do Not Correlate with Mycobacterial Antibiotic Susceptibility

Johanna Raffetseder[◈], Elsje Pienaar[◈¤a], Robert Blomgran, Daniel Eklund, Veronika Patcha Brodin, Henrik Andersson, Amanda Welin[¤b], Maria Lerm*

Division of Microbiology and Molecular Medicine, Department of Clinical and Experimental Medicine, Faculty of Health Sciences, Linköping University, Linköping, SE-58185, Sweden

Abstract

The standard treatment of tuberculosis (TB) takes six to nine months to complete and this lengthy therapy contributes to the emergence of drug-resistant TB. TB is caused by *Mycobacterium tuberculosis* (Mtb) and the ability of this bacterium to switch to a dormant phenotype has been suggested to be responsible for the slow clearance during treatment. A recent study showed that the replication rate of a non-virulent mycobacterium, *Mycobacterium smegmatis,* did not correlate with antibiotic susceptibility. However, the question whether this observation also holds true for Mtb remains unanswered. Here, in order to mimic physiological conditions of TB infection, we established a protocol based on long-term infection of primary human macrophages, featuring Mtb replicating at different rates inside the cells. During conditions that restricted Mtb replication, the bacterial phenotype was associated with reduced acid-fastness. However, these phenotypically altered bacteria were as sensitive to isoniazid, pyrazinamide and ethambutol as intracellularly replicating Mtb. In support of the recent findings with *M. smegmatis,* we conclude that replication rates of Mtb do not correlate with antibiotic tolerance.

Editor: Jordi B. Torrelles, The Ohio State University, United States of America

Funding: The project was funded by the Bill & Melinda Gates Foundation (www.gatesfoundation.org), the Swedish Research Council (grant numbers 2009-3821 and 2012-3349, www.vr.se), the Swedish International Development Cooperation Agency (www.sida.se), the Swedish Heart-Lung Foundation (www.hjart-lungfonden.se), King Oscar II Foundation, Carl Trygger Foundation (www.carltryggersstiftelse.se), and the Clas Groschinsky Foundation (www.groschinsky.org). The funders had no role in study design, data collection and analysis, decision to publish, or preparation of the manuscript.

Competing Interests: The authors have declared that no competing interests exist.

* Email: maria.lerm@liu.se

¤a Current address: Department of Microbiology and Immunology, Department of Chemical Engineering, University of Michigan, Ann Arbor, Michigan, 48109-2136, United States of America
¤b Current address: Phagocyte Research Laboratory, Department of Rheumatology and Inflammation Research, Sahlgrenska Academy, University of Gothenburg, Gothenburg, SE-41346, Sweden

◈ These authors contributed equally to this work.

Introduction

Tuberculosis (TB) is caused by *Mycobacterium tuberculosis* (Mtb), which primarily infects alveolar macrophages. Depending on the host immune status, the infection has different outcomes. In immunocompetent hosts, the bacterium may be controlled through innate immune mechanisms and/or by adaptive immunity [1,2]. In some individuals, the immune system fails to control the infection and the disease progresses to active TB. Factors that contribute to disease progression include HIV co-infection, malnutrition and predisposing genetic variations [1].

Treatment of TB requires administration of several drugs for at least 6 to 9 months, leading to high costs, side-effects and the emergence of drug-resistant strains associated with patient non-compliance. Therefore, one of the key elements in improved global TB control is a more effective treatment regimen to shorten the time of sterilizing antibiotic therapy by several months. Altered bacterial phenotypes have been suggested to be responsible for

tolerance of Mtb against antibiotics, and the prevailing view is that slow- or non-replicating bacteria in hypoxic granulomas are phenotypically tolerant towards antibiotics and thus responsible for the long time required for TB treatment. The hypoxic conditions in the granuloma have been mimicked *in vitro* by progressive oxygen depletion of cultures, rendering Mtb tolerant to isoniazid (INH) [3]. Although the absence of oxygen could directly affect the efficacy of INH [4,5,6], the tolerance has been attributed to the absence of replication [3].

In the human lung, Mtb can persist without the presence of granuloma [7], or in replicating and non-replicating states in subclinical lesions [8]. In the mouse model, Mtb can persist [9] although mice do not form hypoxic granulomas [10,11,12], and before the onset of adaptive immunity, substantial killing occurs [13]. Altogether, this speaks for a major role for innate immunity, at least during the early phase of infection and raises the question of tolerant Mtb being located outside of granulomas. Macrophages, constituting the primary target of infection and the first line of host defense, exert a range of pressures on the bacilli, forcing them to adapt to the harsh intracellular conditions and to shift phenotype, as shown earlier in different macrophage-based models [14,15,16] and mimicked in broth models [17]. We have shown that primary human macrophages are able to control bacterial net growth through mechanisms dependent on phagolysosomal functionality [18]. So far, Mtb replication and death rates have been difficult to determine and often neglected, although considerable evidence exists for divergent numbers of live and dead (or non-culturable) bacteria *in vivo* and *in vitro* [13,19,20,21]. The link between mycobacterial replication and drug tolerance is still not clear, and a recent study in a non-virulent mycobacterium, *Mycobacterium smegmatis*, elegantly showed that tolerance correlates with expression fluctuations of katG (a mycobacterial catalase-peroxidase which protects Mtb from oxidative stress but also transforms INH into its active form [22]) and is independent of replication rate [23]. Furthermore, asymmetrical division of *M. smegmatis* resulting in phenotypically heterogeneous siblings growing at different rates did not cause any differences in antibiotic susceptibilities [24].

With this study, we take these findings into a more physiological setting and evaluate drug susceptibility of phenotypically different, virulent Mtb inside human monocyte-derived macrophages (hMDM). We observed that hMDMs are able to restrict intracellular Mtb net growth for at least 10 days, provided that the initial bacterial burden was low. During growth restriction, Mtb displayed a phenotype that was rich in lipid bodies, but negative for acid-fast staining, both of which are features that have been linked to persistent Mtb. A higher bacterial burden, on the other hand, promoted an actively replicating phenotype that was positive for acid-fast staining. Finally, we tested whether the susceptibility towards first- and second-line TB drugs was different in the characterized phenotypes. Consistently with the findings obtained with *M. smegmatis*, we demonstrate that an altered replication rate of Mtb did not influence the susceptibility of the bacterium to antibiotics.

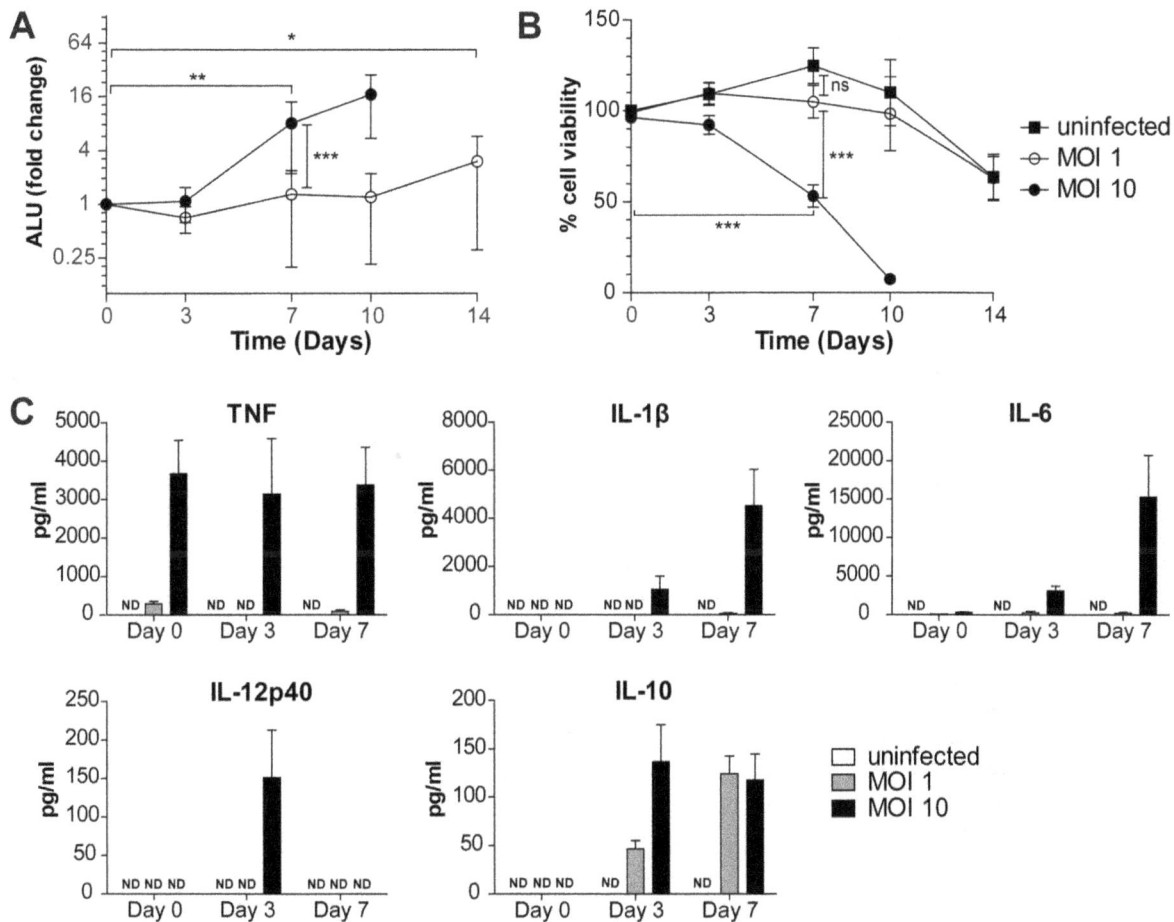

Figure 1. Kinetics of Mtb growth, macrophage cell death and cytokine secretion. Bacterial fold change (A; normalized to day 0 values of the respective MOI) and percentage of macrophage survival (B; normalized to uninfected controls on day 0) were measured during 14 days of H37Rv infection using luminometry for bacterial numbers and calcein-AM for macrophage viability. Arbitrary luminescence units (ALU) for medium supernatant and lysate (Figure S1) measurements were added to give totals (A). n = 7–32 and symbols and error bars represent means and 95% confidence intervals. Comparisons between MOI 1 and MOI 10 (and uninfected controls for (B)) at different time points were done using unmatched 2-way ANOVA of normalized values and Bonferroni post-hoc test for multiple comparisons. Significant changes compared to day 0 were determined using 1-way ANOVA of normalized values and Dunnett's test, and only the first time point significantly different from day 0 is indicated with asterisks (A and B). *p<0.05, **p<0.01, ***p<0.001. (C) For cytokine analysis, medium supernatants were saved on the respective days of infection and analyzed by cytokine bead array for the indicated cytokines. n = 5–7 and bars and error bars depict means and SEMs, respectively. ND: Not detected.

Results

Macrophages control Mtb net growth during a low burden infection

In order to establish whether unstimulated hMDMs were able to restrict growth of virulent Mtb for an extended period of time, we performed infection experiments through 14 days of infection. We found that infection of hMDMs with Mtb H37Rv at a multiplicity of infection (MOI) of 1 did not result in any significant net increase in bacterial numbers for at least 10 days, a period during which cell viability of infected cells was similar to uninfected cells (Figure 1A and B). On the contrary, infection with a higher MOI (MOI 10) resulted in significant bacterial growth by day 7 (Figure 1A), coinciding with extensive cell death (Figure 1B) and release of Mtb from dying cells causing an increase in the extracellular fraction, but not in the cell-associated fraction (Figure S1).

The different outcomes of MOI 1 and MOI 10 infection prompted us to map the inflammatory response of the cells to the different bacterial loads. Cells infected with MOI 10 released high amounts of TNF at day 0, and of IL-1β, IL-6, IL-12p40 and IL-10 starting from day 3. Cells infected with MOI 1 initially secreted TNF at levels corresponding to approximately 10% of the amount secreted from the MOI 10-infected cells. However, at day 3, there was no detectible TNF secretion from MOI 1-infected cells, followed by a slight increase by day 7. The other investigated pro-inflammatory cytokines were low (IL-6 on day 3 and IL-1β and IL-6 on day 7) or undetectable (IL-1β and IL-12p40 on day 3) during MOI 1 infection. On the other hand, the levels of the anti-inflammatory cytokine IL-10 increased by day 3 and were equal for both MOIs by day 7. Uninfected cells did not release any of the cytokines measured (Figure 1C), and hMDMs exhibited a heterogeneous phenotype at the time of infection, expressing both M1 and M2 macrophage makers (Figure S2 and Table S1), corresponding to a more dynamic classification of macrophages, as proposed by Mosser and Edwards [25], rather than the conventional IFNγ-/IL-4-induced M1/M2 phenotypes.

During the course of infection, bacterial numbers were measured using a H37Rv strain carrying a luciferase-encoding plasmid with a hygromycin resistance marker (pSMT1). To rule out the possibility of changes in luciferase expression after infection, we routinely correlated arbitrary luminescence units (ALU) to bacterial CFU. During extended macrophage infection, ALU and CFU correlated well and most importantly, bacterial numbers were not underestimated when using luminometry (Figure S3A). Furthermore, the luminescent signal did not diminish when hygromycin is absent indicating that plasmid loss does not occur during a time period of at least 14 days (Figure S3B). This is further supported by earlier publications on the same plasmid showing that CFU and ALU correlate well for at least 60 days in a murine infection model [26].

Bacterial replication rates are dependent on the initial bacterial burden

To investigate whether the absence of intracellular net growth during MOI 1 infection reflects bona fide non-replicating bacteria or a dynamic equilibrium (growth balanced by killing by macrophages), we used the replication clock plasmid [13]. Briefly, this low copy plasmid is lost from each generation at a constant rate, and together with the proportion of plasmid-containing Mtb, this rate can be used to derive the replication (r) and death (d) rates of the bacteria in a given population at a given time point.

Analysis of plasmid loss from intracellular bacteria revealed that during the initial phase of MOI 1 infection, there was no significant loss of plasmid (Figure 2A, estimated generation time of 6.5 days or 158 h). Between day 7 and 14 of MOI 1 infection, a shorter generation time of 1.5 days (38 h) was accompanied by an increase in bacterial death rate ($r = 0.43$ and $d = 0.40$, Figure 2B), suggesting growth balanced by killing during the later phase of infection. Both phases are consistent with the absence of net growth as observed in Figure 1A. For the time span between day 0 and day 7 during MOI 10 infection, Mtb was estimated to replicate once every 3 days (76 h, $r = 0.22$, $d = 0.07$, Figure 2B). The larger difference between r and d during MOI 10 compared to MOI 1 infection is reflected in the observed net growth during MOI 10 infection (Figure 1A). The method likely underestimates the replication rate (and hence overestimates the generation time) of the MOI 10 infection, since dying macrophages release replicating bacteria into the supernatant. For comparison, we determined the generation time in broth for H37Rv to be 37 hours.

Intracellular bacterial phenotype

Next, we characterized whether the slow-growing Mtb during MOI 1 infection displayed an altered phenotype, as compared to the actively growing Mtb during MOI 10 infection. Persistent Mtb

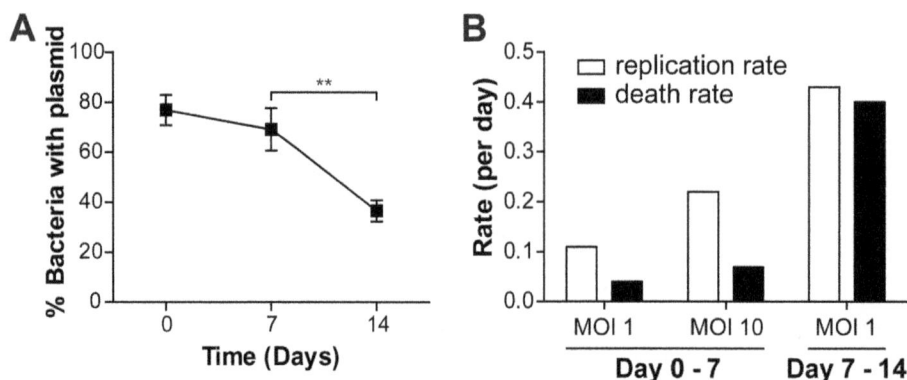

Figure 2. Loss of clock plasmid in the intracellular fraction and estimated mycobacterial replication and death rates. (A) CFU counts from cell lysates during MOI 1 infection on kanamycin-containing plates normalized to total CFU counts on plates without kanamycin. Differences in percentage of bacteria containing the clock plasmid was analyzed using 1-way ANOVA and Tukey's post-hoc test. n = 8–11. **p<0.01. (B) Estimated replication and death rates (per day) for intracellular Mtb were calculated from clock plasmid CFU data. Rates for MOI 10 infection between day 7 and 14 could not be determined due to extensive cell death.

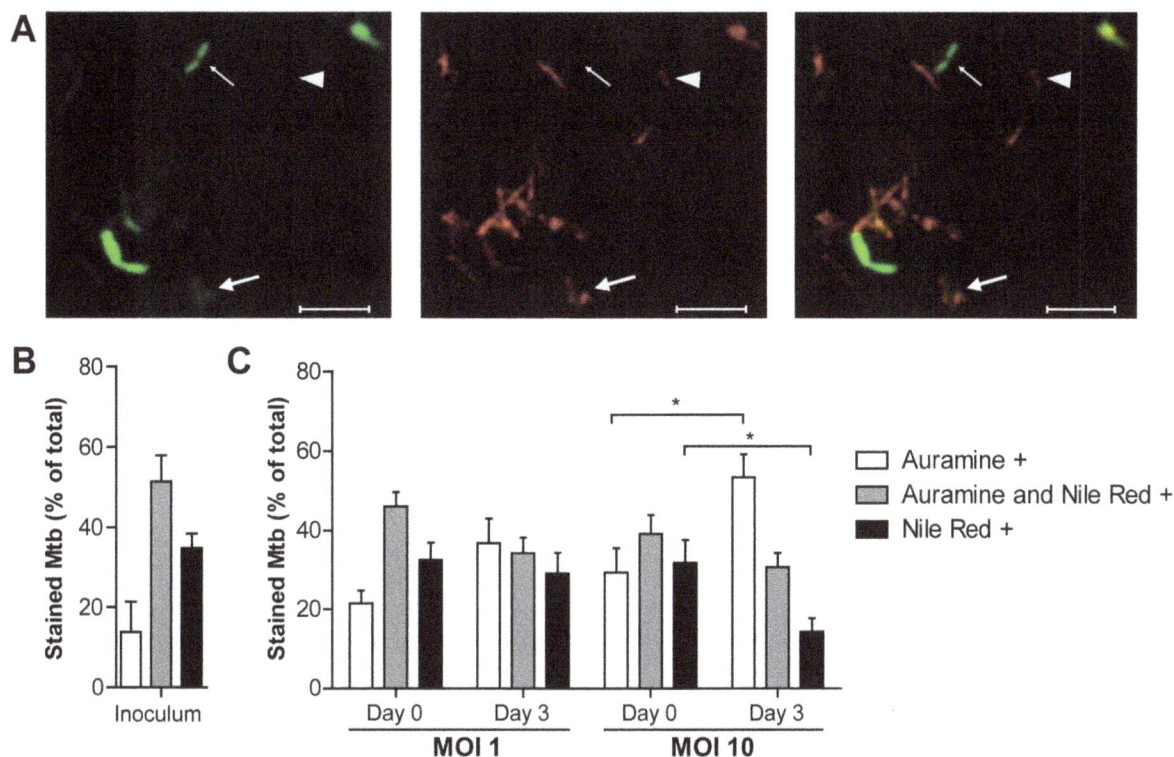

Figure 3. Phenotypic characteristics of Mtb inoculum and during macrophage infection. (A) Representative image of intracellular Mtb stained with Auramine O and Nile Red, counted as either Auramine O-positive (thin arrow), Nile Red-positive (arrowhead) or as positive for both stainings (thick arrow). Scale bar: 5 μm. (B) Percentage of bacteria stained with Auramine O, Nile Red or both in the inoculum. n = 3. (C) Percentage of Auramine O/Nile Red-positive intracellular bacteria, using hMDMs from 20 different donors. Significant changes were determined using 1-way ANOVA comparison followed by Bonferroni's multiple comparison test. *p<0.05. Bars and error bars represent means and SEMs respectively.

are characterized by reduced acid-fastness and intrabacterial accumulation of lipid bodies, as observed *in vivo* in the lungs of latently infected individuals, in sputum from TB patients and in an *in vitro* multiple-stress dormancy model [17,27,28]. In order to determine the phenotype of intracellular Mtb, we implemented a combined acid-fast (Auramine O) and lipid body (Nile Red) staining technique [17]. Representative images of stained intracellular Mtb are shown in Figure 3A. The inoculum displayed a mixed phenotype, with 14% of bacteria being positive for Auramine O only, 51% positive for both Auramine and Nile Red, and 35% positive for Nile Red only (Figure 3B).

One hour after infection (Day 0) with either MOI 1 or MOI 10, the staining pattern of intracellular Mtb resembled the inoculum (Figure 3B and C), indicating that no phenotypic shift occurred during the first hour of infection. This phenotype was not altered during MOI 1 infection by day 3 (Figure 3C), suggesting that the macrophages were able to maintain the initial bacterial phenotype. In contrast, there was a shift in the staining pattern of Mtb infecting the macrophages at the MOI 10 ratio, with a significant increase in Auramine-positive and a significant decrease in Nile Red-positive bacteria (Figure 3C). This phenotypic shift coincided with the bacterial replication observed during MOI 10 infection.

Antibiotic susceptibility of different Mtb phenotypes

Having established that our primary human macrophages were able to maintain an altered, slow-growing and lipid-rich phenotype of Mtb, we tested whether the sensitivity of these bacteria towards some first- and second-line antimycobacterial drugs was different

from the sensitivity of actively replicating, acid-fast bacteria in the same system.

To this end, antibiotics at concentrations based on human peak serum levels were added 1 hour after infection with either MOI 1 or MOI 10 and the number of intracellular bacteria was measured 4 days later. At this time point, no replication had taken place in the MOI 1 situation, whereas one replication had occurred in the MOI 10 situation (as determined by the clock plasmid experiment), thus reflecting situations with non-replicating and replicating bacteria, respectively (schematically outlined in Figure 4A). Significant reduction of bacterial numbers was seen after treatment with three of the first-line drugs ethambutol (EMB), INH and pyrazinamide (PZA) (Figure 4B), but there was no difference between the two MOIs. One possible interpretation of this result may be that the bacteria need time to shift to a different phenotype in the MOI 10 situation. To test this, we performed an additional experiment, in which the antibiotics were added 3 days after infection. Again, bacterial numbers were determined 4 days after addition of antibiotics, and INH was found to significantly kill intracellular bacteria (Figure 4C), but without any difference in antibiotic susceptibility between MOI 1 and MOI 10 infection. The same set of experiments was carried out with second-line drugs (amikacin, capreomycin, kanamycin, metronidazole and streptomycin), but none of the tested drugs caused any significant reduction in bacterial numbers as compared to untreated controls. As observed with the first-line drugs, there was no difference between the two MOIs (Figure S4), and none of the tested first- and second-line drugs rescued cell viability as compared to the untreated controls (Figure S5).

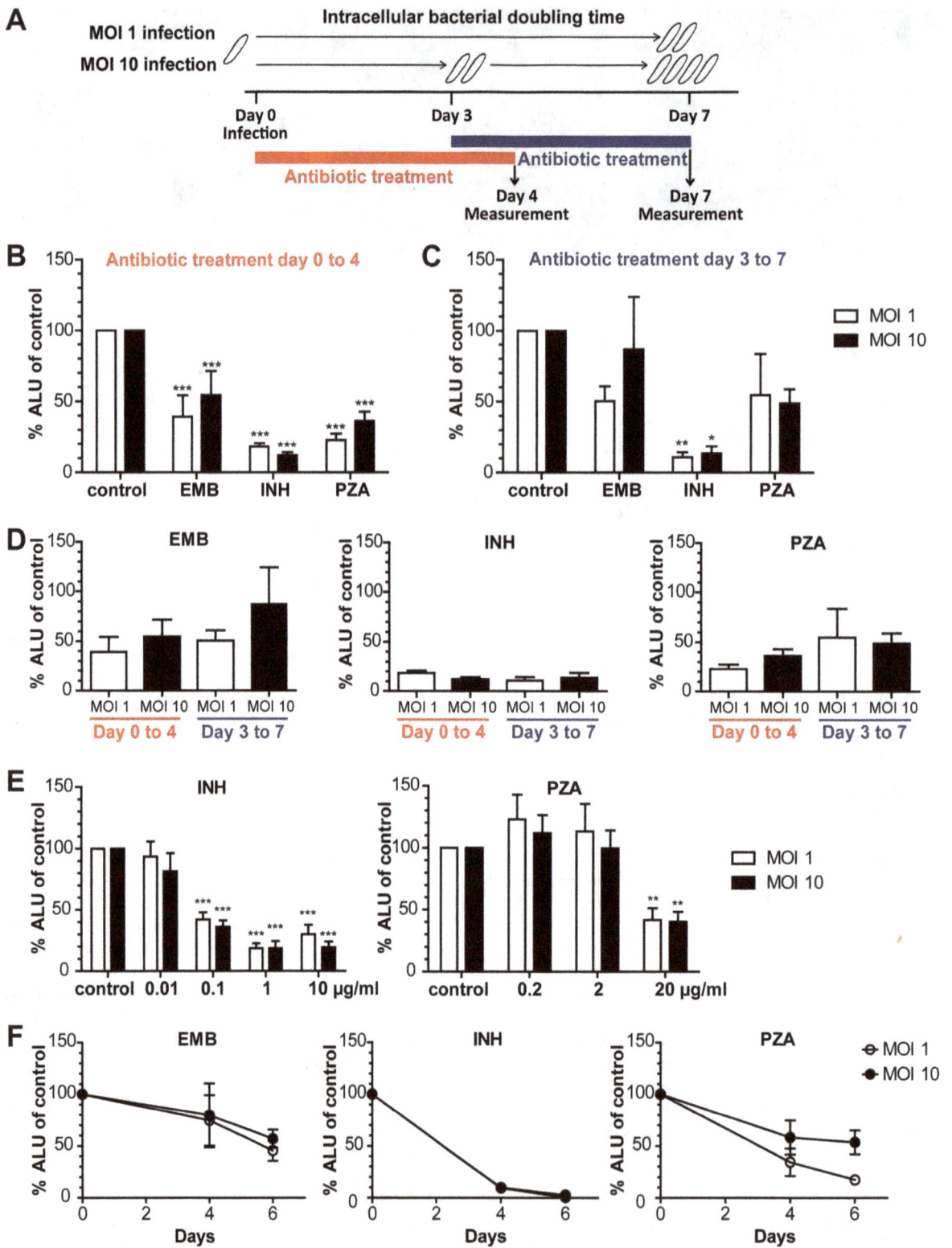

Figure 4. Antibiotic susceptibility of intracellular Mtb. (A) Schematic outline of the experiments, with antibiotics being added either 1 h (B) after infection or after 3 days (C) when Mtb in MOI 10-infected cells already had replicated once. Intracellular bacteria were quantified 4 days later on day 4 or 7, respectively. Antibiotics were used at the following concentrations derived from human peak serum levels: 1 µg/ml ethambutol (EMB), 10 µg/ml isoniazid (INH) and 20 µg/ml pyrazinamide (PZA). ALU as a measure of bacterial numbers were normalized against untreated controls of the same donor. Significant changes were determined using 2-way ANOVA followed by Bonferroni's post-hoc test comparing treated samples to untreated control. Differences between MOIs were not significant. (D) contains the data from (B) and (C). Groups were compared using 1-way ANOVA and Tukey's post-hoc test. No significant differences were found. (E) Lower concentrations of INH and PZA were added than in B–D (0.01, 0.1 and 1 µg/ml INH and 0.2 and 2 µg/ml PZA), and intracellular bacteria were measured on day 4. Significant differences between treated samples and untreated control were determined using 2-way ANOVA followed by Bonferroni's post-hoc test (indicated with asterisks). No significant difference

between MOIs was found. (F) Antibiotics were added 1 h after infection as in (B) and (E), but infection was extended beyond day 4 with another measurement on day 6. n = 5–10 in (B–D), n = 3 in (E) and n = 3 in (F). Bars and error bars represent means and SEMs, respectively. *p<0.05, **p<0.01 and ***p<0.001.

Using the data presented in Figure 4B and 4C, we made a statistical comparison of the percentage of bacteria remaining after 4 days of antibiotics treatment independently of whether the antibiotics were added at day 0 or day 3. The efficacy of EMB, INH and PZA did not differ between MOIs and time points (Figure 4D), indicating that the antibiotic susceptibility was not dependent on the replicative state of the bacteria. In order to rule out that the concentrations of drugs are too high to discriminate between tolerant and susceptible bacteria, we tested lower concentrations of the drugs with the best intracellular effect, INH and PZA, but again, no significant differences between MOIs could be observed (Figure 4E). Treatment with antibiotics was also extended beyond 4 days, showing that bacterial numbers can be further diminished (Figure 4F), which speaks against a residual tolerant population.

The antibiotics were demonstrated to be effective against H37Rv in 7H9 broth cultures (Figure S6), with the exception of PZA that requires acidic pH for activity and was not expected to have any effect in broth, as well as metronidazole which requires anaerobic conditions [3,29]. All the other first and second line drugs effectively killed bacteria in broth, indicating that the bacteria used are genotypically susceptible to those antibiotics.

We also assessed whether treatment with INH, PZA and EMB affected the two studied phenotypes differently using the Auramine O/Nile Red staining protocol on intracellular Mtb after 4 days of infection. The activities of the studied drugs did not affect a certain phenotype more than the other, however, the reliability of the method could have been influenced by the fact that also antibiotic-killed bacteria were stained as indicated by the fragmented appearance of many bacteria (not shown).

Discussion

Aiming to understand how Mtb phenotypes relate to the lengthy treatment required for TB, we investigated antibiotic susceptibility of Mtb inside macrophages. Two major findings guided the investigation: first, evidence has accumulated that not only necrotic granulomas but also macrophages can harbor altered phenotypes of Mtb [14,15,16,30,31,32], and second, that antibiotic susceptibility might not necessarily be coupled to mycobacterial replication rate [23].

We found that unstimulated primary human macrophages harbored an altered phenotype of Mtb during low-burden infection, characterized by slow replication, lipid bodies and reduced acid-fast staining and that this phenotype exhibited similar antibiotic susceptibility as did actively replicating, acid-fast Mtb. The control of bacterial net growth in human macrophages infected with a low bacterial burden was earlier shown to depend on effective phagosomal acidification [18], while macrophages infected with Mtb at higher MOIs undergo necrotic cell death coinciding with intracellular replication [33]. In the present study, long-term infection experiments showed that the balance between macrophages and Mtb at the low MOI could be maintained for at least 10 days. While restriction of mycobacterial growth has been described in other macrophage-based systems, this was dependent on manipulation of the macrophages via factors such as IFN-γ, TNF, GM-CSF or hypoxic conditions [14,15,16]. These studies only describe the absolute numbers of intracellular bacteria and do not provide information about bacterial replication and killing

rates. We included the clock plasmid replication rate analysis [13] in order to distinguish lack of replication from coincident replication and death, both of which would result in unchanged bacterial numbers over time. During low MOI infection, an early phase of slow replication was followed by a phase of faster replication and compensatory killing. The fact that a period of bacterial turnover follows the initial phase suggests that Mtb dynamically cycles between actively replicating and non- or slow-replicating states. A possible explanation for the low bacterial death rate during the early stage of MOI 1 infection is that macrophage effector functions are ineffective against this phenotype of Mtb, which would provide a rationale for its existence. Our observation of an initial stage of slow-replicating bacteria is contrary to previous findings with Mtb CDC1551 infection of murine bone marrow-derived macrophages [19], which showed higher replication (and death) rates associated with a net decrease in bacterial load in the initial phase of the infection, followed by lower replication (and death) rates coincident with a net increase in bacterial numbers. The divergent results may be explained by differential inherent ability of murine and human macrophages to control Mtb infection, by strain variability and possibly also by factors affecting the phenotype(s) of the Mtb inoculum.

The absence of cytokine release from uninfected cells confirms that the cells were not pre-activated and suggests that factors acting inside the cells rather than mediators acting via auto- or paracrine routes contribute to the restriction of intracellular Mtb growth. Analyzing the Mtb inoculum, we found that both replicating (acid-fast-positive) and persister-like (lipid-rich/acid-fast-negative) bacteria were present, probably due to our unagitated Mtb culture conditions. This inoculum phenotypically resembles Mtb found in sputum from TB patients [28], thus constituting a physiologically relevant source of Mtb. Characterization of intracellular Mtb at the higher MOI revealed a significant shift towards the acid-fast-positive, lipid body-negative phenotype, which correlated with a higher replication rate. In contrast, the mixed Mtb phenotypes observed during initial infection were maintained throughout the experiment at the low MOI. Cell wall alterations leading to decreased acid-fastness are features of Mtb persistence in vivo [27,34], and both lost acid-fastness and accumulation of lipid bodies can be induced in a multiple-stress dormancy model [17], and in hypoxic macrophages [14]. In contrast, we show that unstimulated macrophages can harbor an altered Mtb phenotype under normal oxygen pressure. We were unable to quantify whether simultaneous bacterial replication and persistence occurs within the same cell since we could not distinguish the borders of individual cells using this staining protocol. However, the frequent appearance of Auramine O-positive bacteria in the vicinity of Nile Red-positive bacilli suggests that both phenotypes can exist in the same cell. Since the inoculum used in this study contained a mixture of Mtb phenotypes, we cannot make conclusions regarding the ability of the macrophages to induce a phenotypic shift from actively replicating to a lipid-body-rich and acid-fast-negative phenotype. Although this question needs further attention, previous studies have reported induction of stress-regulated genes in Mtb upon uptake into macrophages [19,35], suggesting that the pathogen alters its phenotype to endure the stressful intracellular environment.

Phenotypic drug tolerance has been attributed to the absence of replication, e.g. in *E. coli* [36] and recently also in intracellular *Salmonella* [37]. We found that EMB, INH and PZA efficiently killed intracellular bacilli, and the extent of killing was independent of the MOI, i.e. of the bacterial replication rate. Regardless of MOI and of the time point of addition of antibiotics, the susceptibility pattern was similar, suggesting that antibiotic tolerance of intracellular bacteria does not correlate with bacterial replication rates. Our results provide two possible explanations to the enigmatic fact that INH, which has been traditionally viewed as a drug that is ineffective against non-replicating Mtb [3], is successfully used to treat latent TB [38]. First, we show that it is possible that a macrophage population can balance growth by killing, housing actively replicating bacteria without a net increase of bacterial load, which has been shown to be the case in a mouse model for chronic TB [13]. More importantly, we show that independently of replication rates, INH is as effective in killing Mtb. Our data are supported by the recent study by Wakamoto et al. [23], in which INH tolerance of *Mycobacterium smegmatis* correlates with fluctuations in *katG* expression rather than replication rate. Another study led to a different conclusion and showed THP1 macrophage-induced tolerance to INH in replicating *Mycobacterium marinum* bacteria [30]. In the present study, we cannot exclude the possibility that activated or immunosuppressed macrophages would have rendered Mtb tolerant to INH or the other drugs tested here. We did not measure KatG fluctuations, and to our knowledge, this has not been studied inside macrophages. Furthermore, our model does not necessarily account for the Mtb phenotype found in the hypoxic core of granulomas, where Mtb might undergo a truly non-replicating state (Wayne) or be tolerant to INH due to other factors like oxygen inavailability [4,5]. The absence of activity of second-line drugs in our study is most likely explained by limited intracellular activity of these drugs [39]. Other studies, showing good intracellular effect of these antibiotics, did not investigate macrophage viability [40].

Although first-line drugs effectively killed the bacteria in MOI 10-infected cells in the present study, the treatment did not significantly rescue macrophage viability. The finding suggests that the initial bacterial load rather than the absolute numbers of bacteria determines cell death, however, the reason for this needs further investigation.

To conclude, unstimulated human macrophages were able to maintain phenotypically altered Mtb exhibiting some characteristics of persisters, which supports a role for innate immune cells in latent TB. Being based on infected primary human macrophages as opposed to broth cultures, our model provides a physiological environment in which altered Mtb phenotypes can be studied. Furthermore, we challenge the view that Mtb replication rates determine antibiotic susceptibility inside macrophages.

Materials and Methods

Ethics statement

Blood, collected at the blood bank at Linköping University Hospital, was obtained from healthy donors, who had given written consent for research use of the donated blood in accordance with the Declaration of Helsinki. Since blood donation is classified as a negligible risk to the donors and since de-identified samples were delivered to the researchers, this study did not require a specific ethical approval according to paragraph 4 of the Swedish law (2003:460) on Ethical Conduct in Human Research.

Bacteria

Mtb H37Rv (ATCC) carrying the luciferase-encoding pSMT1 plasmid [26] or both the pSMT1 and the "clock plasmid" pBP10 [13] were grown in Middlebrook 7H9 broth (Difco, Becton Dickinson) supplemented with glycerol, Tween-80 and albumin-dextrose-catalase (ADC, Becton Dickinson) as described earlier [41], and reseeded into fresh medium 7 days before infection. Bacteria carrying the plasmids were selected with 100 µg/ml hygromycin (Sigma) for pSMT1 and 75 µg/ml kanamycin (Sigma) for pBP10.

Human monocyte-derived macrophages

For the preparation of hMDMs from heparinized whole blood or buffy coats, isolation of the mononuclear cell fraction using LymphoPrep (Axis Shield) and differentiation of monocytes were performed as described [18,33]. Monocytes were allowed to differentiate into hMDMs for 5–8 days in Dulbecco's Modified Eagle Medium (DMEM, Gibco) containing 80 µM L-Glutamine (Gibco) and 10% non-heat inactivated human serum (from blood bank at Linköping University Hospital) pooled from 5 donors. The day before infection, cells were trypsinized and re-seeded in serum-containing medium: 1×10^5 cells/well in triplicates in black 96-well plates (Greiner) for determination of bacterial growth and cell viability, and 2.5×10^5 cells/coverslip for staining.

Macrophage characterization

For staining of intracellular macrophage markers, cells were treated with Cytofix/Cytoperm™ (BD Pharmingen) before staining with antibodies. Antibody manufacturers and concentrations used are given in Table S2. Samples stained with fluorophore-conjugated secondary antibodies only served as background controls for intracellularly stained samples. Isotype antibody-treated cells were used as background controls and single- and non-stained cells for color compensation. 10,000 events/sample were acquired using a Gallios Flow Cytometer (Beckman Coulter) and data was analyzed using Kaluza or Flowjo.

Experimental infection

For infection, bacteria were passaged through a 27 gauge needle to remove aggregates and diluted in serum-free medium as described earlier [41], then added to the macrophages at an MOI of 1 or 10. After 1 hour of incubation, the medium was replaced by fresh DMEM containing human serum. For long-term infections, medium was changed on day 3, 7 and 10. For antibiotic susceptibility experiments, antibiotics (all from Sigma Aldrich) were added 1 hour or 3 days after infection. Intracellular bacterial numbers and cell viability were evaluated 4 days after addition of antibiotics as described below. Uninfected and untreated controls were included for all time points. Day 0 measurements were done 2 to 4 hours after infection.

Antibiotic susceptibility in broth

Mtb H37Rv expressing luciferase from the pSMT1 plasmid were prepared from the same culture as used for infection and diluted in Middlebrook 7H9 broth supplemented with Tween 80 and ADC, with or without antibiotics, to a concentration of 10^5 CFU/ml. Antibiotic concentrations used were the same as in the macrophage experiments. After 4 days of incubation, bacterial numbers were determined using the luminescence-based method described below.

Measurement of bacterial numbers and cell viability

Bacterial numbers were determined by a luminescence-based method published previously [41]. Aliquots of medium supernatants and lysates containing luciferase-expressing bacteria were transferred to white 96-well plates (Greiner), and flash luminescence after injection of the luciferase substrate (1% decanal, Sigma Aldrich) was measured in a plate reader (GloMax-Multi+ Detection System with Instinct Software, Promega). The remaining supernatants were pooled, spun down and frozen at $-80°C$ for cytokine analysis, and cell viability was determined as described below prior to subjecting the cells to hypotonic lysis. Arbitrary luminescence units (ALU) obtained from supernatant and lysate measurements were corrected for background luminescence using ALU values from uninfected cells. In order to calculate the total values for each well (intracellular and extracellular bacteria), the ALUs of the supernatant and lysates were standardized for dilutions and summed up. For bacterial growth, the median value of each triplicate of all time points was normalized to the day 0 median of the same experiment (fold change) or normalized to medians of untreated controls of the same day in the antibiotics experiments.

To determine cell viability, cells were washed three times with PBS, followed by 30 min incubation with 4 μM calcein-AM (Molecular Probes). Fluorescence was measured in a plate reader. Arbitrary fluorescence units of infected samples were normalized to those of uninfected cells measured on day 0.

Correlating arbitrary luminescence units to CFU

In order to ensure the stable expression of the luciferase-encoding pSMT1 plasmid in Mtb H37Rv after macrophage infection and to exclude the possibility of underestimating the actual bacterial load due to plasmid loss, ALU measured in medium supernatants and lysates were repeatedly correlated to CFU obtained by traditional plating of the same samples. To do so, supernatant and lysate samples from triplicate wells were pooled, serially diluted and plated in triplicates on Middlebrook 7H10 agar supplemented with ADC. CFUs were counted after two and three weeks of incubation at 37°C, ALU und CFU calculated per well and the median CFU value was correlated to the mean ALU value (since triplicates had been pooled). In order to check for plasmid loss when bacteria are maintained without hygromycin, H37Rv expressing luciferase were grown in 7H9 broth supplemented with ADC in the presence and absence of the selecting antibiotic hygromycin. Every few days, ALUs were measured.

Cytokine analysis

Cytokine analysis was performed using the human flex sets for TNF, IL-1β, IL-6, IL-10 and IL-12p40 for Cytokine Bead Array (Becton Dickinson), according to the manufacturer's instructions followed by an additional fixation step (4% paraformaldehyde for 30 min). Samples were measured using a Gallios Flow Cytometer (Beckman Coulter) and data were analyzed using Kaluza software (Beckman Coulter).

Evaluation of replication rates

The loss of the clock plasmid from intracellularly replicating bacteria was determined by CFU plating of cell lysates on Middlebrook 7H10 plates supplemented with ADC with and without 75 μg/ml kanamycin. Bacteria containing the plasmid grow on both plates, whereas CFUs of bacteria without the plasmid appear only on kanamycin-free plates. The rate of plasmid loss (segregation rate) was determined in logarithmic

phase cultures to be 0.2. Bacterial replication and death rates can be calculated from the segregation rate, total CFU and plasmid containing fractions as outlined elsewhere [13].

Staining of Mtb

Staining of the inoculum and intracellular Mtb was adapted from Garton et al. [42]. Inoculum was streaked on microscope slides, dried and heat-fixed. hMDMs infected on glass coverslips were fixed with 4% paraformaldehyde either 1 hour or 3 days after infection. Microscope slides and coverslips were treated with Auramine O solution (TB Auramine M by Becton Dickinson), acid alcohol and Nile Red (Sigma Aldrich). Between all steps, slides were washed with water. Samples were mounted with fluorescence mounting medium (DAKO). Microscopy was performed using a Zeiss LSM 700 confocal microscope, taking Z-stacks and using the Zen software (Zeiss) for image projection. Bacteria were evaluated for staining with Auramine O and Nile Red.

Supporting Information

Figure S1   Kinetics of Mtb growth in the extracellular and cell-associated fraction. Bacterial fold-change in the macrophage supernatant (A) and lysate (B) during the long-term infection experiments shown in Figure 1A and 1B. Bacterial numbers were measured using luminometry and expressed ALU normalized to Day 0 values. n = 7–32 and symbols and error bars represent means and 95% confidence intervals. Comparisons between MOI 1 and MOI 10 at different time points were done using unmatched 2-way ANOVA of normalized values and Bonferroni post-hoc test for multiple comparisons. Significant changes compared to day 0 were determined using 1-way ANOVA of normalized values and Dunnett's test, and only the first time point significantly different from day 0 is indicated with asterisks. $**p<0.01$, $***p<0.001$.

Figure S2   Macrophage characterization. Surface (CD206, CD 163, DC-SIGN, CD86 and CD14) and intracellular (iNOS2, arginase I, and CD119) staining of hMDMs differentiated for 8 days. Plots show representative expression in one of six donors. Dashed lines show background fluorescence.

Figure S3   Correlation of arbitrary luminescence units to CFU, and plasmid stability. (A) ALUs from Mtb expressing luciferase were measured in aliquots of the cell lysates, and aliquots of the same samples were used for CFU plating. ALU/well and CFU/ well are shown over time from one representative donor of four. (B) Mtb expressing luciferase were grown in the presence and absence of the selecting antibiotic hygromycin and bacterial numbers were quantified by luminometry. One representative experiment of two is shown.

Figure S4   Intracellular susceptibility of Mtb to second-line TB drugs. Antibiotics were added either 1 h after infection (A) or on day 3 (B) after infection. Intracellular bacterial numbers were measured 4 days later, on day 3 or day 7, respectively. Antibiotics were used at the following concentrations: 1 μg/ml amikacin (AMI), 30 μg/ml capreomycin (CAP), 10 μg/ml kanamycin (KAN), 10 μg/ml metronidazole (MTZ) and 10 μg/ml streptomycin (STR). Bacterial numbers were normalized against untreated controls of the same donor. Significant differences were determined using 2-way ANOVA followed by Bonferroni's multiple comparison test comparing treated samples to untreated control. n = 3–6 and bars and error bars represent means and SEMs, respectively.

Figure S5 Cell viability of infected macrophages treated with first- and second-line TB drugs. First-line drug treatments in (A) and (B) correspond to the bacterial growth data shown in Figure 4B and 4C, and second-line drug treatments in (C) and (D) correspond to Figure S4. Antibiotics were added 1 h after infection (A) and (C) or on day 3 (B) and (D), and cell viability was measured at the same time point as intracellular bacterial numbers were determined, on day 4 or 7, respectively and normalized against the cell viability of uninfected cells from the same day. Significant differences were determined using 2-way ANOVA followed by Bonferroni's multiple comparison test comparing treated samples to untreated but infected control. Bars and error bars represent means and SEMs, respectively. $*p < 0.05$, $**p < 0.01$ and $***p < 0.001$.

Figure S6 Antibiotic susceptibility of H37Rv in 7H9 broth. Luciferase-expressing Mtb H37Rv were inoculated in 7H9 broth and exposed to first- and second-line drugs or left untreated (Control) for 4 days. The antibiotic concentrations used were the same as in Figure 4 and Figure S4. The number of bacteria in the samples was then assessed using luminometry and normalized to untreated controls. Bars depict means from four (EMB, INH, PZA) or two (AMI, CAP, KAN, MTZ, STR) independent experiments and error bars represent SEM.

Table S1 Macrophage markers on hMDMs from cells from six independent donors.

Table S2 Antibodies used for macrophage characterization.

Acknowledgments

We are grateful to Professor David Sherman for providing us with the replication clock plasmid, and to Professor Jan Ernerudh and Judit Svensson for advice on macrophage characterization.

Author Contributions

Conceived and designed the experiments: JR EP RB DE VPB AW ML. Performed the experiments: JR EP RB DE VPB AW. Analyzed the data: JR EP RB DE HA VPB AW. Contributed reagents/materials/analysis tools: HA. Contributed to the writing of the manuscript: JR EP RB DE AW ML.

References

1. Lawn SD, Zumla AI (2011) Tuberculosis. Lancet 378: 57–72.
2. Schon T, Lerm M, Stendahl O (2013) Shortening the 'short-course' therapy-insights into host immunity may contribute to new treatment strategies for tuberculosis. J Intern Med 273: 368–382.
3. Wayne LG, Hayes LG (1996) An *in vitro* model for sequential study of shiftdown of *Mycobacterium tuberculosis* through two stages of nonreplicating persistence. Infect Immun 64: 2062–2069.
4. Youatt J (1960) The uptake of isoniazid and related compounds by Mycobacteria. Aust J Exp Biol Med Sci 38: 331–337.
5. Zabinski RF, Blanchard JS (1997) The Requirement for Manganese and Oxygen in the Isoniazid-Dependent Inactivation of *Mycobacterium tuberculosis* Enoyl Reductase. J Am Chem Soc 119: 2331–2332.
6. Magliozzo RS, Marcinkeviciene JA (1996) Evidence for Isoniazid Oxidation by Oxyferrous Mycobacterial Catalase—Peroxidase. J Am Chem Soc 118.
7. Hernandez-Pando R, Jeyanathan M, Mengistu G, Aguilar D, Orozco H, et al. (2000) Persistence of DNA from *Mycobacterium tuberculosis* in superficially normal lung tissue during latent infection. Lancet 356: 2133–2138.
8. Young DB, Gideon HP, Wilkinson RJ (2009) Eliminating latent tuberculosis. Trends Microbiol 17: 183–188.
9. McCune RM Jr, McDermott W, Tompsett R (1956) The fate of *Mycobacterium tuberculosis* in mouse tissues as determined by the microbial enumeration technique. II. The conversion of tuberculous infection to the latent state by the administration of pyrazinamide and a companion drug. J Exp Med 104: 763–802.
10. Via LE, Lin PL, Ray SM, Carrillo J, Allen SS, et al. (2008) Tuberculous granulomas are hypoxic in guinea pigs, rabbits, and nonhuman primates. Infect Immun 76: 2333–2340.
11. Aly S, Wagner K, Keller C, Malm S, Malzan A, et al. (2006) Oxygen status of lung granulomas in *Mycobacterium tuberculosis*-infected mice. J Pathol 210: 298–305.
12. Tsai MC, Chakravarty S, Zhu G, Xu J, Tanaka K, et al. (2006) Characterization of the tuberculous granuloma in murine and human lungs: cellular composition and relative tissue oxygen tension. Cell Microbiol 8: 218–232.
13. Gill WP, Harik NS, Whiddon MR, Liao RP, Mittler JE, et al. (2009) A replication clock for *Mycobacterium tuberculosis*. Nat Med 15: 211–214.
14. Daniel J, Maamar H, Deb C, Sirakova TD, Kolattukudy PE (2011) *Mycobacterium tuberculosis* Uses Host Triacylglycerol to Accumulate Lipid Droplets and Acquires a Dormancy-Like Phenotype in Lipid-Loaded Macrophages. PLoS Pathog 7: e1002093.
15. Estrella JL, Kan-Sutton C, Gong X, Rajagopalan M, Lewis DE, et al. (2011) A Novel *in vitro* Human Macrophage Model to Study the Persistence of *Mycobacterium tuberculosis* Using Vitamin D(3) and Retinoic Acid Activated THP-1 Macrophages. Front Microbiol 2: 67.
16. Vogt G, Nathan C (2011) *In vitro* differentiation of human macrophages with enhanced antimycobacterial activity. J Clin Invest 121: 3889–3901.
17. Deb C, Lee CM, Dubey VS, Daniel J, Abomoelak B, et al. (2009) A novel *in vitro* multiple-stress dormancy model for *Mycobacterium tuberculosis* generates a lipid-loaded, drug-tolerant, dormant pathogen. PLoS One 4: e6077.
18. Welin A, Raffetseder J, Eklund D, Stendahl O, Lerm M (2011) Importance of phagosomal functionality for growth restriction of *Mycobacterium tuberculosis* in primary human macrophages J Innate Immun 3: 508–518.
19. Rohde KH, Veiga DF, Caldwell S, Balazsi G, Russell DG (2012) Linking the transcriptional profiles and the physiological states of *Mycobacterium tuberculosis* during an extended intracellular infection. PLoS Pathog 8: e1002769.
20. Muñoz-Elías E, Timm J, Botha T, Chan W-T, Gomez J, et al. (2005) Replication dynamics of *Mycobacterium tuberculosis* in chronically infected mice. Infect Immun 73: 546–551.
21. Lin P, Ford C, Coleman M, Myers A, Gawande R, et al. (2014) Sterilization of granulomas is common in active and latent tuberculosis despite within-host variability in bacterial killing. Nature Med 20: 75–79.
22. Sherman DR, Sabo PJ, Hickey MJ, Arain TM, Mahairas GG, et al. (1995) Disparate responses to oxidative stress in saprophytic and pathogenic mycobacteria. Proc Natl Acad Sci U S A 92: 6625–6629.
23. Wakamoto Y, Dhar N, Chait R, Schneider K, Signorino-Gelo F, et al. (2013) Dynamic persistence of antibiotic-stressed mycobacteria. Science 339: 91–95.
24. Santi I, Dhar N, Bousbaine D, Wakamoto Y, McKinney JD (2013) Single-cell dynamics of the chromosome replication and cell division cycles in mycobacteria. Nat Commun 4: 2470.
25. Mosser DM, Edwards JP (2008) Exploring the full spectrum of macrophage activation. Nat Rev Immunol 8: 958–969.
26. Snewin VA, Gares MP, Gaora PO, Hasan Z, Brown IN, et al. (1999) Assessment of immunity to mycobacterial infection with luciferase reporter constructs. Infect Immun 67: 4586–4593.
27. Seiler P, Ulrichs T, Bandermann S, Pradl L, Jorg S, et al. (2003) Cell-wall alterations as an attribute of *Mycobacterium tuberculosis* in latent infection. J Infect Dis 188: 1326–1331.
28. Garton NJ, Waddell SJ, Sherratt AL, Lee SM, Smith RJ, et al. (2008) Cytological and transcript analyses reveal fat and lazy persister-like bacilli in tuberculous sputum. PLoS Med 5: e75.
29. Zhang Y, Scorpio A, Nikaido H, Sun Z (1999) Role of acid pH and deficient efflux of pyrazinoic acid in unique susceptibility of *Mycobacterium tuberculosis* to pyrazinamide. J Bacteriol 181: 2044–2049.
30. Adams KN, Takaki K, Connolly LE, Wiedenhoft H, Winglee K, et al. (2011) Drug tolerance in replicating mycobacteria mediated by a macrophage-induced efflux mechanism. Cell 145: 39–53.
31. Peyron P, Vaubourgeix J, Poquet Y, Levillain F, Botanch C, et al. (2008) Foamy macrophages from tuberculous patients' granulomas constitute a nutrient-rich reservoir for M. tuberculosis persistence. PLoS Pathog 4: e1000204.
32. Caceres N, Tapia G, Ojanguren I, Altare F, Gil O, et al. (2009) Evolution of foamy macrophages in the pulmonary granulomas of experimental tuberculosis models. Tuberculosis (Edinb) 89: 175–182.
33. Welin A, Eklund D, Stendahl O, Lerm M (2011) Human Macrophages Infected with a High Burden of ESAT-6-Expressing *M. tuberculosis* Undergo Caspase-1- and Cathepsin B-Independent Necrosis. PLoS One 6: e20302.
34. Bhatt A, Fujiwara N, Bhatt K, Gurcha SS, Kremer L, et al. (2007) Deletion of kasB in *Mycobacterium tuberculosis* causes loss of acid-fastness and subclinical latent tuberculosis in immunocompetent mice. Proc Natl Acad Sci U S A 104: 5157–5162.
35. Tailleux L, Waddell SJ, Pelizzola M, Mortellaro A, Withers M, et al. (2008) Probing host pathogen cross-talk by transcriptional profiling of both *Mycobacterium tuberculosis* and infected human dendritic cells and macrophages. PLoS One 3: e1403.

36. Balaban NQ, Merrin J, Chait R, Kowalik L, Leibler S (2004) Bacterial persistence as a phenotypic switch. Science 305: 1622–1625.
37. Helaine S, Cheverton A, Watson K, Faure L, Matthews S, et al. (2014) Internalization of *Salmonella* by macrophages induces formation of nonreplicating persisters. Science 343: 204–208.
38. Zumla A, Atun R, Maeurer M, Mwaba P, Ma Z, et al. (2011) Viewpoint: Scientific dogmas, paradoxes and mysteries of latent *Mycobacterium tuberculosis* infection. Trop Med Int Health TM & IH 16: 79–83.
39. Dhillon J, Mitchison DA (1989) Activity and penetration of antituberculosis drugs in mouse peritoneal macrophages infected with *Mycobacterium microti* OV254. Antimicrob Agents Chemother 33: 1255–1259.

40. Rastogi N, Labrousse V, Goh KS (1996) *In vitro* activities of fourteen antimicrobial agents against drug susceptible and resistant clinical isolates of *Mycobacterium tuberculosis* and comparative intracellular activities against the virulent H37Rv strain in human macrophages. Curr Microbiol 33: 167–175.
41. Eklund D, Welin A, Schon T, Stendahl O, Huygen K, et al. (2010) Validation of a medium-throughput method for evaluation of intracellular growth of *Mycobacterium tuberculosis*. Clin Vaccine Immunol 17: 513–517.
42. Garton NJ, Christensen H, Minnikin DE, Adegbola RA, Barer MR (2002) Intracellular lipophilic inclusions of mycobacteria *in vitro* and in sputum. Microbiology 148: 2951–2958.

14

Comparative Genomic Analysis Shows That Avian Pathogenic *Escherichia coli* Isolate IMT5155 (O2:K1:H5; ST Complex 95, ST140) Shares Close Relationship with ST95 APEC O1:K1 and Human ExPEC O18:K1 Strains

Xiangkai Zhu Ge[1⊙], Jingwei Jiang[2,3⊙], Zihao Pan[1], Lin Hu[1], Shaohui Wang[4], Haojin Wang[1], Frederick C. Leung[2,3], Jianjun Dai[1]*, Hongjie Fan[1]

1 College of Veterinary Medicine, Nanjing Agricultural University, Nanjing, China, 2 Bioinformatics Center, Nanjing Agricultural University, Nanjing, China, 3 School of Biological Sciences, University of Hong Kong, Hong Kong SAR, China, 4 Shanghai Veterinary Research Institute, Chinese Academy of Agricultural Sciences, Shanghai, China

Abstract

Avian pathogenic *E. coli* and human extraintestinal pathogenic *E. coli* serotypes O1, O2 and O18 strains isolated from different hosts are generally located in phylogroup B2 and ST complex 95, and they share similar genetic characteristics and pathogenicity, with no or minimal host specificity. They are popular objects for the study of ExPEC genetic characteristics and pathogenesis in recent years. Here, we investigated the evolution and genetic blueprint of APEC pathotype by performing phylogenetic and comparative genome analysis of avian pathogenic *E. coli* strain IMT5155 (O2:K1:H5; ST complex 95, ST140) with other *E. coli* pathotypes. Phylogeny analyses indicated that IMT5155 has closest evolutionary relationship with APEC O1, IHE3034, and UTI89. Comparative genomic analysis showed that IMT5155 and APEC O1 shared significant genetic overlap/similarities with human ExPEC dominant O18:K1 strains (IHE3034 and UTI89). Furthermore, the unique PAI I$_{5155}$ (GI-12) was identified and found to be conserved in APEC O2 serotype isolates. GI-7 and GI-16 encoding two typical T6SSs in IMT5155 might be useful markers for the identification of ExPEC dominant serotypes (O1, O2, and O18) strains. IMT5155 contained a ColV plasmid p1ColV$_{5155}$, which defined the APEC pathotype. The distribution analysis of 10 sequenced ExPEC pan-genome virulence factors among 47 sequenced *E. coli* strains provided meaningful information for B2 APEC/ExPEC-specific virulence factors, including several adhesins, invasins, toxins, iron acquisition systems, and so on. The pathogenicity tests of IMT5155 and other APEC O1:K1 and O2:K1 serotypes strains (isolated in China) through four animal models showed that they were highly virulent for avian colisepticemia and able to cause septicemia and meningitis in neonatal rats, suggesting zoonotic potential of these APEC O1:K1 and O2:K1 isolates.

Editor: Mikael Skurnik, University of Helsinki, Finland

Funding: This work was supported by the Fundamental Research Funds for the Central Universities (KYZ201326), the Fund of Priority Academic Program Development of Jiangsu Higher Education Institutions (PAPD) and the Fundamental Research Funds for the Central Universities (KYZ201214). The funders had no role in study design, data collection and analysis, decision to publish, or preparation of the manuscript.

* Email: daijianjun@njau.edu.cn

⊙ These authors contributed equally to this work.

Introduction

Escherichia coli generally colonizes the mammalian intestinal tract commensally, but highly adapted *E. coli* clones can become true pathogens called "pathotypes", some of which cause various lethal diseases after acquisition of specific virulent factors [1,2]. These *E. coli* pathotypes can be broadly classified as intestinal pathogenic *E. coli* or extraintestinal pathogenic *E. coli* (ExPEC) based on the pathogenic types [3]. Intestinal pathogenic *E. coli* strains (IPEC) cause infection in the gastrointestinal system, while ExPEC strains cause urinary tract infections, newborn meningitis, abdominal sepsis, and septicemia in the extraintestinal system [2,4]. ExPEC pathotypes are classically divided into four groups,

based on the disease pathology, namely avian pathogenic *E. coli* (APEC), uropathogenic *E. coli* (UPEC), neonatal meningitis *E. coli* (NMEC), and septicemic *E. coli* [5–7].

In order to discriminate ExPEC from commensal and intestinal pathogenic *E. coli*, several molecular epidemiology approaches are used for ExPEC typing. The classical typing method is the identification of *E. coli* (O: K: H) serotypes, and highly virulent ExPEC isolates can be classified as several specific and predominant O1, O2 and O18 serotypes strains, which can express K1 capsule and are popularly isolated from human and avian colibacillosis [6,8–10]. Related to above mentioned three O serotypes, O6 serotype strains are also highly virulent and popular among UPEC isolates [6,11], and APEC O78 serotype strains are also frequently isolated from avian colibacillosis [6,12]. The phylogroup typing method based on multilocus enzyme electrophoresis (MLEE) and several relevant DNA markers are generally used for identification of *E. coli* genetic and evolutionary characteristics. *E. coli* can be classified as four major phylogroups (A, B1, D and B2) in accordance with the studies of Clermont et al. [13–16], and an additional fifth group (E) [17–19]. Most ExPEC isolates belong to the mainly phylogroup B2 and a lesser group D, especially highly virulent ExPEC strains, while intestinal pathogens and commensals *E. coli* mainly belong to group A and B1 [20]. In addition, the phylogroup E contains almost all serotype O157:H7 strains [18,19,21]. Multilocus sequence typing (MLST) is currently most powerful typing system for the discrimination of bacterial population genetics [22]. The molecular epidemiology shows that phylogenetic diversity of *E. coli* isolates are unambiguously differentiated based on *E. coli* MLST data (clonal complexes and sequence types data) [17,23]. ExPEC and IPEC isolates are generally distributed in distinct clonal complexes i.e. sequence type complexes, containing numerous sequence types (ST) for *E. coli* MLST database. The majority of ExPEC isolates are located in several specific ST complexes (95, 73, 131, 127, 141, et al.), which are called ExPEC dominated clonal complexes[24–27]. Phylogroup B2 ExPEC strians of serotypes O1, O2 and O18 are generally located in ST complex 95, and ExPEC isolates of ST complex 95 are popular objects for ExPEC genetic characteristics and pathogenesis in recent years [5,6,19,27–29].

After its entry via inhalation of fecal dust, APEC colonizes at the avian respiratory tract, and causes local infections and then spreads to various internal organs, resulting in systemic infection in poultry. These APEC-associated systemic infections have been proven economically devastating to global poultry industries [6,29–31]. The phylogroup B2 APEC strains isolated from avian colibacillosis mainly belong to O1:K1, O2:K1, and another O78 serotypes [6,9]. The complete genomic sequence of APEC O1 (an O1:K1:H7 strain; ST95) is first determined, which shares high similarities with the genomes of human UPEC isolates [5]. APEC and NMEC ST95 serotype O18 isolates can both cause meningitis in the rat model and disease in poultry, suggesting that they might have no or minimal host specificity [32]. APEC O78 strain χ7122 (ST23) is the second genome that has been sequenced in APEC isolates, which keeps close relationship with human ST23 ETEC than that of APEC O1 and human ExPEC strains. APEC wildtype strain IMT5155 (O2:K1:H5; ST complex 95, ST140; B2 phylogroup) is often used as a classic infection strain of APEC pathogenicity to identify APEC virulence factors [33–35]. Due to close relationship of ExPEC O2:K1 serotype strains with extraintestinal infection between humans and animals, we reported the complete genome sequence of IMT5155 in order to unravel the evolutionary and genomic features of APEC O2 isolates. We further compared IMT5155 genome with other *E. coli* strains to identify APEC/ExPEC genetic characteristics. In addition, virulence and zoonotic potentials of APEC O1:K1 and O2:K1 serotypes isolates were assessed through animal models for pathogenicity testing.

Materials and Methods

APEC strain and the total DNA extraction

The avian pathogenic *E. coli* strain IMT5155 was isolated from a chicken with the typical clinical symptoms of avian colibacillosis at a German chicken farm in the year 2000 and were provided by Lothar H Wieler and Christa Ewers [33]. The IMT5155 cells were cultured in LB media to its exponential growth phase and harvested by centrifuge. The bacteria genomic DNA extraction

was extracted using the Bacterial DNA Kit (Omega Bio-Tek, America).

454 pyrosequencing of the IMT5155 genome and assembly

A whole genome shotgun library was produced with 5 µg of the genomic DNA of IMT5155. The shotgun sequencing procedure followed the instruction of 454 GS Junior General Library Preparation Kit (Roche). In addition, an 8 kb insert paired end library was produced with 15 µg of the genomic DNA of IMT5155. The paired end sequencing procedure followed the instruction of 454 GS Junior Paired-end Library Preparation Kit (Roche). Paired-end reads were used to orientate the contigs into scaffolds. The DNA libraries were amplified by emPCR and sequenced by FLX Titanium sequencing chemistry (Roche). Two shotgun runs and one paired-end runs were performed based on their individual library. After sequencing, the raw data were assembled by Newbler 2.7 (Roche) with default parameters. Primer pairs were designed along the sequences flanking the gap regions for PCR gap filling. The complete sequences of IMT5155 chromosome and two plasmids have been deposited in GenBank (Accession numbers: CP005930, CP005931, and CP005932, respectively).

Genome annotation of IMT5155

Glimmer 3.02 was used for gene prediction of IMT5155 complete genome [36]. The Glimmer results were corrected manually, and pseudogenes were investigated through genome submission check process for GenBank (http://www.ncbi.nlm.nih.gov/genomes/frameshifts/frameshifts.cgi), and small CDSs in intergenic regions were identified by IASPLS (Iteratively adaptive sparse partial least squares) [37]. Then, all the predicted ORF sequences were translated into protein sequences. BLASTp was applied to align all the above protein sequences against the NCBI non-redundant database (January, 2013) [38]. Protein sequences with alignment length over 90% of its own length and over 50% identity were chosen and the name of the best hit will be assigned to the corresponding predicted gene. rRNA operons were annotated by RNAmmer (http://www.cbs.dtu.dk/services/RNAmmer/), tRNA genes tRNAscan-SE Search Server (http://lowelab.ucsc.edu/tRNAscan-SE/), and tmRNA were annotated by tmRNA Database (http://rth.dk/resources/rnp/tmRDB/) with default parameters.

Phylogenomic analysis of IMT5155 with other *E. coli* pathotypes

46 complete genomes and 1 draft genome of *E. coli* strains were downloaded from NCBI GenBank (File A in File S3). The othologous genes were identified by using the predicted genes of IMT5155 to align to all annotated genes of 47 *E. coli* by BLAT (the BLAST-like alignment tool) [39]. Those single copy IMT5155 genes over 90% of alignment length against all other *E. coli* strains were considered as the common genes, which composed the common genome of 47 *E. coli* strains. Then, all the common genes were aligned by MUSCLE and concatenated together [40]. Finally, the concatenated aligned genes were submitted to MrBayes with the GTR+G+I substitution model [41]. The chain length was set to 10,000,000 (1 sample/1000 generations). The first 2,000 samples were discarded as burn in after scrutinizing the trace files of two independent runs with Tracer v1.4 (http://tree.bio.ed.ac.uk/software/tracer/).

Virulence genes and Genomic islands of IMT5155

The annotated genes were submitted to IslandViewer (http://www.pathogenomics.sfu.ca/islandviewer/genome_submit.php) and PAIDB (https://www.gem.re.kr/paidb/about_paidb.php) with default parameters for the identification of genomic islands s, i.e., pathogenecity island-like region [42,43]. Then the annotated genes were submitted to VFDB database (http://www.mgc.ac.cn/VFs/) for the identification of virulence genes [38,44]. Protein sequences with alignment length over 90% of its own length and over 50% identity were chosen from VFDB database, and the name of the best hit will be assigned to the corresponding predicted gene. Through online prediction and manual inspection, we obtained the detailed and precise information for IMT5155 GIs and virulence genes.

Comparative genomic analysis

For comparative studies, common genes in chromosomes of other *E. coli* strains (APEC O1, CFT073, χ7122, MG1655, SE15, O157Sakai, IHE3034, CE10, 83972, NA114, UMN026, UTI89, E2348/69, RM12579, NRG857c, and UM146) shared with *E. coli* IMT5155 were identified and plotted along with all predicted genes in *E. coli* IMT5155 (with >90% alignment length and >50% identity). The similarities and differences of the predicted genes located in IMT5155 genomic islands were highlighted among the other *E. coli* strains.

p1ColV5155 and 5 plasmids (pAPEC-O2-ColV, pAPEC-O1-ColBM, pUTI89, pMAR2, and pO83-CoRR) were used for plasmid comparative analysis and synteny analysis. The common genes in 5 plasmids shared with p1ColV$_{5155}$ were identified and plotted along with all predicted genes in p1ColV$_{5155}$ as well as some functional genes. All genes of 5 plasmids were aligned with all genes predicted in p1ColV$_{5155}$ respectively. Then, the aligned genes (with >90% alignment length and >50% identity) were shown for synteny analysis. The scripts for comparative ORF analysis and GIs distribution between IMT5155 and other *E. coli* strains were shown in File B in File S3.

The distribution analysis of 10 sequenced B2 ExPEC pan-genome virulence genes among all sequenced *E. coli* strains

The homologous and non-orthologous genes in genomes of 10 sequenced B2 ExPEC strains (NA114, UTI89, IHE3034, IMT5155, APEC O1, S88, CFT073, Clone Di14, ABU83972, 536) were identified by this standard: homology genes, gene sequence identity ≥80% and coverage ≥80%, otherwise it was a non-orthologous gene. The total genes of the homologous and non-orthologous genes of those genomes represent the pan-genome of 10 sequenced B2 ExPEC genomes. The genes of pan-genome for 10 sequenced B2 ExPEC were translated into protein, and then protein of 10 sequenced B2 ExPEC pan-genome were submitted to VFDB database (with >90% alignment length and >50% identity) [38,44]. Then all predicted virulence genes were one by one manually verified through a large number of references about ExPEC virulence factors, and the confirmed virulence-associated genes were classified as six categories: adhesins, invasins, toxins, iron acquisition/transport systems, polysialic acid synthesis, and other virulence genes. For distribution analysis of virulence genes, common genes in 46 *E. coli* genomes (selected consistent with phylogenomic analysis) (File A in File S3) shared with virulence genes of 10 sequenced B2 ExPEC pan-genome were identified with >90% alignment length and >50% identity, and highlighted among all 46 sequenced *E. coli* strains expect draft PCN033 genome sequence. The scripts for virulence genes

statistics and heat-map for virulence gene distribution were shown in File B in File S3.

Pathogenicity testing

All animal experimental protocols were approved by the Laboratory Animal Monitoring Committee of Jiangsu Province, China.

(i) Chicken embryo lethality assay (ELA). The ELA model was performed to evaluate lethality in chicken embryos for IMT5155 and other APEC strains, as previously described [5,32]. Briefly, approximately 500 CFU of each cultured bacterial were inoculated into the allantoic cavity of a 12-day-old, embryonated, specific-pathogen-free egg (Jinan SAIS Poultry Co. Ltd.), and 20 eggs were successively inoculated for every experimental group. PBS-inoculated and uninoculated were used as negative controls. The inoculated eggs were checked daily, and embryo deaths were recorded for 4 days.

(ii) Chick colisepticemia model. IMT5155 and other APEC strains to cause avian colibacillosis were assessed for chick lethality, as previously described [5,32]. Briefly, group of 10 1-day-old SPF chicks (QYH Biotech) were inoculated intratracheally with 0.1 ml bacteria suspensions (approximately 10^7 CFU) for APEC and other strains. The groups for chicks inoculated with PBS and MG1655 acted as negative controls. Measuring time for mortality were 7 days after postinfection. Deaths were recorded, and the survivors after 7 days were euthanatized, and all tested chicks in each group were dissected and examined for lesion scores (ranked from 0 to 3 in accordance with the presence of airsacculitis, pericarditis, and perihepatitis). The air sacs, blood in heart, and brain of all tested chicks were picked using inoculation loops, and then plates of MacConkey agar were crossed by inoculation loops and cultured at 37°C overnight.

(iii) Mouse sepsis model. The mouse sepsis model for virulence evaluation of ExPEC isolates was performed on the basis of previously described methods [28,45,46]. Approximately 10^7 CFU (0.2 ml) of bacteria suspensions for APEC and other strains were injected intraperitoneally into 8-week-old imprinting control region (ICR) mice, and every group contained 10 mice. Mice for health status were observed twice daily during 3 days postinfection, which was score on a 5-step scale (1 = healthy, 2 = minimally ill, 3 = moderately ill, 4 = severely ill, 5 = dead) with the worst score as the score for that day, as described by Johnson et al. [28]. The mean of the 3 daily health status scores represented each mouse's infection process during 3 days postinfection. The blood in heart and brain of all tested mouse were picked using inoculation loops, and then plates of MacConkey agar were crossed by inoculation loops and cultured at 37°C overnight.

(iv) Rat neonatal meningitis model. The abilities to induce septicemia and enter the central nerves system (CNS) for APEC strains were assessed by 5 days old, specific-pathogen-free Sprague-Dawley rats, as previously described [28,32]. And *E. coli* MG1655 and NMEC strain RS218 acted as negative and positive controls, respectively. Groups of 12 rat pups were intraperitoneally inoculated with approximately 200 CFU of bacteria suspensions (20 μl) [32]. At 24 h postinoculation, rats were subsequently euthanized, and 25 μl of blood and 10 μl of cerebrospinal fluid (CSF) from each survivor for infected rat pup were obtained for quantitative cultures. The blood and CSF were plated on MacConkey agar to measure the bacteria concentration in the blood and indicate meningitis, respectively.

Results and Discussion

Sequencing and overview of the complete genome of APEC strain IMT5155

The complete genome of APEC strain IMT5155 was determined by initial *de novo* assembly of two shotgun sequencing runs and one paired-end sequencing run (8-kb insert paired-end library) followed by PCR gap-filling. The raw shotgun reads and paired-end reads were assembled into 121 contigs which were further assembled into eight scaffolds. The N50 contig size was 177,509 bp. The largest scaffold size was 4,907,543 bp (containing 56 large contigs). The second largest scaffold size was 191,765 bp (containing 14 large contigs) indicating that our raw assembly was highly continuous and that might be sequence of *E. coli* large plasmids. Primer pairs were designed to amplify the gaps between contigs. The PCR products were directly sequenced using a Sanger sequencer ABI 3730. For the shotgun runs, one run generated 132,755 reads (~53 Mb) and the other generated 108,804 reads (~47 Mb). The average read length of both shotgun runs was approximately 400 bp. The paired-end run generated 90,792 reads (~26 Mb) with an average read length of approximately 300 bp. Over 99% of the total reads were assembled, resulting in approximately 23-fold coverage of the genome of APEC strain IMT5155.

The complete genome of APEC strain IMT5155 comprises 5,126,057 bp, existing as a circular chromosome of 4,929,051 bp and two plasmids of 194,170 bp and 2,836 bp. Glimmer 3.02 annotated 4,804 CDSs covering 87.87% of IMT5155 chromosome. In addition, 27 pseudogenes and 30 small CDSs in intergenic regions were identified (File C in File S3). p1ColV$_{5155}$ contained 270 Glimmer-predicted CDSs (File D in File S3), and 6 CDSs were identified in p2$_{5155}$. Moreover, 88 tRNA genes, 19 rRNA genes, and 1 tmRNA gene were identified in the IMT5155 chromosome (File C in File S3). The GC content of the IMT5155 chromosome is approximately 50.65%, which is similar to other reported *E. coli* genomes. By contrast, the two plasmids have GC% contents of 49.84% (p1ColV$_{5155}$) and 42.21% (p2$_{5155}$). The large plasmid, p1ColV$_{5155}$, was identified as a ColV plasmid, which was widespread in ExPEC pathotypes, particularly in APEC pathotype[47,48]. Table A in File S2 summarizes the general genomic features of IMT5155 genome. Among 5,144 Glimmer-annotated CDSs found in IMT5155 genome, 5,053 (~98.2%) could be matched to genes in the NCBI nr database (December, 2013).

Whole-genome phylogenetic analysis of IMT5155 compared with other *E. coli* pathotypes

Whole-genome-derived phylogeny of common genomes can accurately illustrate evolutionary relationships among different commensal and pathogenic *E. coli* variants [49]. The genomes of IMT5155 and another 46 *E. coli* strains were selected for mapping the whole-genome evolutionary phylogeny, ranging from a commensal K12 strain, through intestinal pathogenic strains, to the highlighted extraintestinal pathogenic strains (Figure 1). MrBayes was used to construct a BMCMC phylogenetic tree to define the evolutionary phylogeny of 47 whole genome sequenced *E. coli* strains, based on *E. coli* common genes. The common genes identified from IMT5155 and the others 46 *E. coli* genomes comprised 1,782 genes and covered approximately 1.61 Mb. The result of phylogeny showed that 47 *E. coli* strains could be clearly divided into six monophyletic groups, which was similar to the whole-genome-based phylogeny by both Rasko and McNally et al. [26,50] (Figure 1). In the phylogenetic tree, APEC strains IMT5155 and APEC O1 were located in B2 ExPEC cluster

(Figure 1), and an APEC O78 strain χ7122 was located in B1 clade (Figure 1). The phylogenomic tree showed that ST complex 95 APEC dominant O1:K1 and O2:K1 serotypes strains (APEC O1 and IMT5155) have the closest evolutionary relationships with human ExPEC dominant O18:K1 (ST95 complex) strains (UTI89 and IHE3034).

Identification of virulence determinants and genomic islands in the IMT5155 genome

Many virulence-associated factors were identified in IMT5155 genome (Table B in File S2). Adhesins, invasins, and iron uptake systems were critical for APEC/ExPEC pathogenesis, which typically promote motility, achieve the capability of adhesion to and invasion of host tissues, and conduct iron uptake for survival [51–53]. The predicted adhesins of IMT5155 genome were listed in Table B in File S2. Six different chaperone-usher adhesion determinants were identified at IMT5155 genome, including *fim*, *yqi*, *yad*, *auf*, *yfc*, and *fml* operons. APEC strains shared common invasion genes with NMEC strains isolated from patients with neonatal meningitis [28,51]. Several microbial invasion determinants, including *Ibe* proteins, *yijP*, *aslA*, K1 capsule, and *Hcp* family proteins (Table B in File S2) which contribute to invasion of brain microvascular endothelial cells (BMECs), were identified at both APEC and NMEC pathotypes [46,54,55]. IMT5155 possessed ferrous iron transporters *FeoABC* and *SitABCD* (Table B in File S2). Unlike widespread siderophore enterobactin, IMT5155 contained three ExPEC specific pathogen-related siderophores for salmochelin, aerobactin, and yersiniabactin, which took important roles in APEC virulence [52,56] (Table B in File S2).

The distinct genomic islands (GIs) of pathogens that encode various virulence factors are called pathogenicity islands (PAIs), which have a significant difference in GC content compared with the core genome, and some PAIs are usually integrated into tRNA genes [57]. In this study, 20 GIs, ranging from 4 to 96-kb, were annotated on the IMT5155 chromosome via PAIDB and IslandViewer (Table C in File S2). 14 GIs contained several potential virulence factors, as predicted by PAIDB forecast and NCBI BLAST analysis, and these islands could be considered as confirmed or presumed PAIs. Moreover, 5 prophage islands (GI-5, -6, -13, -18, and -19) were identified at IMT5155 chromosome. Among the five prophages, it seemed that GI-13 was a P4 family phage and GI-18 was a P2 family member. The coexistence of these two phages (a satellite and helper phage pair) was quite reasonable [58]. It was also likely that the GI-18 phage could produce two types of tail fibers by DNA inversion like phage Mu and several other phages [59,60]. The detailed and precise information for each GI had been elucidated and listed at Table C in File S2. We then focused on a novel APEC O2 PAI (GI-12) and two GIs (GI-8 and GI-22) coding Type VI secretion systems.

A novel APEC O2 PAI (GI-12), termed PAI I$_{5155}$, was identified from the IMT5155 chromosome, which inserted between the *cadC* and *yidC* genes of *E. coli* core genome, was adjacent to tRNA-Phe (Figure 2 and Table C in File S2). The total GC content of this island was 48.76%, below to the average GC content(50.65%)of IMT5155 chromosome. The size of PAI I$_{5155}$ was approximately 94 kb, composed 105 ORFs. Proteins encoded by ORFs of PAI I$_{5155}$ were shown in Figure 2 and Table C in File S2. PAI I$_{5155}$ was absent in APEC O1 and other ExPEC genomes in this study, and only partial CDSs including several virulence/fitness factors (*aatA*, ireA, *fecIRABCDE*, and *pgtABCP*) were identified in pathogenicity islands of other *E. coli* pathotypes. For virulence factors encoded in PAI I$_{5155}$, AatA of APEC autotransporter adhesin, IreA of iron-regulated virulence factor have

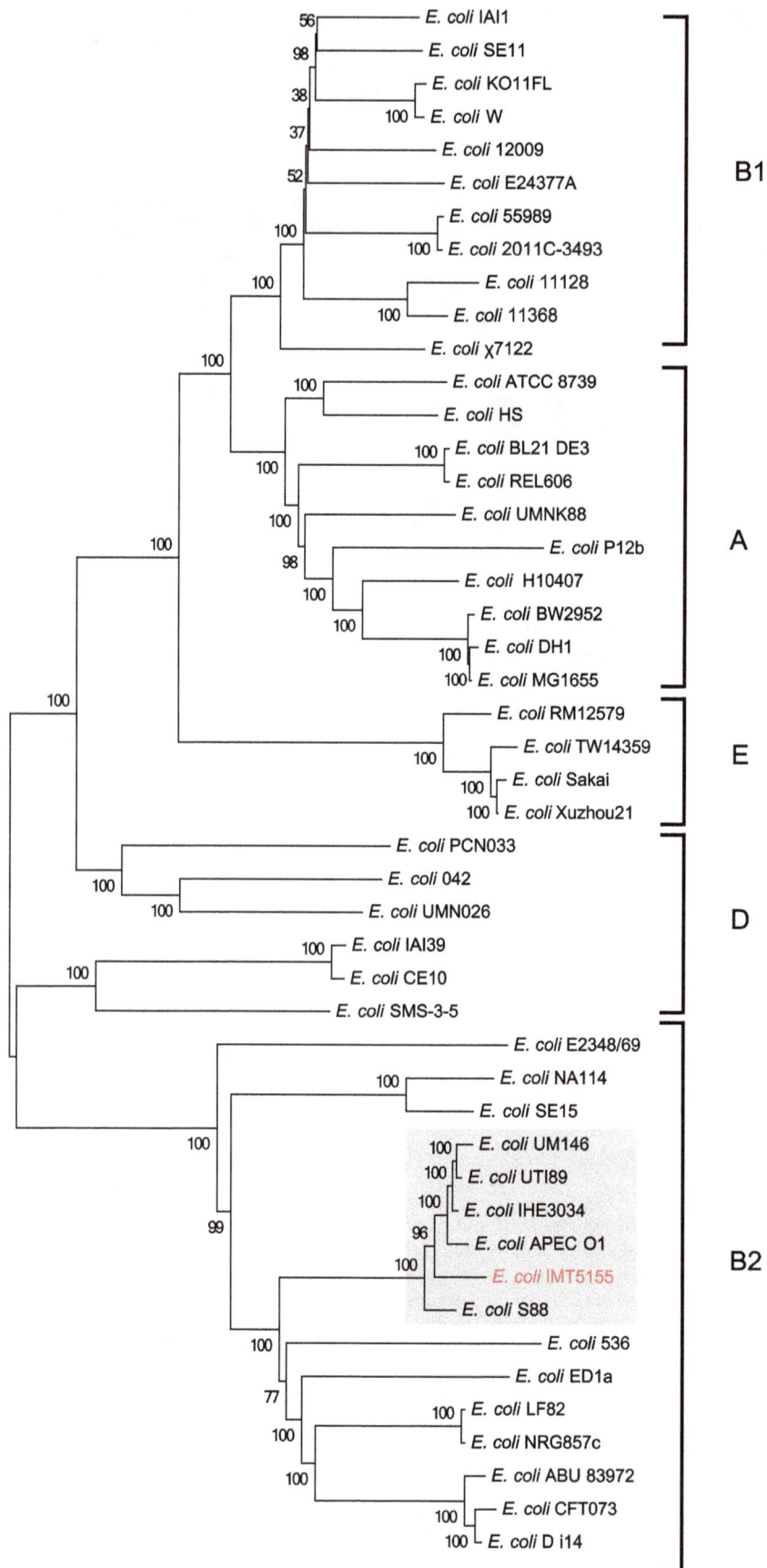

0.001

Figure 1. Phylogenomic tree (1,782 concatenated core genes, 1.61 Mb) of 47 *E. coli* strains. All MrBayes with the GTR+G+I substitution model (BMCMC) was used for the reconstruction of the phylogenomic tree. The chain length was set to 10,000,000 (1 sample/1000 generations). 47 *E. coli* strains clearly divided into monophyletically phylogroups (A, B1, B2, D, and E), and ST complex 95 strains were highlighted in phylogenomic tree. 47 *E. coli* genomes data was listed in File A in File S3.

been confirmed that they were involved in the pathogenicity of APEC/ExPEC [33,61,62], and other putative virulence genes need to be further identified (Figure 2 and Table C in File S2). Unlike other ExPEC, IMT5155 contained the ferric dicitrate transport system, which was previously reported to maintain *E. coli* growth under iron-limited circumstances and widespread among *E. coli* K-12, intestinal pathogenic *E. coli*, and *Shigella* strains [63]. For the putative metabolism/biosynthesis-related systems, those annotated genes of PAI I$_{5155}$ were mainly distributed in ExPEC strains by BLASTN analysis. A putative transketolase-like protein, which was adjacent to a putative ascorbate-specific IIABC component of a PTS system, was also annotated in PAI I$_{5155}$. In addition, like typical PAIs, PAI I$_{5155}$ contained many mobility elements, including four integrases and multiple transposons, suggesting that horizontal gene transfer and genomic recombination were possibly involved in the evolution of PAI I$_{5155}$ (Figure 2 and Table C in File S2). We identified a PAI I$_{5155}$ analogue located in the chromosome of APEC strain DE205B (O2:K1), which was isolated in China (unpublished data) [45]. Therefore, PAI I$_{5155}$ could be considered as a novel arrangement of these virulence factors and metabolism/biosynthesis-related systems. This island currently was only identified in APEC serotype O2 strains. Furthermore, roles of the putative virulence factors and metabolism/biosynthesis-related systems in pathogenicity and fitness of bacterial demands pending further research.

Type VI secretion systems (T6SSs) are distributed widely in many Gram-negative pathogenic bacteria [64]. IMT5155 carried two putative type VI secretion systems, which were located in GI-7 (32.2 kb) and GI-16 (28.2 kb) (Table C in File S2). GI-7, which was inserted between the *mltA* and *serA*-1 genes of B2 ExPEC core genome, was a region (GC content: 52.81%) adjacent to the tRNA-Met. GI-16 (GC content: 51.95%) located directly downstream of a tRNA-Asp, was inserted between the *yafT* and *ramA*-1 genes of *E. coli* core genome. GI-7 and GI-16 were respectively corresponding to T6SS1 and T6SS2, both of which have been recently described by Ma et al. [65]. The genes encoding secretion assembly components, including conserved core components of T6SS and additional unknown proteins [65], were located in GI-7 and GI-16 (Figure A in File S1). The typical T6SS1 (GI-7) was widely prevalent among the B2 and D ExPEC strains, and was elaborated to take roles in pathogenesis of APEC [28,66]. However, it was reported that the T6SS2 was mainly encoded in virulent isolated of B2 ExPEC and might be a potential marker for B2 ExPEC, but not associated with ExPEC virulence [28,65]. In order to identify whether T6SS2 can act as a potential marker for ExPEC dominant serotypes (O1, O2, and O18) strains, we detected almost all of the reported ExPEC O1:K1, O2:K1 and O18:K1 strains (genome sequences available online) and APEC isolates in our laboratory as previously described by Ma et al. [65] (Table D in File S2). We speculated that T6SS2 might be associated with ST95 ExPEC (serotypes O1, O2 and O18) strains, and those B2 phylogroup ExPEC (O1, O2, and O18) strains almost simultaneously contained two T6SSs (T6SS1 and T6SS2) (Table D in File S2).

Comparative genomic analysis of IMT5155 with other *E. coli* pathotypes

Comparative genomic analysis was performed using one by one alignment between IMT5155 genome and other 16 representative

E. coli strains based on their evolutionary relationships and phenotypes. The general comparison of IMT5155 genome content with 16 *E. coli* strains was shown in Table A in File S2. The 16 representative strains encompassed typical commensal *E. coli*, highly pathogenic diarrhoeagenic *E. coli*, and extraintestinal *E. coli* strains. Four of these 16 *E. coli* strains were used as control references for comparative genomic analysis, including the commensal strains (MG1655 and SE15), EHEC strain O157 Sakai, and EPEC strain RM12579. IMT5155 shared different numbers of common chromosomal genes with these strains (Table E in File S2). The comparative chromosomal atlas of IMT5155 with those *E. coli* genomes is shown in Figure 3. The results showed that significant differences in genome content mainly focus on IMT5155 GIs regions (Figure 3). The distribution of IMT5155 GIs among these strains was shown in Table C in File S2. The commensal *E. coli* genomes were usually smaller than *E. coli* pathotypes, and harbored fewer genes, especially accessory genes i.e., genomic islands by genomic recombination than *E. coli* pathotypes [19,49]. As described above, MG1655 harbored merely IMT5155 GIs homology loci (Figure 3 and Table C in File S2). Comparison between B2 phylogroup SE15 and IMT5155 reflected a similar result that only 4 IMT5155 GIs were present in SE15. The EHEC O157:H7 pathotype is a typical highly pathogenic diarrhoeagenic *E. coli* and highlighted the genomic plasticity for lateral gene transfer. EPEC strain RM12579 (O55:H7) is a precursor to O157:H7 pathotype [67,68]. Both E phylogroup Sakai and RM12579 harbored merely IMT5155 GIs homology loci (Figure 3 and Table C in File S2), and Sakai shared the least numbers of chromosomal common genes with IMT5155 (Table E in File S2). The typical EPEC strain E2348/69 (serotype O127:H6) shares close evolutionary relationship with B2 ExPEC pathotypes, but has no common GIs with IMT5155. Two AIEC strains (UM146 and NRG857c) shared relatively largest numbers of common genes with IMT5155. UM146 and NRG857c had12 and 9 common GIs with IMT5155, respectively.

For 9 ExPEC strains in the comparative genomic analysis, APEC O1, IHE3034, and UTI89 exhibited closest phylogenetic relationship with IMT5155 (Figure 1). CFT073, ABU83972 and NA114 were in different subclades of phylogenetic tree relative to IMT5155, respectively (Figure 1). Our phylogenetic tree and previous studies revealed APEC ST23 serotype O78 strain χ7122 arose from distinct lineages with APEC O1 and IMT5155 [12]. In addition, CE10 and UMN026 belong to phylogroup D. The comparative genomic analysis showed that IMT5155 GIs, excepting for PAI I$_{5155}$ and several prophage GIs, were highly conserved in APEC O1, IHE3034, and UTI89 (Figure 3 and Table C in File S2). Furthermore, IMT5155 shared the highest number of common chromosomal genes with IHE3034 (3,948; 83.0% of the total annotated CDSs in IHE3034 genome) (Table E in File S2). In contract, IMT5155 GIs were not widespread among CFT073, ABU83972, NA114, CE10, UMN026, and χ7122 (Table C in File S2). Moreover, 16 of the 20 genomic islands of IMT5155 were absent or poorly conserved in χ7122, and this result further reinforced the fact that ST23 APEC O78 strains lacked of conservation of virulence-associated genomic islands with ST95 APEC serotypes O1 and O2 strains (Figure 3 and Table C in File S2). Interestingly, the results showed that prophage GIs in IMT5155 exhibited partial or no homology among these

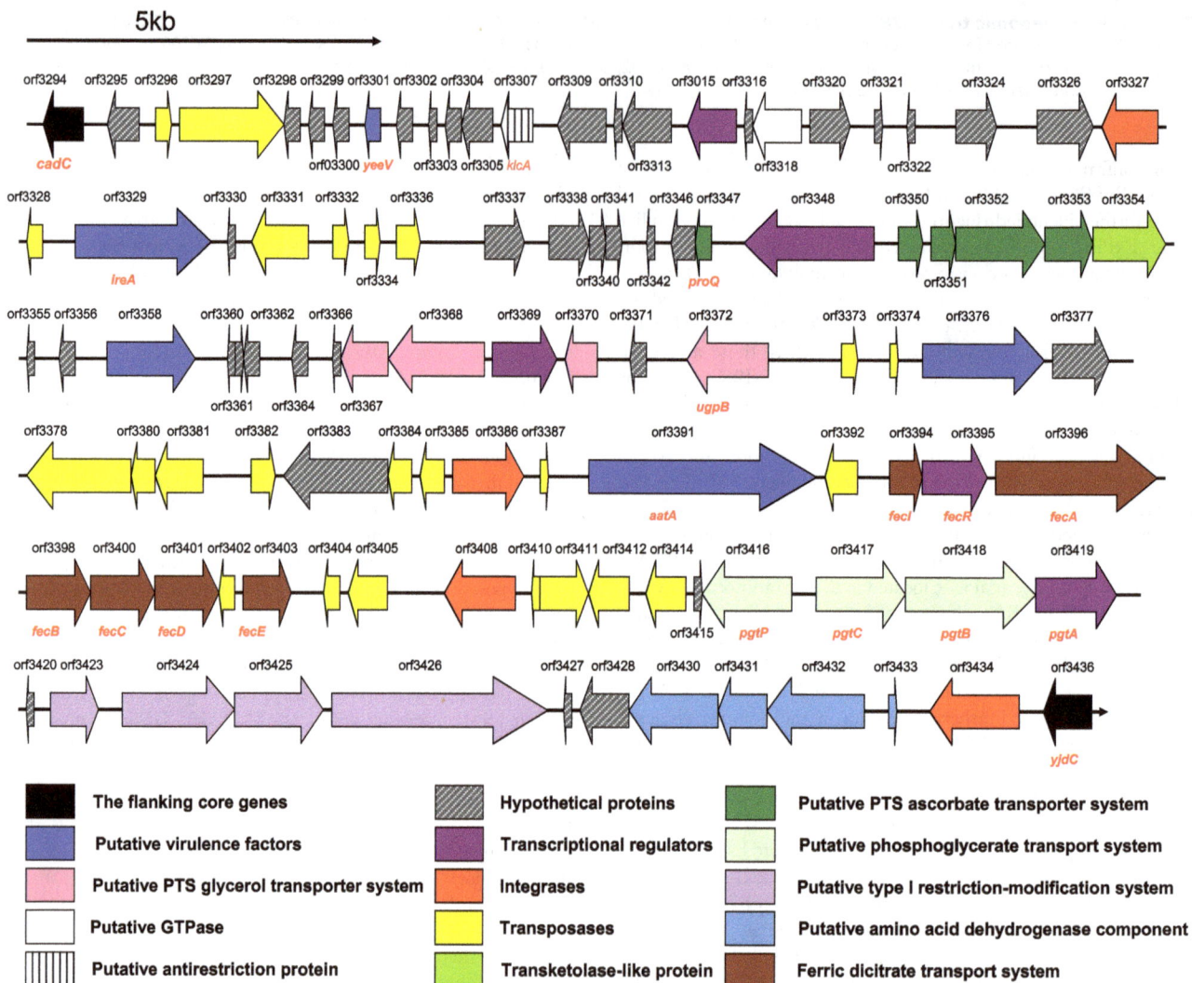

Figure 2. Chimeric feature and genetic context of PAI I$_{5155}$ (GI-12). PAI I$_{5155}$ was inserted between the *cadC* and *yidC* genes of *E. coli* core genome. Proteins encoded by the ORFs of PAI I$_{5155}$ represented by arrows, and the direction of the arrows indicated the direction of transcription. The color keys for functions of these proteins were shown at the bottom.

ExPEC strains. These results showed that genomes of APEC O1 and IMT5155 shared significant genetic overlap/similarities with human ExPEC O18 strains UTI89 and IHE3034. Moreover, those GIs of IMT5155 that were widespread among APEC O1, IHE3034, and UTI89 might be involved in or contribute to the pathogenicity and niche adaptation of ExPEC O1/O2/O18 strains (phylogroup B2; ST complex 95).

Sequence analysis and characterization of IMT5155 ColV plasmid p1ColV5155

(i) Analysis and characterization of the structure of p1ColV$_{5155}$. The IMT5155 strain harbored a 194-kb ColV plasmid, termed p1ColV$_{5155}$, which have been described elsewhere [69]. p1ColV$_{5155}$, which was depicted in a circular map (Figure 4), comprised 214 CDSs, encoding virulence-related proteins, plasmid conjugal transfer proteins, mobile genetic elements, and hypothetical proteins. The number and percentage of common genes of p1ColV$_{5155}$ and the other *E. coli* pathotypes' plasmids were listed in Table F in File S2. p1ColV$_{5155}$ shared more common genes with pAPEC-O2-ColV and pAPEC-O1-

ColBM than the other large plasmids in other *E. coli* pathotypes (Table F in File S2). In an effort to better define p1ColV$_{5155}$ backbone, classical circular genetic map was applied for comparative CDSs analysis of the p1ColV$_{5155}$ with five other large plasmids (pAPEC-O2-ColV, pAPEC-O1-ColBM, pUTI89, pMAR2, and pO83-CoRR), three (pUTI89, pMAR2, and pO83-CoRR) of which acted as references for homology analysis (Figure 4). Plasmids pUTI89, pMAR2, and pO83-CoRR were respectively present in UTI89, E2348/69 and NRG 857C, which shared close evolutionary relationships with IMT5155 in the preceding section. In addition, synteny analysis between CDSs inp1ColV$_{5155}$and the above five plasmids were also performed (Figure B in File S1). For the Tra genes region, we identified the detailed locations of p1ColV$_{5155}$ homologous genes among those five plasmids. The common genes of p1ColV$_{5155}$ with pAPEC-O2-ColV and pAPEC-O1-ColBM were mainly concentrated in virulence and plasmid conjugal transfer regions. The conjugative transfer system regions of pUTI89 and pMAR2 also shared high identity with that regions of p1ColV$_{5155}$. However, the common

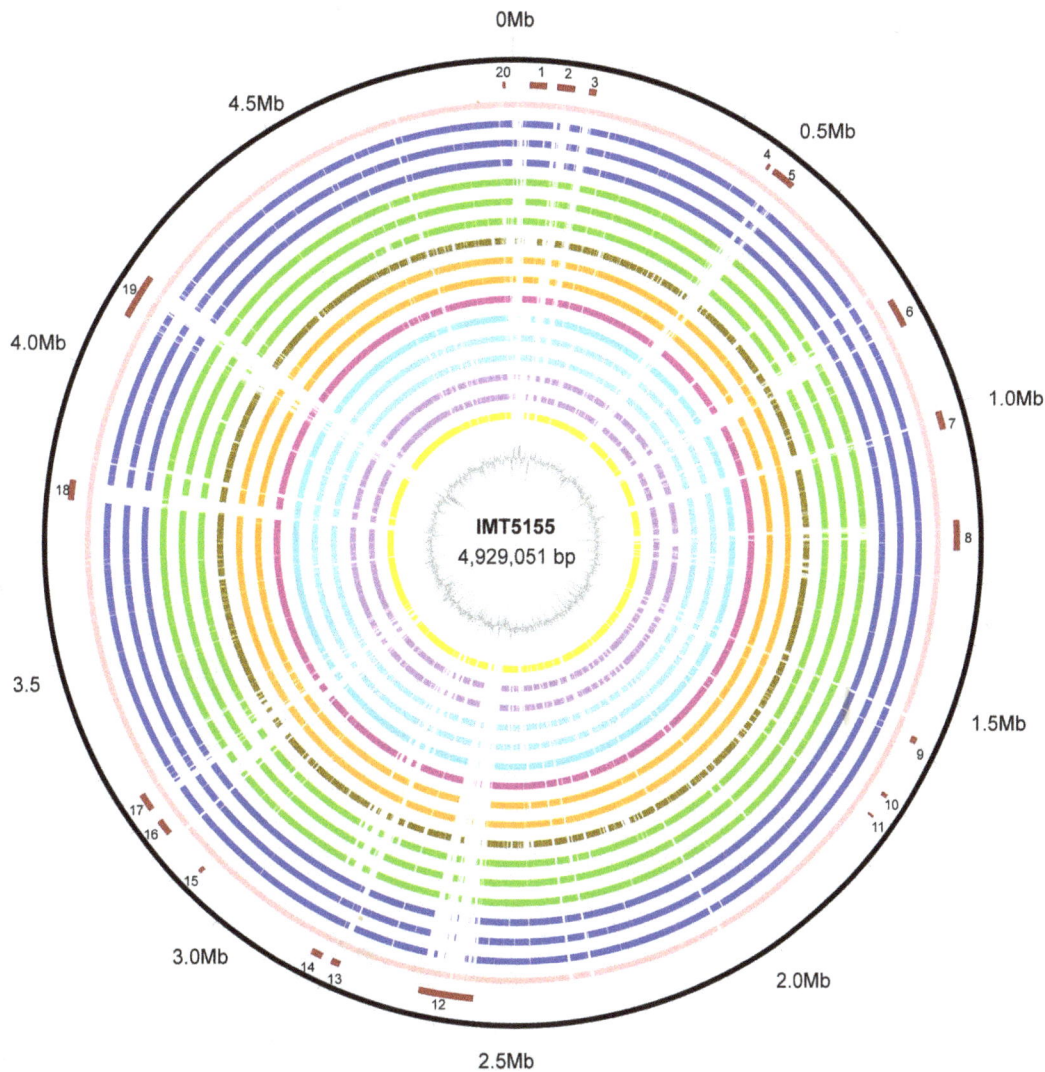

Figure 3. Comparative ORF analysis between IMT5155 and other *E. coli* strains. From outside to inside, the circles represent that: a) coordinate of IMT5155 genome; b) IMT5155 genomic island regions (red); c) IMT5155 (pink); d) APEC O1, IHE304, and UTI189 (blue); e) CFT073, ABU 83972 and NA114 (green); f) χ7122 (olive); g) UM146 and NRG857c (orange); h) SE15 (magenta); i) E2348/69 (cyan); j) CE10 and UMN026 (skyblue); k) O157 Sakai and O55:H7 RM12579 (purple); l) MG1655 (yellow); GC% of IMT5155 (calculated by 500 bp sliding window).

genes between pO83-CoRR and p1ColV$_{5155}$ were mainly located in the virulence region (Figure 4).

(ii) Virulence-associated genes of p1ColV$_{5155}$. ColV plasmids are generally present in ExPEC strains and contain a series of virulence genes [70]. Several genes of ColV plasmids, identified as being involved in APEC virulence and defined the APEC pathotype [47,48,71,72], were found at two virulence regikbons of p1ColV$_{5155}$. The first virulence region with the size of 62.1 kb was from *iroBCDEN* of the salmochelin cluster to *iucABCD* and *iutA* of the aerobactin cluster (Figure 4). The second region was a 24.3-kb virulence gene region from *cvaA* and *cvaB* of the ColV operon to *eitABCD* of a putative iron transport system (Figure 4). In particular, the first virulence region of p1ColV$_{5155}$ was nearly identical to the conserved portion of pAPEC-O2-ColV and pAPEC-O1-ColBM [47,48]. The second virulence region of p1ColV$_{5155}$ was homologous to the variable portion of pAPEC-O2-ColV and pAPEC-O1-ColBM, including *cvaAB*, *tsh*, and *eitABCD* [47,48] (Figure 4). However, the virulence genes' locus in p1ColV$_{5155}$ variable portion was completely inverted to that of

pAPEC-O2-ColV (Figure 4 and Figure B in File S1). Further analysis of variable portion revealed that p1ColV$_{5155}$ contained intact *cvaA* and *cvaB* genes for ColV export, but lacked the *cvaC* gene for ColV synthesis and the *cvi* gene for ColV immunity (Figure 4). Obviously, p1ColV$_{5155}$ neither contained ColB and ColM operons, which were the namesake traits of ColBM plasmids [48] (Figure 4). Therefore, this plasmid named as ColBM plasmid can be excluded, due to the namesake traits of ColBM plasmids. Even though without encoding *cvaC* and *cvi*, p1ColV$_{5155}$ was preferred to be classified as a ColV plasmid, which might lose the intact ColV operon during p1ColV$_{5155}$ evolution. One speculation is that p1ColV$_{5155}$ may be a novel type of ColV plasmid with rearrangements during its evolution. The pathogenic role of the two virulence regions of p1ColV$_{5155}$ might be correspondent to pVM01 of APEC strain E3, which was highly similar to pAPEC-O2-ColV and pAPEC-O1–ColBM [47,48,72]. The conserved section of the pVM01 virulence region was clearly shown to be associated with the virulence of APEC strains. However, the variable sections of this plasmid were not directly

Figure 4. Comparative ORF analysis between p1ColV₅₁₅₅ and other plasmids. From inside to outside, the circles represent that: a) GC% (calculated by 500 bp sliding window); b) common ORFs in pUTI89 (brown); c) common ORFs in pO83_CORR (green); d) common ORFs in pMAR2 (yellow); e) common ORFs in pAPEC-O2-ColV (grey); f) common ORFs in pAPEC-O1-ColBM (purple); g) p1ColV₅₁₅₅ (pink); i) highlighted functional ORFs in the negative strand of p1ColV₅₁₅₅; j) highlighted functional ORFs in the positive strand of p1ColV₅₁₅₅ (orange: RepF IIA, RepF IB, repB; blue: Transfer regions; red: virulence related genes; green: cvaAB locus).

associated with APEC virulence [72]. We speculated that the conserved section of p1ColV₅₁₅₅ virulence region might be involved in virulence of IMT5155.

(iii) Replication and transfer regions of p1ColV₅₁₅₅. Two replication regions were found in the chromosome of p1ColV₅₁₅₅: RepFIIA and RepFIB replicons (Figure 4). The first is a 33.4 kp region encompassing mostly predicted conjugal transfer genes of p1ColV₅₁₅₅, and the second is a 7.8 kp region contained another three conjugal transfer genes adjoining RepFIIA (Figure 4). The

plasmid transfer region of p1ColV₅₁₅₅ was slightly different from that of pAPEC-O2-ColV, which contained a complete plasmid conjugal transfer region [47].

The distribution of 10 sequenced B2 ExPEC pan-genome virulence genes among 46 sequenced *E. coli* strains

E. coli is highly evolved and adapted to the different specific environment. Recent findings show that the frequency of core genome recombination appears a striking decrease from intestinal commensal, through intestinal pathogenic strains, to phylogroup B2 ExPEC strains. Phylogroup B2 ExPEC strains are pathogenic variants, which show highly environmental adaptability with recombination being restricted [26,73]. Comparative genomic analysis of IMT5155 with other *E. coli* pathotypes showed that APEC dominant O1 and O2 serotypes strains (phylogroup B2; ST complex 95) shared significant genetic overlap/similarities with human ExPEC dominant O18 strains (IHE3034, and UTI89), and could be distinguished from APEC O78 strain χ7122, commensal *E. coli*, and IPEC. Accordingly, B2 ExPEC strains should harbor typical ExPEC-specific virulence factors, which could endue ExPEC a selective advantage to adapt/colonize to extraintestinal specific niches during infection relative to intestinal pathogenic strains.

In order to understand the relationship between virulence factors and genetic landscape of B2 ExPEC pathotypes, the distribution of 10 sequenced B2 ExPEC pan-genome virulence genes among 46 sequenced *E. coli* strains was conducted to examine whether B2 ExPEC strains harbored typical ExPEC-specific virulence factors (i.e., determining whether there were significant differences for the distribution of B2 ExPEC virulence genes among different *E. coli* pathotypes) [51]. The pan-genome of sequenced 10 B2 ExPEC strains contained 10,399 orhthologous gene families. The VFDB database predicted 287 virulence genes among these orhthologous genes. 73 virulence-associated genes were manually confirmed among these 287 virulence genes and classified as six categories: adhesins, invasins, toxins, iron acquisition/transport systems, polysialic acid synthesis, and other virulence genes. The details of 73 virulence genes of 10 sequenced B2 ExPEC pan-genome and their distributions among 46 sequenced strains were shown in Figure 5 and Table B in File S2. The distribution diagram showed that 10 sequenced B2 ExPEC pan-genome virulence genes were significant occurring in extraintestinal pathogenic strains compared with commensal and diarrhoeagenic *E. coli*, and several virulence genes were only present among ExPEC strains, such as fimbrial adhesins (*yqi*, *auf*, and *papG*), invasins (*ibeA* and *Hcp*), almost of toxins, and others (Figure 5 and Table B in File S2). The distribution of 10 sequenced B2 ExPEC pan-genome virulence factors provided a meaningful information for ExPEC-specific virulence factors, including several adhesins, invasions, toxins, iron acquisition systems, and others (Figure 5 and Table B in File S2), which were conserved in ExPEC pathotypes and contributed to ExPEC to adapte/colonize extraintestinal specific niches during infection. Moreover, these specific virulence factors might also provide valuable targets for the vaccines design.

Certainly, there may be strain-to-strain variation of the distribution of virulence genes in any specific strains (Figure 5). For example, compared with other B2 ExPEC strains, IMT5155 does not have F1C, P, and S fimbariaes, which are involved in UPEC pathogenesis [53]. We wondered whether there were specific genes or virulence factors to define the APEC pathotype. For 10,399 orhthologous genes of 10 sequenced ExPEC pan-genome, 239 genes were identified in IMT5155 genome relative to the other 9 B2 ExPEC strains (Table G in File S2), and 202 genes were present only in APEC O1, and 24 genes were only common present in APEC strains (IMT5155 and APEC O1) compared with the other 8 B2 ExPEC strains (data not shown). The hypothetical genes and prophage genes were predominant among those specific genes for each APEC strains, and only five virulence genes (*aatA*, *eitA*, *eitB*, *eitC*, and *eitD*) were identified among 24 common genes. Moreover, 600 orhthologous genes were identify as NMEC-specific genes. Similarly, the majority of NMEC-specific genes were prophage genes and hypothetical genes, and no virulence

factors were only present in NMEC (data not shown). Even though 3462 UPEC-specific genes among 10,399 orhthologous genes of 10 sequenced ExPEC pan-genome were identified in six UPEC strains, almost all virulence genes identified in UPEC strains were present among some APEC and UPEC strains. Therefore, there may be slight different distributions of virulence genes for an individual ExPEC strain, but no specific type of virulence genes to define B2 ExPEC subpathotype. The distribution analysis of 10 sequenced B2 ExPEC pan-genome virulence factors were further considered that phylogroup B2 APEC might not be differentiated from group B2 human ExPEC pathotypes (NMEC/UPEC), because two APEC O1 and O2 strains shared ExPEC-specific virulence factors with human ExPEC pathotypes. Furthermore, these results also support the previous findings that phylogroup B2 APEC isolates share remarkable similarities with human ExPEC pathotypes, and might pose a potential zoonosis threat [5,9,10,27,74].

Virulence assessment of APEC O1:K1, O2:K1 and O78 serotypes isolates

The pathogenicity and zoonotic potential of APEC O1:K1 and O2:K1 serotypes isolates, including IMT5155 and several strains isolated in China, were assessed with four animal models [5,28,32,45,46]. In addition, one ST23 APEC O78 strain CVCC1553 and an APEC non-dominant serotype strain Jnd2 (B2; ST95; O39:K1) were also included in the virulence assessment. The strains APEC O1, NMEC RS218, and UPEC CFT073 were used as the positive control, while *E. coli* K-12 MG1655 and CVCC1531 were used as negative control [5,28,32,45,46]. The detail information of these 13 selected strains was shown in Table H in File S2.

The virulence of the selected APEC O1:K1, O2:K1, and O78 strains for natural reservoir were assessed by chicken embryo lethality assay (ELA) and chick colisepticemia model for avian colisepticemia. In ELA assay, the mortalities for un-inoculated, PBS-inoculated, Jnd2, and CVCC1531 inoculated embryos were not obviously different from the negative control MG1655, while seven APEC O1:K1, O2:K1, and O78 strains were significantly different from the negative control MG1655 ($P<0.05$) (Table 1). No significant differences existed among the seven APEC O1:K1, O2:K1, and O78 strains compared to the ELA-positive control strain APEC O1 (high pathogenicity) (Table 1). For chick colisepticemia assay, the mortalities, rates of reisolation from the chick organs, and lesion scores were evaluated. Similarly to ELA results, seven APEC O1:K1, O2:K1, and O78 strains were significantly different from the negative control MG1655 ($P<0.05$) (Table 2) (the original data shown in File E in File S3), while no significant differences were observed among the seven APEC O1:K1, O2:K1, and O78 strains compared to the high-pathogenicity control strain APEC O1 (Table 2). Therefore, based on the results of two models for avian colisepticemia, seven selected APEC O1:K1, O2:K1, and O78 strains was categorized as being highly virulent for natural reservoir. Recent reports show ExPEC isolates of same clonal group could be different for virulence genotypes, because acquisition of accessory virulence traits might be distinct evolutionary paths for strain-to-strain variation [8,9,32]. The virulence genotypes among APEC O1:K1 and O2:K1 strains showed slight differences (Table H in File S2), although the virulence for avian colisepticemia were similar ($P\geq0.05$). Four APEC O2:K1 strains showed almost similar virulence genotypes, and *iucD* and *ironN* were absent in Fy26 and DE205B (Table H in File S2). For the virulence genotypes among three APEC O1:K1 strains, the two O1:K1 isolates (Jnd25 and CVCC249) in China did not harbor *ibeA* (GimA island) and

Figure 5. The distribution diagram of 10 sequenced B2 ExPEC pan-genome virulence genes among 46 *E. coli* strains. The uppermost row showed six classified clusters: 1, adhesins, green; 2, invasins, magenta; 3, iron acquisition/transport systems, blue; 4, polysialic acid synthesis, aquamarine; 5, toxins, purple; 6, others, darksalmon. Right side of the vertical line showed *E. coli* strains that were consistent with phylogenetic tree (Figure 1). The red and black body showed distribution of virulence genes among these strains. A red line meant that the virulence gene of interest was present at a particular strain, while a black line implied the gene was absent.

aatA genes (APEC autotransporter adhesion) compared to APEC O1. The results of ELA assay and chick colisepticemia model showed that Jnd2 was a low-pathogenicity isolate compared to APEC O1 (*P*<0.05), even though previous studies claimed that ST95 B2 strains exhibited enhanced ExPEC virulence [8,75]. There were significant differences between Jnd2 and APEC O1:K1/O2:K1 isolates that Jnd2 genomic did not harbor the typical T6SS1 (GI-7 for IMT5155), *vat*, and *ireA*, which are specifically required for survival and virulence during APEC infection [28,62,66,76] (Table H in File S2). In short, combined pathogenicity tests with comparative genomic analysis, we confirmed that APEC O1:K1 and O2:K1 strains, including IMT5155 and several strains isolated in China, are extraintestinal pathogenic variants for high pathogenicity during infecting avian hosts, which is consistent with previous studies [5,24,26–29,32].

Previous reports put forward the hypothesis that APEC strains have zoonotic potential [6,8,9], and it is confirmed that a subset of APEC ST95 serotype O18 isolates could cause systemic disease in chickens and murine models of human ExPEC-caused septicemia and meningitis [32]. Our comparative genomic analysis further showed that IMT5155 shared significant genetic overlap/similarities with APEC O1 and human ExPEC O18 strains (IHE3034, and UTI89), and O1:K1/O2:K1 strains are common among APEC isolates but which also found among human NMEC and septicemic isolates [6,9]. Certainly, APEC O1 is unable to cause bacteremia or meningitis in the neonatal rat model and keep host specificity by unknown mechanisms [28]. Here, we assessed the zoonotic potential of IMT5155 and the other O1:K1/O2:K1 isolates through two murine models of human ExPEC-caused septicemia and meningitis. For mouse sepsis assay, no mortalities

were observed among mouse intraperitoneally inoculated (approximately 10^7 CFU) with Jnd2, CVCC1531, APEC O1, CFT073, and MG1655 (Table 3) (the original data shown in File F in File S3). The data also showed that six APEC O1:K1/O2:K1 isolates (Jnd25, CVCC249, IMT5155, Fy26, DE164, and DE205B) and O78 strain CVCC1553 were not significantly different from the positive ExPEC reference strain RS218 (rate of mortality:100%)(*P*≥0.05) (Table 3), suggesting that those strains could have its ability to cause sepsis in the mouse through intraperitoneal inoculation. For rat neonatal meningitis assay, CVCC1531 and APEC strain jnd2 were unable to induce bacteremia in blood and CSF in neonatal rats (Table 4) (the original data shown in File G in File S3). The number of bacteria reisolated from the blood and CSF of rats infected with seven strains (Jnd25, CVCC249, IMT5155, Fy26, DE164, DE205B, and CVCC1553) were significantly higher than that of negative control (*P*<0.05) (Table 4). Moreover, IMT515 and five O1:K1/O2:K1 isolates in China showed comparable septicemia and meningitis in neonatal rats, since no significant differences in the blood and CSF counts were observed (*P*≥0.05). Our data demonstrated that IMT515 and five O1:K1/O2:K1 isolates were close to the high-level bacteremia in blood and CSF of RS218-inoculated neonatal rats, suggesting that these APEC O1:K1/O2:K1 isolates were able to cause septicemia and meningitis in neonatal rats. Like the subset of APEC ST95 serotype O18 isolates, our data confirmed that APEC O1:K1 and O2:K1 strains had zoonotic potential.

A subset of APEC ST23 serotype O78 isolates could be acknowledged as APEC-specific pathogens, because APEC O78 strains were clearly differentiated from serotypes O1, O2, and O18 by MLST, phylogroup, and virulence genotypes [9]. The

Table 1. Mortality rates among chick embryos infected with APEC strains.

Strain	Mortality rate[d]	P value vs[a]:	
		MG1655	APEC O1
Uninoculated	0/10	1.0	<0.001
PBS	1/10	0.416	<0.001
MG1655[b]	3/20		<0.001
IMT5155	17/20	<0.001	0.306
Fy26	19/20	<0.001	0.179
DE164	17/20	<0.001	0.306
DE205B	18/20	<0.001	0.271
Jnd25	17/20	<0.001	0.306
CVCC249	16/20	<0.001	0.276
Jnd2	6/20	0.162	<0.001
CVCC1553	16/20	<0.001	0.276
CVCC1531	4/20	0.296	<0.001
APEC O1[c]	25/30	<0.001	

[a]P value measured by Fisher's exact test.
[b]Negative control for the ELA.
[c]Positive control for the ELA.
[d]Data mean the number of dead embryos/total number of embryos tested.

APEC O78 strain χ7122 was used as a classic infection strain of APEC pathogenicity to identify O78-specific virulence genotype [12]. Comparative genomic analysis of IMT5155 with χ7122 was consistent with the description by Dziva et al. that χ7122 were distinct from APEC O1 and IMT5155, and close to human ST23 serotype O78 human ETEC strain [12]. We compared the virulence and zoonotic potential of APEC O78 strain CVCC1553 with ST23 intestinal pathogenic strain CVCC1531. Like APEC O1:K1 and O2:K1 isolates, CVCC1553 was categorized as being highly virulent for natural reservoir, and CVCC1531 was avirulent

in ELA and chick colisepticemia model (Table 1 and Table 2). Meanwhile, both CVCC1553 and χ7122 caused low pathogenicity in the neonatal meningitis mode compared to RS218 and APEC O1:K1/O2:K1 isolates (Table 4) [32]. As discussed by Dziva et al., χ7122 acquired a different virulence gene repertoire via variation in the accessory genome enabling success in avian species, including virulence-associated large plasmids [12]. The virulence genotype of CVCC1553 showed that it also contained the conserved regions of large virulence plasmids (Table H in File S2). Our investigation further confirmed that APEC O78 strains

Table 2. Lethality in 1-day-old chicks for intratracheal inoculation with APEC isolates.

strain	Mortality rate[cf]	Reisolation rate (air sacs)[df]	Reisolation rate (blood) [df]	Reisolation rate (brain) [df]	Mean lesion score[e]
PBS	0/10[f]	1/10[f]	0/10[f]	0/10[f]	0.1 [f]
MG1655[a]	0/10	2/10	0/10	0/10	0.2
IMT5155	6/10	10/10	8/10	8/10	2.3
Fy26	6/10	10/10	9/10	7/10	2.5
DE164	5/10	9/10	9/10	7/10	2.3
DE205B	7/10	10/10	9/10	9/10	2.5
Jnd25	7/10	10/10	10/10	8/10	2.5
CVCC249	6/10	9/10	9/10	7/10	2.3
Jnd2	1/10[f]	5/10	3/10	0/10	0.9
CVCC1553	8/10	10/10	10/10	4/10	2.5
CVCC1531	0/10[f]	3/10[f]	0/10[f]	0/10[f]	0.3
APEC O1[b]	6/10	9/10	8/10	6/10	2.3

[a]Negative control for chick colisepticemia model.
[b]Positive control for chick colisepticemia model.
[c]Data mean the number of dead chicks/total number of chicks tested.
[d]Data mean the number of chicks from which the APEC strain was reisolated/total number of chicks tested.
[e]Mean of lesion scores (ranked from 0 to 3 due to occurrence of airsacculitis, pericarditis, and perihepatitis) for 10 chicks tested.
[f]Values are not significantly different (P≥0.05 by Fisher's exact test) with the negative control.

Table 3. Lethality of ICR mouse for intraperitoneal inoculation with APEC isolates and human ExPEC strains.

strain	Mortality rate[cf]	Reisolation rate (blood) [df]	Reisolation rate (brain) [df]	Mean lesion score[e]
PBS	0/10[f]	0/10[f]	0/10[f]	1.0 [f]
MG1655[a]	0/10	0/10	0/10	1.0
IMT5155	10/10	10/10	10/10	5.0
Fy26	10/10	10/10	10/10	4.9
DE164	10/10	10/10	10/10	4.9
DE205B	10/10	10/10	10/10	5.0
Jnd25	9/10	9/10	9/10	4.7
CVCC249	10/10	10/10	10/10	4.9
Jnd2	0/10[f]	4/10	0/10[f]	1.1 [f]
CVCC1553	10/10	10/10	7/10	5
CVCC1531	0/10[f]	0/10[f]	0/10[f]	1.0 [f]
APEC O1	0/10[f]	7/10	0/10[f]	1.5 [f]
RS218[b]	10/10	10/10	10/10	4.9
CFT073	0/10[f]	8/10	0/10[f]	1.5 [f]

[a]Negative control for mouse sepsis model.
[b]Positive control for mouse sepsis model.
[c]Data mean the number of dead mouse/total number of mouse tested.
[d]Data mean the number of mouse from which the APEC/ExPEC strain was reisolated/total number of mouse tested.
[e]Mean of lesion scores (1 = healthy, 2 = minimally ill, 3 = moderately ill, 4 = severely ill, 5 = dead) for 10 mouse tested.
[f]Values are not significantly different ($P \geq 0.05$ by Fisher's exact test) with the negative control.

could act as avian host-specific extraintestinal pathogenic variant of ST23 lineage to adapt/colonize to extraintestinal specific niches and establish a specific infection by an intratracheal route in avian host.

Conclusions

The study presented here enriches our knowledge of IMT5155 and complements the *E. coli* genome data of O2 serotype and ST140 (ST complex 95). Our phylogeny analyses confirmed that IMT5155 was closest evolutionary relationship with APEC O1

Table 4. Pathogenicities of APEC isolates in the neonatal rat meningitis model.

Strain	Inoculum ($\log_{10}$ CFU per animal)	Mortality rate [c]	Reisolation rate from blood of survivors [d]	Mean $\log_{10}$ CFU ml^{-1} (blood) [e]	Reisolation rate from CSF of survivors [d]	Mean $\log_{10}$ CFU ml^{-1} (CSF) [e]
PBS	0	0/12[f]	0/12[f]	0	0/12	0
MG1655 [a]	2.36	0/12	0/12	0	0/12	0
IMT5155	2.33	1/12[f]	10/11	3.54	10/11	4.02
Fy26	2.34	0/12 [f]	12/12	3.57	10/12	4.10
DE164	2.31	0/12 [f]	12/12	3.41	11/12	3.95
DE205B	2.25	1/12 [f]	10/11	3.51	10/11	4.18
Jnd25	2.35	1/12 [f]	10/11	3.64	10/11	4.3
CVCC249	2.32	0/12 [f]	12/12	3.50	12/12	4.16
Jnd2	2.32	0/12 [f]	2/12 [f]	3.17	0/12 [f]	0
CVCC1553 [f]	2.34	0/12 [f]	7/12	2.85	7/12	3.51
CVCC1531	2.31	0/12 [f]	0/12 [f]	0	0/12 [f]	0
APEC O1 [a]	2.34	0/12 [f]	0/12 [f]	0	0/12 [f]	0
RS218 [b]	2.34	3/12	9/9	3.82	9/9	>4.57

[a]Negative control for the neonatal rat meningitis model.
[b]Positive control for the neonatal rat meningitis model.
[c]Data mean the number of dead rats/total number of rats tested.
[d]Data mean the number of rats from which the APEC/ExPEC strain was reisolated/total number of rat survivors.
[e]Mean number of *E. coli* isolates recovered from the blood and CSF of the rat survivors.
[f]Values are not significantly different ($P \geq 0.05$ by Fisher's exact test) with the negative control.

serotype and human ExPEC O18 serotype strains (APEC O1, IHE3034, and UTI89; ST complex 95), which all belonged to phylogroup B2 and ST complex 95. Comparison of IMT5155 genome with other *E. coli* strains facilitated the identification of APEC/ExPEC genetic characteristics. Our results of comparative genomics showed that APEC dominant O1 and O2 serotypes strains (APEC O1 and IMT5155) shared significant genetic overlap/similarities with human ExPEC dominant O18 strains (IHE3034, and UTI89). The unique PAI I$_{5155}$ (GI-12) was identified and conserved in APEC O2 isolates, and GI-7 and GI-16 encoding two typical T6SSs might be useful markers for the identification of ExPEC dominant serotypes (O1, O2, and O18) strains. IMT5155 contained a ColV plasmid p1ColV$_{5155}$, and virulence genes in p1ColV$_{5155}$ also defined the APEC pathotype. The distribution of 10 sequenced B2 ExPEC pan-genome virulence factors among 47 sequenced *E. coli* provided a meaningful evidence for phylogroup B2 APEC/ExPEC-specific virulence factors, including several adhesins, invasins, toxins, iron acquisition systems, and others, which contributed to ExPEC to adapte/colonize extraintestinal specific niches during infection. The pathogenicity tests of IMT515 and other APEC O1:K1 and O2:K1 serotypes isolates in China through four animal models showed that they were high virulent for avian colisepticemia and able to cause septicemia and meningitis in neonatal rats, suggesting these APEC O1:K1 and O2:K1 isolates had zoonotic potential. Our comparative genomics studies and the pathogenicity tests will promote the investigation of APEC/ExPEC pathogenesis and zoonotic potential of APEC, and pave the way to development of strategies in their prevention and treatment.

Supporting Information

File S1 Figure A. Gene clusters of T6SS1 (GI-7) and T6SS2 (GI-16) in IMT5155 chromosome. Genes encoding conserved domain proteins were represented by the bule colors. And white arrows indicate other unknown proteins, which were not identified as part of the conserved core described by Ma et al. [65]. The flanking core genes were indicated by the black arrows. A) IMT5155 T6SS1 (GI-7); B) IMT5155 T6SS2 (GI-16). Figure B. Synteny analysis based on common ORFs between p1ColV$_{5155}$ and 5 plasmids (pAPEC-O1-ColBM, pAPEC-O2-ColV, pMAR2, pO83_CORR, and pUTI89). Grey ribbons are common ORFs in p1ColV5155 and pAPEC-O2-ColV; Pink ribbons are common

ORFs in p1ColV$_{5155}$ and pAPEC-O1-ColBM; Yellow ribbons are common ORFs in p1ColV5155 and pMAR2; Purple ribbons are common ORFs in p1ColV5155 and PO83-CORR; Green ribbons are common ORFs in p1ColV5155 and PUTI89. Red blocks are repA genes; Purple blocks are *repB* genes; Blue blocks are *Tra* genes.

File S2 Table A. General feature of IMT5155 genome and other *E. coli* strains. Table B. The virulence factors in B2 ExPEC pangenome among 10 *E. coli* strains. Table C. The genomic islands of IMT5155. Table D. The information of 15 ExPEC isolates for simultaneous presence of T6SS1 and T6SS2. Table E. Common genes shared with IMT5155 for 15 *E. coli* strains. Table F. The number and percentage of common genes of other *E. coli* pathotype's plasmids shared with p1ColV$_{5155}$. Table G. The specific genes of IMT5155 relative to other 9 B2 ExPEC strains. Table H. The detail information of the 13 selected strains for pathogenicity testing.

File S3 File A. Detailed description for 47 *E. coli* genomes data. File B. The scripts for comparative genomic analysis. File C. Detailed description for annotated ORFs in the chromosome sequence of IMT5155. File D. Detailed description for annotated ORFs in p1ColV$_{5155}$. File E. The original data for chick colisepticemia assay. File F. The original data for mouse sepsis assay. File G. The original data for rat neonatal meningitis assay.

Acknowledgments

We gratefully acknowledge Lothar H. Wieler (Institute of Microbiology and Epizootics, Freie Universitaet BerlinBerlin, Germany) for the gift of IMT5155 strain. We acknowledge Qiang Li and his colleagues for genome sequencing and analysis at Shanghai Majorbio Bio-pharm Technology Co., Ltd.

Author Contributions

Conceived and designed the experiments: JJD XKZG ZHP. Performed the experiments: XKZG LH SHW HJW. Analyzed the data: XKZG JWJ FCL. Contributed reagents/materials/analysis tools: FCL HJF. Wrote the paper: XKZG. Designed the pathogenicity experiments: JJD XKZG.

References

1. Diard M, Garry L, Selva M, Mosser T, Denamur E (2010) Pathogenicity-associated islands in extraintestinal pathogenic *Escherichia coli* are fitness elements involved in intestinal colonization. J Bacteriol 192: 4885–4893.
2. Kaper JB, Nataro JP, Mobley HL (2004) Pathogenic *Escherichia coli*. Nat Rev Microbiol 2: 123–140.
3. Croxen MA, Finlay BB (2010) Molecular mechanisms of Escherichia coli pathogenicity. Nat Rev Microbiol 8: 26–38.
4. Russo TA, Johnson JR (2000) Proposal for a new inclusive designation for extraintestinal pathogenic isolates of *Escherichia coli*: ExPEC. J Infect Dis 181: 1753–1754.
5. Johnson TJ, Kariyawasam S, Wannemuehler Y, Mangiamele P, Johnson SJ (2007) The genome sequence of avian pathogenic *Escherichia coli* strain O1:K1:H7 shares strong similarities with human extraintestinal pathogenic *E. coli* genomes. J Bacteriol 189: 3228–3236.
6. Ewers C, Li G, Wilking H, Kiessling S, Alt K (2007) Avian pathogenic, uropathogenic, and newborn meningitis-causing *Escherichia coli*: how closely related are they? Int J Med Microbiol 297: 163–176.
7. Ron EZ (2006) Host specificity of septicemic *Escherichia coli*: human and avian pathogens. Curr Opin Microbiol 9: 28–32.
8. Johnson TJ, Wannemuehler Y, Johnson SJ, Stell AL, Doetkott C (2008) Comparison of extraintestinal pathogenic *Escherichia coli* strains from human and avian sources reveals a mixed subset representing potential zoonotic pathogens. Appl Environ Microbiol 74: 7043–7050.
9. Moulin-Schouleur M, Reperant M, Laurent S, Bree A, Mignon-Grasteau S (2007) Extraintestinal pathogenic *Escherichia coli* strains of avian and human

origin: link between phylogenetic relationships and common virulence patterns. J Clin Microbiol 45: 3366–3376.
10. Moulin-Schouleur M, Schouler C, Tailliez P, Kao MR, Bree A (2006) Common virulence factors and genetic relationships between O18:K1:H7 Escherichia coli isolates of human and avian origin. J Clin Microbiol 44: 3484–3492.
11. Brzuszkiewicz E, Bruggemann H, Liesegang H, Emmerth M, Olschlager T (2006) How to become a uropathogen: comparative genomic analysis of extraintestinal pathogenic *Escherichia coli* strains. Proc Natl Acad Sci U S A 103: 12879–12884.
12. Dziva F, Hauser H, Connor TR, van Diemen PM, Prescott G (2013) Sequencing and functional annotation of avian pathogenic *Escherichia coli* serogroup O78 strains reveal the evolution of E. coli lineages pathogenic for poultry via distinct mechanisms. Infect Immun 81: 838–849.
13. Gordon DM, Clermont O, Tolley H, Denamur E (2008) Assigning *Escherichia coli* strains to phylogenetic groups: multi-locus sequence typing versus the PCR triplex method. Environ Microbiol 10: 2484–2496.
14. Wirth T, Falush D, Lan R, Colles F, Mensa P (2006) Sex and virulence in *Escherichia coli*: an evolutionary perspective. Mol Microbiol 60: 1136–1151.
15. Clermont O, Bonacorsi S, Bingen E (2000) Rapid and simple determination of the *Escherichia coli* phylogenetic group. Appl Environ Microbiol 66: 4555–4558.
16. Boyd EF, Hartl DL (1998) Chromosomal regions specific to pathogenic isolates of *Escherichia coli* have a phylogenetically clustered distribution. J Bacteriol 180: 1159–1165.
17. Tenaillon O, Skurnik D, Picard B, Denamur E (2010) The population genetics of commensal *Escherichia coli*. Nat Rev Microbiol 8: 207–217.

18. Escobar-Paramo P, Clermont O, Blanc-Potard AB, Bui H, Le Bouguenec C (2004) A specific genetic background is required for acquisition and expression of virulence factors in *Escherichia coli*. Mol Biol Evol 21: 1085–1094.

19. Touchon M, Hoede C, Tenaillon O, Barbe V, Baeriswyl S (2009) Organised genome dynamics in the *Escherichia coli* species results in highly diverse adaptive paths. PLoS Genet 5: e1000344.

20. Picard B, Garcia JS, Gouriou S, Duriez P, Brahimi N (1999) The link between phylogeny and virulence in *Escherichia coli* extraintestinal infection. Infect Immun 67: 546–553.

21. Kaas RS, Friis C, Ussery DW, Aarestrup FM (2012) Estimating variation within the genes and inferring the phylogeny of 186 sequenced diverse *Escherichia coli* genomes. BMC Genomics 13: 577.

22. Maiden MC, Bygraves JA, Feil E, Morelli G, Russell JE (1998) Multilocus sequence typing: a portable approach to the identification of clones within populations of pathogenic microorganisms. Proc Natl Acad Sci U S A 95: 3140–3145.

23. Jaureguy F, Landraud L, Passet V, Diancourt L, Frapy E (2008) Phylogenetic and genomic diversity of human bacteremic *Escherichia coli* strains. BMC Genomics 9: 560.

24. Kohler CD, Dobrindt U (2011) What defines extraintestinal pathogenic *Escherichia coli*? Int J Med Microbiol 301: 642–647.

25. Tartof SY, Solberg OD, Manges AR, Riley LW (2005) Analysis of a uropathogenic *Escherichia coli* clonal group by multilocus sequence typing. J Clin Microbiol 43: 5860–5864.

26. McNally A, Cheng L, Harris SR, Corander J (2013) The evolutionary path to extraintestinal pathogenic, drug-resistant *Escherichia coli* is marked by drastic reduction in detectable recombination within the core genome. Genome Biol Evol 5: 699–710.

27. Mora A, Lopez C, Dabhi G, Blanco M, Blanco JE (2009) Extraintestinal pathogenic *Escherichia coli* O1:K1:H7/NM from human and avian origin: detection of clonal groups B2 ST95 and D ST59 with different host distribution. BMC Microbiol 9: 132.

28. Johnson TJ, Wannemuehler Y, Kariyawasam S, Johnson JR, Logue CM (2012) Prevalence of avian-pathogenic *Escherichia coli* strain O1 genomic islands among extraintestinal and commensal *E. coli* isolates. J Bacteriol 194: 2846–2853.

29. Rodriguez-Siek KE, Giddings CW, Doetkott C, Johnson TJ, Nolan LK (2005) Characterizing the APEC pathotype. Vet Res 36: 241–256.

30. Antao EM, Glodde S, Li G, Sharifi R, Homeier T (2008) The chicken as a natural model for extraintestinal infections caused by avian pathogenic *Escherichia coli* (APEC). Microb Pathog 45: 361–369.

31. Dho-Moulin M, Fairbrother JM (1999) Avian pathogenic *Escherichia coli* (APEC). Vet Res 30: 299–316.

32. Tivendale KA, Logue CM, Kariyawasam S, Jordan D, Hussein A (2010) Avian-pathogenic *Escherichia coli* strains are similar to neonatal meningitis *E. coli* strains and are able to cause meningitis in the rat model of human disease. Infect Immun 78: 3412–3419.

33. Dai J, Wang S, Guerlebeck D, Laturnus C, Guenther S (2010) Suppression subtractive hybridization identifies an autotransporter adhesin gene of *E. coli* IMT5155 specifically associated with avian pathogenic *Escherichia coli* (APEC). BMC Microbiol 10: 236.

34. Antao EM, Ewers C, Gurlebeck D, Preisinger R, Homeier T (2009) Signature-tagged mutagenesis in a chicken infection model leads to the identification of a novel avian pathogenic *Escherichia coli* fimbrial adhesin. PLoS One 4: e7796.

35. Li G, Laturnus C, Ewers C, Wieler LH (2005) Identification of genes required for avian *Escherichia coli* septicemia by signature-tagged mutagenesis. Infect Immun 73: 2818–2827.

36. Delcher AL, Bratke KA, Powers EC, Salzberg SL (2007) Identifying bacterial genes and endosymbiont DNA with Glimmer. Bioinformatics 23: 673–679.

37. Chen S, Zhang CY, Song K (2013) Recognizing short coding sequences of prokaryotic genome using a novel iteratively adaptive sparse partial least squares algorithm. Biol Direct 8: 23.

38. Altschul SF, Madden TL, Schaffer AA, Zhang J, Zhang Z (1997) Gapped BLAST and PSI-BLAST: a new generation of protein database search programs. Nucleic Acids Res 25: 3389–3402.

39. Kent WJ (2002) BLAT–the BLAST-like alignment tool. Genome Res 12: 656–664.

40. Edgar RC (2004) MUSCLE: multiple sequence alignment with high accuracy and high throughput. Nucleic Acids Res 32: 1792–1797.

41. Ronquist F, Huelsenbeck JP (2003) MrBayes 3: Bayesian phylogenetic inference under mixed models. Bioinformatics 19: 1572–1574.

42. Yoon SH, Park YK, Lee S, Choi D, Oh TK (2007) Towards pathogenomics: a web-based resource for pathogenicity islands. Nucleic Acids Res 35: D395–400.

43. Yoon SH, Hur CG, Kang HY, Kim YH, Oh TK (2005) A computational approach for identifying pathogenicity islands in prokaryotic genomes. BMC Bioinformatics 6: 184.

44. Chen L, Yang J, Yu J, Yao Z, Sun L (2005) VFDB: a reference database for bacterial virulence factors. Nucleic Acids Res 33: D325–328.

45. Zhuge X, Wang S, Fan H, Pan Z, Ren J (2013) Characterization and Functional Analysis of AatB, a Novel Autotransporter Adhesin and Virulence Factor of Avian Pathogenic *Escherichia coli*. Infect Immun.

46. Wang S, Niu C, Shi Z, Xia Y, Yaqoob M (2011) Effects of ibeA deletion on virulence and biofilm formation of avian pathogenic *Escherichia coli*. Infect Immun 79: 279–287.

47. Johnson TJ, Siek KE, Johnson SJ, Nolan LK (2006) DNA sequence of a ColV plasmid and prevalence of selected plasmid-encoded virulence genes among avian *Escherichia coli* strains. J Bacteriol 188: 745–758.

48. Johnson TJ, Johnson SJ, Nolan LK (2006) Complete DNA sequence of a ColBM plasmid from avian pathogenic *Escherichia coli* suggests that it evolved from closely related ColV virulence plasmids. J Bacteriol 188: 5975–5983.

49. Sims GE, Kim SH (2011) Whole-genome phylogeny of *Escherichia coli*/Shigella group by feature frequency profiles (FFPs). Proc Natl Acad Sci U S A 108: 8329–8334.

50. Sahl JW, Steinsland H, Redman JC, Angiuoli SV, Nataro JP (2011) A comparative genomic analysis of diverse clonal types of enterotoxigenic *Escherichia coli* reveals pathovar-specific conservation. Infect Immun 79: 950–960.

51. Logue CM, Doetkott C, Mangiamele P, Wannemuehler YM, Johnson TJ (2012) Genotypic and phenotypic traits that distinguish neonatal meningitis-associated *Escherichia coli* from fecal *E. coli* isolates of healthy human hosts. Appl Environ Microbiol 78: 5824–5830.

52. Gao Q, Wang X, Xu H, Xu Y, Ling J (2012) Roles of iron acquisition systems in virulence of extraintestinal pathogenic *Escherichia coli*: salmochelin and aerobactin contribute more to virulence than heme in a chicken infection model. BMC Microbiol 12: 143.

53. Wright KJ, Hultgren SJ (2006) Sticky fibers and uropathogenesis: bacterial adhesins in the urinary tract. Future Microbiol 1: 75–87.

54. Zhou Y, Tao J, Yu H, Ni J, Zeng L (2012) Hcp family proteins secreted via the type VI secretion system coordinately regulate *Escherichia coli* K1 interaction with human brain microvascular endothelial cells. Infect Immun 80: 1243–1251.

55. Wang S, Shi Z, Xia Y, Li H, Kou Y (2012) IbeB is involved in the invasion and pathogenicity of avian pathogenic *Escherichia coli*. Vet Microbiol 159: 411–419.

56. Schubert S, Picard B, Gouriou S, Heesemann J, Denamur E (2002) Yersinia high-pathogenicity island contributes to virulence in *Escherichia coli* causing extraintestinal infections. Infect Immun 70: 5335–5337.

57. Juhas M, van der Meer JR, Gaillard M, Harding RM, Hood DW (2009) Genomic islands: tools of bacterial horizontal gene transfer and evolution. FEMS Microbiol Rev 33: 376–393.

58. Bobay LM, Rocha EP, Touchon M (2013) The adaptation of temperate bacteriophages to their host genomes. Mol Biol Evol 30: 737–751.

59. Saha RP, Lou Z, Meng L, Harshey RM (2013) Transposable prophage Mu is organized as a stable chromosomal domain of *E. coli*. PLoS Genet 9: e1003902.

60. Harshey RM (2012) The Mu story: how a maverick phage moved the field forward. Mob DNA 3: 21.

61. Li G, Feng Y, Kariyawasam S, Tivendale KA, Wannemuehler Y (2010) AatA is a novel autotransporter and virulence factor of avian pathogenic *Escherichia coli*. Infect Immun 78: 898–906.

62. Russo TA, Carlino UB, Johnson JR (2001) Identification of a new iron-regulated virulence gene, ireA, in an extraintestinal pathogenic isolate of *Escherichia coli*. Infect Immun 69: 6209–6216.

63. Grim CJ, Kothary MH, Gopinath G, Jarvis KG, Beaubrun JJ (2012) Identification and characterization of Cronobacter iron acquisition systems. Appl Environ Microbiol 78: 6035–6050.

64. Shrivastava S, Mande SS (2008) Identification and functional characterization of gene components of Type VI Secretion system in bacterial genomes. PLoS One 3: e2955.

65. Ma J, Sun M, Bao Y, Pan Z, Zhang W (2013) Genetic diversity and features analysis of type VI secretion systems loci in avian pathogenic *Escherichia coli* by wide genomic scanning. Infect Genet Evol 20: 454–464.

66. de Pace F, Nakazato G, Pacheco A, de Paiva JB, Sperandio V (2010) The type VI secretion system plays a role in type 1 fimbria expression and pathogenesis of an avian pathogenic *Escherichia coli* strain. Infect Immun 78: 4990–4998.

67. Kyle JL, Cummings CA, Parker CT, Quinones B, Vatta P (2012) *Escherichia coli* serotype O55:H7 diversity supports parallel acquisition of bacteriophage at Shiga toxin phage insertion sites during evolution of the O157:H7 lineage. J Bacteriol 194: 1885–1896.

68. Eppinger M, Mammel MK, Leclerc JE, Ravel J, Cebula TA (2011) Genomic anatomy of *Escherichia coli* O157:H7 outbreaks. Proc Natl Acad Sci U S A 108: 20142–20147.

69. Böhnke U (2010) Charakterisierung und Bedeutung der Plasmide p1ColV 5155 und p2 5155 für den aviären pathogenen *E. coli*-Stamm IMT5155. Dissertation, Humboldt-Universität zu Berlin, Faculty of Mathematics and Natural Sciences.

70. Johnson TJ, Jordan D, Kariyawasam S, Stell AL, Bell NP (2010) Sequence analysis and characterization of a transferable hybrid plasmid encoding multidrug resistance and enabling zoonotic potential for extraintestinal *Escherichia coli*. Infect Immun 78: 1931–1942.

71. Mellata M, Ameiss K, Mo H, Curtiss R 3rd (2010) Characterization of the contribution to virulence of three large plasmids of avian pathogenic *Escherichia coli* chi7122 (O78:K80:H9). Infect Immun 78: 1528–1541.

72. Tivendale KA, Noormohammadi AH, Allen JL, Browning GF (2009) The conserved portion of the putative virulence region contributes to virulence of avian pathogenic *Escherichia coli*. Microbiology 155: 450–460.

73. Willems RJ, Top J, van Schaik W, Leavis H, Bonten M (2012) Restricted gene flow among hospital subpopulations of Enterococcus faecium. MBio 3: e00151-00112.

74. Rodriguez-Siek KE, Giddings CW, Doetkott C, Johnson TJ, Fakhr MK (2005) Comparison of *Escherichia coli* isolates implicated in human urinary tract infection and avian colibacillosis. Microbiology 151: 2097–2110.

75. Johnson JR, Clermont O, Menard M, Kuskowski MA, Picard B (2006) Experimental mouse lethality of *Escherichia coli* isolates, in relation to accessory traits, phylogenetic group, and ecological source. J Infect Dis 194: 1141–1150.

76. Salvadori MR, Yano T, Carvalho HE, Parreira VR, Gyles CL (2001) Vacuolating cytotoxin produced by avian pathogenic Escherichia coli. Avian Dis 45: 43–51.

15

NAD(P)H-Hydrate Dehydratase- A Metabolic Repair Enzyme and Its Role in *Bacillus subtilis* Stress Adaptation

Miroslava Petrovova[1]⊙, Jan Tkadlec[1,2]⊙, Lukas Dvoracek[1], Eliska Streitova[1,2], Irena Licha[1]*

1 Department of Genetics and Microbiology, Faculty of Science, Charles University, Prague, Czech Republic, **2** Department of Medical Microbiology, 2nd Faculty of Medicine, Charles University, Prague, Czech Republic

Abstract

Background: One of the strategies for survival stress conditions in bacteria is a regulatory adaptive system called general stress response (GSR), which is dependent on the SigB transcription factor in *Bacillus* sp. The GSR is one of the largest regulon in *Bacillus* sp., including about 100 genes; however, most of the genes that show changes in expression during various stresses have not yet been characterized or assigned a biochemical function for the encoded proteins. Previously, we characterized the *Bacillus subtilis*168 osmosensitive mutant, defective in the *yxkO* gene (encoding a putative ribokinase), which was recently assigned *in vitro* as an ADP/ATP-dependent NAD(P)H-hydrate dehydratase and was demonstrated to belong to the SigB operon.

Methods and Results: We show the impact of YxkO on the activity of SigB-dependent P*ctc* promoter and adaptation to osmotic and ethanol stress and potassium limitation respectively. Using a 2DE approach, we compare the proteomes of WT and mutant strains grown under conditions of osmotic and ethanol stress. Both stresses led to changes in the protein level of enzymes that are involved in motility (flagellin), citrate cycle (isocitrate dehydrogenase, malate dehydrogenase), glycolysis (phosphoglycerate kinase), and decomposition of Amadori products (fructosamine-6-phosphate deglycase). Glutamine synthetase revealed a different pattern after osmotic stress. The patterns of enzymes for branched amino acid metabolism and cell wall synthesis (L-alanine dehydrogenase, aspartate-semialdehyde dehydrogenase, ketol-acid reductoisomerase) were altered after ethanol stress.

Conclusion: We performed the first characterization of a *Bacillus subtilis*168 knock-out mutant in the *yxkO* gene that encodes a metabolite repair enzyme. We show that such enzymes could play a significant role in the survival of stressed cells.

Editor: Vasu D. Appanna, Laurentian University, Canada

Funding: The study was supported by projects of Ministry of Education, Youth and Sports of the Czech Republic (LC06066) and by the Charles University by project of Specific University Research (SVV-2014-260081). The funders had no role in study design, data collection and analysis, decision to publish, or preparation of the manuscript.

Competing Interests: The authors have declared that no competing interests exist.

* Email: licha@natur.cuni.cz

⊙ These authors contributed equally to this work.

Introduction

In an effort to understand the global adaptation network that evolved in *Bacillus* sp., several recent studies were carried out, focused on the genome-wide transcriptional profiling of the stress response of *Bacillus subtilis* 168 [1–4]. Several physiological analyses of the *Bacillus subtilis* 168 proteome during the adaptation to various environmental stresses have been published as well [5–7]. These studies identified stress specific regulons that are involved in stress function and confirm that the synthesis of most vegetative proteins is repressed, with the exception of enzymes that take part in adaptive responses.

One of the important strategies for survival in the genus *Bacillus* is a regulatory adaptive system called general stress response (GSR). It occurs as the large expression of stress proteins and is induced by a wide range of stresses, including high and low temperature; osmotic, ethanol, oxidative, and acidic stress; the addition of some antibiotics; starvation for glucose, phosphate, and oxygen; and blue or red light [2], [8–12] It is also induced on the transition into the stationary phase [13] and provides cells unspecified, multiple, and preventive resistance and gives the cells sufficient time for the induction of specific stress responses.

The general stress regulon, dependent on the SigB factor, is one of the largest operons in *Bacillus* sp., including about 100 genes [4]. However, most of the genes that show changes in expression during various stresses have not yet been characterized or assigned a biochemical function for the encoded proteins, and the evidence of the contribution of individual proteins from the general stress regulon to stress resistance of *Bacillus subilis* 168 cells is not complete.

Many genes of this regulon are putative regulatory factors, and all are under complex regulation by the control of other sigma factors and other regulatory proteins or RNAs, which allows their complex networking. It is assumed that their role is to protect DNA, proteins, metabolites, and lipids against the harmful effects of stress and to repair them.

Most recently, it was shown by Young [14] that the extent of stress determines response specificity and that the general stress response pathway activates different genes to a variety of stress conditions.

With the aim of elucidating the mechanism of adaptation of *Bacillus subtilis* to limited concentrations of potassium, we previously isolated a mutant with reduced salt tolerance only at a limited potassium concentration [15] in which the *yxkO* gene was interrupted. The product of this gene was formerly predicted to have a ribokinase activity based on sequence and structural homologies and the presence of ATP- and Mg^{2+}-binding sites [16]. Most recently, while experiments of this work were completed, the biochemical activity of the YxkO protein was assigned *in vitro* as an ADP/ATP-dependent NAD(P)H-hydrate dehydratase (EC 4.2.1.93). This enzyme convert abnormal metabolite NAD(P)H hydrate (NAD(P)HX) to NAD(P)H and is conserved over the kingdoms [17]. NAD(P)HX is slowly catalyzed from NAD(P)H by glyceraldehyde 3-phosphate dehydrogenase [18] or is produced non enzymatically in the course of the non-physiological conditions respectively [19], [20]. NAD(P)HX is unable to react as cofactor and it inhibits several dehydrogenases with detrimental effect on a cell [20], [21]. Enzymes with such activity are called metabolite repair or metabolite-proofreading enzymes and play a role similar to the proofreading activities of DNA polymerases and aminoacyl-tRNA synthetases [22].

The increased transcriptional activity of this gene after osmotic, heat, and ethanol stress was observed in the transcriptomic study of Petersohn [3], as well as in a recent extensive systematic and quantitative exploration of transcriptome changes in *Bacillus subtilis* [4]. The mutant in this gene was included in a phenotype screening study determining the contribution of individual SigB-dependent genes of unknown function to stress resistance, showing a lower survival rate following severe ethanol, heat, and osmotic stresses [23]. Most recently, it was shown to be under exclusive SigB regulation [24].

In accordance with our prior study and in addition to the reduced tolerance to the environmental stresses mentioned above, the mutant in the *yxkO* gene exhibits reduced growth under potassium limitation and altered motility under hyperosmotic conditions. This multiple effect of the gene disruption on phenotype led us originally to the hypothesis that the product of the *yxkO* gene has a regulatory function.

The present study aimed to determine the contribution of the *yxkO* gene product to stress adaptation by estimating the transcriptional activity of the Pctc promoter, as the Ctc protein is considered a marker of general stress response [11], and by discovering the changes of the cytoplasmic protein level pattern in the mutant and a wild-type strain when exposed to salt and ethanol stress.

Materials and Methods

Bacterial strains

The *Bacillus subtilis* strains and plasmids used in this study are listed in Table 1. *Escherichia coli* DH5α strain (*deoR endA1 gyrA96 hsd R17* (r_k^-, m_k^-) *recA1 relA1 supE44 thi-1* (*lacZYA-arg F*) *U169 φ80lacZ M15* F$^-$λ$^-$) (Clontech) was used for propagation of plasmid constructs.

Growth conditions

For genetic manipulations, *Escherichia coli* and *Bacillus subtilis* strains were cultivated routinely in LB medium.

For growth rate measurements, the *Bacillus subtilis* strains (WT, LD1 and MP2) were cultivated under vigorous agitation at 37°C and synchronized in exponential growth by reinoculation from overnight cultures (grown in LB medium) to LB medium for ethanol stress experiments or to MM medium with 0.5 mM K$^+$ (as described previously [15]) for both ethanol and osmotic stress experiments, respectively. The salt stress conditions were performed by the exposure of exponentially growing cells (OD$_{600}$ of ~0.3) to 0.6 M NaCl. For the ethanol stress setup, ethanol to a final concentration of 4% (v/v) was added at the same growth condition as for the salt stress. For the MP2 strain, erythromycin (0.3 µg/ml) was added.

For transcriptional activity, *Bacillus subtilis* strains (WT and MP2) were cultivated in LB or MM medium with the defined concentration of potassium as described above for growth rate measurements; the cell samples were collected in intervals before and after stress exposure, as indicated in the particular experiment (see *Results and Discussion*).

For 2DE analysis, *Bacillus subtilis* strains (WT and MP2) were cultivated in MM medium with 0.5 mM K$^+$ as described above for growth rate measurements and harvested 60 min after stress exposure.

B. subtilis yxkO knock-out mutant strain construction

For transfer of the insertional mutation of the mini-Tn10 transposon to the *yxkO* gene from an asporogenic genetic background from a previously prepared mutant [15], the PBS1 lysate was prepared from the L-42 mutant, and *Bacillus subtilis* 168 was transduced. The transductants were selected on LB chloramphenicol plates (5 µg/ml). The classical transduction protocol was used [25]. Insertion of mini-Tn10 into the *yxkO* gene in a particular clone was confirmed by PCR using primers designed for the *yxkO* gene and the transposon region, as well. The respective PCR product was confirmed by sequencing, and the mutant was named LD1.

For inactivation of the *yxkO* gene, an integrative vector, pMUTIN4 from *Bacillus* Genetic Stock Center (BGSC - ECE 139), was used. A fragment of the *yxkO* allele with a ribosome binding site was generated by PCR with the primers 5′-GGAGGATCCATAACAGGACAATCAGCC-3′ and 5′-CCCGAATTCAGGAAAAGAAAGCAGAGGAG-3′ (the *Eco*RI and *Bam*HI restriction sites for direct cloning into pMUTIN4 are underlined) and ligated into *Eco*RI-*Bam*HI-digested pMUTIN4. The ligation mixture was transformed into *E. coli* DH5α, and clones with pMP2 plasmid were selected on LB ampicillin plates (100 µg/ml). Presence of the *yxkO* allele fragment in the plasmid was confirmed by sequencing.

Circular pMP2 plasmid was then used for transformation into *Bacillus subtilis* SG64, and single-crossover recombinants were selected on LB plates supplemented with erythromycin (0.3 µg/ml). Correct insertion into the chromosome was confirmed in a selected isolate by PCR and sequencing and named MP2.

Bacillus subtilis Pctc promoter probe mutant strain construction

The promoter region of the *ctc* gene [26] was generated by PCR with the primers 5′- CATAGAATTCCCATTTTTCGAGGTTTAAATCCTT-3′ and 5′- TTTAGGATCCCGAGTAAAGTCCGTTCTTTCTT-3′ (the *Eco*RI and *Bam*HI restriction sites for direct cloning into plasmid are underlined) and

Table 1. *Bacillus subtilis* strains, phage, and plasmids used in this study.

Name or code	Relevant genotype or description	Reference, Source, or construction
Strains		
B.s. 168 (WT)	*trpC2*	BGSC (1A1)
B.s. SG4 (WT)	*xglA1, xglR1*	BGSC (1A680)
L-42	*pheA1, spo0F221, trpC2, mini-Tn10::yxkO*	[15]
LD1	*trpC2, mini-Tn10::yxkO*	this study
MP2	Mutin4::*yxkO, trpC, xglA1, xglR1*	this study
WT/Pctc	(*Pctc*Φ *spoVG-lacZ, cat*)::*amy, trpC, xglA1, xglR1*	this study
MP2/Pctc	Mutin4::*yxkO*, (*Pctc*Φ spoVG-*lacZ, cat*)::*amy, trpC, xglA1, xglR1*	this study
Phage PBS1	Bacillusphage PBS1	BGSC (1P1)
Plasmids		
pDG1661	*spoVG-lacZ, spc, cat, bla, amy* 5' and 3' segment	BGSC (ECE112)
pMUTIN	*spoVG-lacZ, Pspac, lacI, erm, bla*	BGSC (ECE139)
pJT2	pDG1661 with *Pctc* insert transcriptionally fused to *spoVG-lacZ*	this study
pMP2	pMUTIN with 5' end segment of *yxkO* with SD site fused to *spoVG-lacZ*	this study

ligated to the same sites of the suicide pDG1661 plasmid, which possesses an insertion site to the *amy* locus (BGSC – ECE112).

The ligation mixture was transformed into *E. coli* DH5 α strain, yielding pJT2 plasmid by selecting on LB plates supplemented by ampicillin (100 μg/ml).

For *Bacillus subtilis* mutant strain (WT/Pctc and MP2/Pctc) construction, the plasmid was linearized by *Xho*I restriction enzyme (RE) and transformed into the *Bacillus subtilis* SG4 strain and the MP2 mutant strain prepared as above. Correct insertion into the *amy* locus and right orientation towards the reporter *lacZ* gene were confirmed by PCR and sequencing in double-crossover recombinants selected on LB plates supplemented with chloramphenicol (5 μg/ml).

β - Galactosidase assay of *lacZ* transcriptional fusions

Samples were cultivated and collected at specified intervals. Cell samples were permeabilized with lysozyme, and galactosidase activities were measured at OD_{420} and expressed in Miller units (M.U.) according to the protocol from the *Bacillus* Genetic Stock Center (BGSC) Catalog of Integration Vectors (http://www.bgsc.org/_catalogs/Catpart4.pdf), page 15.

2DE sample preparation

For one experiment, WT and MP2 mutant were grown and stressed in parallel, and cells were cultivated as specified in the *Growth condition* section. Cells were harvested by centrifugation (4000×g, 4°C, 5 min), washed with 0.01 M Tris/KCl buffer pH 8.0, resuspended in 1 ml of the same buffer with protease inhibitors (Sigma), and disrupted by sonication. The cell debris was removed by centrifugation (15,000×g, 4°C, and 10 min), and protein concentration in the supernatant was estimated with the BCA Protein Assay Kit – Reducing Agent Compatible (Pierce). The cell lysates were then treated with the Bio-Rad Ready Prep 2-D Cleanup Kit and diluted in an appropriate volume of rehydration buffer up to 100 μg of total protein per sample. From each growth condition (WT, WT stressed, MP2, MP2 stressed), samples in technical triplicates were prepared.

2D polyacrylamide gel electrophoresis and protein visualization

Each sample prepared above was loaded onto an IPG strip pH 4–7 (Bio-Rad). Isoelectric focusing (IEF) was performed after passive overnight rehydration at room temperature using voltage that linearly increased to steady state [100 V for 2 h (slow), 300 V for 2 h (slow), 8000 V for 2 h slow, 8000 V for 7 h (rapid), and finished at 500 V] using the Bio-Rad Protean IEF system. After IEF, the strips were washed in equilibration solution (50 mM, Tris-HCl (pH 6.8), 6 M urea, 30% glycerol and 2% SDS) containing 0.02 M DTT for 10 min, followed by a second 10-min wash in equilibrium buffer containing 0.025 M iodoacetamide and bromophenol blue for gel staining.

The separation in the second dimension was carried out using precast gradient gels (Criterion Precast Gel (10.5%–14%) Bio-Rad). Gels were run on the Bio-Rad Criterion Dodeca Cell device at 5 V for 30 min and at 100 V for 2 hours at room temperature. Gels from one independent experiment in technical triplicates for each stress condition were run together (12 gels).

After staining with Colloidal Coomassie G-250 (Simply Blue Safestain, Invitrogen Life technologies, Paisley, UK), the gels were scanned with a GS-800 calibrated densitometer.

Digitalization of gel images and data analysis

The 2DE image analysis was performed using PDQuest 8.0 software (Bio-Rad, Hercules, CA). For matching and quantification, raw images were smoothed to remove noise, background was subtracted, and a spot-by-spot visual validation of automated analysis was done from then on to increase the reliability of the matching. Only spots that exhibited similar intensity in each gel of the technical triplicates in particular growth condition and in both biological replicates were taken for further analysis. Data from PDQuest analysis are published in Table S1. Identified protein spots were manually cut out of the gels and analyzed by MS.

MS analysis

The bands of interest were cut out of the gels and chopped into $1 \times 1 \times 1$-mm pieces. The pieces were destained; to reduce and block cysteines, DTT and iodoacetamide were applied. The samples were trypsinized as described previously [27]. The dried-droplet method of sample preparation was employed, and spectra were acquired on a 4800 Plus MALDI TOF/TOF analyzer (AB Sciex). The data were analyzed using in-house running Mascot server 2.2.07 and matched against the current release of the NCBI protein sequence database: taxonomy: Bacillus (*Taxonomy ID:* 1386), 833049 sequences, and protein scores greater than 72 were significant (p<0.05). Cysteine carbamidomethylation, methionine oxidation, and N, Q deamination were set as fixed or variable modifications, respectively. One missed cleavage site was allowed. Precursor accuracy was set to 50 ppm, and the accuracy for MS/MS spectra was set to 0.25 Da. Detailed results are collected in Table 2.

Results and Discussion

Confirmation of *yxkO* knock-out mutant phenotype

Firstly, to verify the *yxkO* gene phenotype previously described in a sporulation-proficient strain (*spo0F221*) [15], we transduced the cassette of mini-Tn*10* from the L-42 mutant using PBS1 bacteriophage to *Bacillus subtilis* 168, and in parallel, we constructed an insertional mutant using the pMUTIN4 plasmid, as well (see *Materials and Methods*). Both mutants revealed the previously described phenotype, an increase in generation time from 55 min to 115 min and osmosensitivity in low K^+ conditions (15) but also lower viability, even under stress conditions, as we demonstrated in the long-term growth measurements (Figure 1A).

By doing so, we confirmed that the previously described phenotypic changes in mutant L-42 corresponded to the inactivation of the *yxkO* gene and not to the *spo0F* mutation. Both types of the mutant revealed prolonged lag phase (Figure 1B), and lower viability when exposed to ethanol and salt

Table 2. MALDI-TOF peptide mapping identification.

Protein	Protein Accession	Mascote Score	Best Protein Description
Osmotic stress			
Pgk	gi|16080446	104	phosphoglycerate kinase [Bacillus subtilis subsp. subtilis str. 168]
CitC	gi|16079965	885	isocitrate dehydrogenase [Bacillus subtilis subsp. subtilis str. 168]
Mdh	gi|16079964	105	malate dehydrogenase [Bacillus subtilis subsp. subtilis str. 168]
Mdh deg	gi|16079964	150	malate dehydrogenase [Bacillus subtilis subsp. subtilis str. 168]
AtpD iso1	gi|16080734	93	F0F1 ATP synthase subunit beta [Bacillus subtilis subsp. subtilis str. 168]
AtpD iso 2	gi|16080734	215	F0F1 ATP synthase subunit beta [Bacillus subtilis subsp. subtilis str. 168]
AtpD iso 3	gi|16080734	335	F0F1 ATP synthase subunit beta [Bacillus subtilis subsp. subtilis str. 168]
FrlB	gi|16080314	97	hypothetical protein BSU32610 [Bacillus subtilis subsp. subtilis str. 168]
Hag iso1	gi|16080589	83	flagellin [Bacillus subtilis subsp. subtilis str. 168]
Hag iso2	gi|16080589	78	flagellin [Bacillus subtilis subsp. subtilis str. 168]
Hag iso 3	gi|16080589	83	flagellin [Bacillus subtilis subsp. subtilis str. 168]
GroEL	gi|16077670	74	chaperonin GroEL [Bacillus subtilis subsp. subtilis str. 168]
GlnA	gi|16078809	123	glutamine synthetase [Bacillus subtilis subsp. subtilis str. 168]
Ethanol stress			
Pgk	gi|16080446	329	phosphoglycerate kinase [Bacillus subtilis subsp. subtilis str. 168]
CitC	gi|16079965	118	isocitrate dehydrogenase [Bacillus subtilis subsp. subtilis str. 168]
Mdh	gi|16079964	129	malate dehydrogenase [Bacillus subtilis subsp. subtilis str. 168]
Mdh deg	gi|16079964	209	malate dehydrogenase [Bacillus subtilis subsp. subtilis str. 168]
Atp iso1	gi|16080734	125	F0F1 ATP synthase subunit beta [Bacillus subtilis subsp. subtilis str. 168]
Atp iso 2	gi|16080734	155	F0F1 ATP synthase subunit beta [Bacillus subtilis subsp. subtilis str. 168
AtpD iso 3	gi|16080734	341	F0F1 ATP synthase subunit beta [Bacillus subtilis subsp. subtilis str. 168
FrlB	gi|16080314	119	fructoselysine-6-P-deglycase [Bacillus subtilis subsp. subtilis str. 168]
Hag iso1	gi|16080589	157	flagellin [Bacillus subtilis subsp. subtilis str. 168]
Hag iso2	gi|16080589	175	flagellin [Bacillus subtilis subsp. subtilis str. 168]
Hag iso 3	gi|16080589	89	flagellin [Bacillus subtilis subsp. subtilis str. 168]
GroEL	gi|16077670	140	chaperonin GroEL [Bacillus subtilis subsp. subtilis str. 168]
Ilvc iso1	gi|16079881	194	ketol-acid reductoisomerase [Bacillus subtilis subsp. subtilis str. 168]
Ilvc iso2	gi|16079881	183	ketol-acid reductoisomerase [Bacillus subtilis subsp. subtilis str. 168]
Asd	gi|16078738	81	aspartate-semialdehyde dehydrogenase [Bacillus subtilis subsp. subtilis str. 168]
Ald	gi|16080244	79	L-alanine dehydrogenase [Bacillus subtilis subsp. subtilis str. 168]

Detailed data from MALDI-TOF peptide mapping identification of proteins, that differed in levels when exposed to osmotic and ethanol stress.

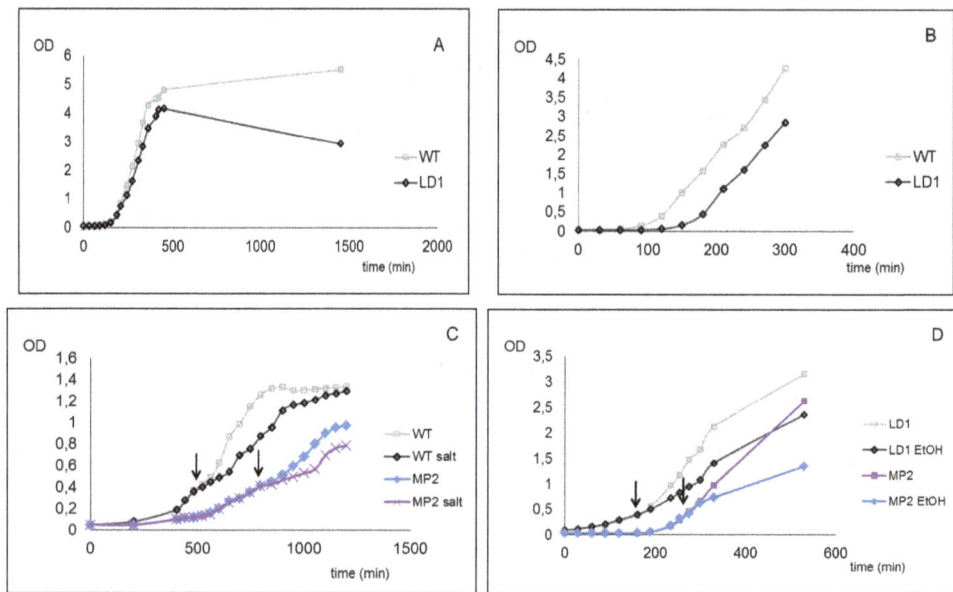

Figure 1. Growth characterization of WT, LD1, and MP2 mutants of *Bacillus subtilis* in long-term cultivation and in response to stress. For long-term growth measurements, the cells were grown in LB medium (A). Effect of MM medium and K⁺ limitation (0.5 mM K⁺) to growth rate of mutant (LD1) (B). Osmotic stress performed only for MP2 mutant in MM medium and K⁺ limitation is demonstrated (C). Effect of ethanol stress was measured in MM medium and K+ limitation for both mutants (LD1, MP2) (D). (WT - *Bacillus subtilis* 168 or SG4, LD1 and MP2– *yxkO* knock-out mutants, for details see Table 1). Exposure of stress is marked by arrows. Measurements were done with cells synchronized in exponential growth, and stress conditions were set up as is described in *Material and Methods*. The typical growth rate curve is shown from measurements made in triplicate for each condition.

stress when cultivated in media with a potassium concentration lower than 1 mM (0.5 mM) (Figure 1C, D). Both types of mutants revealed a similar course of the growth curves.

Characterization of *yxkO* gene promoter transcriptional activity under stress conditions

As the next routine step in characterization of the *yxkO* mutant, we tested activity of the P*yxkO* promoter when exposed to several stress conditions. For this experiment, we used the previously prepared MP2 mutant, as the plasmid pMUTIN4 was constructed to be used for gene knock-out and as a promoter probe with the *lacZ* reporter, as well, with the benefit of complementation of the WT allele. Surprisingly, no significant increase of transcription was observed under any cultivation condition and induction of the WT allele with IPTG, respectively (low, high concentration of K⁺, salt stress), except for ethanol stress, when a slight increase of transcription was detected after the stress exposure (data not shown). The same result was observed previously, when transcriptional activity was tested from solely promoter cloned to pDG1661promoter probe vector (unpublished data). These data indicate that the protein is expressed only at low levels or undergoes more complex regulation of transcription.

Testing of YxkO's impact on activation of general stress response

When direct estimation of the promoter activity of the P*yxkO* promoter had failed, we tested the impact of the YxkO on the induction of SigB-dependent genes. For this purpose, we generally used the well-characterized SigB-dependent P*ctc* promoter, which is transcribed under the most studied stress conditions, and the P*ctc* promoter fused to *lacZ* (P*ctc-lacZ*) beneficially monitors SigB activity in the WT promoter context, as was shown previously [11]. The promoter-probe strains in both genetic backgrounds

were prepared as described in the *Materials and Methods* section, and the promoter activity of P*ctc* was monitored. The tested conditions were ethanol stress in complex media and salt stress when cultivated in high (10 mM) and low (0.5 mM) concentrations of K⁺, respectively. Results are shown in Figure 2.

When exposed to ethanol stress, the promoter activity of P*ctc* in the MP2 mutant was markedly lower (Figure 2A). In the high K⁺ concentration condition assay, the activity of P*ctc* increased in WT and the MP2 mutant after the salt stress, as well, albeit at a lower level in the mutant (Figure 2B). When cells were cultivated in medium with low K⁺ concentration, the transcription from P*ctc* revealed a completely different course. The transcription level was high in WT, even before salt stress, and remained low in the mutant, even after the stress exposure (Figure 2C).

Therefore, we monitored the promoter activity of P*ctc* when cells were cultivated first in the medium with high K⁺ concentration and then shifted to medium with low K⁺ concentration. In both WT and the mutant strain, an increase in the P*ctc* activity occurred after the K⁺ concentration shift but in the mutant the promoter activity was systematically lower at all-time points (Figure 2D). As mentioned above, a 1 mM concentration of K⁺ limits the potassium KtrAB transport system of *Bacillus subtilis* 168, which is predicted to be involved in osmoadaptation [28].

These results indicated that in *Bacillus subtilis* 168, a K⁺ concentration in the medium below 1 mM is a stress condition and causes the general stress response.

The different level of P*ctc* transcription in the presented experiments is in agreement with the recent observation of Young et al. that the extent of the stress causes a differential effect on the adaptation response in subpopulations of *Bacillus subtilis* [14]. It is obvious that potassium deficiency coupled with osmotic stress causes a distinct adaptive response than the osmotic stress itself. As the product of the *yxkO* gene is assumed to be NADH hydrate

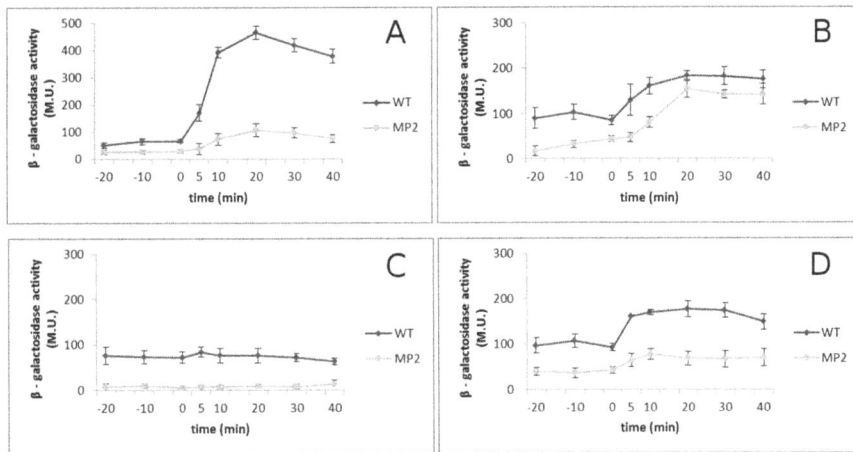

Figure 2. Transcription level of P*ctc* on genetic background of WT and MP2 mutant. P*ctc* activation measurements of WT and MP2 were performed under ethanol stress in LB medium (A), osmotic stress in MM medium with 10 mM K$^+$ concentration (B), osmotic stress in MM medium with 0.5 mM K$^+$ concentration (C), shift from MM medium with 10 mM K$^+$ concentration to MM medium with 0.5 mM K$^+$ concentration (D). Time 0 indicates application of stress. Details of transcription activity measurements, growth, and stress conditions are described in *Material and Methods*.

dehydratase, it is tempting to speculate about the role of energy imbalance in the activation of the SigB regulon and SigB itself.

Comparative 2DE analysis of YxkO impact on an adaptive response to stress

We compared protein levels of cytoplasmic proteins in the pI range of 4–7 in exponentially growing cells of WT and the yxkO knock-out mutant (MP2) of *Bacillus subtilis* in medium with limited potassium and exposed to osmotic and ethanol stress, respectively. Cells were grown and analyzed by comparative 2DE as described in *Materials and Methods*.

The results, summarized in Figures 3–6, show that disruption of the *yxkO* gene caused changes in the level of seven proteins belonging to various metabolic pathways upon both stresses (Figure 3 and Figure 4), and we detected only one protein with a different intensity solely under osmotic stress (Figure 5), three enzymes revealed an altered pattern merely under ethanol stress (Figure 6). Figure S1 illustrates the pattern of the 2DE spot distribution.

Proteins with levels that differ under both stresses

Metabolic enzymes. It is described elsewhere that after osmotic stress, a prompt decrease in expression of most of the glycolytic and citrate cycle enzymes occurs and, in intervals to 60 min, increases to initial levels after resumption of growth arises [5]. Our results are in agreement with this, as we determined the different levels of phosphoglycerate kinase, an enzyme representing the glycolytic pathway, and as we detected changes in the protein level of isocitrate dehydrogenase and malate dehydrogenase from citrate cycle enzymes, as well. In the mutant, 60 minutes after stress exposure, the protein level remained low in the case of all three mentioned proteins, and we detected degradation products of malate dehydrogenase (Mdh-deg, MP2 stress sample) in the case of EtOH stress (Figure 3). There is no study in the literature about the mechanism that is involved in the cessation and recovery of expression of these three vegetative proteins after osmotic stress, but the role of the *yxkO* gene is evident from our experiment.

This effect highlights the recently recognized biochemical activity of YxkO protein as an enzyme that regenerates the hydrated form of NAD(P)H emerging in cells during stress [17].

The absence of dehydration of NADHX and the regeneration of activities of both isocitrate dehydrogenase and malate dehydrogenase may result in energetic imbalance and, subsequently, growth retardation and reduced stress adaptation in the MP2 mutant, as we documented by extension of the lag phase of stressed cultures of mutant cells.

We also detected changes in the protein level of the ATPase subunit, similar to Höper's results [5], and we determined three isoforms of the beta subunit, AtpD. The protein levels of two isoforms (iso1, iso2) manifested the same pattern as in the case of metabolic enzymes, and recovery in the mutant did not occur, either. After stress, there was a significant decrease in the protein level of both isoforms in the mutant (Figure 3).

Stress adaptation. Another protein that was determined to be differentially accumulated under both stresses was fructosamine-6-phosphate deglycase (FrlB, formerly YurP); the protein level increased after the stress in the WT and this increase was absent in the mutant, regardless of stress exposure (Figure 4). This gene was also detected to be upregulated under most stress conditions at the transcription level [4] and was shown to be under direct regulation of the CodY general regulator [29]. This enzyme catalyzes the cleavage of fructosamine-6-phosphate to glucose-6-phosphate and the corresponding amines in *Escherichia coli* and *Bacillus subtilis* and is involved in enzymatic deglycation of Amadori products [30]. Intracellular glycation has been described to occur in *Escherichia coli* [31], and it is very likely to occur in *Bacillus subtilis*, as well. DNA-binding activity of CodY is regulated by the intracellular level of GTP [32], and it was evidenced that during nutrient starvation, the ratio of ATP/GTP is changed [33]. In environmental stress adaptation, this phenomenon has not yet been studied, but it could be supposed that inactivation of NADH-synthesizing enzymes gives rise to an energy imbalance, changing the ATP/GTP ratio, as well.

Changes in the protein value caused by *yxkO* disruption were also recorded for GroEL (Figure 4). In the mutant, the increase of protein level occurred in non-stressed conditions, as well when compared to the WT. This can be explained by extension of the lag phase and the decline of the renewing of isocitrate dehydrogenase levels in the mutant according to WT after stress exposure, which denotes to failure of stress adaptation and triggers

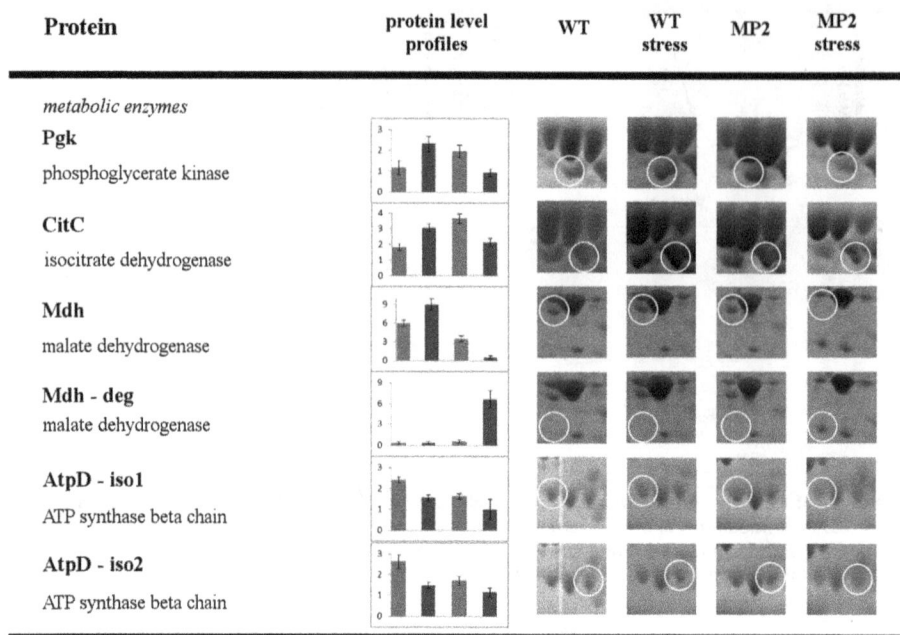

Figure 3. Comparative 2DE analysis of WT versus MP2 mutant exposed to osmotic and ethanol stress – metabolic enzymes. Proteins with protein level profiles that aresimilar for both stresses. For experimental conditions and data evaluation, see *Material and Methods*. Separate columns of the bar charts show the protein level of respective proteins, as calculated from the quantification of the spot volume by PDQuest 8.0 software; y-axes are scaled in intensity for each particular protein. Bars represent each strains and conditions, and there are in the same order as the protein level profiles are presented.

increased levels of GroEL as a result of the devastating effects of both stresses on cellular proteins.

Motility. Another protein that differed in protein level in WT versus MP2 and the changes of which corresponded to the mutant phenotype that the *yxkO* gene disruption affected is flagellin (Figure 4). When the cells were cultivated under a limited

concentration of potassium, there was massive expression of flagellin protein. This corresponds to the rapid movement and substantial flagellation of cells observed in light and electron microscopy, respectively. In native microscopic preparates, we also observed different motility of WT and mutant cells before and after being subjected to salt stress. It has been described that after

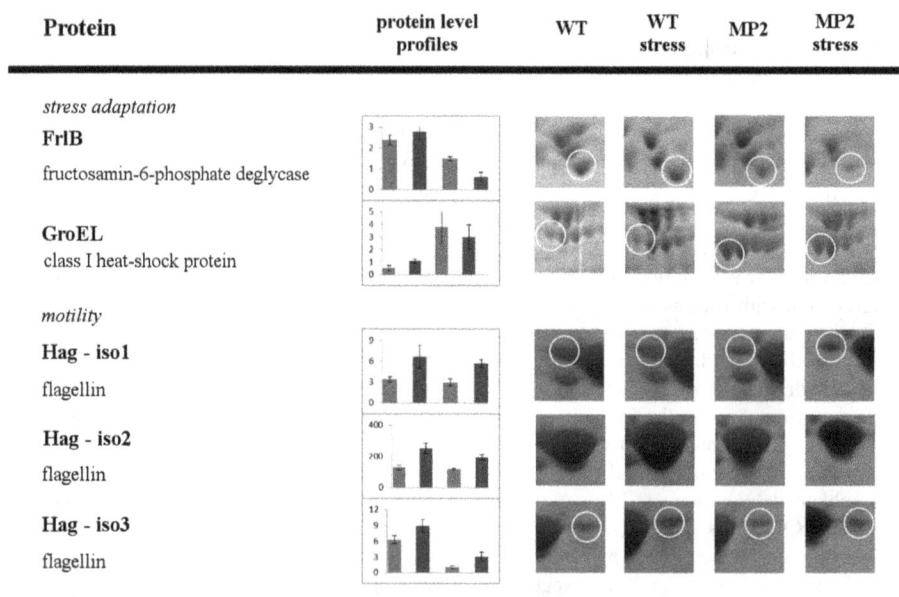

Figure 4. Comparative 2DE analysis of WT versus MP2 mutant exposed to osmotic and ethanol stress – stress adaptation and motility. Proteins with protein level profiles that are similar for both stresses. For experimental conditions and data evaluation, see *Material and Methods*. The picture description is same as for Figure 3.

Protein	protein level profiles	WT	WT stress	MP2	MP2 stress
nitrogen assimilation **GlnA** glutamine synthetase					

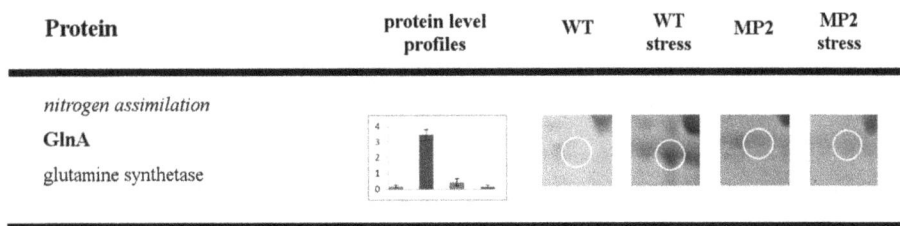

Figure 5. Comparative 2DE analysis of WT versus MP2 mutant exposed to osmotic stress. Proteins with protein level profiles that are unique for osmotic stress. For experimental conditions and data evaluation, see *Material and Methods*. The picture description is same as for Figure 3.

osmotic shock, a decrease in the transcription of genes coding for motility apparatus components occurs [1]. In agreement with Steils results, we have observed that after the shock, the WT cells ceased to move in the adaptation phase, whereas the mutant cells did not show any significant changes in their motility, regardless of the stress (data not shown).

On our gels, we detected three isoforms of flagellin, with the evidence that the formation of quantitatively less represented isoforms (iso1 and iso3) was affected by the disruption of the *yxkO* gene. A drop in other proteins of the flagellar apparatus, such as basal body components, was not detected. This could indicate that the product of the *yxkO* gene does not influence the intracellular amount of flagellin protein but causes its modification, which allows or prevents the transport of monomers out of the cell and its degradation in the cytoplasm, respectively. The origins and nature of these isoforms of flagellin have not been described in the literature to date and require further work to be elucidated. We can speculate that non-enzymatic glycation (see FrlB above) is involved.

Protein with levels that differ solely under osmotic stress

Nitrogen assimilation. Glutamine synthetase, the key enzyme of nitrogen assimilation, is a protein, the level of which was affected by mutation in the *yxkO* gene only after osmotic stress (Figure 5). We determined a massive increase in the protein amount of this enzyme in WT after the salt stress, while in the MP2 mutant, the protein stayed at a low level, even after the stress. Our results differ from those of Höper [5], which can be explained by the fact that we used a mineral medium with ammonium

sulphate as the sole source of nitrogen and a limited concentration of potassium, conditions under which nitrogen metabolism are also significantly influenced by CodY and ATP assistance, as was recently reported by Gunka [34].

Proteins with a different level under ethanol stress

A comprehensive proteomic study of *Bacillus subtilis* under ethanol stress with which we could compare our results has not yet been published. After the ethanol stress, we observed changes in a similar set of proteins as in the salt stress experiment, with a comparable intensity profile (see above). However, we identified proteins, the accumulation of which was influenced only after the ethanol stress (Figure 6). Changes were detected in the protein amount of three enzymes: ketol-acid reductoisomerase (IlvC), aspartate-semialdehyde dehydrogenase (Asd) and amino acid L-alanine dehydrogenase (Ald).

Valin, leucine, and isoleucine metabolism. The first enzyme, ketol-acid reductoisomerase (IlvC), is known to be involved in the biosynthesis of valine, leucine, and isoleucine, and its encoding gene is a part of the *ilv-leu* operon [35]. We identified that the cellular level of IlvC decreased in the mutant after the stress, while in WT, it remained unchanged. Valine, leucine, and isoleucine are the precursors for the biosynthesis of iso- and anteiso-branched fatty acids, which represent the major fatty acid species of membrane lipids in *Bacillus* sp. The increase in their expression was proposed as a long-term adaptation mechanism to cold and ethanol stresses [36]. It is worth mentioning that as already reported by Molle [37], the *ilv-leu*

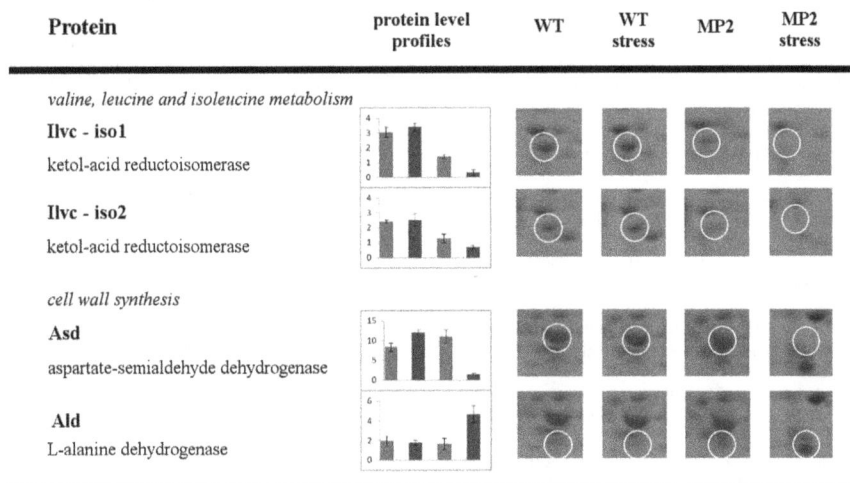

Protein	protein level profiles	WT	WT stress	MP2	MP2 stress
valine, leucine and isoleucine metabolism **Ilvc - iso1** ketol-acid reductoisomerase					
Ilvc - iso2 ketol-acid reductoisomerase					
cell wall synthesis **Asd** aspartate-semialdehyde dehydrogenase					
Ald L-alanine dehydrogenase					

Figure 6. Comparative 2DE analysis of WT versus MP2 mutant exposed to ethanol stress. Proteins with protein level profiles that are unique for ethanol stress. For experimental conditions and data evaluation, see *Material and Methods*. The picture description is same as for Figure 3.

operon was shown to be controlled by the global transcriptional regulator CodY.

Cell wall synthesis. The second enzyme, aspartate-semialdehyde dehydrogenase (Asd), the expression of which was lower after the stress in the mutant, is involved in the metabolism of alanine, aspartate, and glutamate. The third enzyme, amino acid L-alanine dehydrogenase (Ald), catalyzes deamination of alanine to pyruvate. Its level in the mutant strain was markedly higher after the stress, while in WT, it was not affected. Both are the precursors for the synthesis of diaminopimelic acid, which is a constituent of peptidoglycan. It has been reported that ethanol stress increases the expression of proteins involved in cell wall synthesis [38], and phosphorylation of aspartate-semialdehyde dehydrogenase by PtkA kinase was described [39]. An imbalance in expression of this protein may cause a defect in cell wall synthesis in the mutant strain and consequently decrease its adaptability to ethanol stress.

Conclusions

In this study, we performed the first characterization of a knock-out mutant in the *yxkO* gene, which biochemical activity was classified *in vitro* as an NAD(P)H hydrate dehydratase, and provided evidence that it is involved in stress adaptation (to osmotic and ethanol stress and under potassium limitation).

Our results correspond to the observed phenotype of the mutant and show that stress adaptation reactions in the mutant are merely suppressed or not activated, as reflected by the observed decline in the growth and extension of the lag phase, inhibition of recovery of NADH-dependent proteins during the lag phase after stress exposure (Idh, Mdh, etc.), energy imbalance and repression of CodY-regulated proteins (FrlB, IlvC), and inhibition of activation of a general stress protein (Ctc) in the mutant. We have also shown that potassium limitation is a stress condition for *Bacillus subtilis*

that activates the general stress response. This reflects the complexity and interconnection of stress adaptation processes and is evidence that either adaptation mechanism is important in stress resistance. Thus metabolite repair enzymes play a significant role in the survival of stressed cells.

Supporting Information

Figure S1   The 2-DE image of cytoplasmic proteins from WT *Bacillus subtilis* 168 cultivated in the ethanol stress condition, illustrating the pattern of the 2DE spot distribution. The patterns of the 2DE spot distributions in the studied culture conditions of WT and MP2 mutant were similar. Proteins that exhibited changes in protein abundances and were identified by MS MALDI-TOF are pointed out. Details of identified proteins are described in *Results and Discussion*, Figures 3–6, and Table 2.

Acknowledgments

Mass spec measurements were done in the Service Laboratories of the Biology Section, Laboratory of Mass Spectrometry, Charles University in Prague, Faculty of Science, Vinicna 7, 128 44 Praha 2. https://www.natur.cuni.cz/biologie/servisni-laboratore/laborator-hmotnostni-spektrometrie.

Author Contributions

Conceived and designed the experiments: MP JT LD IL. Performed the experiments: MP JT LD ES. Analyzed the data: MP JT LD IL. Contributed reagents/materials/analysis tools: MP JT LD ES IL. Contributed to the writing of the manuscript: MP JT LD IL.

References

1. Steil L, Hoffmann T, Budde I, Volker U, Bremer E (2003) Genome-wide transcriptional profiling analysis of adaptation of Bacillus subtilis to high salinity. J Bacteriol 185: 6358–6370.
2. Price CW, Fawcett P, Ceremonie H, Su N, Murphy CK, et al. (2001) Genome-wide analysis of the general stress response in Bacillus subtilis. Mol Microbiol 41: 757–774.
3. Petersohn A, Brigulla M, Haas S, Hoheisel JD, Volker U, et al. (2001) Global analysis of the general stress response of Bacillus subtilis. J Bacteriol 183: 5617–5631.
4. Nicolas P, Mäder U, Dervyn E, Rochat T, Leduc A, et al. (2012) Condition-Dependent Transcriptome Reveals High-Level Regulatory Architecture in Bacillus subtilis. Science 335: 1103–1106.
5. Hoper D, Bernhardt J, Hecker M (2006) Salt stress adaptation of Bacillus subtilis: a physiological proteomics approach. Proteomics 6: 1550–1562.
6. Hecker M, Pane-Farre J, Volker U (2007) SigB-dependent general stress response in Bacillus subtilis and related gram-positive bacteria. Annu Rev Microbiol 61: 215–236.
7. Hahne H, Mader U, Otto A, Bonn F, Steil L, et al. (2010) A comprehensive proteomics and transcriptomics analysis of Bacillus subtilis salt stress adaptation. J Bacteriol 192: 870–882.
8. Mascher T, Margulis NG, Wang T, Ye RW, Helmann JD (2003) Cell wall stress responses in Bacillus subtilis: the regulatory network of the bacitracin stimulon. Mol Microbiol 50: 1591–1604.
9. Gaidenko TA, Kim TJ, Weigel AL, Brody MS, Price CW (2006) The blue-light receptor YtvA acts in the environmental stress signaling pathway of Bacillus subtilis. J Bacteriol 188: 6387–6395.
10. Boylan SA, Redfield AR, Brody MS, Price CW (1993) Stress-induced activation of the sigma B transcription factor of Bacillus subtilis. J Bacteriol 175: 7931–7937.
11. Benson AK, Haldenwang WG (1992) Characterization of a regulatory network that controls sigma B expression in Bacillus subtilis. J Bacteriol 174: 749–757.
12. Avila-Perez M, van der Steen JB, Kort R, Hellingwerf KJ (2010) Red light activates the sigmaB-mediated general stress response of Bacillus subtilis via the energy branch of the upstream signaling cascade. J Bacteriol 192: 755–762.
13. Boylan SA, Redfield AR, Price CW (1993) Transcription factor sigma B of Bacillus subtilis controls a large stationary-phase regulon. J Bacteriol 175: 3957–3963.
14. Young JW, Locke JC, Elowitz MB (2013) Rate of environmental change determines stress response specificity. Proc Natl Acad Sci U S A 110: 4140–4145.
15. Ulanova D, Holanova V, Prenosilova L, Naprstek J, Licha I (2007) Mutation of a gene encoding a putative ribokinase leads to reduced salt tolerance under potassium limitation in Bacillus subtilis. Folia Microbiol (Praha) 52: 203–208.
16. Zhang RG, Grembecka J, Vinokour E, Collart F, Dementieva I, et al. (2002) Structure of Bacillus subtilis YXKO-a member of the UPF0031 family and a putative kinase. J Struct Biol 139: 161–170.
17. Shumilin I, Cymborowski M, Chertihin O, Jha KN, Herr JC, et al. (2012) Identification of unknown protein function using metabolite cocktail screening. Structure 20: 1715–1725.
18. Oppenheimer NJ, Kaplan NO (1974) Glyceraldehyde-3-phosphate dehydrogenase catalyzed hydration of the 5–6 double bond of reduced beta-nicotinamide adenine dinucleotide (beta NADH). Formation of beta-6-hydroxy-1,4,5,6-tetrahydronicotinamide adenine dinucleotide. Biochemistry 13: 4685–4694.
19. Marbaix AY, Noel G, Detroux AM, Vertommen D, Van Schaftingen E, et al. (2011) Extremely conserved ATP- or ADP-dependent enzymatic system for nicotinamide nucleotide repair. J Biol Chem 286: 41246–41252.
20. Yoshida A, Dave V (1975) Inhibition of NADP-dependent dehydrogenases by modified products of NADPH. Arch Biochem Biophys 169: 298–303.
21. Prabhakar P, Laboy JI, Wang J, Budker T, Din ZZ, et al. (1998) Effect of NADH-X on cytosolic glycerol-3-phosphate dehydrogenase. Arch Biochem Biophys 360: 195–205.
22. Van Schaftingen E, Rzem R, Marbaix A, Collard F, Veiga-da-Cunha M, et al. (2013) Metabolite proofreading, a neglected aspect of intermediary metabolism. J Inherit Metab Dis 36: 427–434.
23. Hoper D, Volker U, Hecker M (2005) Comprehensive characterization of the contribution of individual SigB-dependent general stress genes to stress resistance of Bacillus subtilis. J Bacteriol 187: 2810–2826.
24. Young JW, Locke JC, Elowitz MB (2013) Rate of environmental change determines stress response specificity. Proc Natl Acad Sci U S A 110: 4140–4145.

25. Cutting SM, Vander Horn PB (1990) Genetic analysis, PBS1 Geneneralized transduction. In: Harwood CR, Cutting SM editors. Molecular Biological Methods for Bacillus subtilis: John Wiley and sons.

26. Ray C, Hay RE, Carter HL Jr., Moran CP Jr. (1985) Mutations that affect utilization of a promoter in stationary-phase Bacillus subtilis. J Bacteriol 163: 610–614.

27. Shevchenko A, Tomas H, Havlis J, Olsen JV, Mann M (2006) In-gel digestion for mass spectrometric characterization of proteins and proteomes. Nat Protoc 1: 2856–2860.

28. Holtmann G, Bakker EP, Uozumi N, Bremer E (2003) KtrAB and KtrCD: two K^+ uptake systems in Bacillus subtilis and their role in adaptation to hypertonicity. J Bacteriol 185: 1289–1298.

29. Belitsky BR, Sonenshein AL (2008) Genetic and biochemical analysis of CodY-binding sites in Bacillus subtilis. J Bacteriol 190: 1224–1236.

30. Deppe V, Bongaerts J, O'Connell T, Maurer KH, Meinhardt F (2011) Enzymatic deglycation of Amadori products in bacteria: mechanisms, occurrence and physiological functions. Appl Microbiol Biotechnol 90: 399–406.

31. Mironova R, Niwa T, Hayashi H, Dimitrova R, Ivanov I (2001) Evidence for non-enzymatic glycosylation in Escherichia coli. Mol Microbiol 39: 1061–1068.

32. Ratnayake-Lecamwasam M, Serror P, Wong KW, Sonenshein AL (2001) Bacillus subtilis CodY represses early-stationary-phase genes by sensing GTP levels. Genes Dev 15: 1093–1103.

33. Zhang S, Haldenwang WG (2005) Contributions of ATP, GTP, and Redox State to Nutritional Stress Activation of the Bacillus subtilis σB Transcription Factor J Bacteriol 7554–7560.

34. Gunka K, Commichau F (2012) Control of glutamate homeostasis in Bacillus subtilis: a complex interplay between ammonium assimilation, glutamate biosynthesis and degradation. Mol Microbiol 85: 213–224.

35. Mader U, Hennig S, Hecker M, Homuth G (2004) Transcriptional organization and posttranscriptional regulation of the Bacillus subtilis branched-chain amino acid biosynthesis genes. J Bacteriol 186: 2240–2252.

36. De Mendoza D, Schujman GE, Aguilar PS (2002) Biosynthesis and function of membrane lipids. In: A. L. Sonenshein JAH, and R Losick editor. Bacillus and its closest relatives. Washington, D.C.: ASM Press.

37. Molle V, Nakaura Y, Shivers RP, Yamaguchi H, Losick R, et al. (2003) Additional targets of the Bacillus subtilis global regulator CodY identified by chromatin immunoprecipitation and genome-wide transcript analysis. J Bacteriol 185: 1911–1922.

38. Eiamphungporn W, Helmann JD (2008) The Bacillus subtilis sigma(M) regulon and its contribution to cell envelope stress responses. Mol Microbiol 67: 830–848.

39. Jers C, Pedersen MM, Paspaliari DK, Schutz W, Johnsson C, et al. (2010) Bacillus subtilis BY-kinase PtkA controls enzyme activity and localization of its protein substrates. Mol Microbiol 77: 287–299.

16

Role of ARF6, Rab11 and External Hsp90 in the Trafficking and Recycling of Recombinant-Soluble *Neisseria meningitidis* Adhesin A (rNadA) in Human Epithelial Cells

Giuseppe Bozza[1][◐][¤a], Mirco Capitani[2][◐], Paolo Montanari[1], Barbara Benucci[1], Marco Biancucci[1][¤b], Vincenzo Nardi-Dei[1], Elena Caproni[1], Riccardo Barrile[1][¤c], Benedetta Picciani[2][¤d], Silvana Savino[1], Beatrice Aricò[1], Rino Rappuoli[1], Mariagrazia Pizza[1], Alberto Luini[3], Michele Sallese[2]*, Marcello Merola[1,4]*

[1] Novartis Vaccines, Siena, Italy, [2] Unit of Genomic Approaches to Membrane Traffic, Fondazione Mario Negri Sud, S. Maria Imbaro (CH), Italy, [3] Institute of Protein Biochemistry, CNR, Naples, Italy, [4] Department of Biology, University of Naples "Federico II", Naples, Italy

Abstract

Neisseria meningitidis adhesin A (NadA) is a meningococcus surface protein thought to assist in the adhesion of the bacterium to host cells. We have previously shown that NadA also promotes bacterial internalization in a heterologous expression system. Here we have used the soluble recombinant NadA (rNadA) lacking the membrane anchor region to characterize its internalization route in Chang epithelial cells. Added to the culture medium, rNadA internalizes through a PI3K-dependent endocytosis process not mediated by the canonical clathrin or caveolin scaffolds, but instead follows an ARF6-regulated recycling pathway previously described for MHC-I. The intracellular pool of rNadA reaches a steady state level within one hour of incubation and colocalizes in endocytic vesicles with MHC-I and with the extracellularly labeled chaperone Hsp90. Treatment with membrane permeated and impermeable Hsp90 inhibitors 17-AAG and FITC-GA respectively, lead to intracellular accumulation of rNadA, strongly suggesting that the extracellular secreted pool of the chaperone is involved in rNadA intracellular trafficking. A significant number of intracellular vesicles containing rNadA recruit Rab11, a small GTPase associated to recycling endosomes, but do not contain transferrin receptor (TfR). Interestingly, cell treatment with Hsp90 inhibitors, including the membrane-impermeable FITC-GA, abolished Rab11-rNadA colocalization but do not interfere with Rab11-TfR colocalization. Collectively, these results are consistent with a model whereby rNadA internalizes into human epithelial cells hijacking the recycling endosome pathway and recycle back to the surface of the cell via an ARF6-dependent, Rab11 associated and Hsp90-regulated mechanism. The present study addresses for the first time a meningoccoccal adhesin mechanism of endocytosis and suggests a possible entry pathway engaged by *N. meningitidis* in primary infection of human epithelial cells.

Editor: Matthew Seaman, Cambridge University, United Kingdom

Funding: This work is entirely funded by Novartis Vaccines; no external entities have contributed to cover project expenses. The funders had no role in study design, data collection and analysis, decision to publish, or preparation of the manuscript.

Competing Interests: PM, VND, SS, BA, RR, and MP are employees of Novartis Vaccines, whose company funded this study. GB, RB, and MB were students working at Novartis and BB is a student still working at Novartis. EC was a post-doctoral fellow working at Novartis. MM is a Novartis Scientific collaborator and was responsible for this project. The antigen of this study is one of the components of Bexsero, a commercially available vaccine against meningococcus B. There are no further patents, products in development, or marketed products to declare.

* Email: sallese@negrisud.it (MS); marcello.merola@novartis.com (MM)

◐ These authors contributed equally to this work.

¤a Current address: European Medicines Agency, London, United Kingdom
¤b Current address: Northwestern University of Chicago, Evanston, IL, United States of America
¤c Current address: Wyss Institute for Biologically Inspired Engineering, Harvard University, Boston, MA, United States of America
¤d Current address: Department of Diagnostic and Laboratory Pathology, IRCCS Foundation, "Istituto Nazionale Tumori", Milan, Italy

Introduction

Neisseria meningitidis (meningococcus) is a Gram-negative diplococcus that causes severe invasive disease and represents one of the most devastating bacterial infections. Although fatal if not treated on time, meningococcus invasion appears to be more an undesirable event of a usually commensal bacterium, probably due to a combination of host susceptibility and strain specific propensity to invasiveness [1,2]. The pathophysiology of the bacterium *N. meningitidis* is a process that requires several steps: penetration of the epithelial or mucosal barrier, reaching and surviving the bloodstream, crossing the blood-brain barrier, and eventually causing meningitis through extracellular proliferation [3]. Epithelia are the primary target of bacterial colonization, an event basically asymptomatic and common to both non-virulent and virulent strains. Disease is a rare event compared to the extent of meningococcal nasopharynx colonization. [2,4,5]. Experimental data support attachment of the bacterium to nonciliated cells of the respiratory epithelium [6] and transcellular route of passage through this barrier [6–9]. A recent report shows that bacterial capsule and type 4 pili are important for epithelial cell transcytosis [9] but host and pathogen players involved in this process are far from being defined.

Neisseria meningitidis adhesin A (NadA) was identified through genome wide analysis and afterward proved to be expressed on the bacterial surface [10–12]. NadA is a trimeric outer membrane protein whose gene is present in three out of four known hypervirulent lineages of serogroup B strains [13,14]. NadA is classified as a trimeric autotransporter adhesion (TAA), a family of outer membrane adhesin, which are present only in Gram negative bacteria. TAAs differ from classical autotrasporter protein by the homotrimeric structure hanging on the bacterial surface and the Type Vc secretion system [10,11,15–17]. Like other TAA proteins, NadA has a common modular organization, composed of i) a conserved C-terminal membrane anchor through which the protein is translocated to the cell surface, ii) a long central alpha helical domain (stalk) with high propensity to form coiled-coil structures, iii) an N-terminal globular head that has been associated with binding to specific cellular receptors and iiii) a cleavable signal sequence [18,19]. Generally, TAAs (previously classified as Oligomeric coiled-coil Adhesins (Oca) family [20]) mediate bacterial interaction with host cells or extracellular matrix (ECM) proteins or induce invasion into target cells [21–29]. The specificity of the biological function is thought to reside in the head, while the stalk is thought to ensure adequate exposure of the head from the outer membrane [19].

In previous reports, we have used a NadA heterologous expression system to demonstrate the invasive ability acquired by an Escherichia coli strain exposing surface NadA. Our data support the role of NadA in the uptake of bacteria by Chang cells, a human epithelial cell line [10,30]. A recombinant NadA (rNadA), expressed in E. coli and purified in a soluble form in absence of the anchor (translocator) domain, preserves its immunogenic properties and is included in a multicomponent vaccine against meningococcus B (Bexsero) [31,32]. A peculiar feature of rNadA, perhaps unique among all members of the TAA family, is the ability to preserve a stable trimeric structure in solution [10,13,33]. This recombinant soluble homo-trimer still binds eukaryotic cells [10,13,33–35]. The gain-of-function phenotype acquired by heterologous bacteria expressing NadA and the conserved binding characteristics shown by the recombinant protein provide an opportunity to dissect the function of this adhesin in host-pathogen interaction(s).

When expressed on the surface of E. coli, NadA promotes bacterial internalization [10], by an undefined mechanism. Eukaryotic cells internalize a variety of extracellular substances as well as plasma membrane proteins via a remarkable diversity of endocytic pathways generally classified as clathrin-dependent or independent according to the involvement of clathrin [36–38]. Endocytic vesicles converge into early endosomes (EEs), the main sorting compartment from which most of the cargos are rapidly recycled back to the plasmamembrane (PM), while some are sorted into late endosomes and destined either to lysosomes or to the trans Golgi network (TGN) [39,40]. While lysosome targeting is achieved by clathrin-coated endosomes, mechanisms of clathrin-independent endocytosis have been identified more recently and are emerging as complex, multi-pattern processes of internalization still poorly characterized [41,42]. In particular, ADP ribosylation Factor 6 (ARF6) has been involved in internalization and sorting of cargos targeted to recycling endosome [43]. Initially

described for the major histocompatibility complex (MHC) class I trafficking [44], the list of membrane proteins requiring this factor for internalization and recycling is growing [42,43]. While ARFs coordinate vesicle formation [45], key regulators of endosome trafficking include the Rabs, a family of small GTPase proteins associated to all kind of vesicles [46]. The more than 60 Rabs found in eukaryotic cells communicate with each other through recruitment of effectors that regulate the correct destination of vesicles [39,47].

In our attempt to identify eukaryotic cell proteins involved in the NadA-mediated interaction of bacteria, we found that such processes are sensitive to inhibition of Hsp90 activity, and the host chaperone itself was found to bind NadA in vitro [30]. Hsp90 is a ubiquitous molecular chaperone crucial for maintenance of cellular homeostasis [48]. The Hsp90 family of chaperones includes organelle specialized isoforms and two cytoplasm/nucleus localized factors, the inducible Hsp90α and the constitutive Hsp90β. More recently, several lines of evidence suggest that Hsp90α is also secreted by normal cells in response to stress or insults, and by several tumor cell lines (reviewed in [49]). The major functions of this extracellular pool have been identified as protective response to several kinds of stress, assistance to antigen presentation and promoting tissue invasion by cancer cells [49–52]. Results from our previous study also indicated a protective role of Hsp90 toward NadA mediated invasion, although we did not differentiate the possible role of extracellular and intracellular pools of the chaperone [30].

In this report we have used the soluble recombinant NadA$_{\Delta351-405}$ (rNadA) and demonstrated that it was able to internalize upon binding to the Chang human epithelial cell line. Evidence is provided on the stability of the rNadA trimer at 37°C and on the temperature dependence of NadA binding to Chang cells. In our analysis of the internalization pathway exploited by NadA, we found that the process triggered by NadA was clathrin independent. As reported for MHC class I internalization and recycling to cell surface, ARF6-dependence and recruitment of Rab 11 on rNadA vesicles suggested that the adhesin is sorted in a recycling endosomes pathway. The effect of two Hsp90 inhibitors on the internalization and recycling of the adhesin indicated that extracellular Hsp90 is involved in the trafficking of rNadA.

Materials and Methods

Cell culture and transfection

Chang epithelial cells (Wong-Kilbourne derivative, clone 1-5c-4, human conjunctiva, ATCC CCL-20.2) were grown in Dulbecco's Modified Eagle Medium (DMEM) supplemented with 15 mM glutamine, 100 U/ml penicillin, 100 μg/ml streptomycin, 10% FCS and maintained at 37°C in a controlled humidified atmosphere containing 5% CO_2. Transfection of Chang cells was performed using Hiperfect (Qiagen), jetPEI (Polyplus transfection) or lipofectamine (Invitrogen) according to the manufacturer's instructions.

Antibodies, plasmid vectors and reagents

Anti-MHC-I (Biolegends); anti-EEA1 (Novus Biologicals); anti-ARF6 (Santa Cruz Biotechnology); anti-Rab5 (BD transduction Laboratories); anti clathrin (Sigma); anti-M6PR kindly provided by J. Gruenberg University of Geneva, Switzerland; anti-HSP90 (SPA-830, Stressgene); Alexa-488-conjugated Phalloidin, Cy3-conjugated transferrin, Alexa488- and 543-conjugated secondary antibodies (Invitrogen); Alexa488-conjugated anti-MHC-I (Santa cruz Biotechnology). Antibodies against rNadA were previously described in [53]. Alexa-488-conjugated and Alexa-633 conjugated rNadA were obtained by labelling with Alexa Fluor Microscale Protein Labeling Kit (Molecular Probes-Invitrogen) following manufacturer instructions. cDNA expressing the wild type and ARF6-Q67L mutant were kindly provided by J. Donaldson (to M.S. and A.L.), National Institutes of Health, Bethesda, USA. 17-AAG (17-N-allylamino-17-demethoxygeldanamycin) and Wortmannin were from Sigma. FITC-GA was synthesized by labogen (Milan - Italy). The myc-AP180C plasmid containing the C-terminal portion of the AP180 gene fused in frame with N-terminal myc tag sequence was generated as described elsewhere [54]. Generation of plasmids coding for C-terminal EGFP fusion Rabs was performed as described previously [55].

Binding of rNadA to human Chang epithelial cells

Chang cells were non-enzymatically detached from the support by using cell dissociation solution (Sigma), harvested and resuspended in DMEM medium supplemented with 1% FBS.

Then 3×10^5 cells were mixed with recombinant rNadA (200 μg/ml) for 30 minutes at four different temperatures (RT, 4°C, 30°C and 37°C). After two washes with 1% FBS in phosphate-buffered saline, cells were incubated with the mouse monoclonal anti-NadA antibody 9F11 (1.25 μg/ml to 160 μg/ml) or IgG mouse isotype control (Sigma) for 1 hr at 4°C. Samples were washed twice and then incubated for 30 minutes at 4°C with Allophycocyanin-conjugated goat F(ab)₂ antibody to mouse IgG (1:100, Jackson ImmunoResearch) and finally the cells were analyzed with the FACS-Scan flow cytometer Canto II. The mean fluorescence intensity for each sample was calculated.

To analyze the time dependence of NadA binding to Chang, 3×10^5 cells were incubated with recombinant rNadA (200 μg/ml) at 37°C for different period of time (1 minute to 2 hours), washed and incubated with the mouse monoclonal anti-NadA antibody 9F11 (20 μg/ml) or IgG mouse isotype control for 1 hr at 4°C. FACS analysis was performed as described above.

Size Esclusion-HPLC–multiangle laser light-scattering (MALLS) analysis

Four identical aliquots (100 μl) of rNadA (1,7 mg/ml in 10 mM NaH₂PO₄ 150 mM NaCl pH 7.2) were heated at 37°C for 30 min, 1, 2 and 4 hrs respectively. For each time point, sample was loaded on an analytical size exclusion TSK Super SW3000 (4.6×300-mm column of 4 μm and 300 Å) with a separation range on globular proteins from 10 to 500 kDa (Tosoh Bioscience, Tokyo, Japan). Samples were eluted isocratically in 0.1 M NaH₂PO₄, 0,1 M NaHSO₄ buffer at pH 7.0. MALLS analyses were performed in real time using a multi-angle light-scattering detector Dawn TREOS (Wyatt Corp., Santa Barbara, CA, USA), with an incident laser of 658 nm; intensity of the scattered light was measured at 3 angles simultaneously. Data elaboration was performed using Astra V software (Wyatt). Zimm formalism was used to determine the weight-average molecular mass (MW) in daltons and polydispersity index (MW/Mn) for each oligomer present in solution.

Full length trimeric NadA purification on size exclusion chromatography

A 300 μl aliquot of rNadA (1,7 mg/ml Buffer 10 mM NaH₂PO₄ 150 mM NaCl pH 7.2) was heated at 37°C for 4 hours. At end of this time period, sample was loaded on the gel filtration column. Size exclusion was performed using a Superdex 75 10–300 GL (GE-Healtcare) and elution was performed isocratically in 0.1 M NaH₂PO₄ pH 7.2, 0.15 M NaCl. Protein containing fractions were identified by OD at 280 nm and collected.

Immunofluorescence staining and confocal microscopy

Chang cells were cultured on coverslips to reach 70–80% of confluence, washed three times in medium without serum and then incubated with rNadA (200 μg/ml) in the same medium supplemented with 1% FCS. Incubation was performed at 37°C for the indicated times. Cells were then fixed in paraformaldehyde for 8–10 min at 37°C, permeabilized and incubated with primary antibodies for 1 hr at room temperature. Afterwards, cells were washed three times with PBS, incubated with fluorophore-conjugated secondary antibody for 30–45 min at room temperature, washed again three times in PBS and mounted on slides.

In vivo staining of MHC-I was performed as follows: cells were washed three times in medium without serum and incubated with a mouse monoclonal antibody against MHC-I (10–30 μg/ml) for 1 hr at 4°C. Afterwards, cells were washed 3 times and incubated with recombinant rNadA (200 μg/ml) for 1 hr at 37°C in a medium supplemented with 1% FCS, then fixed and permeabilized. rNadA was stained following the standard procedure while MHC-I was revealed using a fluorescence-conjugated secondary antibody directed against the mouse monoclonal primary antibody.

In vivo staining of HSP90 was performed as follows: cells were washed three times in medium without serum and incubated with a rabbit polyclonal HSP90 antibody (50 μg/ml) for 2–4 hrs at 4°C. Afterward, cells were washed 3 times and incubated with recombinant rNadA (200 μg/ml) for 1 h at 37°C in a medium supplemented with 1% FCS, then fixed and permeabilized. rNadA was stained following the standard procedure while HSP90 was detected using a Alexa543-conjugated secondary antibody directed against the rabbit polyclonal primary antibody.

Samples were analyzed by confocal microscopy (LSM 510, Zeiss) using a 60× oil-immersion objective, maintaining the pinhole of the objective at 1 airy unit. Images were scanned using an Argon 488 laser, a HeNe 543 laser and a HeNe 633 laser, under non-saturating conditions (pixel fluorescence below 255 arbitrary units).

The colocalization analysis and the quantification of immunofluorescence (IF) intensity of rNadA in the cells was performed with LSM510-3.2 software (Zeiss). To assess the colocalization we removed the background immunofluorescence by adjusting the threshold levels and used the histo and colocalization functions of the above software. This software provides two colocalization coefficients that ranges from 0 (no colocalization) to 1 (complete colocalization). The colocalization coefficients indicate the amount of pixels of the channel A that colocalizes with pixels from channel B and viceversa. Finally, we expressed the colocalization extent as a percentage over the total immunofluorescence per channel. The immunofluorescence (IF) intensity was calculated as total immunofluorescence of rNadA in the cell divided by the area of the cell and expressed as arbitrary units (A.U.).

rNadA uptake in the presence of Hsp90 inhibitors

Internalization was performed by adding rNadA to the culture medium at a final concentration of 200 μg/ml and incubating at 37°C for the indicated period of time. Chang cells grown at about 50% confluence were pre-treated overnight with 0.5 μM 17-AAG and the then incubated with recombinant NadA (200 μg/ml) at 37°C for 1, 4 or 16 hrs in presence of the same concentration of 17-AAG. When 10 μM 17-AAG or FITC-GA were used, cells were grown to 70–80% confluence and pretreated for 1 hr with the inhibitors before adding rNadA (200 μg/ml) and subsequently incubated at 37°C for 1,4 or 16 hrs in the continued presence of the inhibitors. To temporarily permeabilize cells, growth medium was substituted with 0.01% saponin in PBS for 30 seconds at room temperature. Cells were then washed 3 times with PBS before adding fresh medium and incubate at 37°C for the indicated time. Afterwards, cells were fixed and labeled as detailed above.

AKT detection in Chang cell lysates

Chang cells were seeded on 24-well tissue culture plates (6–8×10⁴ cells per well) and incubated at 37°C overnight. The following day, cells were treated with HSP90 inhibitors as described above. Total cell extracts were prepared in RIPA buffer (Sigma) supplemented with complete protease inhibitor (Roche).

Equal amounts of proteins were prepared in NuPAGE SDS Buffer under reducing conditions and separated on NuPAGE polyacrylamide gels. Proteins were transferred to nitrocellulose membranes for Western blot analysis. Membranes were then blocked with PBS containing 0.1% Tween 20 (PBST)+10% dried skim milk at room temperature for 1 h. After extensive washings in PBST, proteins bound on nitrocellulose membranes were detected with specific anti-Akt rabbit antibody (Cell Signaling Technology) followed by HRP-conjugated goat anti-rabbit secondary antibody.

Time-lapse Microscopy

Chang cells at 70–80% of confluence on glass-bottom Petri dishes (MatTek Corporation-USA) were incubated overnight at 37°C. The day after, cells were incubated with Alexa488-conjugated anti-MHC-I antibody and 2.5 µg/ml Alexa633-conjugated rNadA, or 5 µg/ml of Cy3-conjugated transferrin and 2.5 µg/ml Alexa-488-conjugated rNadA, or pretreated overnight with 0.5 µM of 17AAG and then with Alexa-488-conjugated rNadA for 60 min at 37°C. Cells were washes three times and supplemented with the medium buffered with 20 mM Hepes pH 7.2. The Petri dish was then placed on a stage of LSM-510 confocal microscope (Zeiss, Germany) equipped with a thermoregulation device at 37°C. Images were acquired every 2 (for MHC-I and NadA; movie S1) or 4 seconds (for transferrin and rNadA, movie S2; rNadA in 17AAG treated or untreated cells, movie S3 and S4 respectively) at 25% laser power intensity.

Results

Purified rNadA binds Chang cells in a time and temperature dependent manner

In previous reports, rNadA binds Chang epithelial cells more than other human cell lines [10]. A temperature dependence has also been suggested, as binding to human monocytes/macrophages was detectable at 37°C but not at 0°C [35]. We sought to investigate the temperature dependence and kinetics of NadA binding to Chang epithelial cells. Non permeabilized Chang cells were incubated with rNadA and, after washing, the bound fraction was revealed by FACS analysis with anti-NadA antibody.

At first this analysis was performed after 30 minutes of incubation at different temperatures. The result is shown in figure 1A. The rNadA binding to Chang cells was temperature dependent, giving the highest value at 37°C while dropping to about 60% at 30°C (figure 1A). Lowering the incubation temperatures at 20°C caused a decrease of the binding ability at less than 20% of the value obtained at 37°C. However, even at 4°C the amount of cell-bound rNadA was measurably higher than the background (independent IgG used as control) Setting incubation temperature at 37°C, we then analyzed the kinetics of rNadA binding to Chang cells (figure 1B). The maximum level of surface bound rNadA was observed at 30 minutes of incubation and decreased at longer incubation times. The time-dependent decrease of surface localized rNadA was obtained with various

monoclonal and polyclonal antibodies used for detection (data not shown).

To exclude a destabilization of the trimeric rNadA structure at 37°C over time, we incubated the recombinant adhesin for 30 min, 1, 2 and 4 hrs and analyzed the products by analytical size exclusion chromatography and MALLS (Multi Angle Laser Light Scattering). The results of such analysis, presented in figure S1 and table S1, showed that the native trimer is the predominant species (about 95%) although the rNadA preparation include C-terminus deleted forms previously characterized [33]. The 4 hour treatment at 37°C completely removed these truncated species and the full length homogenous trimeric population can be purified as shown in figure 1, panels C–E.

The size exclusion chromatography analysis of the sample before and after thermal treatment is shown in figure 1C and 1D, respectively. The SDS-PAGE of the eluted samples confirmed that all contaminants were removed from the fractions containing the majority of the trimeric rNadA following 4 hour incubation at 37°C (fractions 12 and 13 of figure 1C and 1D). The purified full length trimeric rNadA eluted in peak II (lane B of figure 1E) was used at least once in all our experiments and challenged with the crude rNadA preparations. Both samples gave identical results. Thus, trimers of rNadA form a stable structure in solution which is resistant to the temperature conditions used in our studies.

rNadA internalization into Chang epithelial cells

Having excluded that the time-dependent reduction of rNadA binding to Chang cells was caused by changes in the trimeric structure or stability of the protein, we investigated a possible internalization process by confocal immunofluorescence microscopy (figure 2).

Following 15 minutes of incubation with rNadA, cells showed minor amounts of detectable rNadA apparently distributed on the cell surface (figure 2). Upon 30 minutes of incubation, rNadA appeared mostly as dot-like structures. At a longer time of incubation the fluorescence of the dot-like structures slightly increased (figure 2A and B). Hypothesizing that the dot-like structures are indicative of rNadA internalization, this phenomenon could explain, at least partially, the observed loss of rNadA fluorescence observed by FACS analysis at prolonged incubation times. However, the confocal microscopy also indicated that the intracellular pool of rNadA reaches a steady state level at about 30–60 minutes incubation time, possibly because it recycles towards the cell exterior or because of intracellular degradation.

Phosphoinositide 3-kinase (PI3K) control of rNadA internalization

Src phosphorylation of several components of the clathrin dependent endocytosis, including cortactin, arrestin, dynamin and clathrin itself, is essential for internalization [56]. In addition, Src activity is also required for the alternative caveolar endocytosis [57]. Protein kinase C is involved in the endocytosis of several G protein coupled receptor through clathrin or caveolae pathways [58,59]. The phosphoinositide 3-kinase (PI3K) is important in the early steps of endocytosis and and in the regulation of fusion and maturation dynamics of endocytic vesicles [60]. Here we used inhibitors of the different kinases to better understand the internalization pathway(s) exploited by rNadA.

Chang cells were incubated with or without kinase inhibitorsfor 1 hour with vehicle, 100 nM wortmannin (PI3K inhibitor more active on class I enzymes), 100 µM genistein (wide spectrum tyrosine kinase inhibitor known to prevent the caveolae mediated endocytosis [61]), 10 µM Bisindolylmaleimide I (BSMI, a general PKC inhibitor), and 10 µM SU6656 (Src family kinase inhibitor)

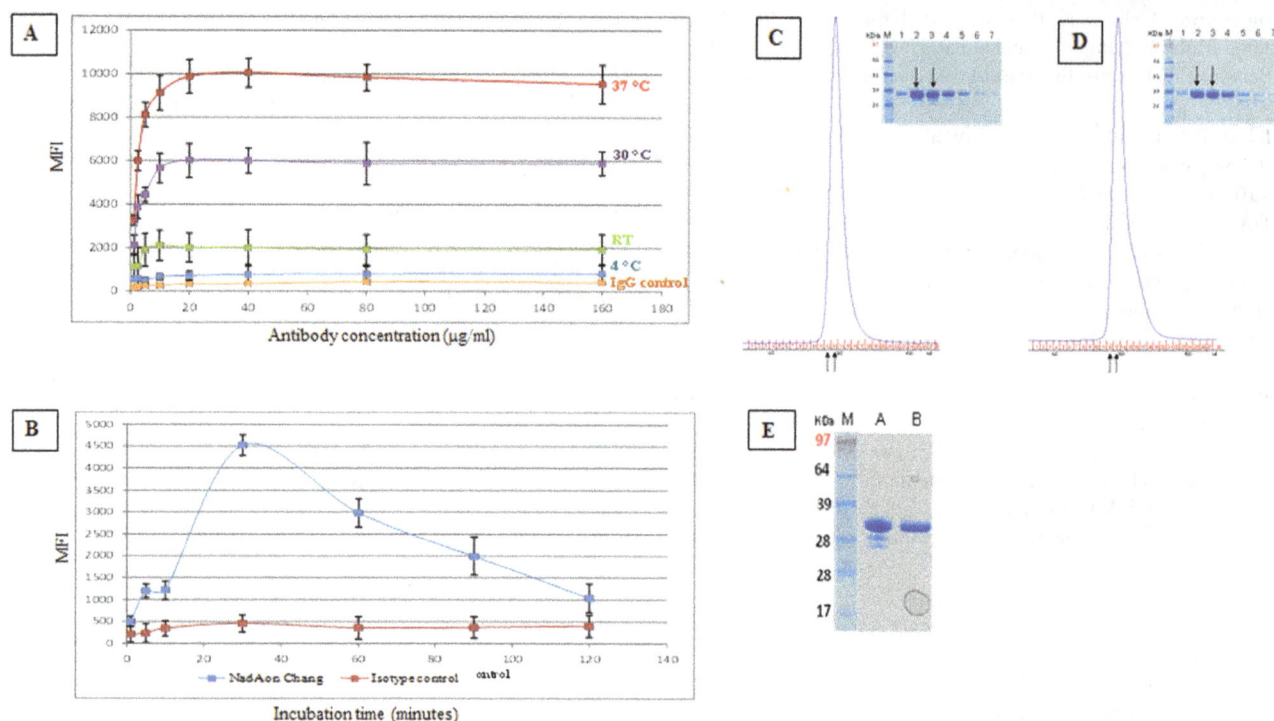

Figure 1. Purified rNadA binds Chang cells in a time- and temperature dependent manner. A and B. Chang cells were incubated with 200 µg/ml rNadA and then washed with PBS – 1% FBS. Binding was detected using anti-NadA 9F11 mouse mAb and Allophycocyanin-conjugated goat anti-mouse secondary antibody. Analysis was performed with Canto II instrument, and mean fluorescence intensity is reported (MFI). *A*: Binding of rNadA to cells for 30 minutes at the indicated temperatures, and using varying concentrations of the primary antibody *B*: Binding of rNadA was performed at 37°C for a period of time ranging from 1 minute to 2 hours as indicated. *C*: SE-HPLC profile of rNadA at room temperature. *D*: SE-HPLC profile of rNadA after heating period of 4 hrs at 37°C. *E*: SDS-Page of collected fractions. Black arrows in **D** indicate the fractions pooled.

and then for an additional hour with rNadA. The internalization of rNadA was unaffected by the action of the tyrosine kinase inhibitors (data not shown) whereas wortmannin treatment reduced the internalization of rNadA (figure 3A and B). The cellular localization of MHC-I was unaffected by exposure to rNadA, genistein, BSMI and SU6656 (not shown), whereas wortmannin shifted the localization of MHC-I towards the PM fraction (figure 3A and C).

In these experiments, rNadA showed a certain extent of endosomal colocalization with MHC class I and we decided to further explore this observation. Antibody labelling of MHC-I following cell permeabilization would show the total pool of MHC-I (PM plus endosomally localized MHC-I). In order to visualize only the recently internalized MHC-I we decided to pre-label the PM-localized MHC-I. Chang cells were treated with an antibody against MHC-I for 1 hr at 4°C. This incubation allowed binding of the antibody to MHC-I at a temperature that prevents MHC-I internalization. The cells were then treated with rNadA for 60 minutes at 37°C, a temperature at which internalization is restored. As shown in figure 3D and E, rNadA appeared to colocalize with MHC-I in endosomes.

Confocal microscopy analysis of rNadA internalization pathway

In an attempt to elucidate the cellular entry mechanism of rNadA, we performed a confocal microscopy analysis in search of rNadA colocalization with different endocytic markers belonging to the clathrin-dependent and -independent pathways.

Chang cells were incubated with rNadA for 60 minutes at 37°C to allow internalization of the adhesin and then stained for rNadA and the endocytic markers, early endosome antigen 1 (EEA1), small GTPase Rab5, M6PR. As shown in figure 4A–C and F, rNadA showed a weak colocalization with early endosomal markers (EEA1, Rab5), as well as with the endo-lysosomal marker M6PR. We also analyzed the colocalization of internalized rNadA with clathrin itself and the transferrin receptor, a prototype cargo of the clathrin-dependent pathway. Remarkably, rNadA containing carriers did not associate with the scaffold protein clathrin, and the colocalization with transferrin receptor was negligible (figure 4D–F). As a control, we showed that transferrin receptor colocalized with clathrin in Chang cells (figure S2). Since these data suggested that the main rNadA internalization pathway does not rely on the clathrin- pathway, we sought to verify this aspect by the analysis of rNadA internalization in presence of an inhibitor of clathrin vesicles formation. The AP180 protein is a crucial adaptor involved in the early stage of chlatrin-dependent endocytosis [54].Expression of the C-terminus moiety of this adaptor, known as AP180C, inhibits transferrin and EGF receptors clathrin-dependent endocytosis by preventing coat formation and leading to clathrin redistribution [54,62]. AP180C transfected Chang cells were incubated with rNadA for 60 minutes at 37°C before to be fixed and treated for confocal analysis. As control, we followed TfR internalization both in absence and presence of the truncated AP180. Results are shown in figure 4G and quantification of rNadA, TfR and AP180C colocalization reported in graph. As expected, TfR internalization was severely impaired by AP180C expression while the intracellular vesicular pattern of rNadA was not affected (figure 4G). Furthermore, the absence of rNadA

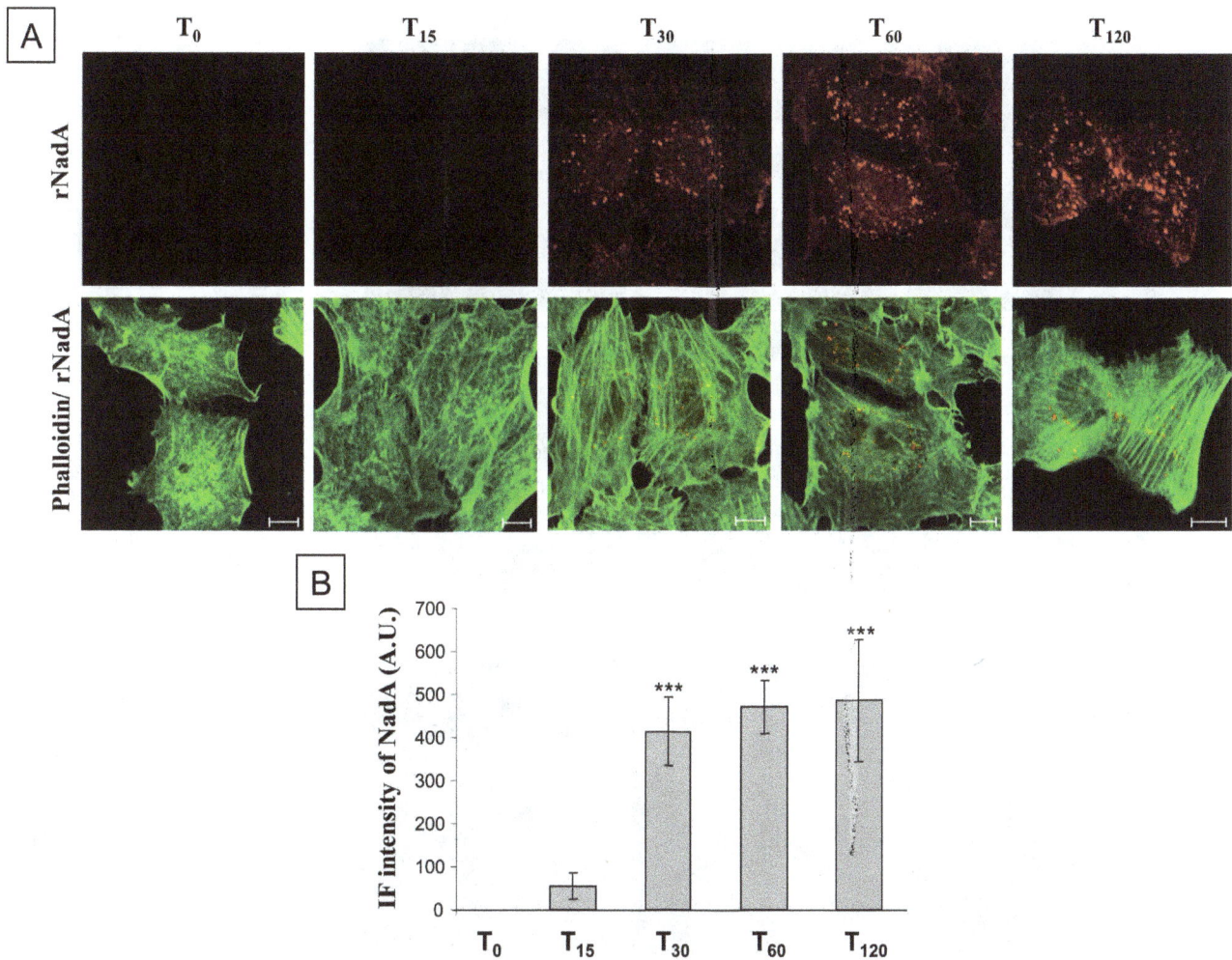

Figure 2. Time course of rNadA internalization. *A:* Chang cells were incubated with 200 µg/ml rNadA at 37°C for 0–120 min and then fixed, permeabilized and double stained for rNadA (upper panels; red) and Alexa488-conjugated phalloidin (green). Merged images of rNadA and phalloidin are shown in the lower panel. Scale bar is 10 µm. *B:* Quantification of IF intensity of rNadA in the experiment illustrated in panel A. Data are mean ± s.e.m representative of two independent experiments, each assessing 10–15 cells, and expressed as Arbitrary Units (A.U.). ***p<0.001, compared to T_0 treated cells (t-test).

colocalization with either TfR or AP180C confirmed that the two proteins follow different internalization pathways.

To further explore the internalization pathway used by rNadA, we analyzed the presence of rNadA in the lysosomal compartment. rNadA showed a very minor colocalization with the lysosomal marker LAMP1 upon 60 minute of incubation as well as at longer incubation times up to 24 hours (not shown). This observation is consistent with a small presence of the adhesin in the early endosomal compartment and indicates that internalized rNadA is not directed towards the degradative lysosomal pathway.

Finally, we analyzed the involvement of the clathrin-independent ARF6-regulated pathway. To verify whether endosomal sorting of internalized rNadA involves the small GTPase ARF6 [60] we monitored the internalization/recycling of rNadA in Chang cells expressing wild-type, and GTP-locked ARF6 (ARF6-Q67L). Previous studies demonstrated that ARF6-Q67L transfection leads to the intracellular accumulation of MHC-I [60]. Indeed, when mutant cells overexpressing ARF6-Q67L were incubated with rNadA overnight there was an accumulation of intracellular rNadA, while in cells overexpressing wild-type ARF6 the classical spotted distribution of rNadA was observed (figure 5).

In conclusion, this indicates that the rNadA endosomal trafficking involves the clathrin independent ARF6-regulated pathway. The carriers generated along this pathway can fuse with the early endosome, or take a more direct route towards the PM [60,63,64]. Indeed, our data suggest that rNadA may use a more direct recycling pathway without involvement of the early endosomal compartment. This hypothesis would be consistent with the negligible colocalization with the lysosomal compartment [64].

rNadA, MHC-I and transferrin internalization dynamics in live cells

In order to validate that rNadA uses the MHC-I internalization/recycling route we investigated their co-internalization in live cells. Chang cells were incubated for 1 hour with fluorescent Alexa633-conjugated rNadA and fluorescent Alexa488-conjugated anti-MHC-I antibody. Cells were rapidly washed with PBS and imaged for 5 minutes by confocal microscope. The movies showed the presence of rapidly moving carriers, containing both rNadA and MHC-I (figure 6A and movie S1). A minor fraction of carriers

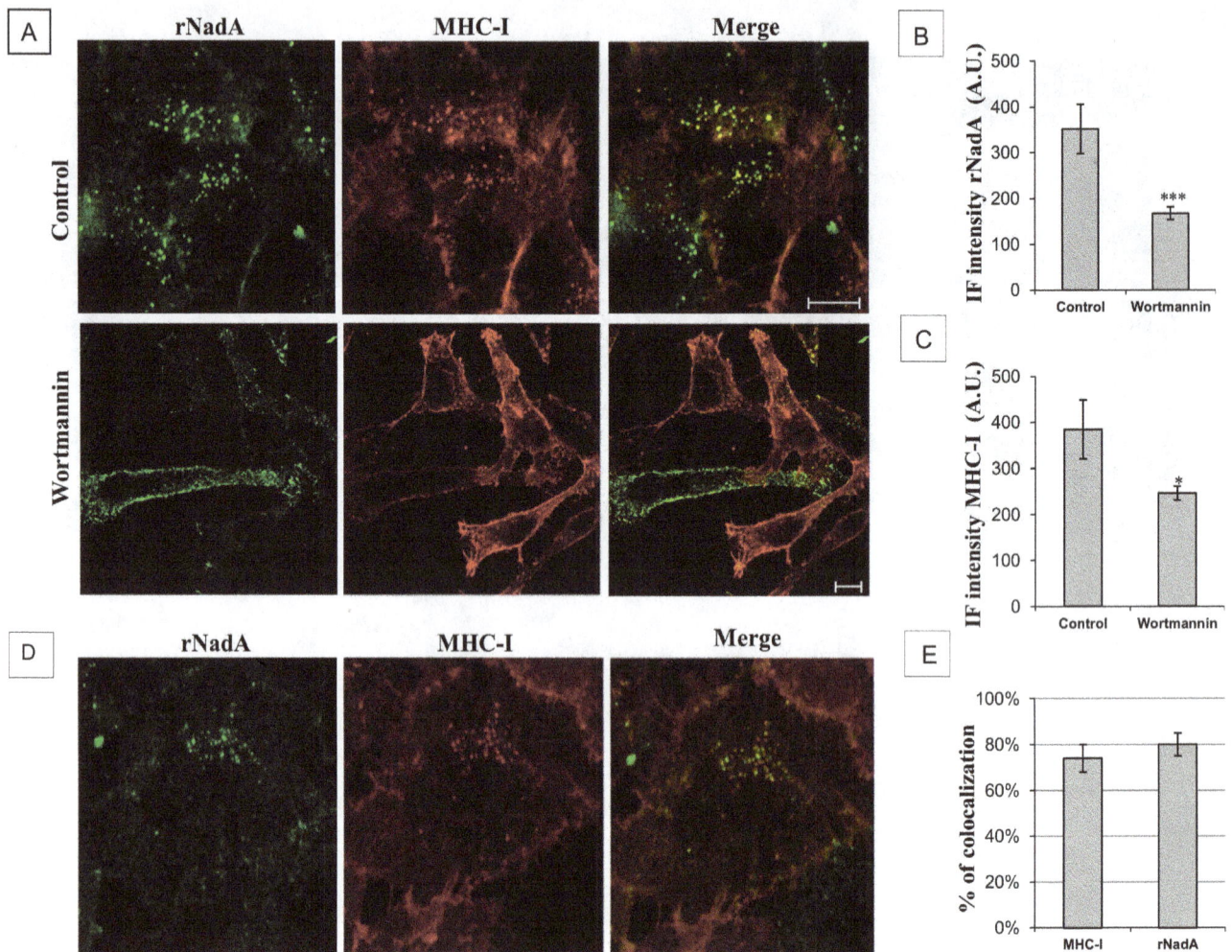

Figure 3. PI3K inhibition impairs rNadA internalization and colocalization of rNadA with MHC-I. *A:* Chang cells were treated with either vehicle (upper panel) or 100 nM Wortmannin (lower panel) for 1 h at 37°C, and then incubated with 200 μg/ml rNadA for an additional hour at 37°C in presence of the inhibitor or vehicle. Cells were then fixed, permeabilized and double stained for rNadA (green) and MHC-1 (red). Merged images are also shown. Scale bar 10 μm. *B:* Quantification of IF intensity of rNadA in the experiment illustrated in panel A. Data are mean ± s.e.m representative of two independent experiments, each assessing 20–25 cells, and expressed as Arbitrary Units (A.U.). ***p<0.001, compared to control cells (t-test). *C:* Quantification of IF intensity of MHC-I in the experiment illustrated in panel A. Data are mean ± s.e.m representative of two independent experiments, each assessing 20–25 cells, and expressed as Arbitrary Units (A.U.). *p<0.05, compared to control cells (t-test). *D:* Chang cells were treated with anti-MHC-I antibody for 1 h at 4°C and then incubated at 37°C for 1 h with 200 μg/ml rNadA. Cells were fixed and double stained for rNadA (green) and MHC-1 (red). Scale bar is 10 μm. *E:* Quantification of MHC-I and rNadA colocalization. MHC-I column indicate the percentage of MHC-I immunofluorescent pixels colocalizing with rNadA immunofluorescent pixels. Conversely, rNadA column indicate the percentage of rNadA immunofluorescent pixels colocalizing with MHC-I immunofluorescent pixels. Data are mean ± s.e.m representative of two independent experiments, each assessing 20–25 cells.

contained either rNadA or MHC-I. This could be due to the previously described dual targeting of surface MHC class I [65].

To cross-check that the rNadA internalization route is clathrin-independent, we tested the overlap between endocytosed rNadA and transferrin, a cargo internalized via the clathrin-dependent pathway. Chang cells were incubated for 1 hour with fluorescent Alexa488-conjugated rNadA and fluorescent Cy3-conjugated transferrin. Cells were rapidly washed with PBS and imaged as above. The movies showed the presence of rapidly moving rNadA or transferrin containing carriers indicating that the intracellular trafficking of rNadA is essentially different from that of transferrin (figure 6B and movie S2). As expected, only a minor fraction of carriers showed the presence of both rNadA and transferrin, perhaps as a result of a sorting/recycling compartment where the clathrin-dependent and independent pathways converge. Taken together, these observations confirm that the intracellular trafficking route of rNadA is clathrin independent and mostly overlaps with that of MHC-I.

Functional involvement of intracellular and external Hsp90 in internalization and trafficking of rNadA

In our recent report [30], we identified Hsp90 as a cellular factor interfering with rNadA-mediated bacterial infection. Although Hsp90 is an intracellular chaperone, its presence in the extracellular space has also been reported [50]. Here, we sought to verify the possible relationship between extracellular Hsp90 and rNadA internalization.

To this end, we selectively labeled the external Hsp90 in live cells by incubating them with an antibody against Hsp90 for 1 hr

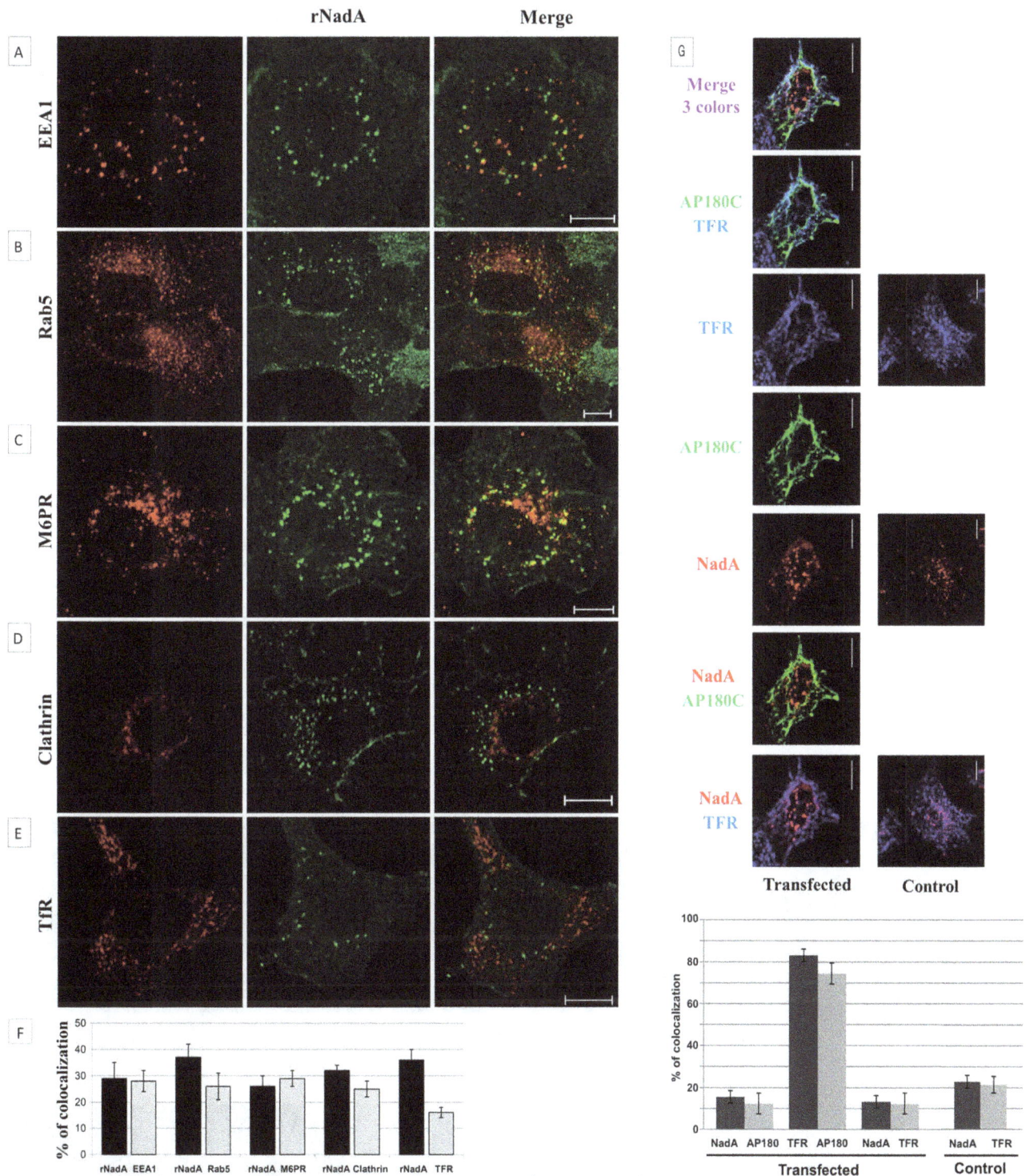

Figure 4. Colocalization of rNadA with endosome markers. *A:* Chang cells were incubated with 200 µg/ml rNadA for 1 h at 37°C, then fixed, permeabilized and double stained for rNadA (green) and EEA1 (*A*), Rab5 (*B*), M6PR (*C*), Clathrin (*D*) and TfR (*E*) (red). Merged images of the red and green signals are also shown. Images are representative of two independent experiments. Scale bar is 10 µm. *F:* Quantification of rNadA colocalization with endosome markers. The black columns indicate the percentage of rNadA immunofluorescent pixels colocalizing with the endosomal marker (EEA1, RAb5, M6PR, Clathrin, TfR as indicated) immunofluorescent pixels. Viceversa the grey columns indicate the percentage of the endosomal marker immunofluorescent pixels colocalizing with rNadA immunofluorescent pixels. Data are mean ± s.e.m representative of two independent experiments, each assessing 20–25 cells. *G:* Chang cells were transfected with an EGFR-AP180C expressing plasmid or with empty vector and incubated 24 hours at 37°C to recover. Cells were then incubated with rNadA, fixed and stained as indicated in point A. EGFR-AP190C is shown in green, rNadA in red and TfR in blu. Quantification was performed as indicated in point F.

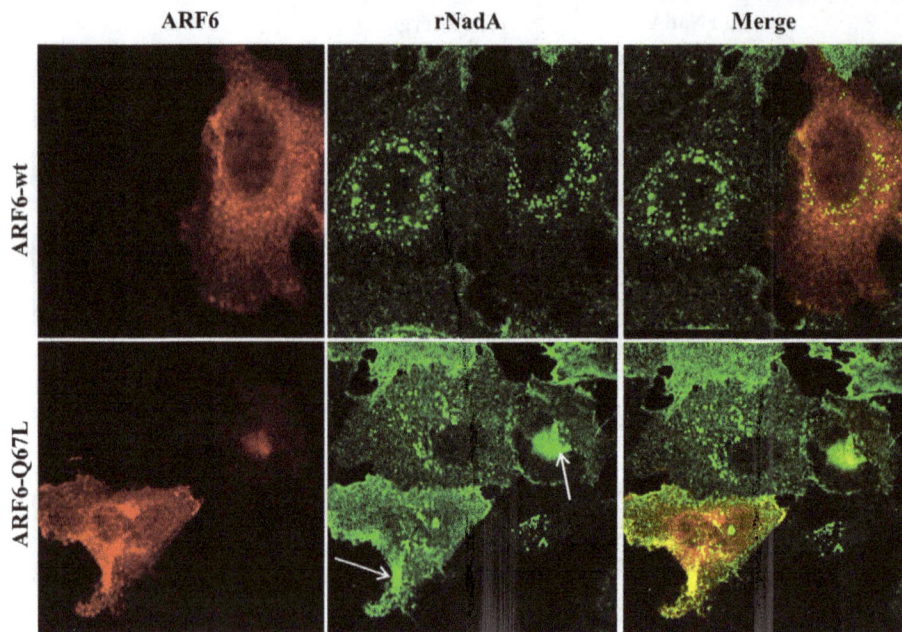

Figure 5. rNadA intracellular distribution is affected by ARF6. Chang cells were transfected in order to overexpress wild type ARF6 (ARF6-wt) or dominant-negative ARF6 (ARF6-Q67L) and then incubated overnight with 200 µg/ml rNadA at 37°C. Afterwards, cells were fixed, permeabilized and double stained for rNadA (green) and ARF6 (red). Merged images are also shown. Scale bar 10 µm. Images are representative of two independent experiments.

at 4°C. This approach allows the binding of the antibody to Hsp90 present on the external side of the PM, while the low temperature prevents internalization. Excess unbound anti-hsp90 antibodies were removed and the cells were treated with rNadA for 60 minutes at 37°C. Finally, the cells were fixed and rNadA was labeled with the appropriate primary and secondary antibodies. The anti-Hsp90 antibodies were revealed using fluorescence-conjugated secondary antibodies. As shown in figure 7A, externally labelled Hsp90

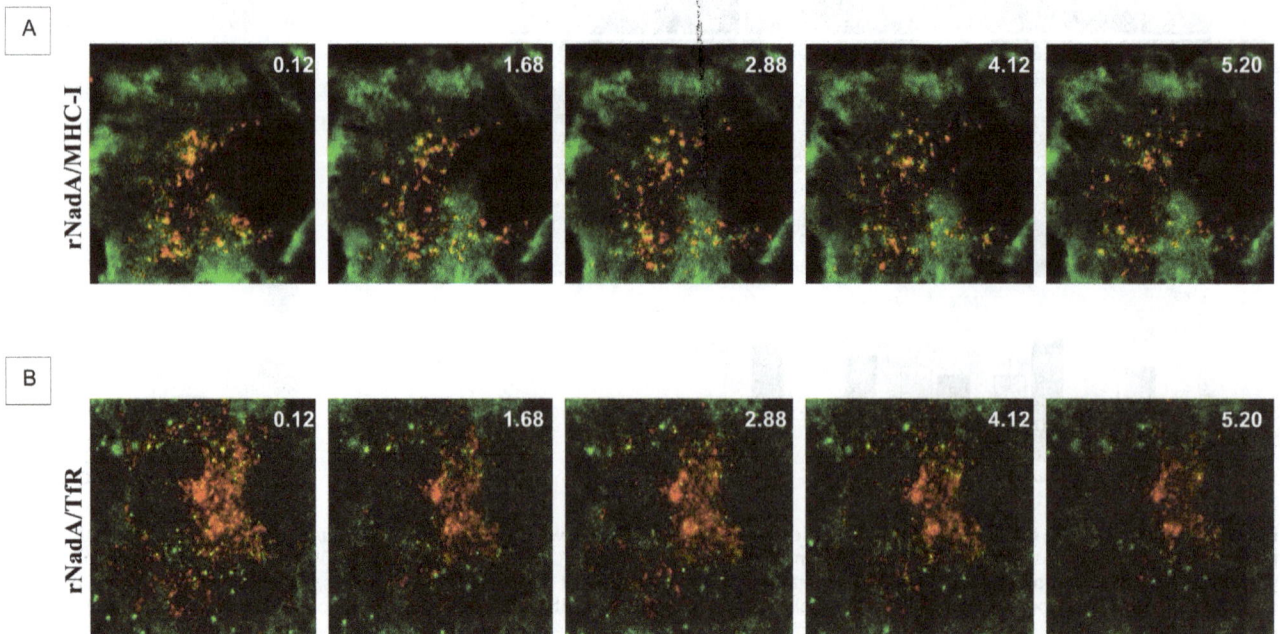

Figure 6. Time lapse rNadA internalization. *A:* Chang cells were incubated for 1 h with Alexa633-conjugated rNadA (shown in red) and Alexa488-conjugated anti-MHC-I antibody (green). Cells were then washed and live images were recorded every 2 seconds by confocal microscopy. Five frames from the Movie S1 are shown. Time (seconds) is indicated in the top right of each panel. *B:* Chang cells were incubated for 1 h with Alexa488-conjugated rNadA (green) and Cy3-conjugated transferrin (red). Cells were then washed and live images were recorded every 4 seconds by confocal microscope. Five frames from the Supplementary Movie S2 are shown. Time (seconds) is indicated in the top right of each panel.

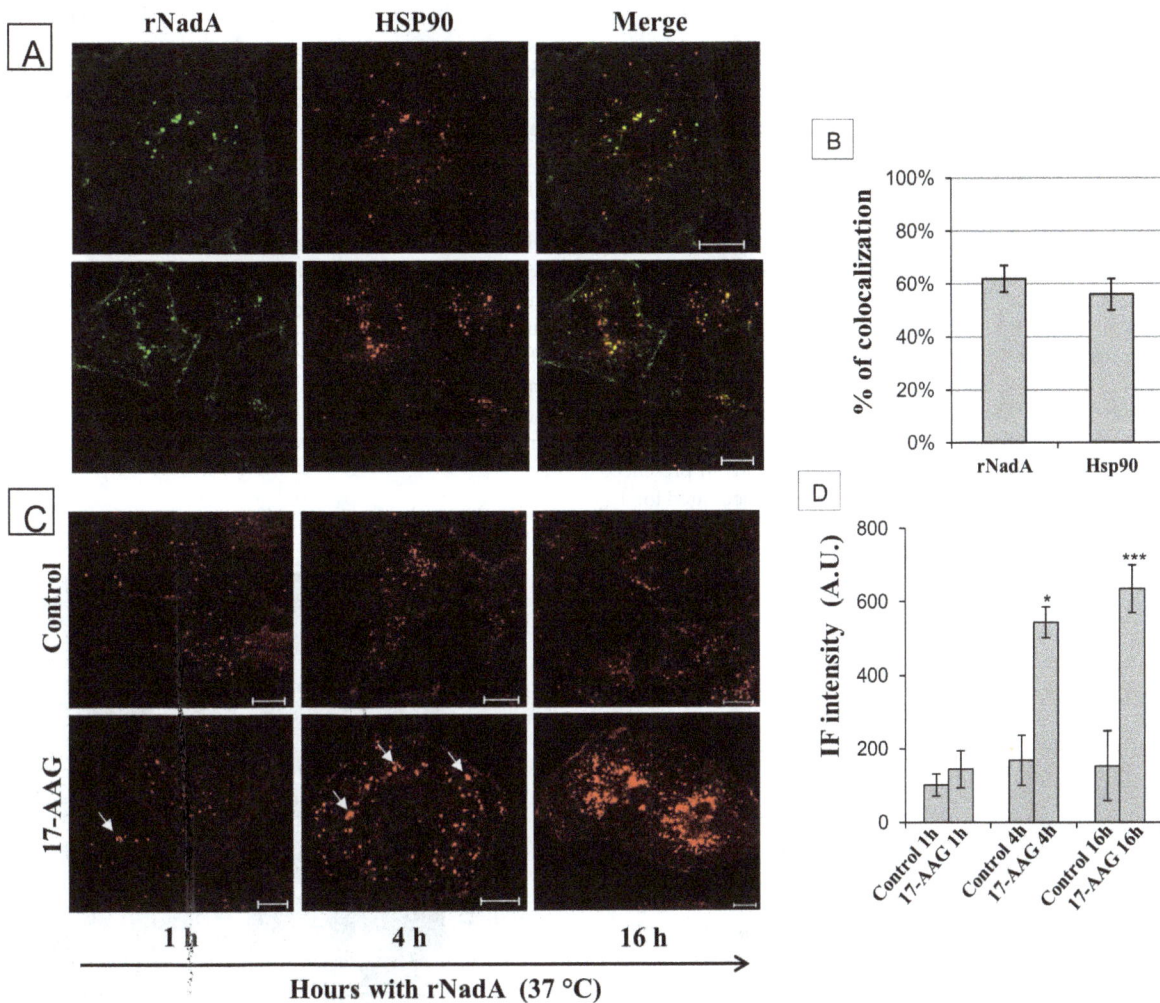

Figure 7. Colocalization of rNadA with HSP90. *A:* Chang cells, stained *in vivo* with a rabbit polyclonal anti-HSP90, were incubated with rNadA for 1 h at 37°C, then fixed, permeabilized and double stained for rNadA (green) and HSP90 (red). Panels are taken from two different experiments. Merged images are also shown. Scale bar 10 µm. *B:* Quantification of rNadA and Hsp90 colocalization. rNadA column indicate the percentage of rNadA immunofluorescent pixels colocalizing with HSP90 immunofluorescent pixels. Conversely, HSP90 columns indicate the percentage of HSP90 immunofluorescent pixels colocalizing with rNadA immunofluorescent pixels. Data are mean ± s.e.m representative of two independent experiments, each assessing 20–25 cells. *C:* Chang cells pre-treated overnight with either vehicle (upper panel) or 0.5 µM 17-AAG (lower panel), were incubated with 200 µg/ml rNadA for 1 h, 4 h and 16 hrs as indicated, then fixed, permeabilized and stained for rNadA (red). Intracellular accumulation of rNadA clusters are indicated by arrows. Scale bar 10 µm. *D:* Quantification of IF intensity of rNadA in the experiment illustrated in panel C. Data are mean ± s.e.m representative of two independent experiments, each assessing 20–25 cells, and expressed as Arbitrary Units (A.U.). ***p<0.001, compared to 1 h of 17-AAG treated cells (t-test). *p<0.05, compared to 1 h of 17-AAG treated cells (t-test).

accumulated intracellularly in compartments containing rNadA, strongly suggesting that Hsp90 present on the external side of the PM is internalized and colocalizes with the rNadA (figure 7B).

The functional role of Hsp90 in rNadA internalization was further investigated using the specific Hsp90 inhibitor 17-AAG. Chang cells pre-treated overnight with 0.5 µM of 17-AAG were then incubated with rNadA for 1, 4 and 16 hours. After 1 hour of rNadA incubation in 17-AAG-treated cells, clusters of rNadA appeared as dot-like structures as compared to control inhibitor-untreated cells (figure 7C, left panels and D). At longer times, clusters of rNadA-containing structures increased in size and number resulting in a remarkable accumulation of rNadA within the cells (figure 7C, central panels and D). At 16 hours the intracellular rNadA amount was roughly double in 17-AAG treated cells compared to control cells (figure 7C, right panels and D). Finally, we investigated the rNadA dynamics in 17-AAG

treated cells. Chang cells were pre-treated overnight with 0.5 µM 17-AAG, incubated for 1 hour with fluorescent Alexa488-conjugated rNadA, rapidly washed with PBS and a movie recorded as above. In contrast to the rapidly moving rNadA containing carriers of the control cells, in 17-AAG treated cells, carriers containing rNadA appeared less mobile. In addition, their fusion to form larger structures or the emergence of small rNadA carriers from these aggregate appeared less frequent in 17-AAG than in controls (movie S3 and S4).

The functional role of external Hsp90 in rNadA internalization was further explored by means of a synthetic membrane-impermeable inhibitor, the FITC-GA, that specifically target the extracellular population of the chaperone. To verify that FITC-GA did not enter the cells we used the cellular AKT levels as a marker for the functionality of the intracellular chaperone [66]. Hsp90 inhibitors (e.g. 17-AAG) promote the proteasomal degradation of this intracellular kinase. Cells were incubated with medium containing 2.5 and 10 µM of either 17-AAG or FITC-GA for 1, 4, 24 and 48 hrs. At the end of each period, cell lysates were prepared and equal amount of proteins were analyzed by electrophoresis and Western blot to reveal the amount of intracellular AKT. As shown in figure 8A, the levels of AKT did not change until 4 hrs of treatment with the inhibitors (figure 8A, lanes 2–7). At 24 and 48 hrs the intracellular kinase was degraded in cells incubated in presence of 17-AAG at both concentrations (figure 8A, lanes 9–10, 13–14) while its amount was not modified in FITC-GA treated cells (figure 8A, lanes 11–12, 15–16). This result strongly suggested that the Hsp90 inhibitor FITC-GA do not exert any activity on the intracellular population of the chaperone.

To verify that the membrane-impermeable inhibitor was an efficient Hsp90 inhibitor, we transiently permeabilized the cells. Chang cells were treated with saponin for 30 seconds in presence of FITC-GA and, once cell integrity was restored, incubated for 1,

4, 24 and 48 hrs. The amount of intracellular AKT was then detected as previously described. As shown in figure 8B, intracellular FITC-GA was revealed to be an efficient Hsp90 inhibitor leading to AKT degradation within 24 hours (figure 8B, lanes 11–18).

Chang cells were preincubated with either 10 µM 17-AAG or 10 µM FITC-GA at 37°C for 1 hr before addition of rNadA, and were further incubated for 1 and 4 hrs in presence of the inhibitors. Cells were then fixed, stained and subjected to confocal microscopy analysis (figure 8C). Compared to the control (upper panel of figure 8C), both inhibitors led to a similar accumulation of intracellular rNadA.

Taken together, the use of these inhibitors confirmed that the rNadA intracellular trafficking is somehow dependent upon functionally active Hsp90 and emphasized the role of the extracellular pool of this chaperone.

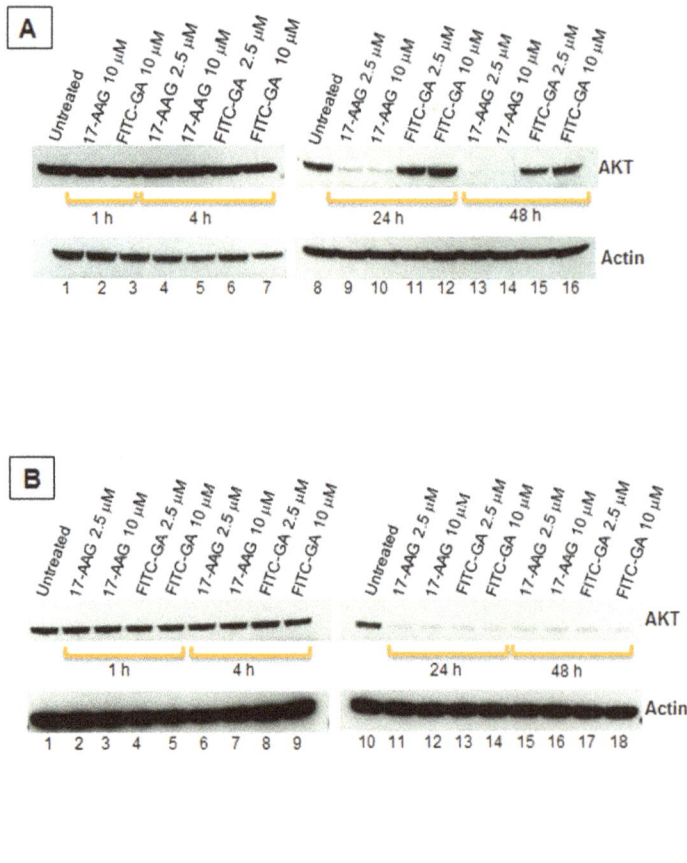

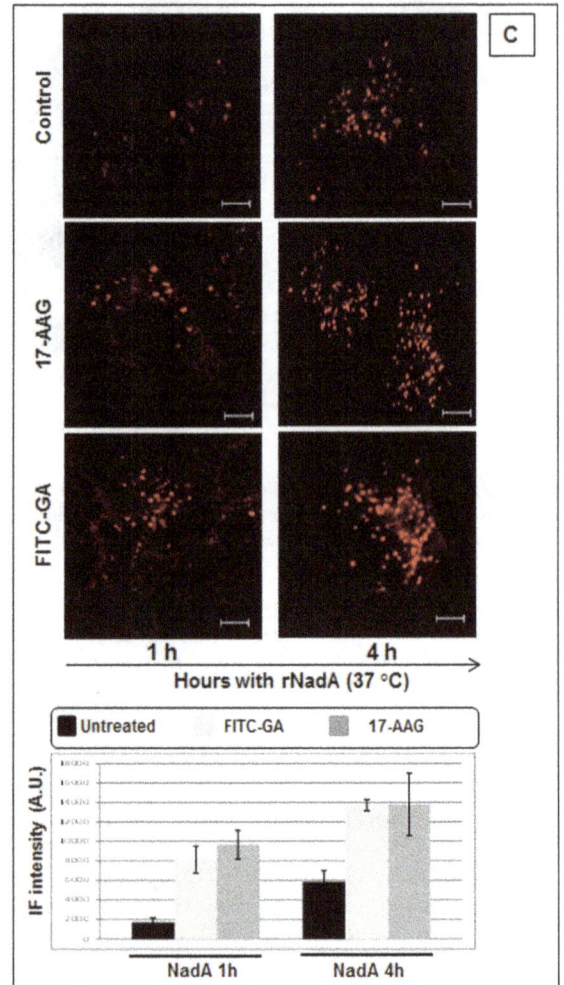

Figure 8. Inhibition of Hsp90 influences AKT degradation and intracellular accumulation of rNadA. *A:* Chang cells were incubated for 1, 4, 24 and 48 hours with either 17-AAG or FITC-GA at two different concentrations. At the end of each period cells were washed, trypsinized and RIPA-buffer total lysate prepared. Proteins from each extract were separated on NuSieve gel, blotted, and AKT or Actin (loading control) were detected using the respective primary antibodies. *B:* Chang cells were pre-treated with 0.01% saponin in PBS for 30 seconds at room temperature, washed three times with PBS and then handled as described in A. *C:* Chang cells were pre-treated for 1 hour with either vehicle (upper panels), 10 µM 17-AAG (middle panel) or 10 µM FITC-GA (lower panel), and then incubated with 200 µg/ml rNadA at 37°C for 1 (left panels) or 4 hours (right panels). Cells were then fixed, permeabilized and stained for rNadA. The drugs were present during the entire incubation period. IF intensity was calculated as mean ± s.e.m in two independent experiments, each assessing 10–15 cells and expressed as Arbitrary Units (A.U.). Scale bar 10 µm.

Colocalization of rNadA with Rab11 and effect of inhibitors

Our data strongly suggested that rNadA is recycled back to the plasma membrane and possibly released back to the cell culture medium. Although several Rabs are known to regulate cellular membrane traffic [43,67], Rab11 has been particularly associated with recycling endosomes and vesicular exocytosis at the plasma membrane [68]. We hypothesized that rNadA was colocalizing with Rab 11. Chang cells were incubated with rNadA for 1 hr and then stained for rNadA and Rab11. Confocal microscopy revealed that vesicles containing rNadA were also associated with Rab11 (figure 9A, upper panel) strongly suggesting that the adhesin moved into a recycling pathway.

We also investigated the Rab11-rNadA colocalization in presence of the two Hsp90 inhibitors. Chang cells were pre-treated with 17-AAG, FITC-GA for 1 hr before adding rNadA and allow internalization for one more hr at 37°C in presence of the inhibitors. Confocal microscopy analysis revealed a different cellular distribution of Rab11 in the presence of either drug (figure 9A, middle and lower panels). In the absence of inhibitors, Rab11 was mostly localized in vesicular structures (figure 9A, upper panel). In presence of 17-AAG or FITC-GA the Rab11 cellular pattern was more distributed, with a tendency to perinuclear accumulation (figure 9, middle and lower panels). Interestingly, both 17-AAG (membrane permeable Hsp90 inhibitor) and FITC-GA (membrane-impermeable inhibitor) treatment resulted in the complete loss of Rab11-rNadA colocalization (figure 9, compare right panels) whereas intracellular accumulation of rNadA was observed. Treatment with the Hsp90 inhibitors did not interfere with the co-localization of TfR with Rab11 positive endosomes (figure S3).

Taken together, these observations indicated that Hsp90 activity was not essential for the formation of vesicles containing rNadA. Rather, it may affect the recruitment of effector molecules necessary for the formation of recycling endosomes.

Due to the high level of colocalization between internalized rNadA and Rab11, we reasoned that it was surprising not to have found rNadA-TfR colocalization in our previous experiment (figure 4E–F). Indeed, TfR recycling, has been reported to rely on Rab11 endosomes [43]. To decipher this apparent contradiction, we performed a simultaneous staining of TfR and Rab11 during rNadA internalization. Chang cells were transfected with an EGFP-Rab11 expressing plasmid and 24 hours later incubated with rNadA for 2 hours. Cells were then fixed and revealed for rNadA and TfR by using specific primary antibodies. Results from confocal microscopy analysis are shown in figure 9B in which, to better discriminate colocalization between pair of proteins, two colors at the time are also shown separately. Furthermore, percentages of colocalization are reported in the graphs of figure 9B.

Confocal microscopy revealed a significant colocalization of rNadA in Rab11 vesicles (figure 9B, upper panel and left graph) although the value was lower than previous experiment. This could be due to the different method used to reveal Rab11 in the experiments of figure 9A and 9B. Whereas physiological levels of Rab11 were detected in figure 9A by specific antibody, overexpression of a fluorescent tagged protein increase the amount of endogenous GTPase and may justify the lower percentage of association. Colocalization of TfR with Rab11 was significantly higher, consistent with data from the literature (figure 9A upper panel and middle graph), but rNadA and TfR did not show contemporary presence in Rab11 vesicles (figure 9B upper panel and right graph).

These data suggest that the rNadA and TfR enter distinct population of Rab11 endosomes. Although surprising, the existence of separate sub-population of Rab11 vesicles has already been observed in a different cell type [69].

Discussion

In this report we have shown that a purified bacterial adhesin binds human epithelial cells and triggers its own internalization. Once internalized, the intracellular trafficking of rNadA avoids lysosome targeting by entering an endosome recycling pathway and is likely transported back to the extracellular space. The extracellular pool of the molecular chaperone Hsp90, a cellular chaperone previously shown to bind rNadA in vitro [30], has been identified as a crucial factor for the intracellular trafficking of the meningococcal adhesin.

The early steps of rNadA endocytosis remain poorly understood and, in particular, we still have not defined the cellular ligand on the membrane. Binding of rNadA to Chang cells is strongly temperature dependent and it dramatically drops below 30°C. Internalization of the adhesin is a slow event (about 30 min) compared to general mechanisms of endocytosis. In the last decade, the endocytic process has shown a surprising diversity of parallel mechanisms in a single cell [38,70]. The rNadA entry process in human epithelial cells is clathrin-independent, insensitive to general tyrosine kinases or specific Src inhibitors, and thus excludes the involvement of a caveolin dependent entry pathway [38]. Treatment with the PI3K inhibitor wortmannin, in contrast, led to rNadA accumulation at the cell periphery, consistent with the role played by this kinase in the early steps of vesicle formation [71,72]. Endocytic vesicles containing rNadA enter an ARF6-regulated trafficking pathway and over-expression of a constitutively activated form of this protein (Q67L) led to the rNadA accumulation in vacuolar structures. Activated ARF6 is described to have a dual role in cells: it promotes internalization at the level of the plasma membrane and is a targeting factor for recycling of cargo to the cell surface [43,73]. Our data suggested that rNadA could avoid the lysosomes by steering into a recycling route. Thus far, a limited number of proteins have been described to use an ARF6-regulated pathway [43,63], the list including β1-integrin and MHC-I that are continuously trafficking from endocytic vesicles back to the plasma membrane. Consistent with our results, expression of ARF6 Q67L mutant has been shown to induce vacuolar structures and blocks recycling to the cell surface of both β1-integrin and MHC-I [74]. The hypothesis that rNadA enters a recycling pathway was further reinforced by its significant colocalization with Rab11 coated vesicles. The small GTPase Rab11 is a recognized marker of recycling endosomes, and a crucial factor regulating the exocytic fusion event [43,67,68,75]. Interestingly, stimulation-dependent recycling of β1-integrin follows an Arf-6 and Rab11 dependent pathway [76]. This integrin has been recently indicated as a putative ligand for NadA by using a heterologous system of invasion [77]. In our hands, the anti-β1 antibody failed to impair both binding and internalization of rNadA in Chang cells (GB and MM unpublished results) but, due to the different experimental system used, a role for this integrin in NadA-mediated binding and/or internalization cannot be ruled out. Intriguingly, we found that Rab11 positive endosomes containing rNadA are distinct from Rab11 positive endosomes carrying TfR, although formation of both populations is PI3K dependent. Although we have not further characterized the identity of these vesicles, the existence of diversified Rab11 endosomes has been recently reported in the rat adrenal phaeochromocytoma cell line PC12 [69].

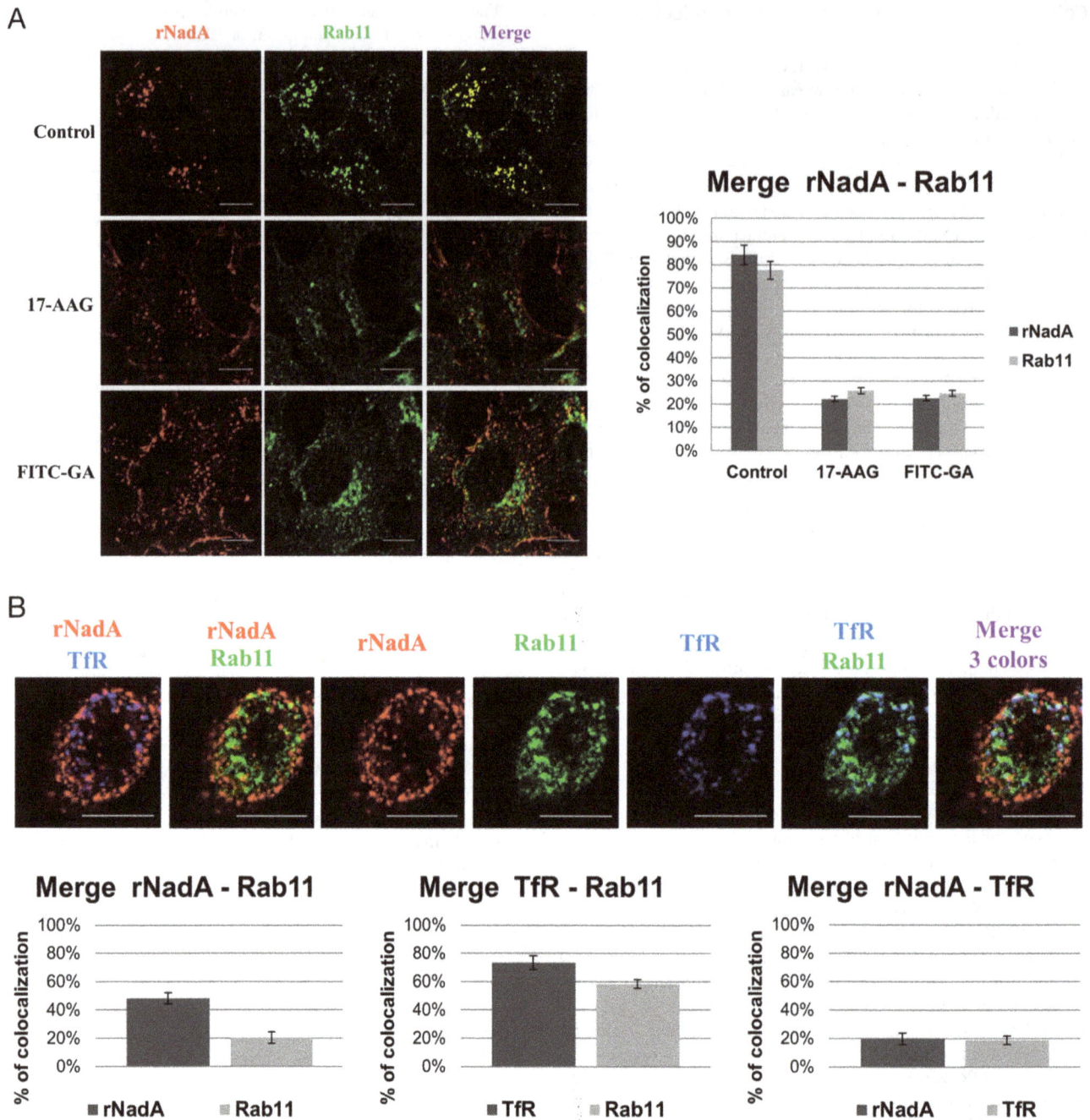

Figure 9. rNada internalization and colocalization with Rab11 and TfR. *A: rNadA and Rab11 colocalization in presence of inhibitors.* The control untreated Chang cells are shown in the upper panel. Pre-treatment of cells was performed for 1 hour with 10 μM 17-AAG (middle panel) or 10 μM FITC-GA (bottom panel). Cells were then incubated with 200 μg/ml rNadA at 37°C for 1 hour. Afterward, cells were fixed, permeabilized and stained for NadA. The drugs were present during the entire incubation period. *Graph*: rNadA columns indicate the percentage of rNadA immunofluorescent pixels colocalizing with Rab11 immunofluorescent pixels in untreated, 17-AAG and FITC-GA cells (from left to right) respectively. Conversely, Rab11 columns indicate the percentage of Rab11 immunofluorescent pixels colocalizing with rNadA immunofluorescent pixels in untreated, 17-AAG and FITC-GA cells (from left to right) respectively. *B: Rab11, rNadA and TfR colocalization.* A: Chang cells were transfected with EGFP-Rab11 plasmid, incubated 24 hours at 37°C. and then incubated with 200 μg/ml of rNadA at 37°C for 2 hours. Afterward, cells were fixed, permeabilized and stained for NadA (Red) and TfR (blu). Rab11 is colored in green. *Graphs*: rNadA columns indicate the percentage of rNadA immunofluorescent pixels colocalizing with Rab11 fluorescent pixels (left) or with TfR immunofluorescent pixels (right). Rab11column indicate the percentage of fluorescent pixels colocalizing with rNadA immunofluorescent pixels (left) or TfR immunofluorescent pixels (right). TfR columns indicate the percentage of TfR immunofluorescent pixels colocalizing with Rab11 fluorescent pixels (middle) or with rNadA immunofluorescent pixels (left). Data are mean ± s.e.m representative of two independent experiments, each assessing 20–25 cells.

Our attempts to identify a rNadA cell surface ligand led us to extracellular Hsp90 [30], a finding that was independently supported by another laboratory [53]. In the present paper, we have shown that external Hsp90 internalizes and colocalizes with the rNadA containing endocytic vesicles in live cells, strongly suggesting that the initial interaction between rNadA and Hsp90 takes place in the extracellular milieu. The extracellular presence of Hsp90 has been controversial for a long time, but it is now accepted that secretion of this chaperone occurs in both tumor and normal cells. Moreover, the functions of extracellular Hsp90 do not rely on the classical chaperone ATPase activity [49,51]. In our previous study we demonstrated a direct binding in vitro between rNadA and a complex of Hsp90 and ADP or 17-AAG. Meanwhile, in the presence of ATP, no such binding was observed [30]. Although the internalization of rNadA is not impaired by interfering with Hsp90, the ensuing intracellular trafficking is affected. Here we show that rNadA uptake in presence of the membrane-impermeable Hsp90 inhibitor FITC-GA led to intracellular accumulation of vesicles that contained the adhesin but were unable to engage Rab11. Interaction of rNadA with external Hsp90, thus, would be required to recruit a pattern of effector molecules for the targeting of the adhesin to the recycling endosomes. On the other hand, involvement of Hsp90 activity in intracellular trafficking has been poorly documented in literature. Its activity is required for activation of the Rho family of small G-protein that are known to regulate cytoskeleton rearrangement and endosomal trafficking [78,79]. Very recently, Cortese et al. reported that recycling of ErbB2 is perturbed by cell treatment with the Hsp90 inhibitor geldanamycin, leading to targeting of the receptor to mixed endosomal compartments [80]. To our knowledge, however, there are no reports on the influence of external Hsp90 on intracellular trafficking. The data presented in this work led us to hypothesize that the binding of rNadA to extracellular Hsp90 may influence a selection of the signaling involved in internalization, but further studies are required to elucidate this aspect.

The role of adhesins in meningococcal pathogenesis is thought to be confined to securing the bacteria to the host cell surface, whereas their contribution to bacterial trafficking is unaddressed [4]. We believe that the model exploited in this work, through the dissection of the pathways employed by an individual adhesin, has shed new light on the possible role that "minor" adhesion factors may have in the pathogenesis of bacterial invasion and, in the context of N. meningitidis, in transcellular passage through the epithelial barrier.

Supporting Information

**Figure S1 Size Exclusion - HPLC coupled with a MALLS (Multi Angle Laser Light Scattering) instrument of rNadA heated at 37°C for different times. *Panel A:* SE-HPLC profile of rNadA at room temperature. *Panel B, C,D, E:* SE-HPLC profiles of rNadA after heating period at 37°C of 30 min, 1, 2 and 4 hrs respectively. *Panel F:* Comparison of the different SE-HPLC profiles of rNadA heated at 37°C for the indicated times. In the panel A, the diverse species eluted at different elution time can be unequivocally identified having been previously characterized [33]. Equilibrated at room temperature, rNadA preparation showed a predominant molecular weight corresponding to the native trimer (peak II) and three minor peaks corresponding to aggregates (peak I of figure 2A), monomer (peak III of figure 2A) and C-deleted monomer (peak IV of figure 2A). The NadA preparation, includes a certain number of C-deleted forms that are able to form trimers and are eluted in the peak II [33]. Incubation at 37°C did not provoke any difference in the SE-

HPLC retention time and the MALLS measured MW of NadA peaks revealed at room temperature but it induced changes in their relative percentages.

Figure S2 Colocalization of Transferrin receptor with clathrin. Chang cells were fixed, permeabilized and double stained for clathrin (red) and Transferrin receptor (green). Merged image is also shown.

Figure S3 Colocalization of Transferrin receptor with Rab11 in presence of Hsp90 inhibitors. The control untreated Chang cells are shown in the upper panel. Pre-treatment of cells was performed for 1 hour with 10 μM 17-AAG (middle panel) or 10 μM FITC-GA (bottom panel). Chang cells were then fixed, permeabilized and double stained for Rab11 (green) and transferrin receptor (blu). Merged images are also shown. Graph report the percentage of colocalization obtained by 3 independent experiments.

Table S1 Relative percentages on rNadA species heated at 37°C for different times as revealed by Size Exclusion - HPLC coupled with a MALLS (Multi Angle Laser Light Scattering) in figure S1.

Movie S1 Dynamics of rNadA and MHC-I internalization in live cells. Chang cells were incubated for 1 h with Alexa488-conjugated anti-MHC-I antibody (green) and Alexa633-conjugated rNadA (red). Cells were then washed and imaged at the confocal microscope. Images were taken every 2 seconds.

Movie S2 Dynamics of rNadA and transferrin internalization in live cells. Chang cells grown overnight were incubated for 1 h with Alexa488-conjugated rNadA (green) and cy3-conjugated transferrin (red). Cells were then washed and imaged at the confocal microscope. Images were taken every 4 seconds.

Movie S3 Dynamics of rNadA in 17-AAG treated cells. Chang cells, pre-treated overnight with 0.5 μM 17-AAG (17-AAG), were incubated for 1 h with Alexa488-conjugated rNadA (green), in presence of the the drug. Cells were then washed and imaged at the confocal microscope. Images were taken every 4 seconds.

Movie S4 Dynamics of rNadA in untreated cells. Chang cells, pre-treated overnight with vehicle, were incubated for 1 h with Alexa488-conjugated rNadA (green). Cells were then washed and imaged at the confocal microscope. Images were taken every 4 seconds.

Acknowledgments

We thank Mirko Cortese (Novartis Vaccines and Diagnostics) for technical help and Elena Fontana for artwork preparation. We are particularly grateful to Robert Janulczyk for editing and critical reading of the manuscript.

Author Contributions

Conceived and designed the experiments: BA SS RR MP AL MS MM. Performed the experiments: GB MC PM BB MB VND EC RB. Analyzed the data: GB MC BB BP MS MM. Wrote the paper: GB MS MM.

References

1. Emonts M, Hazelzet JA, de Groot R, Hermans PW (2003) Host genetic determinants of Neisseria meningitidis infections. Lancet Infect Dis 3: 565–577.

2. Stephens DS (2009) Biology and pathogenesis of the evolutionarily successful, obligate human bacterium Neisseria meningitidis. Vaccine 27 Suppl 2: B71–77.

3. Hill DJ, Griffiths NJ, Borodina E, Virji M (2010) Cellular and molecular biology of Neisseria meningitidis colonization and invasive disease. Clin Sci (Lond) 118: 547–564.

4. Virji M (2009) Pathogenic neisseriae: surface modulation, pathogenesis and infection control. Nat Rev Microbiol 7: 274–286.

5. Carbonnelle E, Hill DJ, Morand P, Griffiths NJ, Bourdoulous S, et al. (2009) Meningococcal interactions with the host. Vaccine 27 Suppl 2: B78–89.

6. Stephens DS, Hoffman LH, McGee ZA (1983) Interaction of Neisseria meningitidis with human nasopharyngeal mucosa: attachment and entry into columnar epithelial cells. J Infect Dis 148: 369–376.

7. Birkness KA, Swisher BL, White EH, Long EG, Ewing EP Jr, et al. (1995) A tissue culture bilayer model to study the passage of Neisseria meningitidis. Infect Immun 63: 402–409.

8. Pujol C, Eugene E, de Saint Martin L, Nassif X (1997) Interaction of Neisseria meningitidis with a polarized monolayer of epithelial cells. Infect Immun 65: 4836–4842.

9. Sutherland TC, Quattroni P, Exley RM, Tang CM (2010) Transcellular passage of Neisseria meningitidis across a polarized respiratory epithelium. Infect Immun 78: 3832–3847.

10. Capecchi B, Adu-Bobie J, Di Marcello F, Ciucchi L, Masignani V, et al. (2005) Neisseria meningitidis NadA is a new invasin which promotes bacterial adhesion to and penetration into human epithelial cells. Mol Microbiol 55: 687–698.

11. Scarselli M, Serruto D, Montanari P, Capecchi B, Adu-Bobie J, et al. (2006) Neisseria meningitidis NhhA is a multifunctional trimeric autotransporter adhesin. Mol Microbiol 61: 631–644.

12. Pizza M, Scarlato V, Masignani V, Giuliani MM, Arico B, et al. (2000) Identification of vaccine candidates against serogroup B meningococcus by whole-genome sequencing. Science 287: 1816–1820.

13. Comanducci M, Bambini S, Brunelli B, Adu-Bobie J, Arico B, et al. (2002) NadA, a novel vaccine candidate of Neisseria meningitidis. J Exp Med 195: 1445–1454.

14. Comanducci M, Bambini S, Caugant DA, Mora M, Brunelli B, et al. (2004) NadA diversity and carriage in Neisseria meningitidis. Infect Immun 72: 4217–4223.

15. Cotter SE, Surana NK, St Geme JW 3rd (2005) Trimeric autotransporters: a distinct subfamily of autotransporter proteins. Trends Microbiol 13: 199–205.

16. Linke D, Riess T, Autenrieth IB, Lupas A, Kempf VA (2006) Trimeric autotransporter adhesins: variable structure, common function. Trends Microbiol 14: 264–270.

17. Surana NK, Cutter D, Barenkamp SJ, St Geme JW 3rd (2004) The Haemophilus influenzae Hia autotransporter contains an unusually short trimeric translocator domain. J Biol Chem 279: 14679–14685.

18. Lyskowski A, Leo JC, Goldman A (2011) Structure and biology of trimeric autotransporter adhesins. Adv Exp Med Biol 715: 143–158.

19. Dautin N, Bernstein HD (2007) Protein secretion in gram-negative bacteria via the autotransporter pathway. Annu Rev Microbiol 61: 89–112.

20. Roggenkamp A, Ackermann N, Jacobi CA, Truelzsch K, Hoffmann H, et al. (2003) Molecular analysis of transport and oligomerization of the Yersinia enterocolitica adhesin YadA. J Bacteriol 185: 3735–3744.

21. Yang Y, Isberg RR (1993) Cellular internalization in the absence of invasin expression is promoted by the Yersinia pseudotuberculosis yadA product. Infect Immun 61: 3907–3913.

22. McMichael JC, Fiske MJ, Fredenburg RA, Chakravarti DN, VanDerMeid KR, et al. (1998) Isolation and characterization of two proteins from Moraxella catarrhalis that bear a common epitope. Infect Immun 66: 4374–4381.

23. Laarmann S, Cutter D, Juehne T, Barenkamp SJ, St Geme JW (2002) The Haemophilus influenzae Hia autotransporter harbours two adhesive pockets that reside in the passenger domain and recognize the same host cell receptor. Mol Microbiol 46: 731–743.

24. Zhang P, Chomel BB, Schau MK, Goo JS, Droz S, et al. (2004) A family of variably expressed outer-membrane proteins (Vomp) mediates adhesion and autoaggregation in Bartonella quintana. Proc Natl Acad Sci U S A 101: 13630–13635.

25. O'Rourke F, Schmidgen T, Kaiser PO, Linke D, Kempf VA (2011) Adhesins of Bartonella spp. Adv Exp Med Biol 715: 51–70.

26. Serruto D, Spadafina T, Scarselli M, Bambini S, Comanducci M, et al. (2009) HadA is an atypical new multifunctional trimeric coiled-coil adhesin of Haemophilus influenzae biogroup aegyptius, which promotes entry into host cells. Cell Microbiol 11: 1044–1063.

27. El Tahir Y, Skurnik M (2001) YadA, the multifaceted Yersinia adhesin. Int J Med Microbiol 291: 209–218.

28. Hill DJ, Virji M (2003) A novel cell-binding mechanism of Moraxella catarrhalis ubiquitous surface protein UspA: specific targeting of the N-domain of carcinoembryonic antigen-related cell adhesion molecules by UspA1. Mol Microbiol 48: 117–129.

29. Riess T, Andersson SG, Lupas A, Schaller M, Schafer A, et al. (2004) Bartonella adhesin a mediates a proangiogenic host cell response. J Exp Med 200: 1267–1278.

30. Montanari P, Bozza G, Capecchi B, Caproni E, Barrile R, et al. (2012) Human heat shock protein (Hsp) 90 interferes with Neisseria meningitidis adhesin A (NadA)-mediated adhesion and invasion. Cell Microbiol 14: 368–385.

31. Bowe F, Lavelle EC, McNeela EA, Hale C, Clare S, et al. (2004) Mucosal vaccination against serogroup B meningococci: induction of bactericidal antibodies and cellular immunity following intranasal immunization with NadA of Neisseria meningitidis and mutants of Escherichia coli heat-labile enterotoxin. Infect Immun 72: 4052–4060.

32. Litt DJ, Savino S, Beddek A, Comanducci M, Sandiford C, et al. (2004) Putative vaccine antigens from Neisseria meningitidis recognized by serum antibodies of young children convalescing after meningococcal disease. J Infect Dis 190: 1488–1497.

33. Magagnoli C, Bardotti A, De Conciliis G, Galasso R, Tomei M, et al. (2009) Structural organization of NadADelta(351-405), a recombinant MenB vaccine component, by its physico-chemical characterization at drug substance level. Vaccine 27: 2156–2170.

34. Mazzon C, Baldani-Guerra B, Cecchini P, Kasic T, Viola A, et al. (2007) IFN-gamma and R-848 dependent activation of human monocyte-derived dendritic cells by Neisseria meningitidis adhesin A. J Immunol 179: 3904–3916.

35. Franzoso S, Mazzon C, Sztukowska M, Cecchini P, Kasic T, et al. (2008) Human monocytes/macrophages are a target of Neisseria meningitidis Adhesin A (NadA). J Leukoc Biol 83: 1100–1110.

36. McMahon HT, Boucrot E (2011) Molecular mechanism and physiological functions of clathrin-mediated endocytosis. Nat Rev Mol Cell Biol 12: 517–533.

37. Mayor S, Pagano RE (2007) Pathways of clathrin-independent endocytosis. Nat Rev Mol Cell Biol 8: 603–612.

38. Doherty GJ, McMahon HT (2009) Mechanisms of endocytosis. Annu Rev Biochem 78: 857–902.

39. Huotari J, Helenius A (2011) Endosome maturation. EMBO J 30: 3481–3500.

40. Saftig P, Klumperman J (2009) Lysosome biogenesis and lysosomal membrane proteins: trafficking meets function. Nat Rev Mol Cell Biol 10: 623–635.

41. Sandvig K, Pust S, Skotland T, van Deurs B (2011) Clathrin-independent endocytosis: mechanisms and function. Curr Opin Cell Biol 23: 413–420.

42. Donaldson JG, Porat-Shliom N, Cohen LA (2009) Clathrin-independent endocytosis: a unique platform for cell signaling and PM remodeling. Cell Signal 21: 1–6.

43. Grant BD, Donaldson JG (2009) Pathways and mechanisms of endocytic recycling. Nat Rev Mol Cell Biol 10: 597–608.

44. Radhakrishna H, Donaldson JG (1997) ADP-ribosylation factor 6 regulates a novel plasma membrane recycling pathway. J Cell Biol 139: 49–61.

45. D'Souza-Schorey C, Chavrier P (2006) ARF proteins: roles in membrane traffic and beyond. Nat Rev Mol Cell Biol 7: 347–358.

46. Jean S, Kiger AA (2012) Coordination between RAB GTPase and phospho-inositide regulation and functions. Nat Rev Mol Cell Biol 13: 463–470.

47. Hutagalung AH, Novick PJ (2011) Role of Rab GTPases in membrane traffic and cell physiology. Physiol Rev 91: 119–149.

48. Li J, Soroka J, Buchner J (2012) The Hsp90 chaperone machinery: Conformational dynamics and regulation by co-chaperones. Biochim Biophys Acta 1823: 624–635.

49. Li W, Sahu D, Tsen F (2012) Secreted heat shock protein-90 (Hsp90) in wound healing and cancer. Biochim Biophys Acta 1823: 730–741.

50. Tsutsumi S, Neckers L (2007) Extracellular heat shock protein 90: a role for a molecular chaperone in cell motility and cancer metastasis. Cancer Sci 98: 1536–1539.

51. Schmitt E, Gehrmann M, Brunet M, Multhoff G, Garrido C (2007) Intracellular and extracellular functions of heat shock proteins: repercussions in cancer therapy. J Leukoc Biol 81: 15–27.

52. Binder RJ, Vatner R, Srivastava P (2004) The heat-shock protein receptors: some answers and more questions. Tissue Antigens 64: 442–451.

53. Cecchini P, Tavano R, Polverino de Laureto P, Franzoso S, Mazzon C, et al. (2011) The soluble recombinant Neisseria meningitidis adhesin NadA(Delta351-405) stimulates human monocytes by binding to extracellular Hsp90. PLoS One 6: e25089.

54. Ford MG, Pearse BM, Higgins MK, Vallis Y, Owen DJ, et al. (2001) Simultaneous binding of PtdIns(4,5)P2 and clathrin by AP180 in the nucleation of clathrin lattices on membranes. Science 291: 1051–1055.

55. Rojas R, van Vlijmen T, Mardones GA, Prabhu Y, Rojas AL, et al. (2008) Regulation of retromer recruitment to endosomes by sequential action of Rab5 and Rab7. J Cell Biol 183: 513–526.

56. Delom F, Fessart D (2011) Role of Phosphorylation in the Control of Clathrin-Mediated Internalization of GPCR. Int J Cell Biol 2011: 246954.

57. Shajahan AN, Tiruppathi C, Smrcka AV, Malik AB, Minshall RD (2004) Gbetagamma activation of Src induces caveolae-mediated endocytosis in endothelial cells. J Biol Chem 279: 48055–48062.

58. Cha SK, Wu T, Huang CL (2008) Protein kinase C inhibits caveolae-mediated endocytosis of TRPV5. Am J Physiol Renal Physiol 294: F1212–1221.

59. Alvi F, Idkowiak-Baldys J, Baldys A, Raymond JR, Hannun YA (2007) Regulation of membrane trafficking and endocytosis by protein kinase C:

emerging role of the pericentrion, a novel protein kinase C-dependent subset of recycling endosomes. Cell Mol Life Sci 64: 263–270.

60. Naslavsky N, Weigert R, Donaldson JG (2003) Convergence of non-clathrin- and clathrin-derived endosomes involves Arf6 inactivation and changes in phosphoinositides. Mol Biol Cell 14: 417–431.

61. Sharma DK, Brown JC, Choudhury A, Peterson TE, Holicky E, et al. (2004) Selective stimulation of caveolar endocytosis by glycosphingolipids and cholesterol. Mol Biol Cell 15: 3114–3122.

62. Zhao X, Greener T, Al-Hasani H, Cushman SW, Eisenberg E, et al. (2001) Expression of auxilin or AP180 inhibits endocytosis by mislocalizing clathrin: evidence for formation of nascent pits containing AP1 or AP2 but not clathrin. J Cell Sci 114: 353–365.

63. Eyster CA, Higginson JD, Huebner R, Porat-Shliom N, Weigert R, et al. (2009) Discovery of new cargo proteins that enter cells through clathrin-independent endocytosis. Traffic 10: 590–599.

64. Eyster CA, Cole NB, Petersen S, Viswanathan K, Fruh K, et al. (2011) MARCH ubiquitin ligases alter the itinerary of clathrin-independent cargo from recycling to degradation. Mol Biol Cell 22: 3218–3230.

65. Zagorac GB, Mahmutefendic H, Tomas MI, Kucic N, Le Bouteiller P, et al. (2012) Early endosomal rerouting of major histocompatibility class I conformers. J Cell Physiol 227: 2953–2964.

66. Powers MV, Workman P (2006) Targeting of multiple signalling pathways by heat shock protein 90 molecular chaperone inhibitors. Endocr Relat Cancer 13 Suppl 1: S125–135.

67. Stenmark H (2009) Rab GTPases as coordinators of vesicle traffic. Nat Rev Mol Cell Biol 10: 513–525.

68. Takahashi S, Kubo K, Waguri S, Yabashi A, Shin HW, et al. (2012) Rab11 regulates exocytosis of recycling vesicles at the plasma membrane. J Cell Sci 125: 4049–4057.

69. Kobayashi H, Fukuda M (2013) Arf6, Rab11 and transferrin receptor define distinct populations of recycling endosomes. Commun Integr Biol 6: e25036.

70. Kumari S, Mg S, Mayor S (2010) Endocytosis unplugged: multiple ways to enter the cell. Cell Res 20: 256–275.

71. Amyere M, Payrastre B, Krause U, Van Der Smissen P, Veithen A, et al. (2000) Constitutive macropinocytosis in oncogene-transformed fibroblasts depends on sequential permanent activation of phosphoinositide 3-kinase and phospholipase C. Mol Biol Cell 11: 3453–3467.

72. Araki N, Egami Y, Watanabe Y, Hatae T (2007) Phosphoinositide metabolism during membrane ruffling and macropinosome formation in EGF-stimulated A431 cells. Exp Cell Res 313: 1496–1507.

73. Schweitzer JK, Sedgwick AE, D'Souza-Schorey C (2011) ARF6-mediated endocytic recycling impacts cell movement, cell division and lipid homeostasis. Semin Cell Dev Biol 22: 39–47.

74. Brown FD, Rozelle AL, Yin HL, Balla T, Donaldson JG (2001) Phosphatidy-linositol 4,5-bisphosphate and Arf6-regulated membrane traffic. J Cell Biol 154: 1007–1017.

75. Ullrich O, Reinsch S, Urbe S, Zerial M, Parton RG (1996) Rab11 regulates recycling through the pericentriolar recycling endosome. J Cell Biol 135: 913–924.

76. Powelka AM, Sun J, Li J, Gao M, Shaw LM, et al. (2004) Stimulation-dependent recycling of integrin beta1 regulated by ARF6 and Rab11. Traffic 5: 20–36.

77. Nagele V, Heesemann J, Schielke S, Jimenez-Soto LF, Kurzai O, et al. (2011) Neisseria meningitidis adhesin NadA targets beta1 integrins: functional similarity to Yersinia invasin. J Biol Chem 286: 20536–20546.

78. Amiri A, Noei F, Feroz T, Lee JM (2007) Geldanamycin anisimycins activate Rho and stimulate Rho- and ROCK-dependent actin stress fiber formation. Mol Cancer Res 5: 933–942.

79. Yang W, Jansen JM, Lin Q, Canova S, Cerione RA, et al. (2004) Interaction of activated Cdc42-associated tyrosine kinase ACK2 with HSP90. Biochem J 382: 199–204.

80. Cortese K, Howes MT, Lundmark R, Tagliatti E, Bagnato P, et al. (2013) The HSP90 inhibitor geldanamycin perturbs endosomal structure and drives recycling ErbB2 and transferrin to modified MVBs/lysosomal compartments. Mol Biol Cell 24: 129–144.

17

Aedesin: Structure and Antimicrobial Activity against Multidrug Resistant Bacterial Strains

Sylvain Godreuil[1,9], Nadia Leban[2,9], André Padilla[3], Rodolphe Hamel[4], Natthanej Luplertlop[5], Aurélie Chauffour[6], Marion Vittecoq[7], François Hoh[3], Frédéric Thomas[4], Wladimir Sougakoff[6], Corinne Lionne[2], Hans Yssel[6], Dorothée Missé[4]*

1 Centre Hospitalier Régional Universitaire de Montpellier, Hôpital Arnaud de Villeneuve, Département de Bactériologie-Virologie, Montpellier, France, 2 Centre d'études d'agents Pathogènes et Biotechnologies pour la Santé, CNRS-UMR 5236/UM1/UM2, Montpellier, France, 3 Centre de Biochimie Structurale Inserm U1054, CNRS UMR5048, Montpellier, France, 4 Laboratoire MIVEGEC, UMR 224 IRD/CNRS/UM1, Montpellier, France, 5 Department of Microbiology and Immunology, Faculty of Tropical Medicine, Mahidol University, Bangkok, Thailand, 6 Centre d'Immunologie et des Maladies Infectieuses, Inserm U1135, Sorbonne Universités, UPMC, APHP Hôpital Pitié-Salpêtrière, Paris, France, 7 Centre de Recherche de la Tour du Valat, le Sambuc, Arles, France

Abstract

Multidrug resistance, which is acquired by both Gram-positive and Gram-negative bacteria, causes infections that are associated with significant morbidity and mortality in many clinical settings around the world. Because of the rapidly increasing incidence of pathogens that have become resistant to all or nearly all available antibiotics, there is a need for a new generation of antimicrobials with a broad therapeutic range for specific applications against infections. Aedesin is a cecropin-like anti-microbial peptide that was recently isolated from dengue virus-infected salivary glands of the *Aedes aegypti* mosquito. In the present study, we have refined the analysis of its structural characteristics and have determined its antimicrobial effects against a large panel of multidrug resistant bacterial strains, directly isolated from infected patients. Based on the results from nuclear magnetic resonance spectroscopy analysis, Aedesin has a helix-bend-helix structure typical for a member of the family of α-helix anti-microbial peptides. Aedesin efficiently killed Gram-negative bacterial strains that display the most worrisome resistance mechanisms encountered in the clinic, including resistance to carbapenems, aminoglycosides, cephalosporins, 4th generation fluoroquinolones, folate inhibitors and monobactams. In contrast, Gram-positive strains were insensitive to the lytic effects of the peptide. The anti-bacterial activity of Aedesin was found to be salt-resistant, indicating that it is active under physiological conditions encountered in body fluids characterized by ionic salt concentrations. In conclusion, because of its strong lytic activity against multidrug resistant Gram-negative bacterial strains displaying all types of clinically relevant resistance mechanisms known today, Aedesin might be an interesting candidate for the development of alternative treatment for infections caused by these types of bacteria.

Editor: Chiaho Shih, Academia Sinica, Taiwan

Funding: This work was supported by grants from the French Research Agency Agence Nationale de la Recherche (ANR-12-BSV3-0004-01) and the French Infrastructure for Integrated Structural Biology (FRISBI) ANR-10-INSB-05-01. Nadia Leban was supported by a fellowship of the Infectiopôle Sud foundation. The funders had no role in study design, data collection and analysis, decision to publish, or preparation of the manuscript.

Competing Interests: The authors have declared that no competing interests exist.

* Email: dorothee.misse@ird.fr

9 These authors contributed equally to this work.

Introduction

Antibiotics have saved millions of lives worldwide by significantly decreasing the mortality associated with infectious diseases. However, these drugs are losing their effectiveness because of increasing antimicrobial resistance, as their massive and repetitive use in human and veterinary medicine has resulted in the emergence of multidrug-resistant (MDR) strains of bacteria that has become a serious global problem without any signs of abating. The propensity of microbes to develop multidrug-resistance is a natural trait following billions of years of evolution. Indeed, widespread resistance against several types of modern synthetic antibiotics has been discovered among bacterial strains that had been geologically isolated from the surface of the earth for more than 4 millions years [1], demonstrating that mechanisms of antibiotic modification and inactivation are part of the highly specific evolutionary adaptations of these microorganisms to evade the cytotoxic action of antibiotics, even those they have yet to encounter.

Particularly worrisome is the emergence of methicillin-resistant *Staphylococcus aureus* (MRSA) and *Enterococcus faecium*, glycopeptide-resistant Enterococcus (GRE), as well as MDR Gram-negative enterobacteria, in particular *Escherichia coli*, *Klebsiella pneumonia*, *Acinetobacter baumanii* and *Pseudomonas aeruginosa*, that, because of their production of broad-spectrum β-lactamases, i.e. AmpC cephalosporinase overproduction and extended spectrum β-lactamase, have become resistant to the third generation of cephalosporins [2]. The more recent emergence and expansion of the so-called carbapenemases determining resistance to carbapenems, a class of antibiotics of last resort for many bacterial infections, is also a cause of concern since these enzymes are presently found in the four known classes of β-lactamases (class A, B, C and D) and are determined by genes frequently harbored on highly transferable plasmids, in particular those coding for the carbapenemases KPC (class A), VIM and NDM (class B), and OXA-48 (class D) (see Table S1 for the corresponding resistance profiles). As resistance towards antibiotics becomes more common, there is an increased need for alternative treatments. However, novel antibiotics are not being developed at anywhere near the pace necessary to keep ahead of the natural ability of bacteria to evolve and defend themselves against antibiotics and, in addition, there has been a continued decline in the number of newly approved drugs [3]. Therefore, in addition to better management of antibiotic use, there is an urgent need for the development of novel therapeutic approaches to treat infections with MDR bacterial strains.

Ubiquitous in nature, antimicrobial peptides (AMP) are a unique and diverse group of molecules that were initially identified in insects and that form an important component of the innate immune system in all living organisms [4,5]. AMP typically have broad spectrum activity against pathogenic bacteria and fungi, with various modes of action that may differ among bacterial species. Based on structure-function relationship, these peptides with a length between 12 and 50 amino acids can be divided in three classes based on their secondary structure: α-helical peptides, β sheet peptides - or mixed structures - and so-called extended peptides that do not fold into regular secondary structure elements and that often contain high proportions of certain amino acids, specifically Arg, Trp or Pro residues [6]. Most AMP carry a cationic charge that promotes selective interaction with negatively charged bacterial membranes, rather than zwitterionic mammalian cell surfaces. In addition, they contain amphipathic domains which facilitate their interaction with fatty acyl acids, thereby enabling them to associate with membranes, which is a definite property of these peptides. Many linear AMP are unstructured in aqueous solution and require a membranous environment to adopt such a stable, amphipathic, conformation. As most bacterial surfaces are anionic, the initial contact between the peptide and the target organism is electrostatic. Their amino acid composition, amphipathicity, cationic charge and size allow the AMP to attach to and insert into membrane bilayers to form pores by 'barrel-stave', 'carpet' or 'toroidal-pore' mechanisms [7]. In contrast to many conventional antibiotics, AMP appear to be bactericidal [8] instead of bacteriostatic, although in many cases, the exact mechanism of killing is not known [7]. Because of their particular mode of action, the antimicrobial properties of AMP have raised clinical attention and research interest over the past years [9]. Importantly, natural AMP have co-evolved with bacterial strains and their ability to permeabilize cytoplasmic membranes is less prone to the development of resistance, such as changes in the molecular charge of cell surface proteins or proteolytic cleavage following the release of extracellular proteases. The latter processes will not only take much longer periods of time, as compared to resistance induced by conventional antibiotics, but also have the potential to compromise cell wall integrity and are therefore detrimental to bacterial survival.

Recently, we have reported the identification of a cecropin-like AMP from the dengue virus-infected salivary glands of *Aedes aegypti* [10] for which the term Aedesin is coined. The chemically synthesized form of this peptide with a length of 36 amino acid residues was found to possess antibacterial activity against *E. coli*. In the present study, we have refined the analysis of Aedesin

structural characteristics using nuclear magnetic resonance spectroscopy analysis and have furthermore determined its antimicrobial effects against a large panel of multidrug resistant clinical bacterial isolates and susceptible control reference strains.

Materials and Methods

Peptide synthesis

The identification of the cecropin-like peptide AAEL000598 peptide was recently described [10]. The peptide, with the following sequence $^{26}GGLKKLGKKLEGAGKRVFKASEKALP-VVVGIKAIGK^{61}$ and referred to as Aedesin in the present study, was chemically synthesized by Proteogenix (Schiltigheim, France) using FMOC (N-(9 fluorenyl)methoxycarbonyl) chemistry. The peptide is numbered starting from ^{26}G till K^{61}, the first 25 residues not being included as they correspond to the leader sequence. In addition, a peptide of identical amino acid composition, but with a scrambled sequence (*VAKGLIKGVKAKGELPAKGVFKGLKE-SIGKRAVLKG*) and referred to as VG26-61, was synthesized and used as a negative control. The peptides were purified by reverse-phase preparative HPLC on a C18 column (20×250 mm; Shimpack) using an appropriate 0-90% water/acetonitrile gradient in the presence of 0.05% trifluoroacetic acid. The purity of both peptides was checked by mass spectrometry and was more than 95% (data not shown). The molecular mass of both peptides was determined by matrix-assisted laser desorption ionization time-of-flight mass spectrometer (Axima-CFR Plus; Shimadzu). The concentration of the peptides was determined using an UV spectrometer.

Nuclear magnetic resonance spectroscopy

The NMR sample was prepared by dissolving the Aedesin in a mixture of 50% PBS pH 7.4/50% TFE at a concentration of 784 μM in a 3 mm tube. TFE-d3 was purchased from Euriso-top. For the experiment in D_2O the sample was lyophilized and dissolved in a mixture of 50% D_2O/50% TFE. Spectra were acquired on 700 MHz Avance Bruker spectrometer equipped with triple-resonance (^1H, ^{15}N, ^{13}C) z-gradient cryo-probe. Experiments were recorded using the Bruker TOPSPIN pulse sequence library (v.2.1). For all experiments, the recycling delay was 1.5 sec. 2D-Nuclear Overhauser effect spectroscopy (NOESY) experiment with excitation sculpting water suppression were acquired at 283K and 302K, with 48 scans and 2048 (t2) ×512 (t1) data size, and 10.2 ppm spectral width. The NOE mixing time was 200 msec. 2D- Total correlation spectroscopy (TOCSY) experiments with excitation sculpting water suppression was acquired at 283K with 32 scans and 2048 (t2) ×512 (t1) data size, and 10.2 ppm spectral width. The mixing time was 60 msec. 2D-^{15}N-^1H HSQC with binomial water suppression was acquired at 283K with 1024 scans and 1500 (t2) ×128 (t1) data size, and 10.2 ppm for the ^1H and 40 ppm for the ^{15}N spectral width. 2D-^{13}C-^1H HSQC was acquired with a D_2O/TFE (50/50%) sample at 283K with 512 scans and 2048 (t2) ×182 (t1) data size, and 10.2 ppm for the ^1H and 80 ppm for the ^{13}C spectral width. All spectra are referenced to the internal reference DSS (4,4-dimethyl-4-silapentane-1-sulfonic acid) [11].

NMR data were processed using Topspin software and were analyzed using strip-plots. Side chain assignments were carried out using 2D-NOESY and 2D-TOCSY experiments with D_2O/TFE samples. The side chain ^1H resonances were assigned, with the exception of Hδ-Hε of Lys residues, the Hζ of Phe43 and the Hγ of Leu50. The NH of the first Gly residue remained unassigned. ^{15}N assignments were derived from the 2D-^{15}N-^1H HSQC, however, due to NH superimposition, the ^{15}N resonances of

Glu36, Ala38, Val42, Phe43, Ser46, Val54 and Ile56 were not assigned. ^{13}C assignments were derived from the 2D-^{13}C-^{1}H HSQC but the Cα of residues Gly26, Leu31, Lys34, Leu35, Lys40 and Ile56, and the Cβ of residues Lys29, Lys30, Lys33, Lys34, Lys40 and Lys44 could not be assigned.

Structure calculation

Structure calculations were carried out by using the programs CYANA and CNS. From the NOESY at 283K, NOEs were classified from strong, medium and weak, corresponding to 2.8, 3.6 and 4.4 Å upper bound constraints, respectively. Structure calculations were performed with CYANA (v. 2.1) [12] using the 372 distance restraints from 2D- NOESY experiments. The NH, Hα, ^{15}N, ^{13}Cα and ^{13}Cβ chemical shifts were converted into 52 Φ/Ψ dihedral angle constraints using TALOS+ (v. 1.2).

CYANA was used to calculate 100 structures, of which the 20 conformers with the lowest target function were refined by CNS (v. 1.2) [13] using 1000 steps of torsion angle dynamics at 250 K and 1000 steps of slow cooling to 100K, followed by 200 steps of Powell minimization. The final 20 conformers were selected with the lowest NOE and dihedral angle violations, and are the structures discussed herein and deposited (PDBs). The final 20 structures contained no NOE violations greater than 0.3 Å and no dihedral angle constraint violations greater than 2°. Structures were validated using PROCHECK [14]. The structure of Aedesin has been deposited at the Protein Data Bank (www.rcsb.org), under the entry assigned accession code: 2MMM.

Circular Dichroism (CD) analysis

CD spectroscopy was used to investigate the secondary structure adopted by Aedesin in membrane-mimetic environments (1, 5 and 100 mM sodium dodecyl sulfate (SDS)). CD analysis was performed using a Chirascan Circular Dichroism Spectrophometer (Applied photophysics, Surrey, United Kingdom) with a polarized selected quartz cuve of 0,5 mm path length at 20°C. Wavelength from 180 to 260 were measured with a step of 0,5 nm and a bandwith of 2 nm. CD spectra were generated from an average of five scans of each sample. The peptide concentration was 45 μM in phosphate buffer (pH 7.4) containing 137 mM NaF for all experiments. Percentage of helicity was calculated using CONTIN Software (http://dichroweb.cryst.bbk.ac.uk).

Bacterial strains

Five susceptible reference (E. coli ATCC 25922, A. baumannii ATCC 17978, P. aeruginosa ATCC 27853, E. faecalis ATCC 700802 and S. aureus ATCC 25923) and nineteen human clinical multidrug-resistant (MDR) or extensively drug-resistant (XDR) [15] strains commonly involved in human infections were used for MIC determination for Aedesin (Table S1). Fifteen and four clinical MDR/XDR isolates were collected at the Department of Bacteriology of the Montpellier University Hospital (DBUH) and Paris Salpêtrière University hospital respectively from 2012 to 2014. Among these bacteria, we have selected three A. baumannii, three P. aeruginosa, five E. coli, two K. pneumonia, three S. aureus and three E. faecium isolates. According to routine procedures, species identification was performed using matrix-assisted laser desorption ionization–time of flight (MALDI-TOF) mass spectrometry (MS) system methods (Bruker Biotyper) and the phenotypes of resistance to antibiotics were determined by using the disk (Bio-Rad, Marne-la-Coquette, France) diffusion method according to guidelines edited by the European Committee on Antimicrobial Susceptibility Testing (http://www.eucast.org). Zone diameter results were interpreted based on breakpoints established for each bacteria species by the Antibiogram Committee of the French Society of Microbiology (http://www.sfm-microbiologie. org). The definition of MDR, XDR and pandrug-resistant (PDR) came from international consensus Multidrug-resistant, extensively drug-resistant and pandrug-resistant bacteria: an international expert proposal for interim standard definitions for acquired resistance [15].

Antibacterial activity

The antimicrobial activity of antibiotics, Aedesin and the scrambled control peptide VG26-61 against bacterial strains was determined by measuring the minimal inhibitory concentration (MIC) which represents the lowest concentration of drug or peptide that inhibits bacterial growth, using a broth microdilution method in 96-well plates (Microtest Tissue Culture plate, FALCON). In brief, pre-cultures were prepared by inoculation of 3 mL Mueller-Hinton (MH) browth and incubation at 37°C overnight under shaking. The pre-cultures were diluted to 1/100 in 3 mL MH and incubated for an additional 4 h at 37°C. The first column of the plate was a negative growth control, containing only 0.1 mL of MH. Columns 2 and 11 contained each 0.05 mL of peptide with a final concentration range of 0.0625 to 32 μg/mL, obtained by successive dilution of the peptide in the MH medium. The diluted peptides were prepared in the plate at concentrations 2 times higher than the desired final concentrations followed by the addition of the same volume of inoculum (total volume 0.1 mL/well). Inocula were prepared to obtain a final OD of 0.001 at a wavelength of 600 nm (Infinite F200 PRO, TECAN, Lyon, France), corresponding to 10^6 CFU/mL. The last column of the plate was a positive growth control (without peptide), containing 0.05 mL of inoculum plus 0.05 mL of MH. The plates were incubated at 37°C for 22–24 h prior to the determination of the MIC, corresponding to the lowest concentration of drug or peptide necessary for preventing bacterial growth as visually observed (no growth viewed from the back of the plate against a dark background illuminated with reflected light) and confirmed by OD measurement at 600 nm in a plate reader (Infinite F200 PRO, TECAN).

Gentamicin and tobramycin were used for the susceptibility testing with Escherichia coli ATCC 25922 strain as an internal control. The breakpoints were determined using the European Committee on Antimicrobial Susceptibility Testing (EUCAST).

Bactericidal activity

Non-treated bacteria were cultured to mid-log phase at 37°C in MH medium, spun for 10 minutes at 1000 g, resuspended in MH medium and diluted to an OD at 600 nm of 0.30 corresponding to 10^8 CFU/ml. The bacteria were then grown for an additional 13 h in the absence or the presence of Aedesin followed by the determination of the OD at 600 nm.

Analysis of the results

To evaluate the reproducibility of the assay, independent tests were performed using the susceptible referent strains (three tests) and the 19 clinical isolates (three tests). The reproducibility value was defined as the percentage of strains which gave the same MIC±1 $\log_2$ dilution at each test. The lecture of the MIC was performed by two independent operators.

Transmission Electron Microscopy

Treated bacterial pellets were washed in phosphate buffered saline, fixed overnight in 2.5% glutaraldehyde (Electron Microscopy Sciences, Hatfield, US) and in 0.1 M sodium phosphate buffer at 4°C. Cells were post-fixed in 1% osmic acid (Electron

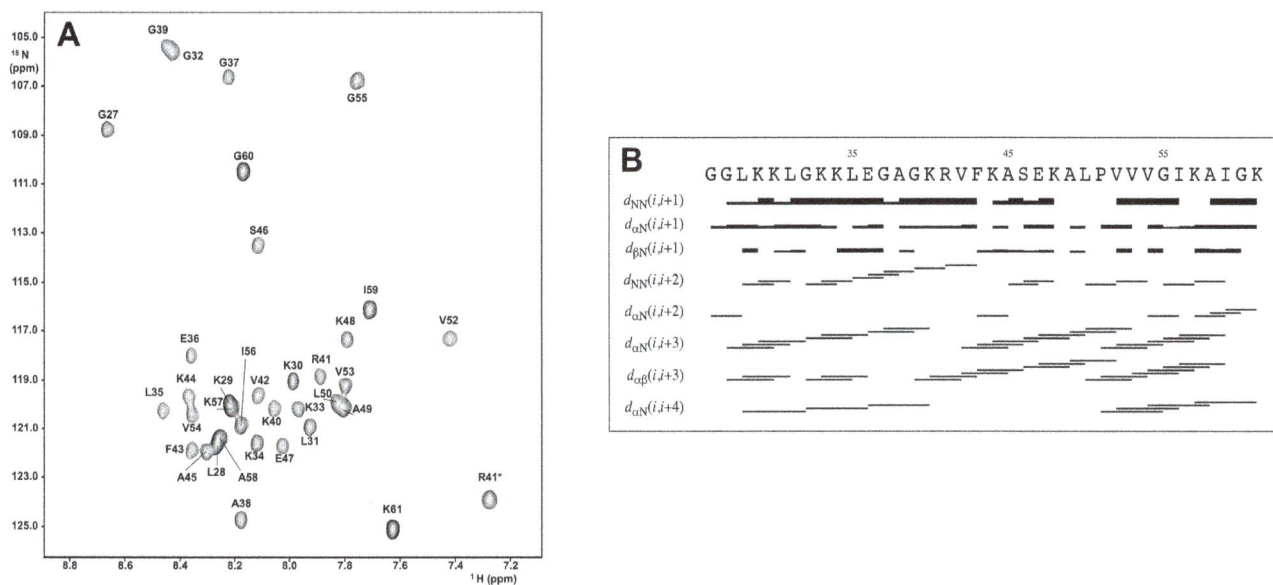

Figure 1. NOESY spectrum and NOE connectivities of Aedisine. (A) ^{15}N-^{1}H HSQC spectrum of Aedesin (G21- K61) at 50% TFE, pH 7.4 and 283K (mixing time, 200 ms). * indicates side chain NεH. (B) Schematic representation of NOE connectivities for Aedesin in 50% TFE. The intensity of the connectivity is reflected by the thickness of the bars.

Microscopy Sciences) for 1 hour at 4°C and with 0.5% tannic acid (Merck-Millipore, Darmstad, Germany) at 4°C for 30 min. Bacteria were dehydrated in a graded series of ethanol solutions (70/90/100%) for 30 min, embedded into resin and left to polymerize at ambient temperature for 1 h. Resins were sectioned by cutting an 80 nm film at 25°C using an ultramicrotome Ultracut Reichert (Leica Microsystemes SAS, Nanterre, France). Imaging was carried out using a Hitachi H1700 transmission electron microscope (Hitachi, Verrières-le-Buisson, France).

Scanning Electron Microscopy

The morphological changes of bacterial cells, either untreated or incubated with Aedesin or the scrambled control peptide VG26-61, were determined by Scanning Electron Microscopy. Bacteria were spun at 300 g for 30 min after which the pellets were washed three times with phosphate buffered saline and deposited in 12 well plates. Samples were observed using a Hitachi S4000 electron microscope.

Results

NMR structure of GK 26-61

Both NOESY and TOCSY spectra were collected for Aedesin at 283K, pH 7.4 in 50% TFE. The spin systems were identified based on the TOCSY spectrum with a mixing time of 60 ms and sequential assignments were obtained using the NOESY spectrum

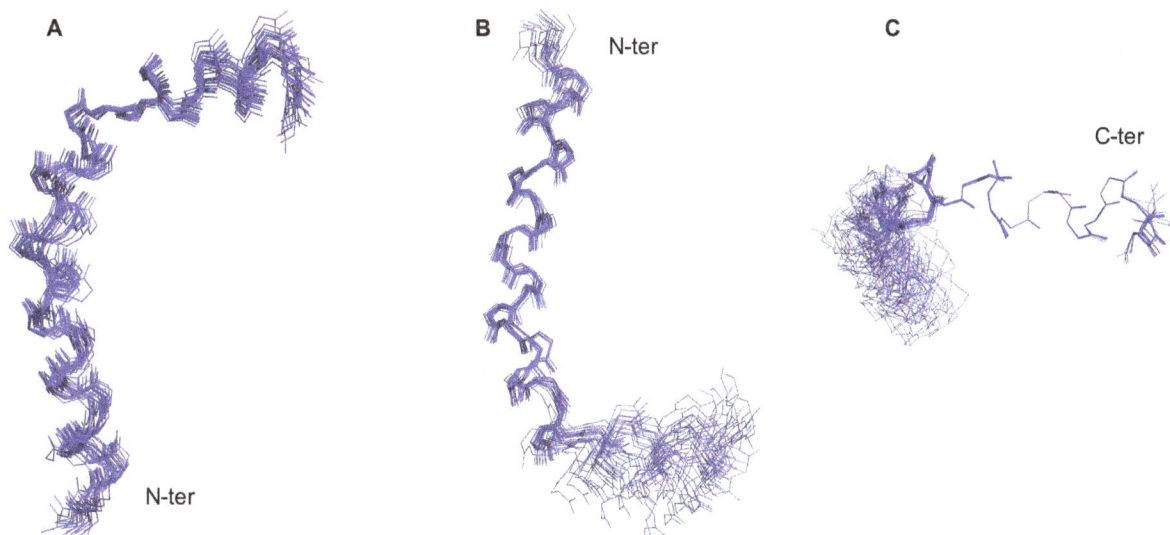

Figure 2. Calculated structures of Aedesin. (A) Superimposition of the 20 structures of Aedesin using backbone atoms. (B,C) the structures were aligned by two sections which are helix 1 from residues Lys^{30} to Lys^{48}, and helix 2 from residues Val^{52} to Ile^{59}, respectively.

Table 1. Summary of structural constraints and structure statistics.

NOE constraints	
Intraresidues (|i-j| = 0)	75
Sequential (|i-j| = 1)	142
Medium range (2≤|i-j|≤4)	155
Long range (|i-j|>4)	0
Dihedral angles	52
Structural Statistics (20 Structures)	
NOE violations, number >0.3 Å	0
Dihedral angle violations >2°	0
RMSD for geometrical analysis	
Bond lengths (Å)	0.0026+/−0.00013
Bond angles (degree)	0.4057+/−0.0077
Improper (degree)	0.3294+/−0.0151
RMSD from experimental constraints	
Distance (Å)	0.0293+/−0.0011
Dihedral angle (degree)	0.1659+/−0.0311
Mean total energy (kcal.mol-1)	83.82+/−5.60
Atomic RMSD	
Overall (26–61)	
Backbone	0.846
Heavy atoms	1.597
Helix 1 (30–48)	
Backbone	0.534
Heavy atoms	1.409
Helix 2 (52–59)	
Backbone	0.047
Heavy atoms	0.658

with a mixing time of 200 ms. Figure 1A shows the assignments of the ^{15}N-^{1}H cross peaks for GK 26-61. The ^{15}N resonances of Glu36, Ala38, Val42, Phe43, Ser46, Val54 and Ile56 could not be unambiguously assigned and are represented according to standard amino-acid ^{15}N chemical shifts. The (i, i+3) NOE connectivities denote an α-helical structure. Two stretches of dαN(i, i+4) NOE connectivities indicate the presence of regular α-helical conformation in the Leu[28]-Lys[40] and Pro[51]-Lys[61] regions. A summary for the sequential and medium range distance constraints for Aedesin in 50% TFE is shown in Figure 1B.

The solution structure of Aedesin was calculated using 372 NOE constraints derived from the NOESY spectrum at 283K. Dihedral angle constraints were obtained from NH, Hα, ^{15}N, ^{13}Cα and ^{13}Cα chemical shifts data converted into 52 Φ/Ψ dihedral angle constraints using TALOS$^+$. The analysis of the 20 overlapping structures of Aedesin (Figure 2) shows that the helical conformation is roughly continuous with a bent at residues 49-51, whereas those of the 20 final structures resulted in a Root-mean-square deviation (RMSD) of 0.846 Å for the backbone atoms and 1.597 Å for the heavy atoms. The structure of Aedesin is depicted as a helix-bent-helix structure with good RMSD statistics for the N-terminal helix (helix 1) and for the C-terminal helix (helix 2) taken separately. Structural statistics and the root mean square deviations for the 20 lowest energy structures of Aedesin are given in Table 1. The Ramachandran plot computed by PROCHECK shows that all the residues fall in the allowed conformational regions. Three amino acids, namely Ala-49, Leu-50 and Pro-51, are in the helical region of the Ramachandran plot. This also supports the NOE data obtained for these residues with the presence of dαN(i, i+3) and dαβ(i, i+3) NOE connectivities. The helical wheel diagram of Aedesin shows the amphipathic character of the first and second α-helices, as well as the opposite localization of their hydrophobic and positively charged residues, respectively (Figure 3). Whereas helix 1 has a prevalence of hydrophilic charged residues and a rather reduced hydrophobic side, the short second helix has a hydrophobic surface that consists of two Val

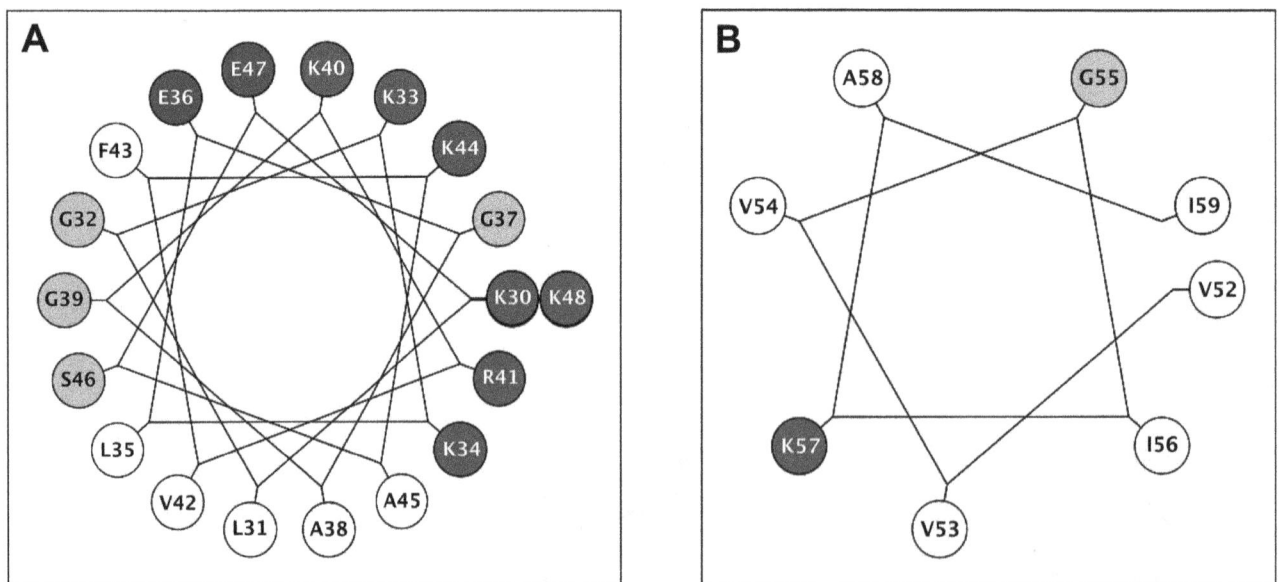

Figure 3. Helical wheel diagrams of Aedesin. (A) N-terminal helix region (helix 1) from Lys[30] to Lys[48] and **(B)** C-terminal helix (helix 2) from Val[52] to Ile[59]. The hydrophobic or charged residues are indicated in black letters within the white circles and white letters within the dark grey circles, respectively. Other residues, including non-polar amino acids, are indicated in the light grey circles.

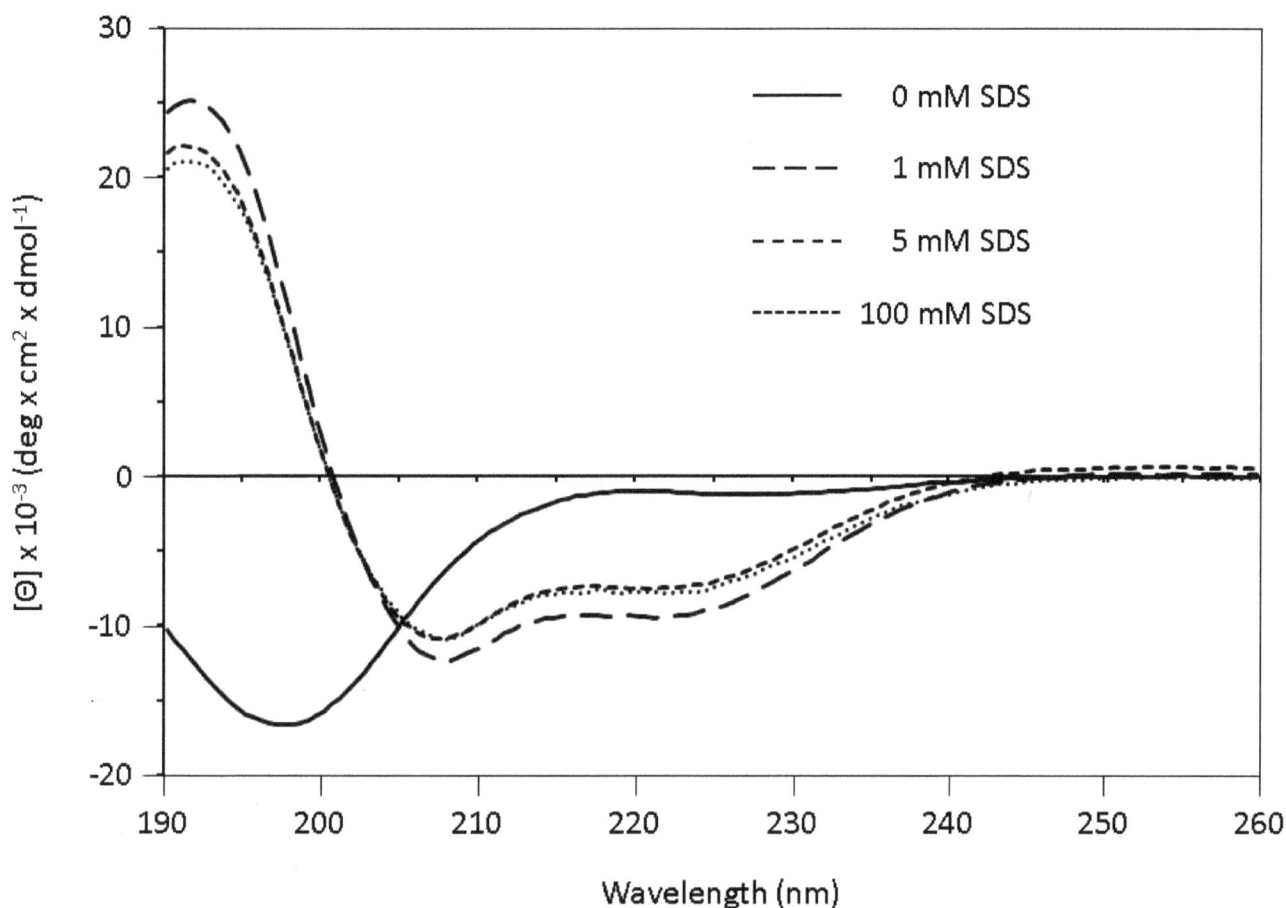

Figure 4. Circular dichroism of Aedesin in the presence of SDS micelles. CD spectra of the peptide were measured in phosphate buffer containing 137 mM NaF at SDS concentrations of 0, 1, 5 and 100 mM.

and two Ile residues, indicating a stronger hydrophobic potential than the first helix.

Circular dichroism measurements

To investigate the secondary structure of Aedesin in a membrane-like environment, we analyzed the CD spectra of the peptide dissolved under increasing concentrations of SDS leading to the formation of micelles. As shown in Figure 4, the CD spectrum of Aedesin exhibited double minimum bands at 208 and 222 nm which indicate that Aedesin adopted a well-defined α-helical structure, already in the presence of 1 mM SDS, with a total helix content of 30% which remained stable also at concentrations of 5 and 100 mM SDS, respectively. In contrast, in the absence of SDS, the peptide was unable to form an α-helical structure.

Antimicrobial activity of Aedesin

The antimicrobial activity of the Aedesin was determined on a comprehensive series of pathogenic, non-resistant, as well as MDR or XDR, Gram-positive and Gram-negative, bacterial isolates, commonly involved in human infections. The resistance pheno-types of each of the MDR or XDR strains, including *S. aureus*, *E. faecalis*, *E. faecium*, *E. coli*, *P. aeruginosa*, *A. baumannii* and *K. pneunomiae* to different classes of antibiotics is recapitulated in Table S1. In particular, the *E. coli* NMD1 and OXA48, as well as the *K. pneunomiae* KPC and VIM strains were selected because of

the serious problems that they cause in the clinic, being resistant to the latest generation of antibiotics. Aedesin displayed strong anti-bacterial activity against all sixteen Gram-negative strains tested, independent of their antibiogram, as demonstrated by the low MIC values ranging between 1 and 4 (Table 2). In contrast, no antibacterial effects of the peptide were observed against different isolates of Gram-positive *S. aureus*, *E. faecalis* and *E. faecium* strains showing MIC values over 32. The scrambled control peptide VG26-61 was totally ineffective, irrespective of the bacterial strain.

High salt concentrations are known to interfere with electro-static contact between AMP and the negatively charged bacterial membrane, thereby potentially inhibiting their anti-microbial effects. To determine the activity of Aedesin in such an environment, the peptide was tested for salt resistance in the presence of either 150 mM NaCl, 1 mM $MgCl_2$, 1 mM $CaCl_2$, or a combination of these salts. Under these experimental conditions, Aedesin still showed a strong antimicrobial effect against all Gram-negative MDR strains, with MIC values between 1 and 2, indicating that its mode of action is maintained in a high salt environment (Table 3).

Aedesin has bactericidal activity

The bactericidal activity of Aedesin was determined against two different MDR bacterial strains by measuring the viability following culture in the presence of either the Aedesin or the

Table 2. Antimicrobial activities of Aedesin against MDR bacterial strains.

Isolates	MIC (µg/mL) of Aedesin	MIC (µg/mL) of VG26-61
E. coli		
ATCC 25922	2 (1–2)	>32
EcESBL1	4 (2–4)	>32
Ec2	4 (2–4)	>32
EcESBL3	4 (2–4)	>32
EcNMD1	2	>32
EcOXA48	2	>32
P. aeruginosa		
ATCC 27853	4 (2–4)	>32
Pa1	1	>32
Pa2	2	>32
Pat3	1	>32
A. baumannii		
ATCC 17978	2	>32
Ab1	2	>32
Ab2	2	>32
Ab3	1	>32
K. pneumoniae		
KpKPC	2	>32
KpVIM	1	>32
S. aureus		
ATCC 25923	>32	>32
MRSA1	>32	>32
MRSA2	>32	>32
MRSA3	>32	>32
Enterococcus		
ATCC 700802	>32	>32
EfmGRE1	>32	>32
EfmGRE2	>32	>32
EfmGRE3	>32	>32

VG26-61 control peptide. Following a 13 h culture of *E. coli* and *P. Aeruginosa* in the presence of Aedesin at a concentration of 2 µg/mL, the OD_{600} diminished from 0.35±0.04 at the onset of the cultures to 0.17±0.01 and 0.11±0.01, respectively. In contrast, culture of the bacteria in the presence of the VG26-61 did not have any effect on their growth, with OD_{600} values of 1.2±0.01 and 1.5±0.01, respectively, at the end of the cultures.

Effect of Aedesin treatment on the morphology of *E. coli*

E. coli treated with phosphate buffer only or with the scrambled control peptide VG26-61 had an intact outer membrane and displayed a regular cytoplasm, as shown by transmission electron microscopy analysis (Figure 5A,B). However, exposure of the bacteria to Aedesin resulted in strong aggregation and an important alteration of their cell membrane (Figure 5C). The strongly altered surface morphology of the bacteria treated with Aedesin was even more evident following analysis by scanning electron microscopy (Figure 5D–F).

Discussion

Infections caused by MDR bacterial strains, resistant to even the latest class of antibiotics, have become a serious and worldwide problem. This is the consequence of a variety of microbial mechanisms, including production of enzymes that modify or destroy the active components of the antibiotic (by far the most prevalent mechanism), modification of the metabolic pathways that are antibiotic targets, as well as reduction of drug accumulation by rendering the bacterial cell wall impermeable for the antibiotic or by increasing active efflux of antibiotics across the cell surface [16]. Because of their particular mechanism of action, which is associated with a decreased tendency to induce bacterial resistance, AMP have gained considerable interest over the past decade as a possible alternative means to combat multidrug-resistance. In the present study, we have determined the antimicrobial capacity of one such AMP, denominated Aedesin, a cecropin-like peptide derived from the saliva of DENV-infected *Aedes aegypti* mosquitos [10]. In insects, cecropins form a large family of cationic α-helical peptides that are active mainly against Gram-negative bacteria [17,18,19,20,21]. Indeed, similar to CecropinA, Aedesin was found to be selective for Gram-negative bacteria and to efficiently kill a wide variety of MDR bacterial strains, including *P. aeruginosa*, *A. baumannii*, *K. pneumoniae* and *E. coli* with MIC values between 1 and 2 µg/mL.

The antimicrobial activity of certain AMPs, such as human β-defensins and the major human cationic host defense peptide LL-37, is strongly antagonized in conditions characterized by high ionic concentrations, which might preclude their therapeutic use in serum or other bodily fluids. For example, human β–defensins, as well as the major human cationic host defense peptide LL-37,

Table 3. Salt resistance of Aedesin.

Added salt	Concentration (mM)	E. coli		P. aeruginosa	
		MIC (µg/mL) of Aedesin	MIC (µg/mL) of VG26-61	MIC (µg/mL) of Aedesin	MIC (µg/mL) of VG26-61
none	-	1	>32	1–2	>32
NaCl	150	1	>32	1	>32
CaCl₂	1	1	>32	1	>32
MgCl₂	1	2	>32	2	>32
NaCl +	150	1	>32	1	>32
CaCl₂ +	1				
MgCl₂	1				

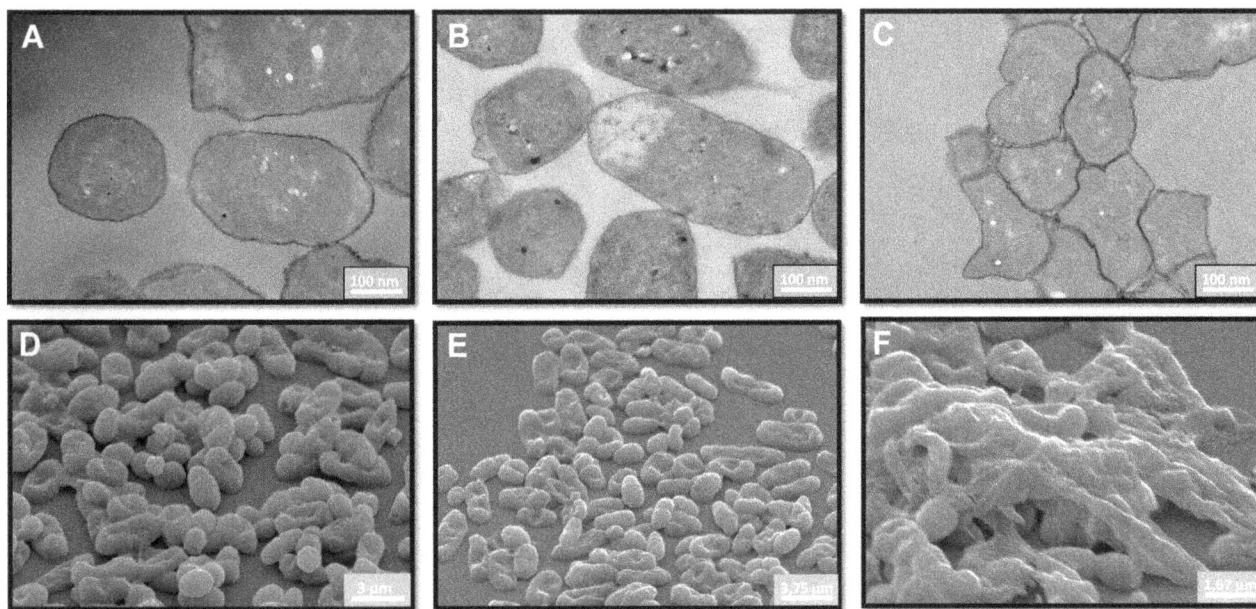

Figure 5. Electron microscopic analysis of Aedesin-treated bacteria. *E. coli* were either untreated (A,D) or incubated with VG26-61 (B,E) or Aedesin (C,F), respectively for 2 h at 37°C, prepared as indicated in Materials and Methods and analyzed by transmission (A–C) and scanning (D–F) electron microscopy.

are rapidly inactivated in the NaCl concentrations present in the airway surface liquid of cystic fibrosis patients [22], whereas interactions between cationic peptides and the outer surface component of Gram-negative bacteria are inhibited in the presence of high concentrations of bivalent ions [23]. However, the strong anti-bacterial activity of Aedesin was not affected by the presence of NaCl, $MgCl_2$, $CaCl_2$, or a combination of these salts, at concentrations similar to those present in human bodily fluids [24], indicating that its killing mechanism is salt-resistant. Moreover, Aedesin is not toxic for human cells, at any of the concentrations used [10], further indicating that this AMP might have potential therapeutic use in a physiological environment. It is to be stressed however that peptides, and in particular AMP, have poor *in vivo* stability, in particular when composed of L-amino acids, and are readily disintegrated by proteolytic enzymes in bodily fluids or recognized and processed by tissue-resident antigen-presenting cells which limits their systemic therapeutic use. Moreover, renal clearance limits the *in vivo* half-life of peptides in the circulation to only a few hours [25]. These considerations notwithstanding, several approaches that impede proteolysis in serum conditions, while retaining the bactericidal activity of the AMP, have been reported, such as substitution of L- by D-amino acids, cyclization of the peptides, use of fluorinated amino acids, beta peptides or conjugation of fatty acids [26]. Another strategy is the substitution of certain residues by unusual amino acids. For example, the replacement of Arg residues within the Oncocin-1 peptide by Orn [27] or Arg substitution within cationic amphiphilic or cationic polypeptides by Aib and Agp residues [28] were shown to confer protection to degradation and improve serum stability. A detailed structure-function analysis of Aedesin therefore needs to be carried out to determine, and possibly ameliorate, the pharmacokinetics and bioavailability of this peptide for systemic use.

The results from CD analysis of the peptide in the presence of increasing concentrations of SDS showed that Aedesin readily adopts a helical structure in a hydrophic environment. This finding was confirmed and extended by the results from NMR analysis demonstrating that Aedesin consists of two regular amphipatic α-helices in the Lys^{30}-Lys^{48} and Val^{52}-Ile^{59} regions, respectively, at the N- and C-terminal part of the peptide. The N-terminal region contains a large stretch of positively charged residues including six Lys residues. In contrast, the C-terminal helix is clearly hydrophobic with two Ile and three Val residues, separated by a single charged Lys. Although the presence of a helix-hinge-helix is a common feature found in many cecropin family members, this property does not guarantee its antibacterial activity. For example, cecropin B1, although sharing a similar conformation with Aedesin, has poor anti-microbial activity [29], underscoring the correct composition and distribution of key amino acid residues in Aedesin that are critical for this function.

Our results from scanning electron microscopic analysis show that Aedesin strongly alters the bacterial morphology, indicating that it exerts its lytic function by disrupting the bacterial outer membrane of Gram-negative bacteria. Similar results have been reported for Cecropins B, D [19], as well as Cecropin A. Indeed, using lipid vesicles with varying phospholipid composition, mimicking mammalian or microbial membranes, Cecropin A was found to preferentially permeate microbial or fungal membranes characterized by the presence of negatively charged phospholipids, rather than zwitterionic phospholipid-containing mammalian membranes [29]. In this respect, it is of note that Aedesin also kills the parasite *Leishmania donovani* [10], which is in agreement with previously published reports that demonstrate the presence of high amounts of lipophosphoglycan molecules in the membrane of this promastigote [30,31], thus forming a protective anionic barrier shielding that is sensitive to cationic molecules or ionizable phospholipid groups that cause destabilization of the membrane [32].

Like other cecropins, Aedesin is ineffective against various MDR *S. aureus* strains, thus corroborating the notion that the cytoplasmic membranes of Gram-positive bacteria are inherently more resistant to these cationic peptides, as compared to Gram-negative microorganisms [19]. Indeed, the interaction between the peptide and the bacterial membrane is determined by the lipid composition of the membrane, its surface charge density and by the presence of an electrochemical potential across the membrane, underscoring the difference between the membrane components, resulting in their differential sensitivity to membrane permeabilization by cationic peptides, between both groups of bacteria.

In conclusion, the results of this study show that the killing of MDR bacterial strains by Aedesin is independent from most mechanisms of bacterial resistance. Although it is unlikely, in its present form, to be used to systemically treat MDR Gram-negative bacterial infections, the topical use of this cationic AMP could be envisaged. For example, polymyxin B and E, while toxic at clinical doses for systemic use as anti-bacterial drugs, have been successfully implemented in the treatment of cutaneous infections caused by *P. aeruginosa* and *A. baumannii* [25], both strains that are highly susceptible to the bactericidal effects of Aedesin, as shown in the present study. Certain AMP have also been formulated in artificial tear solutions, lens preservation fluid and generic wound creams [33]. Of great interest is the application of cationic peptides against biofilm-forming bacterial infections. In particular, their application as nanofilms or other coating materials for surgical devices, including catheters and medical implants are currently under study [34]. Substitution experiments

to determine the essential amino acid residues involved in the lytic function of this peptide, while trying to preserve or ameliorate its stability are currently underway.

Acknowledgments

The authors thank Dr Laurent Chaloin (CPBS, Montpellier) for critical discussion, Dr Guillaume Arlet (St Antoine Hospital, Paris) for providing MDR bacterial strains, and Chantal Cazevieille (CRIC, Montpellier) and Aymeric Neyret (CPBS, Montpellier) for expert help with electron microscopy.

Author Contributions

Conceived and designed the experiments: DM HY CL. Performed the experiments: SG N. Leban AP RH N. Luplertlop AC FH. Analyzed the data: N. Leban SG AP WS CL MV FT HY DM FH. Contributed reagents/materials/analysis tools: SG WS DM CL. Contributed to the writing of the manuscript: DM HY AP SG WS.

References

1. Bhullar K, Waglechner N, Pawlowski A, Koteva K, Banks ED, et al. (2012) Antibiotic resistance is prevalent in an isolated cave microbiome. PLoS One 7: e34953.
2. Diene SM, Rolain JM (2013) Investigation of antibiotic resistance in the genomic era of multidrug-resistant Gram-negative bacilli, especially Enterobacteriaceae, Pseudomonas and Acinetobacter. Expert Rev Anti Infect Ther 11: 277–296.
3. Nigam A, Gupta D, Sharma A (2014) Treatment of infectious disease: Beyond antibiotics. Microbiol Res.
4. Hoffmann JA (2003) The immune response of Drosophila. Nature 426: 33–38.
5. Ferrandon D, Imler JL, Hetru C, Hoffmann JA (2007) The Drosophila systemic immune response: sensing and signalling during bacterial and fungal infections. Nat Rev Immunol 7: 862–874.
6. Nguyen LT, Haney EF, Vogel HJ (2011) The expanding scope of antimicrobial peptide structures and their modes of action. Trends Biotechnol 29: 464–472.
7. Brogden KA (2005) Antimicrobial peptides: pore formers or metabolic inhibitors in bacteria? Nat Rev Microbiol 3: 238–250.
8. Reddy KV, Yedery RD, Aranha C (2004) Antimicrobial peptides: premises and promises. Int J Antimicrob Agents 24: 536–547.
9. Toke O (2005) Antimicrobial peptides: new candidates in the fight against bacterial infections. Biopolymers 80: 717–735.
10. Luplertlop N, Surasombatpattana P, Patramool S, Dumas E, Wasinpiyamongkol L, et al. (2011) Induction of a peptide with activity against a broad spectrum of pathogens in the Aedes aegypti salivary gland, following Infection with Dengue Virus. PLoS Pathog 7: e1001252.
11. Wishart DS, Bigam CG, Yao J, Abildgaard F, Dyson HJ, et al. (1995) 1H, 13C and 15N chemical shift referencing in biomolecular NMR. J Biomol NMR 6: 135–140.
12. Guntert P (2004) Automated NMR structure calculation with CYANA. Methods Mol Biol 278: 353–378.
13. Brunger AT, Adams PD, Clore GM, DeLano WL, Gros P, et al. (1998) Crystallography & NMR system: A new software suite for macromolecular structure determination. Acta Crystallogr D Biol Crystallogr 54: 905–921.
14. Laskowski RA, MacArthur MW, Moss DS, Thornton JM (1993) PROCHECK: a program to check the stereochemical quality of protein structures. J Appl Cryst 26P: 283–291.
15. Magiorakos AP, Srinivasan A, Carey RB, Carmeli Y, Falagas ME, et al. (2012) Multidrug-resistant, extensively drug-resistant and pandrug-resistant bacteria: an international expert proposal for interim standard definitions for acquired resistance. Clin Microbiol Infect 18: 268–281.
16. Walsh C (2000) Molecular mechanisms that confer antibacterial drug resistance. Nature 406: 775–781.
17. Hultmark D, Steiner H, Rasmuson T, Boman HG (1980) Insect immunity. Purification and properties of three inducible bactericidal proteins from hemolymph of immunized pupae of Hyalophora cecropia. Eur J Biochem 106: 7–16.
18. Boman HG, Hultmark D (1987) Cell-free immunity in insects. Annu Rev Microbiol 41: 103–126.
19. Moore AJ, Beazley WD, Bibby MC, Devine DA (1996) Antimicrobial activity of cecropins. J Antimicrob Chemother 37: 1077–1089.
20. Kim JK, Lee E, Shin S, Jeong KW, Lee JY, et al. (2011) Structure and function of papiliocin with antimicrobial and anti-inflammatory activities isolated from the swallowtail butterfly, Papilio xuthus. J Biol Chem 286: 41296–41311.
21. Silvestro L, Weiser JN, Axelsen PH (2000) Antibacterial and antimembrane activities of cecropin A in Escherichia coli. Antimicrob Agents Chemother 44: 602–607.
22. Goldman MJ, Anderson GM, Stolzenberg ED, Kari UP, Zasloff M, et al. (1997) Human beta-defensin-1 is a salt-sensitive antibiotic in lung that is inactivated in cystic fibrosis. Cell 88: 553–560.
23. Piers KL, Brown MH, Hancock RE (1994) Improvement of outer membrane-permeabilizing and lipopolysaccharide-binding activities of an antimicrobial cationic peptide by C-terminal modification. Antimicrob Agents Chemother 38: 2311–2316.
24. Cole AM, Darouiche RO, Legarda D, Connell N, Diamond G (2000) Characterization of a fish antimicrobial peptide: gene expression, subcellular localization, and spectrum of activity. Antimicrob Agents Chemother 44: 2039–2045.
25. Marr AK, Gooderham WJ, Hancock RE (2006) Antibacterial peptides for therapeutic use: obstacles and realistic outlook. Curr Opin Pharmacol 6: 468–472.
26. Matsuzaki K (2009) Control of cell selectivity of antimicrobial peptides. Biochim Biophys Acta 1788: 1687–1692.
27. Knappe D, Henklein P, Hoffmann R, Hilpert K (2010) Easy strategy to protect antimicrobial peptides from fast degradation in serum. Antimicrob Agents Chemother 54: 4003–4005.
28. Zikou S, Koukkou AI, Mastora P, Sakarellos-Daitsiotis M, Sakarellos C, et al. (2007) Design and synthesis of cationic Aib-containing antimicrobial peptides: conformational and biological studies. J Pept Sci 13: 481–486.
29. Lee E, Jeong KW, Lee J, Shin A, Kim JK, et al. (2013) Structure-activity relationships of cecropin-like peptides and their interactions with phospholipid membrane. BMB Rep 46: 282–287.
30. Orlandi PA, Turco SJ (1987) Structure of the lipid moiety of the Leishmania donovani lipophosphoglycan. J Biol Chem 262: 10384–10391.
31. McConville MJ, Bacic A (1990) The glycoinositolphospholipid profiles of two Leishmania major strains that differ in lipophosphoglycan expression. Mol Biochem Parasitol 38: 57–67.
32. Yao H, Hatta I, Koynova R, Tenchov B (1992) Time-resolved x-ray diffraction and calorimetric studies at low scan rates: II. On the fine structure of the phase transitions in hydrated dipalmitoylphosphatidylethanolamine. Biophys J 61: 683–693.
33. Cole N, Hume EB, Vijay AK, Sankaridurg P, Kumar N, et al. (2010) In vivo performance of melimine as an antimicrobial coating for contact lenses in models of CLARE and CLPU. Invest Ophthalmol Vis Sci 51: 390–395.
34. Park SC, Park Y, Hahm KS (2011) The role of antimicrobial peptides in preventing multidrug-resistant bacterial infections and biofilm formation. Int J Mol Sci 12: 5971–5992.

18

Purification and Characterization of Plantaricin ZJ5, a New Bacteriocin Produced by *Lactobacillus plantarum* ZJ5

Da-Feng Song[1,2]**, Mu-Yuan Zhu**[2]*****, Qing Gu**[1]*****

1 Key Laboratory for Food Microbial Technology of Zhejiang Province, Department of Biotechnology, Zhejiang Gongshang University, Hangzhou, China, 2 State Key Laboratory of Plant Physiology and Biochemistry, College of Life Sciences, Zhejiang University, Hangzhou, China

Abstract

The aim of this study is to investigate the antimicrobial potential of *Lactobacillus plantarum* ZJ5, a strain isolated from fermented mustard with a broad range of inhibitory activity against both Gram-positive and Gram-negative bacteria. Here we present the peptide plantaricin ZJ5 (PZJ5), which is an extreme pH and heat-stable. However, it can be digested by pepsin and proteinase K. This peptide has strong activity against *Staphylococcus aureus*. PZJ5 has been purified using a multi-step process, including ammonium sulfate precipitation, cation-exchange chromatography, hydrophobic interactions and reverse-phase chromatography. The molecular mass of the peptide was found to be 2572.9 Da using matrix-assisted laser desorption/ionization time-of-flight mass spectrometry (MALDI-TOF MS). The primary structure of this peptide was determined using amino acid sequencing and DNA sequencing, and these analyses revealed that the DNA sequence translated as a 44-residue precursor containing a 22-amino-acid N-terminal extension that was of the double-glycine type. The bacteriocin sequence exhibited no homology with known bacteriocins when compared with those available in the database, indicating that it was a new class IId bacteriocin. PZJ5 from a food-borne strain may be useful as a promising probiotic candidate.

Editor: Indranil Biswas, University of Kansas Medical Center, United States of America

Funding: This project was supported by the National Natural Science Foundation of China (No. 31071513 and 31271821), the Natural Science Foundation of Zhejiang Province (No. Z3110399 and LY13C00002), the National High Technology Research and Development Program "863" Program) of China (2012AA022105B), the International Science and Technology Cooperation Program of China (2013DFA32330), the Scientific Research Project for the Education Department of Zhejiang Province, China (No. Y201120078), the key science and technology innovation team of Zhejiang Province (No. 010R50032), and National Agricultural System Program of China (CARS-05). The funders had no role in study design, data collection and analysis, decision to publish, or preparation of the manuscript.

Competing Interests: The authors have declared that no competing interests exist.

* Email: myzhu@zju.edu.cn (MYZ); guqing2002@hotmail.com (QG)

Introduction

Bacteriocins are ribosomally synthesized peptides that in most cases, exhibit antibacterial activity against bacteria that are closely related to the producing bacteria. Bacteriocins from lactic acid bacteria (LAB) are mostly small, heat-stable, hydrophobic, and cationic peptides [1,2]. These peptides have attracted significant attention because of their possible applications as non-toxic additives for food preservation and prevention of food spoilage by food-borne pathogenic bacteria [3,4,5].

Depending on their structural characteristics, LAB bacteriocins have been classified by Klaenhammer (1993) into three main groups [6]: class I is small (<5 kDa) peptides called lantibiotics, which contains post-translationally modified amino acid residues, such as lanthionine and 3-methyllanthionine; class II is small, heat-stable, nonmodified and membrane-active peptides; and class III is large (>10 kDa) and heat-labile peptides [7]. Most bacteriocins belong to class II, which are divided into four groups, namely class IIa (pediocin-like peptide bacteriocins), IIb (two-peptide bacteriocins), IIc (circular peptide bacteriocins), and IId (nonpediocin linear one peptide bacteriocins).

Nisin is the most typical bacteriocin and is used as a food preservative worldwide [8,9]. However, the application of nisin is limited because it exhibits activity against Gram-positive bacteria [10]. More than 20 plantaricins have been found, and a number of plantaricins have been purified, such as plantaricin JK [11], plantaricin 1.25 [12], plantaricin NC8 [13], Plantaricin C [14] or plantaricin PASM1 [15]. Some plantaricins, such as plantaricin PASM1 and plantaricin 1.25, show a narrow antibacterial spectrum, exhibiting antibacterial activity against bacteria closely relate to the producer microorganism such as *lactobacillus* strains, similar to many bacteriocins. Plantaricin C, plantaricin JK and plantaricin NC8, like nisin, appear to inhibit some Gram-positive bacteria, such as *Lactobacillus sake*, *Enterococcus faecalis*, and *Bacillus subtilis*, but have no activity against Gram-negative bacteria. Therefore, more broad-spectrum antimicrobial bacteriocins are desired.

We screened LAB strains to obtain new bacteriocins that exhibited antibacterial activity not only against Gram-positive bacteria but also against Gram-negative bacteria. We have isolated more than 150 LAB strains from 13 varieties of fermented mustards obtained from supermarkets in China. A LAB strain known as ZJ5, which produced bacteriocins, was obtained. Here, the identification, purification, and characterization bacteriocin of the strain ZJ5 is described. In addition, the gene encoding the bacteriocin was cloned and sequenced.

Materials and Methods

Bacterial Strains and Media

The bacterial strains that served as indicator strains are listed in Table 1. All LAB strains were grown in deMan Rogosa Sharpe (MRS) broth at 30°C without agitation. The other bacteria were grown in LB or YPD broth at 37°C with agitation. The growth of strain ZJ5 was measured turbid metrically at OD600. At least two separate experiments were conducted for each test organism. All media components were purchased from Merck, Germany. Other biochemicals were procured from Sigma-Aldrich, USA.

The Detection of Inhibitors and Taxonomic Identification

Fermented mustards (10 g) were homogenized in 90 mL of saline solution and then planted in serial dilutions in MRS media. Plates were aerobically incubated at 30°C for 48 h. Several colonies were picked at random for bacteriocin screening and replicated onto four sets of agar plates. After 8 hours of culture, three sets of agar plates were overlaid with 10 mL of overnight culture from the indicator strains *Escherichia coli*, *Staphylococcus aureus*, and *Micrococcus luteus*. After incubation at 37°C, all plates were examined for inhibition zones around individual colonies.

The ZJ5 strain was selected because of its strong antibacterial activity against three indicator strains, and it was subsequently subjected to phenotypic and genotypic identification. This strain was identified based on its cell morphology as observed by microscopy, gram staining, and catalase reaction. Total DNA from the bacteriocin-producing strain was obtained by the alkaline lysis method [16]. The 16S rRNA gene was amplified and sequenced [17].

Purification of Bacteriocin

Purification was performed by a multi-step protocol [13,15]. Three liters of culture supernatant from the ZJ5 strain grown in MRS at 30°C for 24 h at a pH of 4.0 was subjected to ammonium sulfate precipitation. Supernatants of *L. plantarum* ZJ5 cultures were precipitated with ammonium sulfate at 80% (wt/vol). The precipitate, collected by centrifugation, was dissolved in 50 mM sodium phosphate buffer, pH 6.0 (buffer A) and desalted by dialysis (100 Da cut-off membrane, Sangon, China). The bacteriocin was then loaded onto an SP-Sepharose Fast Flow cation-exchange column (GE Healthcare, Sweden) in the ÄKTA purifier system (GE Healthcare, Sweden). The column was equilibrated

Table 1. Antimicrobial activity of PZJ5 against indicator strains.

Indicator species	Strain[a]	MIC (μM)[b]
Micrococcus luteus	CGMCC 1.193	0.153±0.008
Staphylococcus aureus	CGMCC 1.879	0.132±0.005
Staphylococcus aureus	CGMCC 1.128	0.218±0.013
Staphylococcus aureus	ATCC 6538P	0.178±0.025
Staphylococcus aureus	CGMCC 1.2386	0.210±0.024
Lactobacillus plantarum	CGMCC 1.551	0.722±0.075[c]
Lactobacillus plantarum	CGMCC 1.124	0.242±0.014
Lactobacillus plantarum	CGMCC 1.11	0.885±0.079
Lactobacillus plantarum	CGMCC 1.511	1.352±0.136
Lactobacillus plantarum	CGMCC 1.556	0.682±0.018
Lactococcus lactis	ATCC 15577	1.225±0.129
Bacillus subtilis	CGMCC 1.1627	0.121±0.006
Enterococcus faecalis	CGMCC 1.125	0.146±0.027
Shigella. flexneri	CGMCC 1.1868	0.135±0.015
Listeria monocytogenes	ATCC 7648	0.112±0.009
Pseudomonas aeruginosa	CGMCC 1.647	0.144±0.010
Shigella dysenteriae	ATCC 9753	0.185±0.011
Escherichia coli	JM109	0.135±0.015
Escherichia coli	CGMCC 1.1580	0.235±0.038
Salmonella spp.	CGMCC 1.1552	0.136±0.012
Pseudomonas putida	CGMCC 1.645	0.075±0.003
Rhodotorula rubra	CGMCC 2.1034	NA[d, e]
Saccharomyces cerevisiae	CGMCC 2.1643	NA[d]

[a]ATCC, American Type Culture Collection, Rockville, MD; CGMCC, China General Microbiological Culture Collection Center, Peking, China.
[b]MIC was determined by the agar-well diffusion method [18].
[c]MRS medium was used instead of LB medium.
[d]YPD medium was used instead of LB medium.
[e]NA, no activity.

with buffer A at a flow rate of 1 ml/min, and the bacteriocin was eluted using a NaCl gradient (0% to 100% of 1 M) in buffer A in 1 ml fractions. The protein concentration was monitored at 215 nm, and activity was determined in terms of AU/ml.

Then the eluted active fraction was subjected to a Resource ETH (GE Healthcare, Sweden) equilibrated with 1 M NaCl in 10 mM sodium phosphate buffer at pH 5.0 (buffer B). The active fraction was then eluted with a NaCl gradient (100% to 0% of 1 M) in buffer B in 1 ml fractions.

For further purification, the active eluted solution was applied to a Zorbax SB-C18 column (Agilent, USA) in an LC-2000 Plus high-performance liquid chromatography (HPLC) system (Jasco, Tokyo, Japan). Then the active fraction was eluted with a gradient (0% to 100%) of Milli-Q water-acetonitrile containing 0.06% trifluoroacetic acid at 1 ml/min.

Bacteriocin Activity Assay

The agar-well diffusion method [18] was utilized to detect the antibacterial spectrum of the ZJ5 fractions, which acquired during the purification process. The samples were adjusted to pH 5.0. As an indicator strain, Escherichia coli was inoculated in LB agar (ca. 10^5 CFU/mL), and wells (8 mm diam) were punched in the plate. The wells were filled with 100 µL of samples, and the plates were incubated overnight at 30°C. The diameter of the inhibition zones (mm) around the wells was measured. The MIC is defined as the minimum PZJ5 concentration that yielded a clear zone of growth inhibition in the indicator lawn. All activity measurements were conducted at least three times.

Polyacrylamide Gel Electrophoresis

To determine the purity and molecular weight, the purified peptide was subjected to tricine-sodium dodecyl sulphate-polyacrylamide gel electrophoresis (tricine-SDS-PAGE). After electrophoresis, the gel was cut into two parts, one of which was stained while the other was fixed in 20% (v/v) isopropanol, 10% (v/v) acetic acid for 1.5 h, rinsed in 0.5% Tween 80 for 16 h and subsequently rinsed in ddH$_2$O overnight. The gel was then overlaid with soft agar containing the indicator strain (10^6 colony-forming units) and was incubated overnight at 37°C.

Stability Against pH, Heat, and Enzyme Conditions

To determine the sensitivity of the purified peptide to pH, plantaricin ZJ5 was re-suspended in different buffer solutions with pH ranging from 2.0 to 8.0 (50 mM HCl-KCl: pH 2.0, and 4.0; 50 mM phosphate buffer: pH 6.0 and 7.0; and 50 mM Tris-Cl: pH 8.0) and incubated for 2 h at 37°C. Thermo stability was determined at different temperatures in experiments where the protein samples were boiled (100°C in a water bath for 30 min) and autoclaved (121°C at 15 psi for 15 min) prior to assaying for activity [15].

Plantaricin ZJ5 was treated with the following enzymes at 1 mg/mL: trypsin (8.0), α-chymotrypsin (8.0), proteinase K (7.5), pepsin (2.0), papain (7.0), lipase (9.0), and α-amylase (6.0) (Sigma-Aldrich, USA). After incubation at 37°C for 2 h, enzyme activity was terminated by heating at 100°C for 5 min. Then residual plantaricin activity was determined using the agar-well diffusion method described above using Escherichia coli as an indicator strain. The area of inhibition was calculated from the diameter of the inhibition zones, and the decrease ratio was displayed as a percentage. Untreated peptide samples were taken as respective controls [19].

N-terminal Amino Acid Sequencing and Molecular Mass Spectrometry

The molecular mass was characterized by using matrix-assisted laser desorption/ionization time-of-flight mass spectrometry (MALDI TOF-Target) (Bruker Daltonics, USA). The HPLC-purified peptide plantaricin ZJ5 was analyzed via automated Edman degradation using a PPSQ-31A Protein Sequencer (Shimadzu Corporation, Japan).

DNA Sequence Analysis

DNA polymerase was purchased from TaKaRa Bio, Tokyo, Japan. Chromosomal DNA from the ZJ5 strain was obtained by the alkaline lysis method [16] and used as a DNA template for PCR amplification. Table 2 listed all the primers used in this study. The degenerate primers AGF1 and AGR1, which were designed from the amino acid sequences obtained, were used to amplify a genomic DNA fragment containing a portion of PZJ5. To amplify the upstream and downstream areas of the complete sequence, thermal asymmetric interlace PCR (Tail-PCR) was performed as described previously [20,21]. Chromosomal DNA was used as a template for PCR with a structural gene-specific primer (LB-TR1) and random primer (AD1-AD5). The second (with primers LB-TR2 and AD1-AD5) and the third (with primers LB-TR3 and AD1-AD5) PCR were performed as described above. The products were purified by a quick PCR purification kit (Sangon, China). Then they were sequenced. The DNA sequence of the downstream region was analyzed using random primers and the following specific primers: RB-TR1, RB-TR2, and RB-TR3. Based on the sequences obtained from the Tail-PCR products, the new specific primers PZJ5-F and PZJ5-R were synthesized and used to confirm the entire gene sequence using the procedures described above.

Analysis of DNA and Amino Acid Sequences

The DNA and amino acid sequences obtained were analyzed using DNAStar software, version 5.0 (DNASTAR, Madison, USA). Database searches were performed using BLAST (NCBI; http://www.ncbi.nlm.nih.gov/BLAST/).

Nucleotide Sequence Accession Numbers

The sequences of the 16S rRNA gene from the ZJ5 strain were submitted to the GenBank database, and its accession number is KF032707. The nucleotide sequence containing the bacteriocin structural gene presented in this paper has been assigned the accession number KF032708.

Results

Identification of Bacteriocin-producing Strain ZJ5

The production of the ZJ5 strain showed strong bacteriocin activity against E. coli. The strain was a Gram-positive and catalase-negative bacillus that did not produce gas. The 16S rRNA sequence from ZJ5 had a similarity value of more than 99.9% with L. plantarum JDM1 (GenBank accession No. NC012984). The carbohydrate fermentation pattern of the ZJ5 strain was also in agreement with that of other L. plantarum strains (data not shown). Therefore, ZJ5 strain could be classified as a member of the L. plantarum genus. The strain was designated L. plantarum ZJ5, and the bacteriocin produced by this strain was named plantaricin ZJ5 (PZJ5).

Table 2. Oligonucleotide primers used to obtain the PZJ5 gene structure.

Primers	Sequences (5'-3')[a]
AGF1	AARACNAARCARCARTT
AGR1	GTRTANCCRAANACYTT
LB-TR1	GTATAACCAAATACTT
LB-TR2	GTGTTTGTGCTTTGAT
LB-TR3	TGATTAAAAATTGTTG
RB-TR1	TTTAATCAAAGCACAA
RB-TR2	ATCAAAGCACAAACAC
RB-TR3	AAGTATTTGGTTATAC
AD Primers1	NTCGASTWTSGWGTT
AD Primers2	NGTCGASWGANAWGAA
AD Primers3	WGTGNAGWANCANAGA
AD Primers4	TGWGNAGSANCASAGA
AD Primers5	AGWGNAGWANCAWAGG
PZJ5-F	AGATTCCAGGCAATG
PZJ5-R	GGAATAAATCAGTTA

In the degenerate primers, N, R, Y, and D indicate A/T/G/C, A/G, T/C, and A/T/G, respectively.

Purification of Plantaricin ZJ5

PZJ5 was purified from the culture supernatant by ammonium sulfate precipitation and a two-step procedure, including cation-exchange and hydrophobic-interaction chromatography (Table 3). The active eluted solution obtained using hydrophobic-interaction chromatography was then subjected to reverse-phase HPLC. It produced one peak, which was eluted at an approximately 60% acetonitrile concentration. The active fraction was subjected again to RP-HPLC (Figure 1). The final yield of the peptide obtained in these purification steps was 1.7% from 3 L of the ZJ5 culture supernatant; the details are summarized in Table 3. PZJ5 was analyzed using Tricine-SDS-PAGE. A single protein of ca. 3 kDa was detected and its activity was confirmed by a gel antimicrobial assay (Figure 2).

Characterization of PZJ5

The inhibitory spectrum of PZJ5 is shown in Table 1. The plantaricin ZJ5 spectrum differed from the spectra of previously reported plantaricins, exerting high activity against a broad range of both Gram-positive and Gram-negative bacteria, with partic-

ularly strong activity being observed against *Lactobacillus plantarum*, *Listeria monocytogenes*, *Bacillus subtilis*, *Micrococcus luteus*, *Pseudomonas putida*, and *Escherichia coli* but no activity against *Rhodotorula rubra* and *Saccharomyces cerevisiae*. 100% stability was recorded in a wide pH range from 2.0 to 6.0. The activity was not recorded at neutral and alkaline pH, but was regained when the pH was reverted to 5.0 (Table S1). Incubation at different temperatures also led to no losses in activity. 100% activity was recorded even after boiling and autoclaving. When a purified preparation was treated with different hydrolytic enzymes, the inhibitory action was significantly reduced by the proteolytic enzymes. The inhibitory action was completely abolished by pepsin and proteinase K (62%) but not affected after treatment with lipase and α-amylase (Table 4).

Molecular Mass Analysis and Amino Acid Sequencing of Plantaricin ZJ5

The molecular mass of the purified peptide was confirmed by MALDI-TOF/MS analysis, and the molecular mass of PZJ5 was determined to be 2572.9 Da (m/z = 2573.9) (Fig. 3). N-terminal

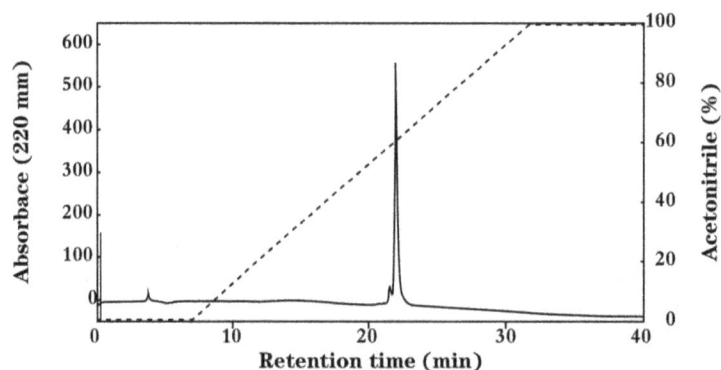

Figure 1. Reverse-phase HPLC analysis of plantaricin ZJ5. Elution was performed with a 0–100% linear gradient of acetonitrile and water containing 0.06% TFA.

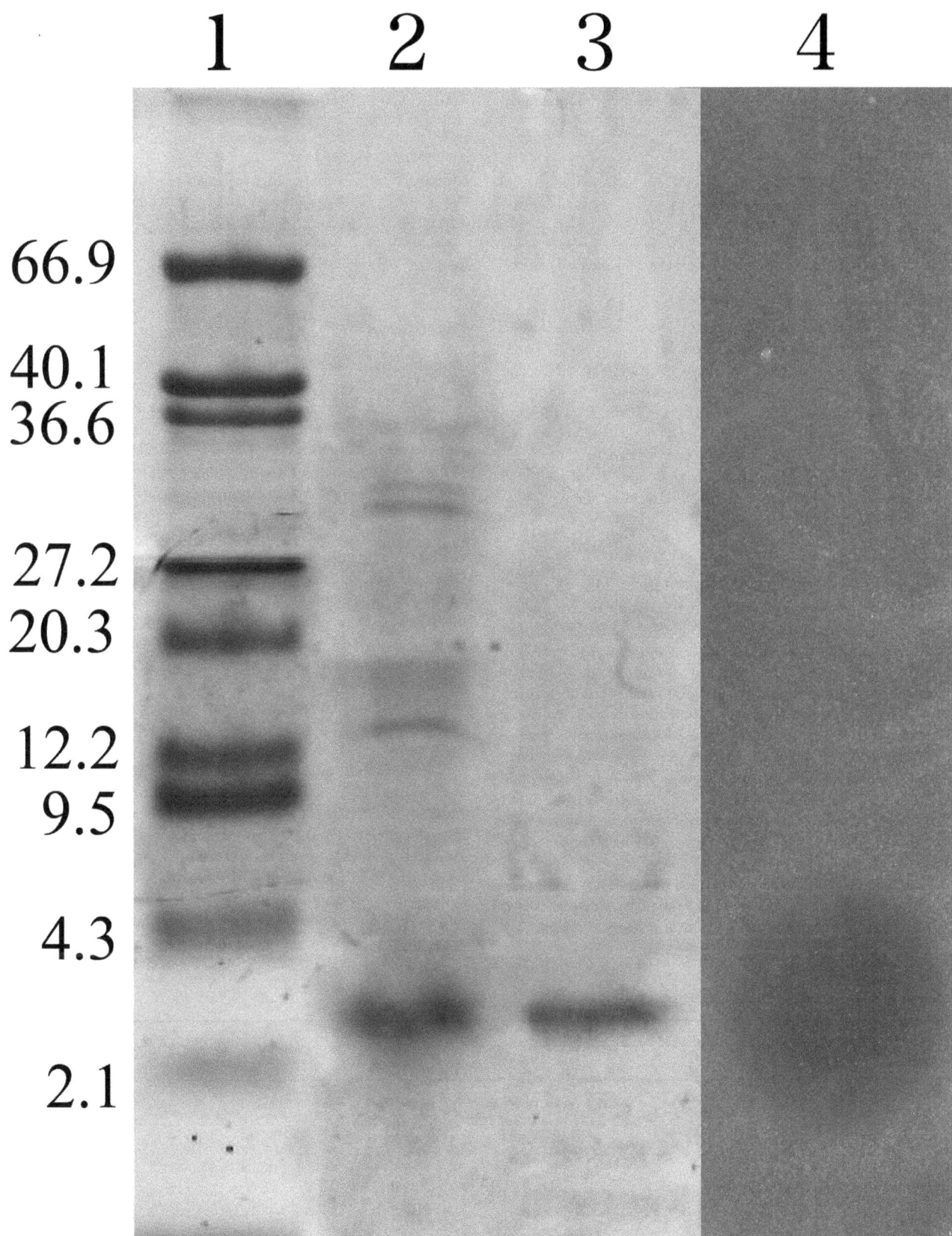

Figure 2. Tricine-Sodium dodecyl sulphate-polyacrylamide gel electrophoresis (tricine-SDS-PAGE) of isolated plantaricin ZJ5. Lane 1, molecular mass marker with the corresponding value in kDa on the left (Tiandz, China); lane 2, PZJ5 precipitated by ammonium sulfate; lane 3, purified PZJ5; lane 4, gel overlaid with indicator strain *Staphylococcus aureus*.

Table 3. Summary of purification index of plantaricin ZJ5 produced by *L. Plantarum* ZJ5.

Purification stage	Volume (ml)	Total protein (mg)	Total activity (AU)	Specific activity[a] (AU/mg)	Purification fold	Yield (%)
Culture supernatant	3,000	8,445	640,000	75.78	1.0	100
Ammonium sulfate precipitation	200	504.5	160,000	317.14	4.2	25
Cation-exchange chromatography	30	39.21	48,000	1,224.17	16.1	7.5
hydrophobic-interaction chromatography	10	8.76	20,000	2,283.11	30.1	3.1
C18 reverse-phase HPLC	5	1.04	11,000	10,576.92	139.5	1.7

[a]The specific activity is the ratio between the total activity and total protein

amino acid sequence analyses of PZJ5 were performed using automated Edman degradation, and 22 amino acids were determined to be N'-KTKQQFLIKAQTQLFKVFGYTL. The unimpaired Edman degradation sequencing suggests that the bacteriocin belongs to a linear, class II bacteriocin. It's estimated that the isoelectric point was 10.18, and its overall charge at pH 7.0 was 4. The amino acid sequence of plantaricin ZJ5 peptide was compared with NCBI Genbank database and Bactibase based on protein-protein homology (Blastp), indicating that the sequenced residues didn't have apparent homology with other known bacteriocins or proteins.

Analysis of the Gene Encoding plantaricin ZJ5

Degenerate primers were based on the amino acid sequence obtained in the manner described above. When PCR was performed with total DNA from *L. plantarum* ZJ5 and the degenerate primers, a 63 bp fragment was produced. The DNA sequence was used to design specific nests and random primers. These were then used for Tail-PCR and to obtain adjacent sequences. The DNA sequence of the structural gene encoding plantaricin ZJ5 was obtained in this way (Figure 4). Sequence analysis revealed a ribosome binding site, a possible leader peptide sequence, and a terminator sequence. This bacteriocin was found to be translated as a 44 amino acid pre-bacteriocin that was then processed to produce mature plantaricin ZJ5, which consists of 22 amino acids.

A double-glycine leader peptide is used to synthesize plantaricin ZJ5. This leader peptide shares 62% homology with plantaricin ASM1, a plantaricin produced by *Lactobacillus plantarum* A-1 (data not shown).

Discussion

In this study, we describe the purification and characterization of a new bacteriocin (PZJ5) produced by an environmental isolate strain LAB *L. Plantarum* ZJ5. Plantaricin ZJ5 shows stability under heating and acidic conditions (pH2.0–6.0). However, the peptide demonstrated reduced activity under alkaline conditions. Being proteinaceous in nature, the complete inactivation or significant reduction in the antimicrobial properties of purified plantaricin was observed after treatment with pepsin and proteinase K, but the same was not observed with lipase or α-amylase treatment. Most plantaricins appear to inhibit some Gram-positive bacteria, exhibiting antibacterial activity against bacteria closely related to the producer microorganism and have no activities against Gram-negative bacteria. Plantaricin ZJ5 not only has high activity against Gram-positive bacteria, but also has strong activity against Gram-negative bacteria, such as *Escherichia coli* and *Salmonella spp*. PZJ5 exhibited activity over a wide range of pH, heat, and bactericidal antimicrobial spectrums, signifying that plantaricin ZJ5 could preserve its structure and bactericidal functions even under extreme conditions, which is an important property in view of its potential use as a biopreservative in foods.

Plantaricin ZJ5 was purified from the culture supernatant of *L. Plantarum* ZJ5 using a four-step purification procedure. Because bacteriocin peptides share many common properties, a purification strategy used by several groups could be applied, such that for plantaricin S [22] and NC8 [13]. Plantaricin ZJ5 is cationic and hydrophobic, which was demonstrated by cation-exchange chromatography, hydrophobic-interaction chromatography and RP-

Table 4. Peptidase sensitivity of plantaricin ZJ5.

Peptidase	Diam (mm)[a]
Control (no peptidase)	25
trypsin	25
Cation-exchange chromatography	25
proteinase K	15
pepsin	0
α-chymotrypsin	25
papain	25
lipase	25
α-amylase	25

[a]Enzyme treatments were performed at 37°C for 2 h, and *Escherichia coli* was used as the indicator strain. Diameters were based on the clear zones of inhibition.

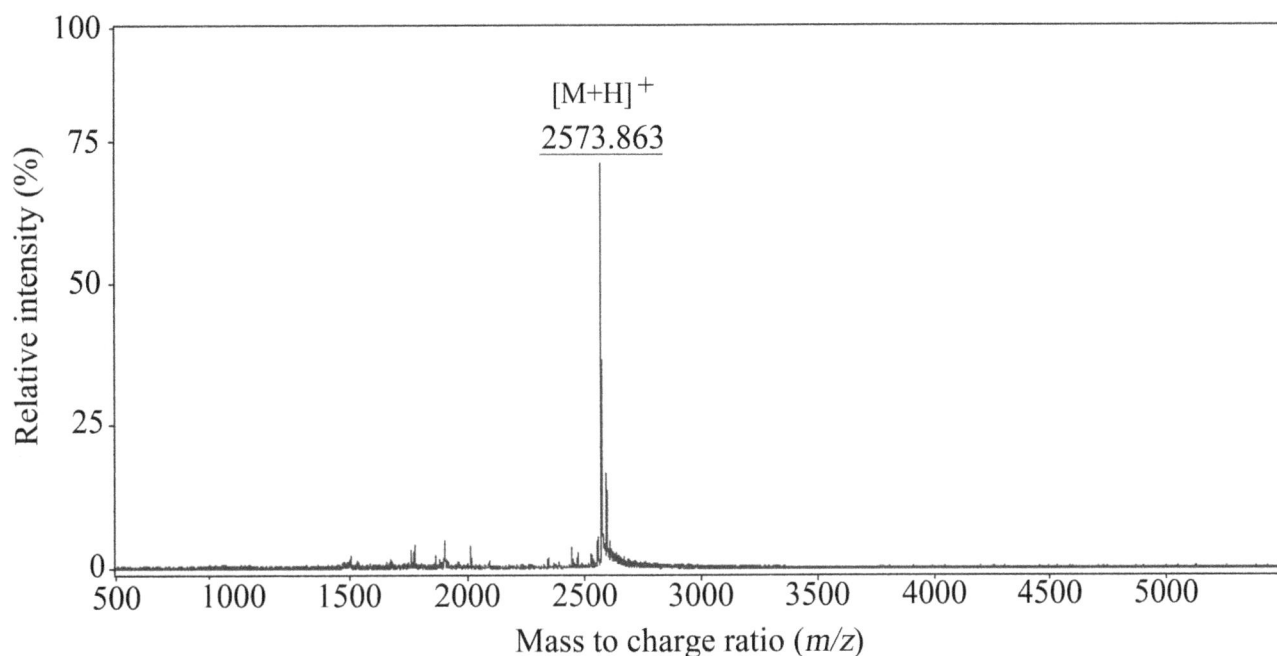

Figure 3. MALDI-TOF mass spectrometry of purified plantaricin ZJ5. A *m/z* 2573.863 monoisotopic peak ([M+H]⁺) is evident.

FPLC. Both properties are in agreement with earlier reports on bacteriocins [13,23,24,25].

Edman degradation analysis, degenerate PCR and Tail-PCR were performed to obtain the gene structure of plantaricin ZJ5 (Figure 4). Comparing the amino acid sequence of the purified plantaricin ZJ5 peptide with other bacteriocins by using Blastp in NCBI showed no apparent homology with other known bacteriocins or proteins. This result shows apparently a lack of similarity and also suggests that plantaricin ZJ5 is a novel bacteriocin. The lack of unusual amino acids in the entire amino acid sequence of plantaricin ZJ5, heat resistance and low molecular mass indicates that plantaricin ZJ5 belongs to the group of class IId bacteriocins [26,27].

The DNA sequence of the codons for a putative leader peptide, ribosome-binding site, terminator sequence and structural gene encoding plantaricin ZJ5 was identified (Figure 4). Although the amino acid sequence of the mature PZJ5 peptide did not show significant homology to any other known bacteriocins, PZJ5 was found to contain a well-known leader peptide characterized by two conserved glycine residues. This leader peptide is found in many bacteriocins. This type of peptide is believed to cleave the N-terminal leader peptide from the bacteriocin to perform at the Gly-Gly site and to transport the resulting mature polypeptide across the plasma membrane [25,28,29,30]. The molecular weight of PZJ5 was calculated to be 2631.1 Da; this value is approximately 57.2 Da higher than the value obtained by mass spectrometry. This result also suggests a post-translational

```
  1    CTG AGA AAA TTC CCT AGA TTC CAG GCA ATG ATC TAA TTC CAA ATT    45
                                                        ‾‾‾‾‾‾‾‾‾‾
                                                          -35

 46    ATA TTG ATG TTA GAA TAT AAT GAC GCA TTT AAA AAG GAG GTA CAT    90
                       ‾‾‾‾‾‾‾‾‾‾                 ‾‾‾‾‾‾‾‾‾‾‾
                          -10                         RBS

 91    ACT ATG AGT TTA GTT AAA GAA AAT AAA ACC TTA GTA GAT GAA CTA   135
  1        M   S   L   V   K   E   N   K   T   L   V   D   E   L    15

136    ATC GCA CAA ATC ACC TTT GGT GGA AAA ACA AAA CAA CAA TTT TTA   180
 16     I   A   Q   I   T   F   G   G   K   T   K   Q   Q   F   L    30
                                       ↑
181    ATC AAA GCA CAA ACA CAA TTA TTT AAA GTA TTT GGT TAT ACA TTA   225
 31     I   K   A   Q   T   Q   L   F   K   V   F   G   Y   T   L    45

226    TAA CTT AGC TAC ATT TAA CTG ATT TAT TCC CGT TTA AAT CGT GAG   270
 46     *                                                            46

271    CTT ACT GAA ATT TAC AAC AAA   291
```

Figure 4. Nucleotide and deduced amino acid sequences in the region containing pzj5. The putative −35 and −10 (Pribnow box) promoter sequences and ribosome binding site (RBS) are underlined. The vertical arrow indicates the Gly-Gly cleavage site of the presequence. The stop codon is indicated with an asterisk.

modification in PZJ5, but the nature of the modification is unknown.

In recent years, a variety of LAB bacteriocins have attracted attention due to their potential application as the next generation of antimicrobial compounds used in food preservation [31]. *L. plantarum* is one of the major LABs, and it can be isolated from a variety of fermented vegetables [32,33]. *L. plantarum* ZJ5 produces the promising novel bacteriocin plantaricin ZJ5, which exhibits activity over a wide range of pH, heat, and bactericidal antimicrobial spectrum while at the same time being safely degradable by human digestive enzymes. Because of these properties, it is considered that plantaricin ZJ5 will be used as a novel biopreservative.

Author Contributions

Conceived and designed the experiments: DFS MYZ QG. Performed the experiments: DFS. Analyzed the data: DFS. Contributed reagents/materials/analysis tools: DFS. Contributed to the writing of the manuscript: DFS.

References

1. Lemos MM, Dias DCA, Ferreira GLS (2008) Inhibition of vancomycin and high-level aminoglycoside-resistant enterococci strains and Listeria monocytogenes by bacteriocin-like substance produced by Enterococcus faecium E86. Curr Microbiol 57: 429–436.

2. Aunpad R, Na-Bangchang K (2007) Pumilicin 4, a novel bacteriocin with anti-MRSA and anti-VRE activity produced by newly isolated bacteria Bacillus pumilus strain WAPB4. Curr Microbiol 55: 308–313.

3. Henderson JT, Chopko AL, van Wassenaar PD (1992) Purification and primary structure of pediocin PA-1 produced by Pediococcus acidilactici PAC-1.0. Arch Biochem Biophys 295: 5–12.

4. Morisset D, Berjeaud JM, Marion D, Lacombe C, Frere J (2004) Mutational analysis of mesentericin y105, an anti-Listeria bacteriocin, for determination of impact on bactericidal activity, in vitro secondary structure, and membrane interaction. Appl Environ Microbiol 70: 4672–4680.

5. Cleveland J, Montville TJ, Nes IF, Chikindas ML (2001) Bacteriocins: safe, natural antimicrobials for food preservation. Int J Food Microbiol 71: 1–20.

6. Klaenhammer TR (1993) Genetics of bacteriocins produced by lactic acid bacteria. FEMS Microbiol Rev 12: 39–85.

7. Nes IF, Holo H (2000) Class II antimicrobial peptides from lactic acid bacteria. Biopolymers 55: 50–61.

8. Lubelski J, Rink R, Khusainov R, Moll GN, Kuipers OP (2008) Biosynthesis, immunity, regulation, mode of action and engineering of the model lantibiotic nisin. Cell Mol Life Sci 65: 455–476.

9. Mangalassary S, Han I, Rieck J, Acton J, Jiang X, et al. (2007) Effect of combining nisin and/or lysozyme with in-package pasteurization on thermal inactivation of Listeria monocytogenes in ready-to-eat turkey bologna. J Food Prot 70: 2503–2511.

10. Naas H, Martinez-Dawson R, Han I, Dawson P (2013) Effect of combining nisin with modified atmosphere packaging on inhibition of Listeria monocytogenes in ready-to-eat turkey bologna. Poult Sci 92: 1930–1935.

11. Anderssen EL, Diep DB, Nes IF, Eijsink VGH, Nissen-Meyer J (1998) Antagonistic Activity of Lactobacillus plantarum C11: Two New Two-Peptide Bacteriocins, Plantaricins EF and JK, and the Induction Factor Plantaricin A. Appl. Envir. Microbiol. 64: 2269–2272.

12. Remiger A, Eijsink VG, Ehrmann MA, Sletten K, Nes IF, et al. (1999) Purification and partial amino acid sequence of plantaricin 1.25 alpha and 1.25 beta, two bacteriocins produced by Lactobacillus plantarum TMW1.25. J Appl Microbiol 86: 1053–1058.

13. Maldonado A, Ruiz-Barba JL, Jimenez-Diaz R (2003) Purification and genetic characterization of plantaricin NC8, a novel coculture-inducible two-peptide bacteriocin from Lactobacillus plantarum NC8. Appl Environ Microbiol 69: 383–389.

14. Gonzalez B, Arca P, Mayo B, Suarez JE (1994) Detection, purification, and partial characterization of plantaricin C, a bacteriocin produced by a Lactobacillus plantarum strain of dairy origin. Appl. Envir. Microbiol. 60: 2158–2163.

15. Hata T, Tanaka R, Ohmomo S (2010) Isolation and characterization of plantaricin ASM1: a new bacteriocin produced by Lactobacillus plantarum A-1. Int J Food Microbiol 137: 94–99.

16. Anderson DG, McKay LL (1983) Simple and rapid method for isolating large plasmid DNA from lactic streptococci. Appl. Envir. Microbiol. 46: 549–552.

17. Weisburg WG, Barns SM, Pelletier DA, Lane DJ (1991) 16S ribosomal DNA amplification for phylogenetic study. J. Bacteriol. 173: 697–703.

18. Tagg JR, McGiven AR (1971) Assay system for bacteriocins. Applied Microbiology 21: 943.

19. Masuda Y, Ono H, Kitagawa H, Ito H, Mu F, et al. (2011) Identification and characterization of leucocyclicin Q, a novel cyclic bacteriocin produced by Leuconostoc mesenteroides TK41401. Appl Environ Microbiol 77: 8164–8170.

20. Liu YG, Whittier RF (1995) Thermal asymmetric interlaced PCR: automatable amplification and sequencing of insert end fragments from P1 and YAC clones for chromosome walking. Genomics 25: 674–681.

21. Singer T, Burke E (2003) High-throughput TAIL-PCR as a tool to identify DNA flanking insertions. Methods Mol Biol 236: 241–272.

22. Jimenez-Diaz R, Ruiz-Barba JL, Cathcart DP, Holo H, Nes IF, et al. (1995) Purification and partial amino acid sequence of plantaricin S, a bacteriocin produced by Lactobacillus plantarum LPCO10, the activity of which depends on the complementary action of two peptides. Appl Environ Microbiol 61: 4459–4463.

23. Morgan SM, O'Connor PM, Cotter PD, Ross RP, Hill C (2005) Sequential actions of the two component peptides of the lantibiotic lacticin 3147 explain its antimicrobial activity at nanomolar concentrations. Antimicrob Agents Chemother 49: 2606–2611.

24. Yamazaki K, Suzuki M, Kawai Y, Inoue N, Montville TJ (2005) Purification and characterization of a novel class IIa bacteriocin, piscicocin CS526, from surimi-associated Carnobacterium piscicola CS526. Appl Environ Microbiol 71: 554–557.

25. Tosukhowong A, Zendo T, Visessanguan W, Roytrakul S, Pumpuang L, et al. (2012) Garvieacin Q, a novel class II bacteriocin from Lactococcus garvieae BCC 43578. Appl Environ Microbiol 78: 1619–1623.

26. Franz CM, van Belkum MJ, Holzapfel WH, Abriouel H, Galvez A (2007) Diversity of enterococcal bacteriocins and their grouping in a new classification scheme. FEMS Microbiol Rev 31: 293–310.

27. Izquierdo E, Wagner C, Marchioni E, Aoude-Werner D, Ennahar S (2009) Enterocin 96, a novel class II bacteriocin produced by Enterococcus faecalis WHE 96, isolated from Munster cheese. Appl Environ Microbiol 75: 4273–4276.

28. Franke CM, Tiemersma J, Venema G, Kok J (1999) Membrane topology of the lactococcal bacteriocin ATP-binding cassette transporter protein LcnC. Involvement of LcnC in lactococcin A maturation. pp.8484–8490.

29. Havarstein LS, Diep DB, Nes IF (1995) A family of bacteriocin ABC transporters carry out proteolytic processing of their substrates concomitant with export. Mol Microbiol 16: 229–240.

30. Kotake Y, Ishii S, Yano T, Katsuoka Y, Hayashi H (2008) Substrate recognition mechanism of the peptidase domain of the quorum-sensing-signal-producing ABC transporter ComA from Streptococcus. Biochemistry 47: 2531–2538.

31. Soomro AH, Masud T, Anwaar K (2002) Role of lactic acid bacteria (LAB) in food preservation and human health—a review. Pakistan Journal of Nutrition 1: 20–24.

32. Lee C (1997) Lactic acid fermented foods and their benefits in Asia. Food Control 8: 259–269.

33. Tanganurat W, Quinquis B, Leelawatcharamas V, Bolotin A (2009) Genotypic and phenotypic characterization of Lactobacillus plantarum strains isolated from Thai fermented fruits and vegetables. J Basic Microbiol 49: 377–385.

19

Chemical Interference with Iron Transport Systems to Suppress Bacterial Growth of *Streptococcus pneumoniae*

Xiao-Yan Yang[1][9], Bin Sun[2][9], Liang Zhang[1], Nan Li[1], Junlong Han[1], Jing Zhang[1], Xuesong Sun[1]*, Qing-Yu He[1]*

1 Key Laboratory of Functional Protein Research of Guangdong Higher Education Institutes, Institute of Life and Health Engineering, College of Life Science and Technology, Jinan University, Guangzhou, China, **2** School of Pharmaceutical Sciences, Southern Medical University, Guangzhou, China

Abstract

Iron is an essential nutrient for the growth of most bacteria. To obtain iron, bacteria have developed specific iron-transport systems located on the membrane surface to uptake iron and iron complexes such as ferrichrome. Interference with the iron-acquisition systems should be therefore an efficient strategy to suppress bacterial growth and infection. Based on the chemical similarity of iron and ruthenium, we used a Ru(II) complex R-825 to compete with ferrichrome for the ferrichrome-transport pathway in *Streptococcus pneumoniae*. R-825 inhibited the bacterial growth of *S. pneumoniae* and stimulated the expression of PiuA, the iron-binding protein in the ferrichrome-uptake system on the cell surface. R-825 treatment decreased the cellular content of iron, accompanying with the increase of Ru(II) level in the bacterium. When the *piuA* gene (SPD_0915) was deleted in the bacterium, the mutant strain became resistant to R-825 treatment, with decreased content of Ru(II). Addition of ferrichrome can rescue the bacterial growth that was suppressed by R-825. Fluorescence spectral quenching showed that R-825 can bind with PiuA in a similar pattern to the ferrichrome-PiuA interaction *in vitro*. These observations demonstrated that Ru(II) complex R-825 can compete with ferrichrome for the ferrichrome-transport system to enter *S. pneumoniae*, reduce the cellular iron supply, and thus suppress the bacterial growth. This finding suggests a novel antimicrobial approach by interfering with iron-uptake pathways, which is different from the mechanisms used by current antibiotics.

Editor: Roy Martin Roop II, East Carolina University School of Medicine, United States of America

Funding: This work was supported by the National Natural Science Foundation of China (21271086, to Q.-Y. H.; 31000373, to X. S.), Guangdong Natural Science Research Grant (32213027, to Q.-Y. H.; S2012010008685 to X. S.), the Fundamental Research Funds for the Central Universities (11610101 to Q.-Y. H.; 21611201, to X. S.), and the Pearl River Rising Star of Science and Technology of Guangzhou City (2011J2200003, to X. S.). The funders had no role in study design, data collection and analysis, decision to publish, or preparation of the manuscript.

Competing Interests: The authors have declared that no competing interests exist.

* Email: tqyhe@jnu.edu.cn (QYH); tsunxs@jnu.edu.cn (XS)

[9] These authors contributed equally to this work.

Introduction

Iron is a critical nutrient for bacterial growth and survival, as a major determinant in the development of infection in host. However, the concentration of free iron in host is extremely low ($<10^{-18}$ M). In order to acquire enough iron from their host environments, bacteria have developed highly specific and effective iron-acquisition systems located on the membrane surface [1–3]. Blocking or interfering with the iron-acquisition systems could disrupt bacterial iron homeostasis and thus suppress bacterial growth.

Streptococcus pneumoniae is a dangerous bacterium responsible for various life-threatening diseases including otitis media, septicemia, pneumonia and meningitis in immuno-compromised individuals [4,5]. In *S. pneumoniae*, there are three ABC transporters as known iron-transport systems including PiaABC, PiuABC and PitABC, respectively responsible for the acquisition of heme, ferrichrome and ferric irons [6–9]. Heme, ferrichrome and ferric irons are firstly bound by lipoproteins PiaA, PiuA and PitA and transferred to the permeases PiaB, PiuB and PitB to go

across the cell membrane using energy provided by the ATP hydrolysis through PiaC, PiuC and PitC, respectively. Since the lipoproteins PiaA, PiuA and PitA work as iron-receptors located on the cellular surface, they are the valuable targets for the design of novel antibacterial agents. It must be pointed out that Cheng *et al* have characterized a ferrichorme-binding PiaA protein encoded by gene SP_1032 in *S. pneumnoniae* TIGR4 (corresponding to gene SPD_0915 in *S. pneumoniae* D39) [9], which is actually the protein PiuA through sequence alignments.

Earlier investigations have suggested that bacteria obtain iron mainly in the forms of complexes, heme-iron and ferrichrome-iron compounds as for *S. pneumoniae*. In this study, we selected our previously synthesized Ru(II) complex R-825 to chemically resemble the heme/ferrichrome compounds. Based on the chemical similarity between Fe(III) and Ru(II), we expected that R-825 could compete with heme/ferrichrome for binding to PiaA/PiuA in *S. pneumoniae*. Our experiments showed that R-825 indeed inhibited the bacterial growth by competing with ferrichrome for PiuA binding. We verified that complex R-825 can get access into the bacterium through ferrichrome-transport systems, reducing the iron supply and thus suppressing the bacterial growth.

Materials and Methods

Materials

Ru(II) complex R-825 was synthesized according to the procedure with minor modifications as described in our previous publication, where the complex was defined as 1a [10]. The structure of R-825 is shown in Figure 1.

Bacterial strains and growth conditions

Single gene deleted *piaA*- (SPD_1652), *piuA*- (SPD_0915) and *pitA*- (SPD_0226) mutant strains of *S. pneumoniae* D39 were constructed by long flanking homology-polymerase chain reaction (LFH-PCR) [11,12], mutants were made by replacing *piaA*, *piuA* and *pitA* genes of D39 with gene encoding resistance to erythromycin (erm)[13]. Briefly, the 500 bp region upstream of *piaA* was amplified using primers piaA–P1 and piaA–P3, while the 500 bp region downstream of *piaA* was amplified using primers piaA–P2 and piaA–P4. The erythromycin gene was amplified using primers erm-F and erm-R. The three PCR fragments generated were joined together by overlap extension PCR using primers piaA–P1 and piaA–P2 to form approximately 2-kb final linear DNA construct the deletion fragment. Then, the linear DNA construct was used for homologous recombination and transformed into *S. pneumoniae* D39. Transformants were selected with 0.25 μg/mL erythromycin selection agar plates, mutants were confirmed by PCR and Western blotting. The *piuA*- and *pitA*- mutant strains were constructed using a similar method. All mutations were stable after six sequential passages in Todd-Hewitt broth supplemented with 0.5% yeast extract (0.5% THY) without antibiotic selection.

These mutant strains together with wild-type *S. pneumoniae* D39 were cultured at 37°C and 5% CO_2 in 0.5% THY or on Columbia agar added with 5% sterile defibrinated sheep blood purchased from Ruite company (Guangzhou, China) in which all animal experiment procedures were conducted in strict accordance with the recommendations in the Guide for the Care and Use of Laboratory Animals of the National Institutes of Health. Disc susceptibility assays were performed as described by CLSI guidelines [14]. In short, filter paper disks (5 mm in diameter) containing 50 μg of R-825 were placed on the plates (10^7 bacteria/plate), and inhibition zones were examined after 24 h of incubation at 37°C. Tests using gentamicin as the control were carried out in parallel.

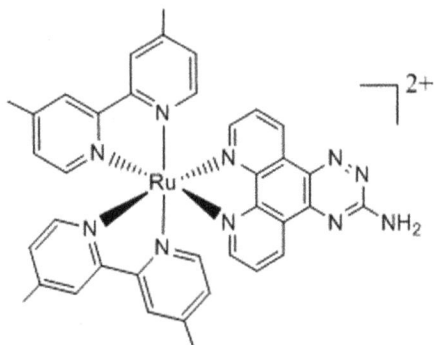

Figure 1. Structure of ruthenium(II) complex R-825 [10].

Minimal inhibitory concentration (MIC) assay

R-825 was dissolved in water to a concentration of 10 mg/mL (12 mM). The MIC assays were carried out in triplicate by using the standard micro-dilution method [14]. Different concentrations of R-825 with 7.5, 15, 30, and 60 μM were added to diluted media containing 10^6 bacteria/mL, followed by incubation in 24-well plates at 37°C for 24 h. The concentration of R-825 corresponding to the well with no visible bacteria growth ($OD_{600} < 0.1$) was taken as the MIC.

Cytotoxicity assay

The human alveolar epithelial cell line A549 was cultivated in DMEM media with 10% fetal bovine serum (FBS) at 37°C in a humidified atmosphere of 5% CO_2. The cytotoxicity of R-825 in A549 cells was assessed using LDH cytotoxicity assay kit by the following procedure as described in the manufacture's manual. Briefly, 5×10^3 A549 cells were seeded into a sterile 96-well plate in triplicate, incubated for 24 h without R-825, followed by a 48 h incubation in the dark with different amounts of R-825 (30, 60, 120, 240 and 480 μM). At the end of the incubation, the cell plate was centrifuged at 400 *g* for 5 min, then 120 μL of the media from each well was transferred to a new plate, and 60 μL of LDH mixture was added to the supernatant and incubated for 30 min in dark at room temperature. Cytotoxicity was assessed by monitoring the absorbance at 490 nm in a microplate reader. Whole cell lysate was used as a positive control.

ICP-MS analysis

Sample preparation for ICP-MS was conducted as a previously described method [15]. Briefly, bacteria were grown in 0.5% THY in the presence or absence of R-825 with sub-MIC concentrations at 37°C and then harvested when the optical density at 600 nm (OD_{600}) reached 0.5–0.6. Cells were pelleted by centrifugation at 8,000 *g* at 4°C for 10 min, and then washed three times with $1 \times PBS$ that had been treated with chelex-100 resin. Subsequently, the wet cell pellets were dried using a Scanvac Freeze Dryer (Labgene Scientific, Switzerland) and the dry weights were calculated. The dry cell mass was disrupted by resuspending pellets in 2 mL of 14% HNO_3, then heated to 95°C for 20 min. Samples were centrifuged at 13,200 *g* for 30 min, the supernatants were collected and an internal standard indium (In) was added to the samples. Metal contents of these samples were analyzed in an iCAP ICP-MS (Thermo Scientific, U.S.A.). Results were showed as ng of Fe and Ru per mg dry weight of cells. All data were evaluated with at least three independent biological experiments.

Real-time quantitative PCR (RT-qPCR)

Total RNA was extracted from *S. pneumoniae* D39 strain with and without R-825 (in sub-MIC) treatment by TRIZOL method according to the manufacturer's manual, and quantified by Nanodrop 2000 spectrophotometer. Any genomic DNA was removed using RNase-free DNase I. cDNA synthesis was performed using 1 μg of total RNA and iScript Reverse Transcriptase (Bio-Rad) according to the manufacture's instruction. RT-qPCRwas carried out using EvaGreen Dye (Bio-Rad) in a Miniopticon RT-qPCR System (Bio-Rad). The cycle threshold (Ct) value was measured; relative quantification of specific gene expression was calculated using the $2^{-\Delta\Delta Ct}$ method, with the 16S rRNA as the reference gene. Genes with a two-fold or greater difference in expression level relative to control were considered significant. The primer sequences used for RT-qPCR are shown in Table 1.

Table 1. The primer sequences used for RT-qPCR experiments.

Primer	Sequence (5′–3′)
16S rRNA-F	5′-CTGCGTTGTATTAGCTAGTTGGTG-3′
16S rRNA-R	5′-TCCGTCCATTGCCGAAGATTC-3′
PiaA-F	5′-TAGTCAGACAGAGACCAGT-3′
PiaA-R	5′-CTTTCATAGAACCAACATT-3′
PiuA-F	5′-ATTTGACGATTTGGATGGACTT-3′
PiuA-R	5′-GATTTGTATGCTGCTACAGGAG-3′
PitA-F	5′-ATGACTGTTGGTCTCTCTT-3′
PitA-R	5′-TTGTTTTAGCATTTTTACG-3′

Cloning and purification of PiuA protein

The *piuA* gene without the N-terminal lipoprotein signal sequence was PCR amplified from *S. pneumoniae* D39 genomic DNA with the forward primer 5′-CCGCCGGAGCTCTCTTC-TAATTCTGTTAAAAA-3′ and the reverse primer 5′-GCCGCCGAATTCTTATTTCGCATTTTTGC-3′, creating the SacI and EcoRI restriction sites (underlined). The PCR product was digested with SacI and EcoRI and ligated into the expression vector pBAD/HisA to generate pBAD-PiuA. The construct was transformed into *E. coli* TOP10 for high-level expression of recombinant protein. The transformants were incubated at 37°C with vigorous shaking in LB medium when the OD_{600} reached at 0.8, followed by induction with 0.05% L-arabinose for 6 h. Harvested cells were lysed by sonication for 30 min (5 s on/5 s off, on ice), the supernatant was collected by centrifugation at 4 °C, 10,000 g for 30 min and the protein was then isolated by using Ni-NTA His-bind Resin (1.5 mL, Qiagen). Fractions containing His6-PiuA protein were harvested and verified with SDS-PAGE and Western blotting. The His-tag was cleaved with entorokinase for 24 h at room temperature, and then removed by Ni-NTA to produce purified PiuA protein. The purity of the purified protein was examined by SDS-PAGE. The identity of the PiuA protein was further confirmed by using ABI-4800 plus MALDI TOF/TOF mass spectrometer according to a previously described method [16]. Proteins were identified by the MASCOT search engine (V2.1) against NCBI *S. pneumoniae* D39 protein database based on the MS and MS/MS spectra, protein identifications with Mascot scores C. I. % >95 were considered significant. PiaA protein (the forward primer piaA-F: 5′-GCGAGCTCGAGACCAGTAGCTCTGCTC-3′, and the reverse primer piaA-R: 5′-CGCCGCGAATTCTTATTT-CAAAGCTTTTTG-3′) and PitA protein (the forward primer pitA-F: 5′-GCGAGCTCATGACTGTTGGTCTCTCTTA-3′, and the reverse primer pitA-R: 5′-CGCCGCGAATTCT-TACTGTTTAGATTGGATAT-3′) were also cloned and purified using the similar procedure.

Immunization experiments and Western blotting

Purified His6-PiaA, His6-PiuA and His6-PitA proteins were used as antigens for the immunization experiments as previously described in the literature to generate multicolon antibodies [17]. The specificities of the antibodies were detected with Western blotting using purified proteins and whole-cell lysates of wild type and *piaA*-, *piuA*- and *pitA*- mutant D39 strains.

For Western blotting analysis, untreated and R-825-treated (sub-MIC) *S. pneumoniae* D39 were harvested by centrifugation at 6,000 g for 10 min at 4°C when the absorbance reading of 0.6 at 600 nm was reached. Then pellets were washed three times with 1×PBS and disrupted by sonication to extract proteins, the concentrations of the cellular proteins were measured by Bradford assay. The protein extracts were separated by 12% SDS–PAGE and then electroblotted onto polyvinylidene fluoride membranes. The protein expressions of PiaA, PiuA and PitA were detected with anti-PiaA, -PiuA and -PitA antibodies and quantified using ImageMaster 2D Platinum 6.0. Total proteins separated by SDS-PAGE and stained with Coomassie brilliant blue R250 were used as the loading control.

Fluorescence spectroscopy of apo-PiuA protein titrated with R-825

Fluorescence measurements were performed in a Hitachi F7000 spectrofluorophotometer. Fluorescence emission spectra were recorded from 290 to 450 nm after exciting at 280 nm. Both slit widths of excitation and emission beams were 5 nm. Spectra were acquired for 2 µM apo-PiuA (20 mM Tris-HCl, 100 mM NaCl, pH 7.4) with varying concentrations of R-825 (from 1.2 to 19.6 µM) and ferrichrome (from 0.4 to 4.0 µM). Vancomycin (from 1.2 to 19.2 µM) was also titrated to the apo-PiuA solution as a negative control. Relative changes in fluorescence emission (ΔF) at 343 nm during the titration verses the concentrations were fitted to a titration curve. The data were analyzed with Hill plot equation in Origin 8.5 to acquire the affinity constants (Ka).

Results

Ru(II) complex R-825 possesses antibacterial activity against *S. pneumoniae*

To investigate the antibacterial activity of R-825, we tested its effects on the growth of *S. pneumoniae* D39 strain in batch cultures, and determined its MIC value for *S. pneumoniae*. As shown in Figure 2A, R-825 can inhibit the bacterial growth in a concentration-dependent manner. At the concentration of 30 µM (MIC), R-825 completely suppressed the growth of *S. pneumoniae*. The sub-MIC, 15 µM, was selected to treat *S. pneumoniae* for the following real-time quantitative PCR and ICP-MS assays. MIC determination for the mutant strains was also performed and the results are listed in Table 2. Notably, *piuA*- mutant has a MIC = 60 µM, double of those for the wild-type and other mutant strains.

R-825 is not toxic to human A549 cells

The cytotoxic activity of R-825 against human A549 cells was evaluated by using LDH assay kit, with whole cell lysate as a positive control. As shown in Figure 2B, the viability of A549 cells had not significant changes under the 48 h treatment of R-825

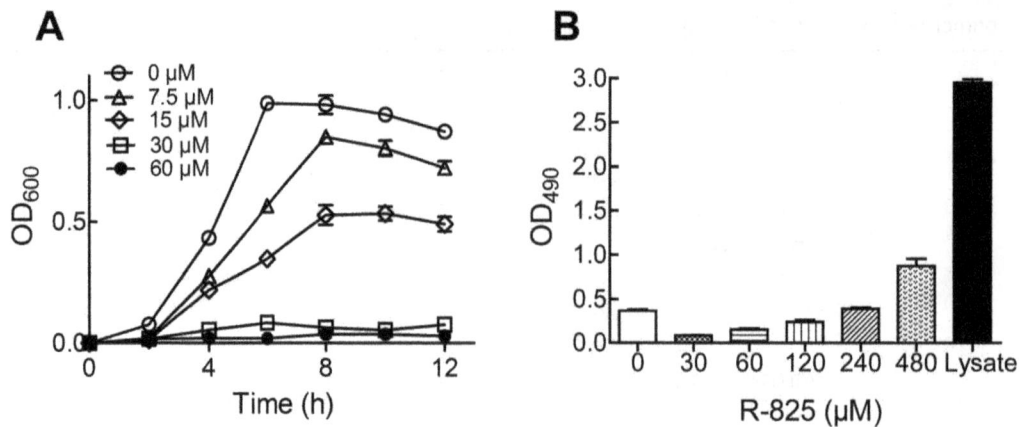

Figure 2. Effect of R-825 on *S. pneumoniae* and human A549 cell growth. (A) R-825 inhibited *S. pneumoniae* growth in a concentration-dependent manner. (B) R-825 showed little cytotoxicity to human cells. Determination of the cytotoxicity against A549 cell line was performed by incubating the cells with R-825 for 48 h using LDH kit, cell lysate was used as positive control. The data shown represent the mean of three experiments; error bars indicate SEM.

with up to 480 µM, a concentration substantially higher than the corresponding MIC value for the bacterium. Results demonstrated that R-825 has very low toxicity to human cells, implicating its high selective toxicity toward bacteria over human host.

R-825 reduces bacterial iron content and up-regulates PiaA and PiuA

To determine whether R-825 can compete with iron for the uptake system by *S. pneumoniae*, we measured the intracellular iron and ruthenium levels in the bacterium with and without R-825 treatment by using ICP-MS analysis. The results are shown in Figure 3A. As compared with control, the treatment with R-825 in its sub-MIC significantly decreased the iron concentration and correspondingly increased the ruthenium concentration in wild-type *S. pneumoniae*.

Previous studies have shown that many metal ion transporters are regulated by their substrates [18–20]. To investigate whether R-825 regulates iron-uptake proteins, we monitored the mRNA expression levels of lipoproteins PiaA, PiuA and PitA, the three surface iron chelates in the known iron-uptake ABC transporter systems in *S. pneumoniae* [6,7]. We observed that R-825 treatment up-regulated *piuA* and *piaA* gene expressions, but exhibited no significant impact on *pitA* expression, as shown in Figure 3B. Western blotting was also performed to measure the protein expression levels after R-825 treatment. As shown in Figure 3C, R-825 stimulated the expression of PiuA and PiaA but exerted no effect on PitA expression.

To confirm the interference of R-825 with the iron uptake system, we also investigated the protein expression level in the

bacteria cultured with iron replete and restricted medium. As shown in Figure S1 in File S1, upon iron starvation, the expression level of PiaA is significantly increased, and PiuA is slightly increased while PitA is unchanged in wild-type D39 strain. However, the expression levels of both PiuA and PitA are evidently up-regulated in *piaA*- mutant strain upon iron starvation. These results are basically consistent with the protein change tendency upon Ru-825 treatment. These consistent observations suggest that R-825 may enter *S. pneumoniae via* PiaABC or PiuABC iron-uptake systems.

R-825 is uptaken *via* ferrichrome-transport system

To determine whether R-825 would be taken up by *S. pneumoniae via* PiaABC (for heme uptake) or PiuABC (for ferrichrome uptake) system, we individually deleted *piaA* and *piuA* genes in the bacterium to construct the *piaA*- and *piuA*- single mutant strains, in which heme- and ferrichrome-binding ability was respectively impaired in the bacterium, and verified the effects using Western blotting (figure S2 in File S1). The sensitivity of the mutant strains to R-825 was compared to that of wild-type *S. pneumoniae* by measuring the growth inhibition zones and MIC values corresponding to R-825 treatment. As comparison, *pitA*-mutant was also constructed and tested (figure S2 in File S1).

Our experimental results are shown in Table 2. As compared with wild-type *S. pneumoniae*, both *piaA*- and *pitA*- mutants exhibited no significant difference in growth inhibition zones and MICs. In contrast, *piuA*- mutant strain displayed much smaller growth inhibition zone with a doubled MIC value, indicating that the *piuA*- strain is resistant to R-825 (Table 2). This suggests that

Table 2. Minimal Inhibitory Concentrations (MICs) and Diameter of Growth Inhibition Zones (mm) of R-825 against WT *S. pneumoniae* D39, *piaA*-, *piuA*- and *pitA*- mutant strains.

Bacteria	MICs (µM)	Diameter of Growth Inhibition Zones (mm)
WT *S. pneumoniae* D39	30	16
piaA- mutant	30	16
piuA- mutant	60	10
pitA- mutant	30	14

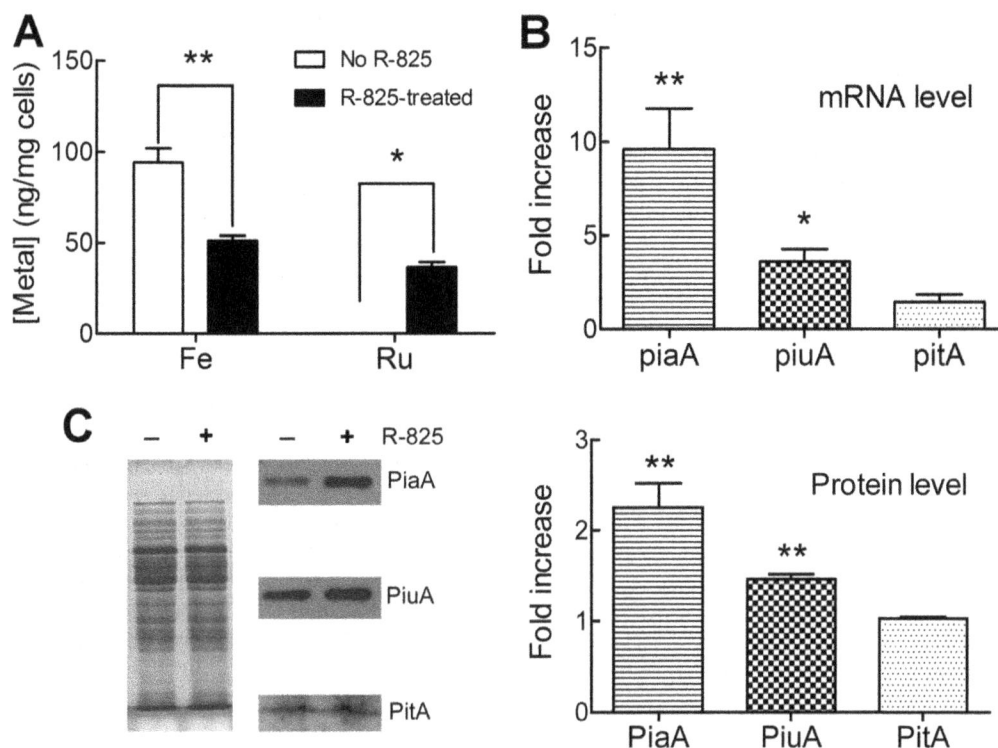

Figure 3. R-825 reduced iron uptake and stimulated the expressions of PiaA and PiuA in *S. pneumoniae*. (A) The cellular iron and ruthenium contents of the wild-type *S. pneumoniae* with and without R-825 treatment. (B) The increase of mRNA expression levels of *piaA*, *piuA* and *pitA* in the presence of 15 μM R-825, as compared with their expressions in the absence of R-825. Data were normalized with housekeeping gene 16S rRNA (mean ± SEM, n = 3, **$p < 0.01$, *$p < 0.05$ versus the untreated control). (C) The comparison of the protein expression levels in the bacterium with and without R-825 treatment. Whole cell proteins were used as loading control for the Western blotting.

the antibacterial activity of R-825 is largely related to PiuA but not PiaA and PitA; without PiuA, the uptake of R-825 may be impaired in the *piuA*- mutant strain. To confirm this hypothesis, we detected the intracellular ruthenium levels in the *piuA*- mutant and wild-type strains under R-825 treatment. As observed in Figure 4A, the level of intracellular ruthenium in *piuA*- mutant strain was substantially reduced to half content of the wild-type bacterium.

Ferrichrome but not iron ion and hemin rescues the bacterial growth suppressed by R-825

The above experiments indicated that R-825 may be taken up by *S. pneumoniae* through PiuABC, the ferrichrome transport system. Accordingly, ferrichrome should be able to compete with R-825 for the uptake and antagonize the antimicrobial activity of R-825. We therefore examined the antimicrobial activity of R-825 against *S. pneumoniae* in the presence of increasing concentrations of ferrichrome, hemin and FeCl₃. As shown in Figure 5,

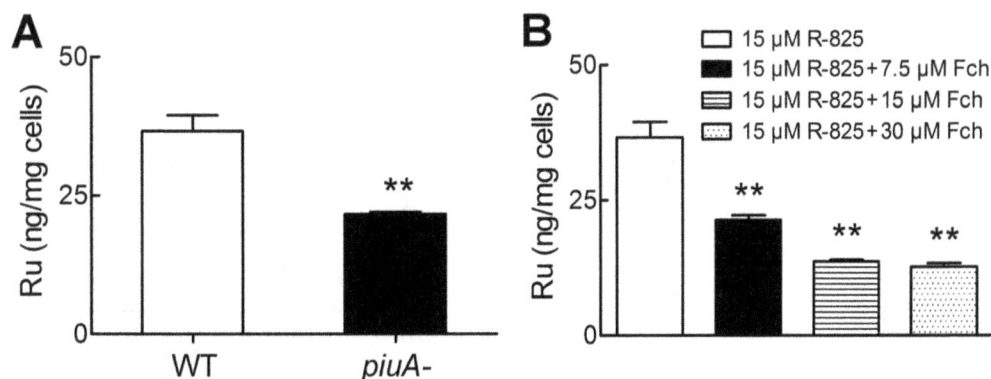

Figure 4. Cellular ruthenium concentrations in R-825 treated *S. pneumoniae* and with ferrichrome (Fch) competition as determined by ICP-MS. (A) Ruthenium contents in wild-type *S. pneumoniae* and in *piuA*- mutant as treated with 15 μM R-825. (B) Ruthenium contents in wild-type *S. pneumoniae* in the addition of increasing ferrichrome. Results are representative of the mean±SEM from three independent experiments (*$p < 0.01$).

ferrichrome addition indeed reversed the growth-inhibitory effects of R-825 in a dose-dependent manner with either sub-MIC or MIC treatments (Fig. 5A and 5B, respectively). When the molar ratio of ferrichrome to R-825 reached higher than 2:1, the maximum OD$_{600}$ of bacterial growth could be restored to near normal level. In contrast, adding either hemin (Fig. 5C) or FeCl$_3$ (Fig. 5D) could not rescue the bacterial growth inhibited by R-825. The complete inhibition observed in Figure 5C with 30 μM hemin addition was due to the toxicity of hemin itself to the bacterium.

We also measured the bacterial ruthenium contents in response to the ferrichrome addition in Figure 5A. As shown in Figure 4B, ferrichrome addition gradually reduced the cellular concentration of ruthenium while rescuing the bacterial growth (Fig. 5A). This observation indicated an actual competition between ferrichrome and R-825 for the bacterial uptake. These results all together confirmed that R-825 enters *S. pneumoniae* indeed *via* ferrichrome-uptake pathway PiuABC, rather than PiaABC and PitABC transport systems.

R-825 can bind to PiuA protein *in vitro*

The binding between R-825 and PiuA protein was tested to validate the interaction of R-825 with PiuABC systems. For comparison, the binding between ferrichrome/vancomycin and

PiuA was also determined. The quenching fluorescence spectra of apo-PiuA upon titration with the chemicals are shown in Figure 6; both R-825 and ferrichrome have a similar binding pattern with PiuA protein, while vancomycin as a negative control almost does not bind to PiuA protein. When the step-wise fluorescence quenching versus chemical concentration data were curve-fitted to Hill plot equation 1, the binding constants were determined to be $(0.30 \pm 0.18) \times 10^6$ M^{-1} and $(1.12 \pm 0.28) \times 10^6$ M^{-1} for R-825 and ferrichrome, respectively. These results revealed that both R-825 and ferrichrome could specifically bind to PiuA, with ferrichrome having a higher binding affinity than R-825 for PiuA.

Discussion

Bacterial resistance to antibiotics is becoming a significant threat to global public health [21]. There is an urgent need to develop novel antimicrobial drugs with an action mode different from the current antibiotics. The bacterial cell membrane demonstrates decreased permeability, serving as a barrier to limit the intracellular access of substances such as antibiotics by passive diffusion. Reduced membrane permeability is one of the main mechanisms for antibiotic resistance (reviewed in [22]). A strategy to circumvent the 'impermeability' resistance problem is to target the iron-transport systems for the drug delivery into the bacterial

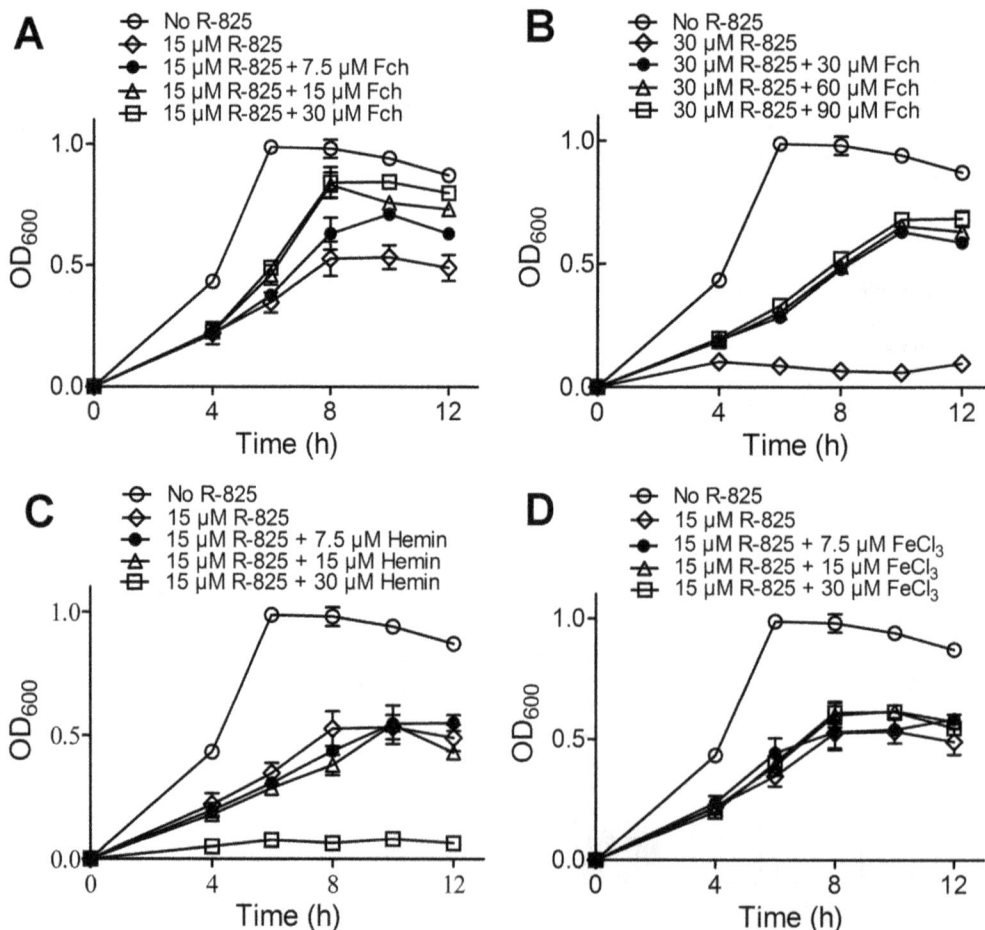

Figure 5. Ferrichrome, but not hemin or FeCl$_3$ rescued the bacterial growth suppressed by R-825. (A) Sub-MIC treated *S. pneumoniae* with increasing amounts of Fch. (B) MIC-treated *S. pneumoniae* with increasing amounts of Fch. (C) Sub-MIC treated *S. pneumoniae* with increasing amounts of hemin. Hemin itself at ≥30 μM has toxicity to the bacterium, causing the further inhibition of the bacterial growth. (D) Sub-MIC treated *S. pneumoniae* with increasing amounts of FeCl$_3$.

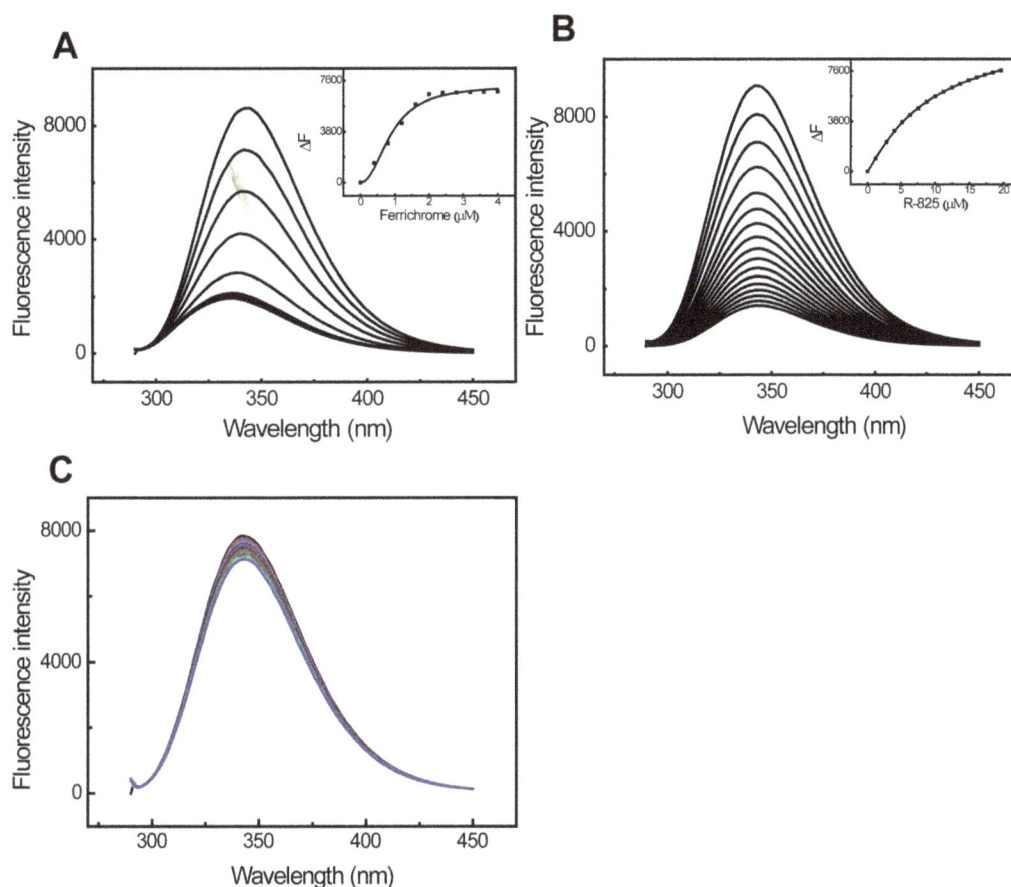

Figure 6. R-825 binds to PiuA protein *in vitro*. Fluorescence spectra of 2 µM apo-PiuA protein in 20 mM Tris-HCl (pH 7.4) titrated with aliquot ferrichrome (A), R-825 (B), or vancomycin (C). The inserted binding isotherm curves were built from the relative change values of fluorescence intensity (ΔF) at 343 nm versus concentrations. The inner panels are the curve-fitting analyses using Hill plot in the program Origin 8.5.

cells. Since the iron-transport systems are widespread among bacteria and work as active channels to transfer iron, some studies had explored antibacterial compounds including Ga(NO$_3$)$_3$ [23], DFO-Ga [24], non-iron metalloporphyrins (MPs) [25], and albomycin [8], enterobactin-cargo conjugates [26] as 'Trojan horses' to enter the bacteria through the iron-uptake systems.

In our long-term research on the mechanism of iron acquisition in bacteria, we understand that most iron obtained by bacteria is in the form of siderophores or heme since free iron ion is severely restricted in the host. Accordingly, blocking or interfering with either the heme- or siderophore-uptake pathway would be an effective approach to limit the acquisition of iron, one of the crucial elements for bacterial growth and survival. Here we selected our previously synthesized ruthenium(II) complex R-825 (Fig. 1) for the experiment, with an expectation that this compound can compete with either heme or ferrichrome to bind to the iron-receptors PiaA or PiuA, reduce the availability of iron to the bacteria and thus suppress the bacterial growth. Both ruthenium and iron are the members of VIII family in the periodic table, sharing chemical similarities in terms of coordination and binding with ligands. More importantly, Ru(II) is not toxic to the human body, making it an excellent candidate for antimicrobial drug development. Our current experiments demonstrated that R-825 indeed has selective antimicrobial activity against *S. pneumoniae*, with no evident toxic effects towards human A549

cells even in a concentration significantly greater than the corresponding MIC value (Fig. 2).

As expected, we detected that R-825 can be internalized by *S. pneumoniae*, accompanying with the decrease of bacterial iron uptake (Fig. 3A). Correspondingly, the expression of lipoproteins PiaA and PiuA in the bacterium was stimulated to compensate for the decreased iron availability under R-825 competition (Fig. 3B&C). Obviously, PiaA and/or PiuA iron-uptake systems were involved and thus we constructed *piaA-* and *piuA-* gene deletion mutant strains for further tests. Based on the MIC and growth inhibition zones assays (Table 2), we observed that only *piuA-* mutant showed an increased resistance to R-825, accompanying with a significant decrease in the content of cellular ruthenium in the mutant (Fig. 4A). This means that, without PiuA, less R-825 can be uptaken into cells and thus the bacterium became resistant to the drug. In other words, R-825 may be internalized by the bacterium mainly *via* PiuABC pathway.

Among the three iron-transport systems in *S. pneumoniae*, PiuABC system mainly conveys ferrichrome and its analogues into cells with the interaction between PiuA and ferrichrome [8]. Our experiments validated that R-825 can also interact with PiuA *in vitro*, in a binding pattern similar to that of PiuA-ferrichrome interaction (Fig. 6), suggesting that an active competition between R-825 and ferrichrome may occur for PiuA binding *in vivo*. Correspondingly, adding ferrichrome to the medium should compete with R-825 for the binding to PiuA, decrease the

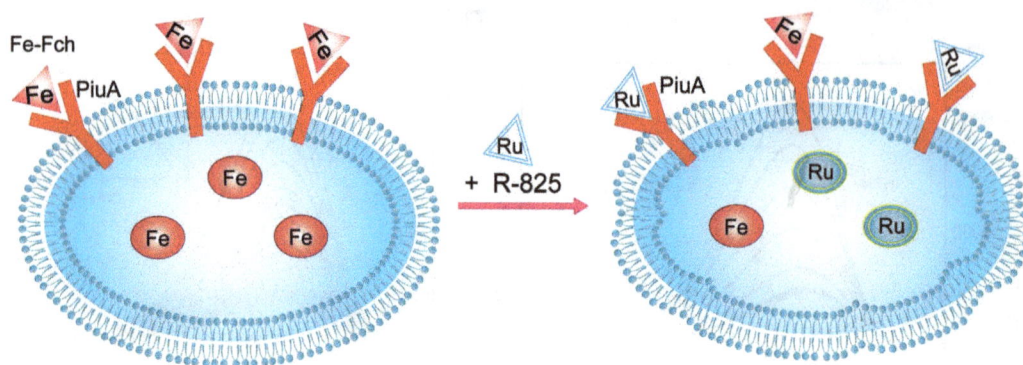

Figure 7. A model to summarize the proposed mechanism of Ru-complex action on the bacteria. R-825 is uptaken by *S. pneumoniae via* ferrichrome-transport pathway, reducing the competitive iron uptake into cells and thus suppressing the bacterial growth.

ruthenium uptake (Fig. 4B) and thus antagonize the antibacterial activity of R-825. Our competition experiments support this observation, in which adding ferrichrome can rescue the R-825-suppressed bacterial growth in a dose-dependent manner while adding hemin or FeCl$_3$ had no affect (Fig. 5). We can therefore conclude that R-825 interferes with the ferrichrome-uptake system by competing with ferrichrome for PiuA binding, reduces the iron internalization and thus suppresses the bacterial growth.

The fact that ruthenium compounds have antimicrobial activities has been recognized recently [27–31]. In particular, Keene and his co-workers characterized the susceptibility and cellular uptake of inert Ru(II) complexes [30,31] and demonstrated that di-nuclear Ru(II) complexes may enter eukaryotic cells by passive diffusion, while mononuclear complexes may have a different mode of action. This observation echoes our current finding that mononuclear Ru(II) complex R-825 is mostly uptaken into cells *via* the active ferrichrome-transport pathway. Certainly, this active transportation may not be the only way for R-825 to enter the bacterium, as attested by the fact that certain amounts of ruthenium were still detected in the *piuA*– mutant strain (Fig. 4A). On the other hand, our R-825 has a lower affinity than ferrichrome for PiuA binding (Fig. 6), suggesting that this compound can be further modified in terms of its structure and lipophilicity to enhance the ability to compete with ferrichrome and thus to optimize its antibacterial activity against *S. pneumoniae*.

In summary, we used a Ru(II) complex R-825 to chemically compete with iron compounds for binding to the iron-binding ligands in the iron-transport systems in *S. pneumoniae*. By performing various experiments, we demonstrated that R-825 can be transported into the bacterium *via* ferrichrome-transport pathway, competitively reducing iron uptake into cells and thus suppressing the bacterial growth (Fig. 7). Since the ferrichrome-uptake pathways are widely spread among bacteria, R-825 and derivatives may represent a new candidate class of antimicrobial agents for further development.

Supporting Information

File S1. Figure S1. The relative levels of PiaA, PiuA and PitA proteins in the iron replete or restricted conditions. **Figure S2.** The constructed *piaA*-, *piuA*- and *pitA*- mutant strains were verified by Western blotting and the purity of PiuA protein was verified by SDS-PAGE.

Author Contributions

Conceived and designed the experiments: XS QYH. Performed the experiments: XYY BS. Analyzed the data: LZ. Contributed reagents/materials/analysis tools: NL JH JZ. Contributed to the writing of the manuscript: XYY XS QYH.

References

1. Ratledge C, Dover LG (2000) Iron metabolism in pathogenic bacteria. Annu Rev Microbiol 54: 881–941.
2. Schaible UE, Kaufmann SH (2004) Iron and microbial infection. Nat Rev Microbiol 2: 946–953.
3. Cassat JE, Skaar EP (2013) Iron in infection and immunity. Cell Host Microbe 13: 509–519.
4. Mitchell TJ (2000) Virulence factors and the pathogenesis of disease caused by *Streptococcus pneumoniae*. Res Microbiol 151: 413–419.
5. Whalan RH, Funnell SG, Bowler LD, Hudson MJ, Robinson A, et al. (2006) Distribution and genetic diversity of the ABC transporter lipoproteins PiuA and PiaA within *Streptococcus pneumoniae* and related streptococci. J Bacteriol 188: 1031–1038.
6. Brown JS, Gilliland SM, Holden DW (2001) A *Streptococcus pneumoniae* pathogenicity island encoding an ABC transporter involved in iron uptake and virulence. Mol Microbiol 40: 572–585.
7. Brown JS, Gilliland SM, Ruiz-Albert J, Holden DW (2002) Characterization of pit, a *Streptococcus pneumoniae* iron uptake ABC transporter. Infect Immun 70: 4389–4398.
8. Pramanik A, Braun V (2006) Albomycin uptake via a ferric hydroxamate transport system of *Streptococcus pneumoniae* R6. J Bacteriol 188: 3878–3886.
9. Cheng W, Li Q, Jiang YL, Zhou CZ, Chen Y (2013) Structures of *Streptococcus pneumoniae* PiaA and its complex with ferrichrome reveal insights into the substrate binding and release of high affinity iron transporters. PLoS One 8: e71451.
10. Sun B, Chu J, Chen Y, Gao F, Ji LN, et al. (2008) Synthesis, characterization, electrochemical and photophysical properties of ruthenium(II) complexes containing 3-amino-1,2,4-triazino[5,6-f]-1,10-phenanthroline. J Mol Struct 890: 203–208.
11. Wach A (1996) PCR-synthesis of marker cassettes with long flanking homology regions for gene disruptions in *S. cerevisiae*. Yeast 12: 259–265.
12. Ong CL, Potter AJ, Trappetti C, Walker MJ, Jennings MP, et al. (2013) Interplay between manganese and iron in pneumococcal pathogenesis: role of the orphan response regulator RitR. Infect Immun 81: 421–429.
13. Lanie JA, Ng WL, Kazmierczak KM, Andrzejewski TM, Davidsen TM, et al. (2007) Genome sequence of Avery's virulent serotype 2 strain D39 of *Streptococcus pneumoniae* and comparison with that of unencapsulated laboratory strain R6. J Bacteriol 189: 38–51.
14. Institute CaLS (2012) Performance Standards for Antimicrobial Susceptibility Testing: Twenty-second Informational Supplement M100-S22.CLSI, Wayne, PA, USA.
15. McDevitt CA, Ogunniyi AD, Valkov E, Lawrence MC, Kobe B, et al. (2011) A molecular mechanism for bacterial susceptibility to zinc. PLoS Pathog 7: e1002357.

16. Wang Y, Cheung YH, Yang Z, Chiu JF, Che CM, et al. (2006) Proteomic approach to study the cytotoxicity of dioscin (saponin). Proteomics 6: 2422–2432.

17. Brown JS, Ogunniyi AD, Woodrow MC, Holden DW, Paton JC (2001) Immunization with components of two iron uptake ABC transporters protects mice against systemic *Streptococcus pneumoniae* infection. Infect Immun 69: 6702–6706.

18. Whitby PW, Sim KE, Morton DJ, Patel JA, Stull TL (1997) Transcription of genes encoding iron and heme acquisition proteins of Haemophilus influenzae during acute otitis media. Infect Immun 65: 4696–4700.

19. Kehres DG, Zaharik ML, Finlay BB, Maguire ME (2000) The NRAMP proteins of *Salmonella typhimurium* and *Escherichia coli* are selective manganese transporters involved in the response to reactive oxygen. Mol Microbiol 36: 1085–1100.

20. Li C, Tao J, Mao D, He C (2011) A novel manganese efflux system, YebN, is required for virulence by *Xanthomonas oryzae* pv. *oryzae*. PLoS One 6: e21983.

21. Ho J, Tambyah PA, Paterson DL (2010) Multiresistant Gram-negative infections: a global perspective. Curr Opin Infect Dis 23: 546–553.

22. Pages JM, James CE, Winterhalter M (2008) The porin and the permeating antibiotic: a selective diffusion barrier in Gram-negative bacteria. Nat Rev Microbiol 6: 893–903.

23. Kaneko Y, Thoendel M, Olakanmi O, Britigan BE, Singh PK (2007) The transition metal gallium disrupts *Pseudomonas aeruginosa* iron metabolism and has antimicrobial and antibiofilm activity. J Clin Invest 117: 877–888.

24. Banin E, Lozinski A, Brady KM, Berenshtein E, Butterfield PW, et al. (2008) The potential of desferrioxamine-gallium as an anti-Pseudomonas therapeutic agent. Proc Natl Acad Sci U S A 105: 16761–16766.

25. Stojiljkovic I, Kumar V, Srinivasan N (1999) Non-iron metalloporphyrins: potent antibacterial compounds that exploit haem/Hb uptake systems of pathogenic bacteria. Mol Microbiol 31: 429–442.

26. Zheng T, Bullock JL, Nolan EM (2012) Siderophore-mediated cargo delivery to the cytoplasm of *Escherichia coli* and *Pseudomonas aeruginosa*: syntheses of monofunctionalized enterobactin scaffolds and evaluation of enterobactin-cargo conjugate uptake. J Am Chem Soc 134: 18388–18400.

27. Dwyer FP, Gyarfas EC, Rogers WP, Koch JH (1952) Biological activity of complex ions. Nature 170: 190–191.

28. Biersack B, Diestel R, Jagusch C, Sasse F, Schobert R (2009) Metal complexes of natural melophlins and their cytotoxic and antibiotic activities. J Inorg Biochem 103: 72–76.

29. Bolhuis A, Hand L, Marshall JE, Richards AD, Rodger A, et al. (2011) Antimicrobial activity of ruthenium-based intercalators. Eur J Pharm Sci 42: 313–317.

30. Li F, Mulyana Y, Feterl M, Warner JM, Collins JG, et al. (2011) The antimicrobial activity of inert oligonuclear polypyridylruthenium(II) complexes against pathogenic bacteria, including MRSA. Dalton Trans 40: 5032–5038.

31. Li F, Feterl M, Mulyana Y, Warner JM, Collins JG, et al. (2012) In vitro susceptibility and cellular uptake for a new class of antimicrobial agents: dinuclear ruthenium(II) complexes. J Antimicrob Chemother 67: 2686–2695.

Permissions

List of Contributors

Monika Garcia
Departments of Cell Biology and Neuroscience, University of California Riverside, Riverside, California, United States of America

Sandeep Dhall and Manuela Martins-Green
Departments of Cell Biology and Neuroscience, University of California Riverside, Riverside, California, United States of America
Bioengineering Interdepartmental Graduate Program, University of California Riverside, Riverside, California, United States of America

Danh Do and Neal Schiller
Division of Biomedical Sciences, University of California Riverside, Riverside, California, United States of America

Jane Kim and Eugene A. Nothnagel
Department of Botany and Plant Sciences, University of California Riverside, Riverside, California, United States of America

Julia Lyubovitsky
Department of Bioengineering, University of California Riverside, Riverside, California, United States of America

Dayanjan Shanaka Wijesinghe and Charles E. Chalfant
Hunter Holmes McGuire Veterans Administration Medical Center, Richmond, Virginia, United States of America
Department of Biochemistry & Molecular Biology, Virginia Commonwealth University, Richmond, Virginia, United States of America
Virginia Commonwealth University Reanimation Engineering Science Center, Richmond, Virginia, United States of America
The Massey Cancer Center, Richmond, Virginia, United States of America

Angela Brandon and Rakesh P. Patel
Department of Pathology, University of Alabama at Birmingham, Birmingham, Alabama, United States of America

Antonio Sanchez and Sean Gallagher
Department of Product Technology, UVP, LLC, an Analytik Jena Company, Upland, California, United States of America

Dominique Wobser, Martin Berthold and Andrea Kropec
Division of Infectious Diseases, Department of Medicine, University Medical Center Freiburg, Freiburg, Germany

Diana Laverde and Felipe Romero-Saavedra
Division of Infectious Diseases, Department of Medicine, University Medical Center Freiburg, Freiburg, Germany
EA4655 U2RM Stress/Virulence, University of Caen Lower-Normandy, Caen, France

Johannes Huebner
Division of Infectious Diseases, Department of Medicine, University Medical Center Freiburg, Freiburg, Germany
Division of Pediatric Infectious Diseases, Dr. von Hauner Children's Hospital, Ludwig-Maximilians-University, Munich, Germany
German Center for Infection Research (DZIF), Partnersite Munich, Munich, Germany

Wouter Hogendorf and Gijsbert van der Marel3 Jeroen Codee
Bio-organic Synthesis Unit, Faculty of Science, Leiden Institute of Chemistry, Leiden University, Leiden, Netherlands

Catherine A. Butler, Stuart G. Dashper, Lianyi Zhang, Christine A. Seers, Helen L. Mitchell, Deanne V. Catmull, Michelle D. Glew, Jacqueline E. Heath, Yan Tan, Hasnah S. G. Khan and Eric C. Reynolds
Oral Health Cooperative Research Centre, Melbourne Dental School, Bio21 Institute, The University of Melbourne, Victoria, Australia

Eduardo Balsanelli, Válter Antonio de Baura, Fábio de Oliveira Pedrosa, Emanuel Maltempi de Souza,Rose Adele Monteiro
Department of Biochemistry and Molecular Biology, Universidade Federal do Paraná, Curitiba, Paraná, Brazil

Tarmo Ketola and Johanna Mappes
Centre of Excellence in Biological Interactions, Department of Biological and Environmental Science, University of Jyväskylä, Jyväskylä, Finland

Ji Zhang and Jouni Laakso
Centre of Excellence in Biological Interactions, Department of Biological and Environmental Science, University of JyväskyläJyväskylä, Finland

Department of Biological and Environmental Science, University of Helsinki, Helsinki, Finland

Anni-Maria ŐrmäläOdegrip
Department of Biological and Environmental Science, University of Helsinki, Helsinki, Finland

Lindsay Aldrich, Yuri Ragoza, Marissa Talamantes, Katharine D. Andrews and Helene L. Andrews-Polymenis
Department of Microbial Pathogenesis and Immunology, College of Medicine, Texas A&M University, Bryan, Texas, United States of America

Lydia M. Bogomolnaya
Department of Microbial Pathogenesis and Immunology, College of Medicine, Texas A&M University, Bryan, Texas, United States of America
Institute of Fundamental Medicine and Biology, Kazan Federal University, Kazan, Russia

Michael McClelland
Department of Microbiology and Molecular Genetics, University of California Irvine, Irvine, California, United States of America

Phuong Thi Mai Nguyen
Institute of Biotechnology, Vietnam Academy of Science and Technology, Hanoi, Vietnam

Megan L. Falsetta and Mireya Gonzalez-Begne
Center for Oral Biology, University of Rochester Medical Center, Rochester, New York, United States of America

Hyun Koo
Center for Oral Biology, University of Rochester Medical Center, Rochester, New York, United States of America
Biofilm Research Labs, Levy Center for Oral Health, Department of Orthodontics, School of Dental Medicine, University of Pennsylvania, Philadelphia, Pennsylvania, United States of America

Geelsu Hwang
Biofilm Research Labs, Levy Center for Oral Health, Department of Orthodontics, School of Dental Medicine, University of Pennsylvania, Philadelphia, Pennsylvania, United States of America

Teresa Koller and Andrew F Bent
Department of Plant Pathology, University of Wisconsin –Madison, Madison, Wisconsin, United States of America

Yuki Yamanaka
Department of Frontier Bioscience, Hosei University, Koganei, Tokyo, Japan,

Akira Ishihama and Kaneyoshi Yamamoto
Department of Frontier Bioscience, Hosei University, Koganei, Tokyo, Japan,
Micro-Nano Technology Research Center, Hosei University, Koganei, Tokyo, Japan

Taku Oshima
Graduate School of Information Sciences, Nara Institute of Science and Technology, Ikoma, Nara, Japan

Yassine Nait Chabane, Sara Marti, Christophe Rihouey, Stéphane Alexandre, Julie Hardouin, Thierry Jouenne and Emmanuelle Dé
Unité Mixte de Recherche 6270 CNRS - Laboratory "Polyméres, Biopolyméres, Surfaces", University of Rouen, Mont-Saint-Aignan, France

Olivier Lesouhaitier
Laboratory of "Microbiologie Signaux et Micro-Environnement" - Equipe d'Accueil 4312, University of Rouen, Evreux, France

Jordi Vila
Department of Microbiology, Hospital Clinic, Barcelona, Spain

Jeffrey B. Kaplan
Department of Biology, American University, Washington, District of Columbia, United States of America

Ya-Wen Chiang and Ting-Kai Lin
Department of Molecular Biology and Human Genetics, Tzu-Chi University, Hualien, Taiwan

Hsin-Hou Chang and Der-Shan Sun
Department of Molecular Biology and Human Genetics, Tzu-Chi University, Hualien, Taiwan
Institute of Medical Sciences, Tzu-Chi University, Hualien, Taiwan

Guan-Ling Lin and You-Yen Lin
Institute of Medical Sciences, Tzu-Chi University, Hualien, Taiwan

Jyh-Hwa Kau
Department of Microbiology and Immunology, National Defense Medical Center, Taipei, Taiwan

Hsin-Hsien Huang and Hui-Ling Hsu
Institute of Preventive Medicine, National Defense Medical Center, Taipei, Taiwan

Jen-Hung Wang
Department of Medical Research, Tzu Chi General Hospital, Hualien, Taiwan

Shixiang Gao
State Key Laboratory of Pollution Control and Resource Reuse, School of the Environment, Nanjing University, Nanjing, 210093, China

Na Wang
State Key Laboratory of Pollution Control and Resource Reuse, School of the Environment, Nanjing University, Nanjing, 210093, China
Nanjing Institute of Environmental Science, Ministry of Environmental Protection of China, Nanjing, 210042, China

Shaojun Jiao
Nanjing Institute of Environmental Science, Ministry of Environmental Protection of China, Nanjing, 210042, China

Xiaohong Yang, Jun Zhang and Boping Ye
School of Life Science and Technology, China Pharmaceutical University, Nanjing, 210009, China

Johanna Raffetseder, Elsje Pienaar, Robert Blomgran, Daniel Eklund, Veronika Patcha Brodin, Henrik Andersson, Amanda Welin and Maria Lerm
Division of Microbiology and Molecular Medicine, Department of Clinical and Experimental Medicine, Faculty of Health Sciences, Linköping University, Linköping, SE-58185, Sweden

Xiangkai Zhu Ge, Zihao Pan, Lin Hu, Haojin Wang, Jianjun Dai and Hongjie Fan
College of Veterinary Medicine, Nanjing Agricultural University, Nanjing, China

Jingwei Jiang and Frederick C. Leung
Bioinformatics Center, Nanjing Agricultural University, Nanjing, China
School of Biological Sciences, University of Hong Kong, Hong Kong SAR, China

Shaohui Wang
Shanghai Veterinary Research Institute, Chinese Academy of Agricultural Sciences, Shanghai, China

Miroslava Petrovova, Lukas Dvoracek and Irena Licha
Department of Genetics and Microbiology, Faculty of Science, Charles University, Prague, Czech Republic
Department of Medical Microbiology 2nd Faculty of Medicine, Charles University, Prague, Czech Republic

Jan Tkadlec and Eliska Streitova
Department of Medical Microbiology 2nd Faculty of Medicine, Charles University, Prague, Czech Republic

Giuseppe Bozza, Paolo Montanari, Vincenzo Nardi-Dei, Barbara Benucci, Marco Biancuccib, Elena Caproni, Riccardo Barrile, Silvana Savino, Beatrice Aricó, Rino Rappuoli and Mariagrazia Pizza
Novartis Vaccines, Siena, Italy

Marcello Merola
Novartis Vaccines, Siena, Italy
Department of Biology, University of Naples "Federico II", Naples, Italy

Mirco Capitani, Benedetta Picciani, Michele Sallese
Unit of Genomic Approaches to Membrane Traffic, Fondazione Mario Negri Sud, S. Maria Imbaro (CH), Italy

Alberto Luini
Institute of Protein Biochemistry, CNR, Naples, Italy

Sylvain Godreuil
Centre Hospitalier Régional Universitaire de Montpellier, Hôpital Arnaud de Villeneuve, Département de Bactériologie-Virologie, Montpellier, France

Nadia Leban and Corinne Lionne
Centre d'études d'agents Pathogénes et Biotechnologies pour la Santé, CNRS-UMR 5236/UM1/UM2, Montpellier, France

André Padilla and François Hoh
Centre de Biochimie Structurale Inserm U1054, CNRS UMR5048, Montpellier, France

Rodolphe Hamel, Dorothée Missé, Frédéric Thomas
Laboratoire MIVEGEC, UMR 224 IRD/CNRS/UM1, Montpellier, France

Natthanej Luplertlop
Department of Microbiology and Immunology, Faculty of Tropical Medicine, Mahidol University, Bangkok, Thailand

Auréie Chauffour, Wladimir Sougakoff and Hans Yssel
Centre d'Immunologie et des Maladies Infectieuses, Inserm U1135, Sorbonne Universités, UPMC, APHP Hôpital Pitié-Salpêtrière, Paris, France

Marion Vittecoq
Centre de Recherche de la Tour du Valat, le Sambuc, Arles, France

Qing Gu
Key Laboratory for Food Microbial Technology of Zhejiang Province, Department of Biotechnology, Zhejiang Gongshang University, Hangzhou, China

Da-Feng Song
Key Laboratory for Food Microbial Technology of
Zhejiang Province, Department of Biotechnology,
Zhejiang Gongshang University, Hangzhou, China
State Key Laboratory of Plant Physiology and
Biochemistry, College of Life Sciences, Zhejiang
University, Hangzhou, China

Mu-Yuan Zhu
State Key Laboratory of Plant Physiology and
Biochemistry, College of Life Sciences, Zhejiang
University, Hangzhou, China

**Xiao-Yan Yang, Liang Zhang, Nan Li, Junlong Han,
Jing Zhang, Xuesong Sun and Qing-Yu He**
Key Laboratory of Functional Protein Research of
Guangdong Higher Education Institutes, Institute of
Life and Health Engineering, College of Life Science
and Technology, Jinan University, Guangzhou, China

Bin Sun
Key Laboratory of Functional Protein Research of
Guangdong Higher Education Institutes, Institute of
Life and Health Engineering, College of Life Science
and Technology, Jinan University, Guangzhou, China
School of Pharmaceutical Sciences, Southern Medical
University, Guangzhou, China

Index